Current Topics in Behavioral Neurosciences

Volume 51

Current Topics in Behavioral Neurosciences provides critical and comprehensive discussions of the most significant areas of behavioral neuroscience research, written by leading international authorities. Each volume in the series represents the most informative and contemporary account of its subject available, making it an unrivalled reference source. Each volume will be made available in both print and electronic form.

With the development of new methodologies for brain imaging, genetic and genomic analyses, molecular engineering of mutant animals, novel routes for drug delivery, and sophisticated cross-species behavioral assessments, it is now possible to study behavior relevant to psychiatric and neurological diseases and disorders on the physiological level. The Behavioral Neurosciences series focuses on translational medicine and cutting-edge technologies. Preclinical and clinical trials for the development of new diagnostics and therapeutics as well as prevention efforts are covered whenever possible. Special attention is also drawn on epigenetical aspects, especially in psychiatric disorders.

CTBN series is indexed in PubMed and Scopus.

More information about this series at http://www.springer.com/series/7854

Grant D. Searchfield • Jinsheng Zhang
Editors

The Behavioral Neuroscience of Tinnitus

Springer

Editors
Grant D. Searchfield
Medical and Health Sciences
University of Auckland
Auckland, New Zealand

Jinsheng Zhang
Department of Communication Sciences and Disorders, Department of Otolaryngology-Head Neck and Surgery
Wayne State University
Detroit, MI, USA

ISSN 1866-3370 ISSN 1866-3389 (electronic)
Current Topics in Behavioral Neurosciences
ISBN 978-3-030-85502-4 ISBN 978-3-030-85503-1 (eBook)
https://doi.org/10.1007/978-3-030-85503-1

This Springer imprint is published by the registered company Springer Nature Switzerland AG.
The registered company address is: Gewerbestrasse 11, 6330 Cham, Switzerland

Preface

Tinnitus is a fascinating topic in behavioral neuroscience. At face value, tinnitus, the false perception of sound, would appear a very specific and punctate topic. But the reality is that tinnitus is a very heterogenous experience and a complex research field. The tinnitus sound is best explained to occur from a cascade of neuropathophysiological events, often commencing with an injury to the peripheral and/or central auditory systems that eventually leads to the perception of a false sound. The disorder of tinnitus is far more than hearing a sound that is not there, it is an unpleasant and aversive experience that is also associated with anxiety, depression, sleep disruption, sometimes with catastrophic effects on quality of life.

Professor Abe Shulman coined the term "Tinnitology" for the science of tinnitus; although this label has not received widespread adoption, the concept behind it – that tinnitus is its own subspecialty of behavioral neuroscience – is strongly supported by this book. To understand such a complex disorder, multiple disciplines need to come together to share knowledge and come up with solutions. This volume is an illustration of efforts to bridge the knowledge of a number of fields including sensory neurophysiology, neuropsychiatry, genetics, neuropharmacology, neuromodulation, psychology, and audiology. Such efforts are not new (for example, early multidisciplinary efforts include the First International Tinnitus Seminar (New York, 1979) [1] and the Ciba Foundation (London 1981) [2]). However, the pace of research has accelerated in the last decade with exciting new discoveries and with the hope for more effective treatments. While some of the chapters in this new book have similar titles to those in earlier publications, the depth of knowledge and understanding of tinnitus have advanced tremendously in the intervening 40 years. New advancements in genetics, biomarkers, big data, functional brain measures, and assessment of behavior have enabled more detailed and conclusive answers than ever before.

The purpose of this book is to provide the most up-to-date and forward looking knowledge about the behavioral neuroscience of tinnitus. If the readers question is: "is there anything new in tinnitus?," this book provides a resounding answer of yes!

Seventeen chapters provide comprehensive up-to-date summaries of the tinnitus experience and scientific methods from benchtop for fundamental mechanisms, clinical diagnoses and treatments, to population studies. In the *Epidemiology of Tinnitus* (Biswas and Hall), the current understanding of the distribution and determinants of tinnitus are reviewed, while the genetics of tinnitus is the focus of *Genetic Inheritance and Its Contribution to Tinnitus* (Amant, Gallego-Martinez, Lopez-Escamez). Pharmacological Evaluation of Drugs in Animal Models of Tinnitus (Zheng, McTavish, Smith) summarizes and compares studies on pharmacological evaluation of tinnitus treatment in different animal models. Modeling electrical sensory and brain stimulation effects on tinnitus in animals is the focus of *Animal Models of Cochlear and Brain Stimulation Effects on Tinnitus* (Zhang, Firestone, Elattma). The development of different forms of biomarkers and mechanisms for tinnitus are reviewed in *Functional Neuroanatomy of Salicylate- and Noise-Induced Tinnitus and Hyperacusis* (Salvi and colleagues), *Neuroinflammation and Tinnitus* (Shulman and colleagues), and through use of large data sets *Using Big Data to Guide Therapy Development* (Schlee and colleagues). Pharmacotherapy and neuromodulation for tinnitus are described and reviewed in chapters on the *Pharmacotherapy of Tinnitus* (Kleinjung and Langguth), sound therapy and its mechanisms *Sense and Sensibility* (Searchfield), invasive and non-invasive *Brain Stimulation* (De Ridder, Adhia, Langguth), and combined use of somatosensory and auditory bimodal stimulation *Bimodal Auditory-Electrical Stimulation for the Treatment of Tinnitus: Pre-Clinical and Clinical Studies* (Riffle and colleagues). Cognitive and affective aspects of tinnitus form another theme within the book: the *Neurobiology of Stress Induced Tinnitus* reviews stress-induced tinnitus and its modulation (Szczepek and Mazurek), the behavioral correlates of tinnitus are summarized in *Psychological Comorbidities of Tinnitus* (Hébert), and a systematic review investigates the *Psychosocial Variables That Predict Chronic and Disabling Tinnitus* (Kleinstäuber and Weise). Various aspects of the clinical assessment of tinnitus are reviewed including the emerging knowledge of the use of momentary evaluation, *Momentary Analysis of Tinnitus: Considering the Patient* (Deutsch and Piccirillo), *Tinnitus Questionnaires for Research and Clinical Use* (Theodoroff), and the psychoacoustic matching of tinnitus, *Principles and Methods for Psychoacoustic Evaluation of Tinnitus* (Vajsakovic, Maslin, Searchfield). The closing chapter is a collaborative effort to describe some of the exciting new developments in tinnitus on the horizon, and has an optimistic message for tinnitus sufferers: more effective and targeted therapies are coming.

We are extremely grateful to the world-leading tinnitus researchers who have contributed very comprehensive and authoritative reviews. While in many ways tinnitus remains an enigma, the chapters in this volume illustrate the tremendous progress that has been made in understanding its sensory and neuropathophysiological underpinnings, its assessment and management. Although the heterogenous nature of tinnitus is a challenge to research, it is a test that has been taken on by the authors.

We believe this book will be of interest to students, researchers, and clinicians with a general interest in neuro-sensory disorders, and in particular tinnitus.

Auckland, New Zealand Grant D. Searchfield
Detroit, MI, USA Jinsheng Zhang

References

1. Proceedings of the First International Tinnitus Seminar (New York, 8–9 June 1979). J Laryngol Otol Suppl 1981;4:1–217
2. Ciba Foundation Symposium 85 (1981) Appendix II guidelines for recommended procedures in tinnitus testing. In: Evered D, Lawrenson G (eds) Tinnitus. Pitman Books, Ltd., London, pp 303–306

Contents

Part I
Tinnitus in Humans

Prevalence, Incidence, and Risk Factors for Tinnitus

Roshni Biswas and Deborah A. Hall

Contents

The original version of this chapter was revised. A correction to this chapter can be found at https://doi.org/10.1007/7854_2020_188.

R. Biswas
Hearing Sciences, Division of Clinical Neuroscience, School of Medicine, University of Nottingham, Nottingham, UK

Department of Environmental Health Sciences, Istituto di Ricerche Farmacologiche Mario Negri IRCCS, Milan, Italy

D. A. Hall (✉)
Hearing Sciences, Division of Clinical Neuroscience, School of Medicine, University of Nottingham, Nottingham, UK

NIHR Nottingham Biomedical Research Centre, Nottingham, UK

Nottingham University Hospitals NHS Trust, Nottingham, UK

University of Nottingham Malaysia, Semenyih, Malaysia
e-mail: deborah.hall@nottingham.edu.my

Curr Topics Behav Neurosci (2021) 51: 3–28
https://doi.org/10.1007/7854_2020_154
Published Online: 26 August 2020

Abstract How many people are affected by tinnitus? Is the risk of developing tinnitus on the rise or has it been declining over time? What modifiable lifestyle factors could help to prevent tinnitus? These population-based questions can be addressed through epidemiological research. Epidemiology refers to the underlying and basic science of public health. It describes the study of the distribution and determinants of health-related states or events in specified populations and the application of this study to the control of health problems. There are two key concepts in epidemiology: (1) measures of frequency and (2) measures of effect. In this chapter, we introduce the two main measures of frequency, prevalence and incidence. We also introduce the notion of risk factors, critical for understanding measures of effect concerning the risk of developing a health condition. In both sections, we provide illustrative examples from the published literature on tinnitus. We end by offering a critical evaluation of the current status of epidemiological research on tinnitus and point to some promising future directions.

Keywords Adults · Epidemiology · Measures of effects · Population · Tinnitus

1 Measures of Frequency

Measures of frequency describe how common a condition is, in reference to the size of the population (i.e., the population at risk of developing the condition) and some defined measure of time.

1.1 Prevalence

Prevalence considers all cases of tinnitus, i.e., both new and preexisting cases (Fig. 1).

$$\text{Percent prevalence} = \frac{\text{All cases (new and preexisting) at a given period of time}}{\text{Total population at the same time period}} \times 100$$

For example, in a sample of 100,000 people studies, 5,000 report tinnitus. This gives a prevalence of 5% in the sample, and when calculated for a nationally representative population, this number can be extrapolated to estimate prevalence in the overall country population. It is to be noted that when calculating prevalence, the numerator is a subset of the denominator. Prevalence values are influenced by incidence (rate of new cases) and duration of the condition. Higher prevalence can be attributed to increased number of new cases and/or increased survival of cases without any cure and vice versa. This is particularly important for chronic

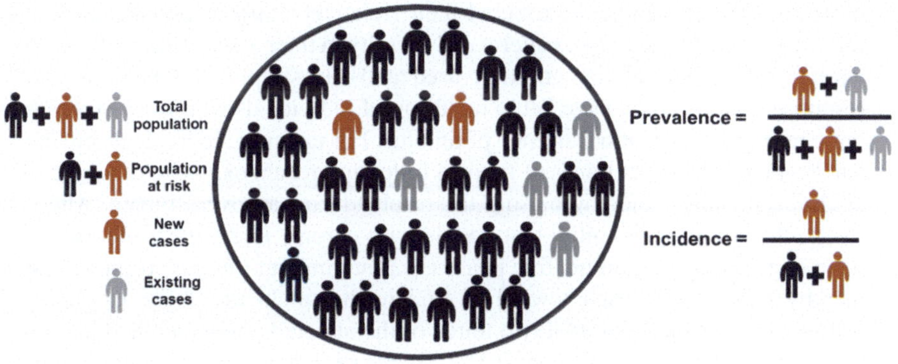

Fig. 1 Graphical representation of prevalence and incidence in a population

conditions, such as tinnitus, where there is no cure and also where it is difficult to pinpoint the date of onset due to the long-standing insidious nature of the condition. Prevalence is often chosen to measure occurrence for chronic conditions, and percent prevalence can be estimated in three different ways, as follows:

1. *Point prevalence* refers to the number of cases of tinnitus at a specific time point. This could be measured using a survey asking individuals if they are currently experiencing tinnitus. The major advantage of assessing point prevalence is that it provides a snapshot of the disease burden. A large sample size can be included, and the one-time assessment method is convenient. In chronic conditions like tinnitus, it is useful for capturing individuals who experience the symptoms most of the time or all of the time.

   Ideally, point prevalence would be data collected at one specific point in time, for example, if patients presenting at a clinic appointment were asked if they were experiencing tinnitus at that particular moment. However, it is usually difficult to assess prevalence of a condition in the entire sample population at one instant, and some amount of flexibility is acceptable in defining time. For all practical purposes, point in time can be defined as an event rather than a distinct calendar date. Therefore, it is appropriate to consider point prevalence as the number of individuals affected by tinnitus at the time of survey, even though the actual data collection maybe conducted on different calendar dates for different participants.

   In a population-based cross-sectional study conducted on adults aged 18 years and above living in the city of Sao Paulo, Brazil, between April and October 2012, the point prevalence of tinnitus was reported to be 22%. This 22% included study participants who responded "yes" to the question "Do you have ringing in your ears?" with binary response options ("yes" or "no") (Oiticica and Bittar 2015). Since the question is not bound by a defined time frame and asks about the presence of the symptom in the current context, we can assume that the prevalence thus measured is an estimate of point prevalence. The authors of the above study acknowledged a prevalence rate that was higher than that reported by some other studies. The authors attributed their (relatively) higher prevalence to the lack of a defined time frame, pattern, and severity in their assessment method. The unrestricted nature of the question used and the simplified ("yes" or "no") response options could have led to inclusion of individuals who were experiencing a rare and transient occurrence of tinnitus. For example, individuals having a single acute episode of transient tinnitus following recent exposure to loud noise could be included, thereby biasing the result toward an overestimate. Alternatively, there is the possibility that people with chronic intermittent tinnitus who were experiencing a tinnitus-free phase (i.e., intermittent tinnitus) respond "no," thereby biasing the result toward an underestimate. In the above example, a follow-up question found that two thirds of the affected population had intermittent tinnitus (Oiticica and Bittar 2015). Therefore, when relying on point prevalence estimates, it is important to consider the impact of the waxing and waning nature of chronic symptoms which can bias the prevalence estimate.

2. *Period prevalence* refers to the number of cases of tinnitus over a specific time period. Often in the tinnitus literature, the distinction between point and period prevalence has not been made clear by authors reporting their study. Perhaps the reason is the practical difficulty in assessing all cases at the exact same instant since surveys tend to have a data collection period that spans weeks or months. Period prevalence is conceptually similar to point prevalence except that it considers a wider time range. While point prevalence provides a single snapshot of burden, period prevalence paints a picture over a longer period. For example, this could be measured using a survey asking individuals if they have experienced tinnitus during the past 12 months. Another difference is that, when calculating period prevalence, the denominator is the average or the mid-interval population, contrasted with point prevalence where the denominator is the population at the same point in time.

   In tinnitus research, some studies have measured tinnitus prevalence at specified time periods. The study by Bhatt et al. (2016), using the nationally representative population-based data from the 2007 US National Health Interview Survey to assess the prevalence of tinnitus, is such an example. This was a cross-sectional study that measured tinnitus in the preceding 12 months on adults aged 18 years and above, using the question "In the past 12 months, have you been bothered by ringing, roaring, or buzzing in your ears or head that lasts for 5 minutes or more?," with the response options "yes" or "no." The study reported a prevalence estimate of 9.6% (Bhatt et al. 2016). Like point prevalence, the potential to recruit a large sample and ease of administration are major advantages. Although, unlike point prevalence, since a longer time period (e.g., 12 months) is considered, individuals experiencing intermittent episodes are less likely to be missed. However, there is a potential risk for overestimation as individuals having had a single acute incident of transient tinnitus in the last 12 months may respond affirmatively. In such situations, it would be difficult to distinguish between cases of chronic, long-term tinnitus and cases of acute or transient tinnitus.

3. *Lifetime prevalence* refers to the number of cases of tinnitus over an individual's lifetime. For example, this could be measured using a survey asking individuals if they have ever experienced tinnitus. Cases of chronic tinnitus are distinctive from cases of acute or transient tinnitus, and so this might be reflected in the form of the question posed. Many cases of tinnitus tend to be chronic in nature, and so one might expect lifetime and period prevalence to yield similar estimates. When assessing tinnitus burden, the debilitating form of chronic tinnitus presents the main public health concern. If a survey asks specifically about "ever" having tinnitus, the resultant estimate would be a mix of acute, transient, and chronic symptoms and not necessarily a true reflection of tinnitus that is bothersome to the population.

   To our knowledge, no tinnitus prevalence study has particularly assessed lifetime prevalence. However, the result from the analysis of the UK Biobank data collected between 2006 and 2010 might potentially include a mix of point and lifetime prevalence (McCormack et al. 2014). The UK Biobank is a large dataset that recruits representative middle-aged UK population from national health registries (Allen et al. 2012). The question on current tinnitus used was "Do you get or have you had noises (such as ringing or buzzing) in your head or in one or both ears that lasts for more than five minutes at a time?" with the response option "yes" or "no." But by including both the "do you" and "have you had" phrases, the question becomes somewhat ambiguous in its interpretation. Respondents experiencing acute or transient or chronic symptoms could both answer affirmatively, potentially leading to overestimation of percent prevalence.

1.2 Incidence

Incidence considers new (or newly diagnosed) cases of tinnitus. Specifically, incidence rate calculates the frequency with which new tinnitus events occur over a particular time frame (e.g., the number of new cases per year) (Fig. 1). Incidence rate is calculated by dividing the number of new cases over a specified period either by the average population (usually mid-period which is the population half-way through the period being evaluated) or person-time which is a measure of the number of persons at risk and the time they were at risk. For example, in a sample of 1,000 non-tinnitus persons, 100 developed tinnitus over 2 years of observation. The incidence proportion is 100 cases per 1,000 persons, i.e., 10% over a 2-year period, or 50 cases per 1,000 person-years (incidence rate), because the incidence proportion (100 per 1,000) is divided by the number of years.

Incidence is assessed using observational study designs which involve observing people without tinnitus in a non-controlled environment without actually interfering or manipulating with other aspects of the study and therefore are non-experimental. The observation can be prospective or retrospective, and here an example of each is given.

The EHLS study measured the incidence proportion of tinnitus in the study population, which is the number of subjects that develop tinnitus anytime during the follow-up period. Of the 3,753 study participants at baseline, 3,429 tinnitus-free participants were included in the incidence study and followed up for a period of 10 years. Of them, 2,922 individuals aged 48 to 92 years provided information on their tinnitus status at least once and up to 4 times. Results indicated that the cumulative incidence of tinnitus was 13% (Nondahl et al. 2010).

Methodological disadvantages of this type of population-based prospective cohort study are that long-term follow-ups are expensive to conduct and resource intensive, and they are prone to loss to follow-up. For example, for the EHLS, there was a 15% dropout rate over survey follow-ups. A methodological advantage is that in addition to estimating incidence the data can be used to assess the association between potential risk factors and tinnitus. For example, in the 10-year EHLS follow-up, hearing loss and head injury were found to increase the risk of developing tinnitus. With respect to modifiable risk factors, smoking was found to increase the risk of developing tinnitus, while moderate intake of alcohol was found to have a protective effect (Nondahl et al. 2010).

As an alternative study design, retrospective population-based studies are those which review existing health records. This design conveys certain methodological advantages. Not only is the study reasonably cost-effective to conduct, but loss to follow-up is not an issue. However, incident reporting is dependent on medical help-seeking behavior. This means that the findings are not directly comparable with prospective surveys on the general population, such as EHLS. Retrospectively considering newly diagnosed cases of tinnitus using hospital records was the method used by Martinez et al. (2015) to measure incident cases of clinically significant tinnitus in England. Records of clinically significant tinnitus were defined in a number of different ways in an attempt to capture all cases. For example, it could be a patient discharged from hospital with a primary discharge diagnosis of tinnitus or a patient having a recording of tinnitus by the family doctor and having either a specific diagnosis or a relevant onward referral. The data was anonymized patient records sourced from two National Health Service databases: the UK Clinical Practice Research Datalink (CPRD) and Hospital Episode Statistics (HES). The observation period (i.e., the time in which a patient from the study population was at risk for developing clinically significant tinnitus) was from 2002 to 2011. This study calculated incidence rate given by the number of new cases of clinically significant tinnitus divided by the total person-years at risk of the study population and reported incidence rate to be 5.4 new cases of clinically significant tinnitus per 10,000 person-years (Martinez et al. 2015).

1.3 Variability in the Estimates of Tinnitus Frequency

Systematic reviews on the prevalence of tinnitus indicate large variability in estimates. For example, a review of 35 prevalence studies by McCormack et al. (2016)

indicated prevalence estimates ranging from 5% to 43% (Gibrin et al. 2013; McCormack et al. 2016; Quaranta et al. 1996). Major contributors to this variability include differences in geographic location of study, differences in population demographics (e.g., studies conducted on specific age groups), the lack of an agreed assessment question, and heterogeneous reporting measures (Gallus et al. 2015; McCormack et al. 2016). Because of this lack of standardization when estimating tinnitus prevalence, it is not possible to pool estimates across countries to understand the global burden of tinnitus nor to examine differences across countries or world regions.

Perhaps the most concerning source of variability relates to the assessment question used for deciding if a person has tinnitus or not. In their systematic review, McCormack et al. (2016) found eight different assessment questions for tinnitus. The two most commonly used terms in the tinnitus definitions were "tinnitus lasting for more than five minutes at a time" and "experiencing in the last one year" (McCormack et al. 2016). The second term defines a period prevalence, and 12 months is a commonly used time frame (National Institute of Mental Health 2017). Approximately 34% (12/35) studies that this review included had the phrase "tinnitus lasting for more than 5 min at a time" in the assessment question (McCormack et al. 2016). For example, a study by Hannula et al. (2011) used the point prevalence assessment question "Nowadays, do you ever get noises in your head or ears (tinnitus) which usually last longer than five minutes?" (Hannula et al. 2011). Another study by Dawes et al. (2014) used the UK Biobank question "Do you get or have you had noises (such as ringing or buzzing) in your head or in one or both ears that lasts for more than five minutes at a time?" (Dawes et al. 2014). From the above examples, it can be noted that although these two questions including the "lasting for five minutes" phrase are apparently similar, there is a key difference. While one of them defines a time frame "nowadays," the other did not. Whether or not "nowadays" defines a point prevalence or period prevalence is somewhat ambiguous. "Nowadays" could mean "now" or "in the last few days" or even "in the last month or so." On the other hand, the inclusion of "do you get or have you had" phrases in the close-ended UK Biobank question with binary response options ("yes" or "no") broadens the scope of the question and could provide a combined estimate for point, period, and lifetime prevalence of tinnitus. Therefore, while it is relatively easy to define whether a population survey is assessing period prevalence, it can be more difficult to determine point prevalence assessments due to ambiguity in interpreting the question.

The second most commonly used phrase "experiencing tinnitus in the last year" was used in 26% (9/35) of studies included in the systematic review by McCormack et al. (2016). This gives a period prevalence estimate (McCormack et al. 2016). Studies by Nondahl et al. (2002) and Spankovich et al. (2017) measured tinnitus prevalence using an assessment question where subjects were asked about tinnitus in the last year (e.g., "In the past year have you had buzzing, ringing or noise in your ears?") (Nondahl et al. 2002; Spankovich et al. 2017). Other studies by Xu et al. (2011) in China, and Bhatt et al. (2016) in the USA, used questions that assessed tinnitus over the last year and lasting for 5 min (e.g. "In the past year have you had noises in your ears or head which lasted longer than 5 minutes?") (Bhatt et al. 2016; Xu et al. 2011). Although both questions include a time frame of 1 year, one specifies a duration for tinnitus, and the other does not. Thus, it can be concluded that studies with broadly similar tinnitus assessment questions might differ by inclusion or omission of key concepts that address crucial information. Moreover, most studies do not provide a justification for the use of their choice of words while framing tinnitus assessment questions (McCormack et al. 2016).

Similar inconsistencies were also noted in "severe tinnitus" definitions. Eight out of the 13 studies (62%) reporting tinnitus severity that McCormack et al. included in their review defined severity or bothersomeness of tinnitus in terms of how bothered, annoyed, or worried it made the individual. However, not all severity assessment questions included all three emotional descriptors. For example, Jun et al. (2015) and Welch and Dawes (2008) both used the term "annoy" to assess the effect of tinnitus severity, although the latter included the term "upset" ("How annoying or upsetting is it?") as well (Jun et al. 2015; Welch and Dawes 2008). On the other hand, the severity question from the UK Biobank uses three emotional descriptors ("How much do these noises worry, annoy or upset you when they are at their worst?"), while the Swedish Longitudinal Occupational Survey of Health (SLOSH) replaced "annoy" with the term "bother" ("How much do you feel that the tinnitus sounds worry, bother or upset you?") (Hasson et al. 2010; McCormack et al. 2014). Although the emotional descriptors address the disturbance of tinnitus, there are underlying differences in their meaning. Since severity is a self-reported perception, these subtle nuances can affect the findings of a study. McCormack et al. (2016) also reported two other, less commonly used concepts for addressing tinnitus severity in published literature. These are severity assessed in terms of its impact on sleep, concentration (3 out of 8 studies) or in terms of an individual's ability to lead a normal life (2 out of 8 studies) (McCormack et al. 2016).

Tinnitus frequency, measured by "how often the tinnitus happens," is another component that has been assessed in some studies to distinguish constant or persistent tinnitus from occasional occurrences (Bhatt et al. 2016; Degeest et al. 2017; Folmer et al. 2011; Shargorodsky et al. 2010; Spankovich et al. 2017). This component is also plagued by the same lack of consensus across the community. Interestingly, this heterogeneity is related more to how constant or persistent tinnitus is defined in the response options, rather than how the frequency question is framed. Shargorodsky et al. (2010) defined frequent tinnitus as occurring "almost always or at least once a day" (Shargorodsky et al. 2010). Folmer et al. (2011) defined constant tinnitus as occurring "almost always," while Spankovich et al. (2017) defined persistent or constant tinnitus as tinnitus occurring "at least once a month or more" (Folmer et al. 2011; Spankovich et al. 2017). In another study, Degeest et al. (2017) defined constant tinnitus as lasting for more than 72 h (Degeest et al. 2017). Bhatt et al. (2016) included nearly constant as a response option in their study (Bhatt et al. 2016).

1.4 Estimates of Frequency for "Any Tinnitus"

Differences in assessment questions and lack of explicit statements on whether the study is assessing point, period or lifetime prevalence are noted across most of the published literature on tinnitus. A call to meet this challenge of developing standardized assessment questions has been raised recently by several authors (Gallus et al. 2015; McCormack et al. 2016).

The impact of the variability in assessment questions can be illustrated by comparing two well-known population-based studies of tinnitus. The Blue Mountain Hearing Study (BMHS) and the Epidemiology of Hearing Loss Study (EHLS) are population-based surveys conducted in Australia and the USA, respectively, to evaluate the characteristics of hearing loss in older adults (Cruickshanks et al. 1998; Sindhusake et al. 2003). The study populations of BMHS and EHLS are similar in terms of age, sex, marital status, pure tone average, ethnicity, occupational history, and cardiovascular profile. Nondahl et al. (2002) analyzed the EHLS data from 1993 to a follow-up of 5 years and reported the prevalence of significant tinnitus as 8.2%. They defined significant tinnitus as "buzzing, ringing, or noise in the ears in the past year that was at least moderate in severity or that caused problems getting to sleep" (Nondahl et al. 2002). In contrast, the findings from BMHS data between 1997 and 1999 as reported by Sindhusake et al. (2003) were tinnitus prevalence of 30%. In BMHS tinnitus was reported as "any prolonged ringing or buzzing in the ears or head within the past one year … lasting for five minutes or longer" (Sindhusake et al. 2003). Thus, despite the relatively similar sociodemographics, these two studies reported quite discrepant estimates of percent period prevalence. In a comment to the article by Sindhusake et al. (2003), Nondahl et al. (2004) attributed this discrepancy to the different reporting measures used (Nondahl et al. 2004; Sindhusake et al. 2003). In their 5-year follow-up analysis, EHLS added the same 5 minutes duration qualifier to their question ("buzzing, ringing or noise in your ears in the past year that usually lasts longer than five minutes"). Following this change, the prevalence estimate rose from 8.2% to 17.9% in the EHLS population (Nondahl et al. 2004). Although this estimate was still lower than the BMHS estimate of 30%, it can be noted that the alignment in definitions was perhaps responsible for bringing the two prevalence estimates closer.

One might expect prevalence estimates for any tinnitus to be reliable, so that findings across studies can be compared and meaningfully interpreted. The issue of reliability introduces two statistical concepts: accuracy and precision. Accuracy is defined as the degree to which a measurement, or an estimate based on measurements, represents the true value of the attribute that is being measured. Precision is defined as the inverse of the variance of a measurement or estimate (Last 1988). In other words, accuracy determines how close the estimate is to the true value, while precision determines how close the repeated estimates are to each other after successive measurements. It is to be noted that accuracy and precision do not always go hand in hand. From the published data, the median range of prevalence estimates in Western Europe and America is between 10% and 15% (Baguley et al. 2013).

This gives a benchmark for reflecting on accuracy and precision. For example, a tinnitus assessment item might give an average prevalence estimate of 13% with a 6–30% range, and then the average may be accurate (because it is close to the expected range of 10–15%, Baguley et al. 2013), but not precise (because the observed range is greater than expected). In another example, a tinnitus assessment item might give repeated estimates around 40%, indicating good precision, but since this value exceeds the expected 10–15% range (Baguley et al. 2013), it is probably not very accurate.

Given the small number of studies that use each of the assessment items, it is difficult to draw formal conclusion about which is the best way to measure prevalence. However, it seems reasonable to posit that estimates of period prevalence using the 12-month time frame and characterizing tinnitus duration as "lasting for more than 5 minutes" seem to give more accurate and precise prevalence estimates compared to other assessment items, with a narrow range of values from 6% to 15% (Bhatt et al. 2016; Gallus et al. 2015;; Xu et al. 2011).

1.5 Distinguishing "Severe Tinnitus" from "Any Tinnitus"

Knowledge about the burden of a condition contributes to understanding the magnitude of its impact on the population and can inform decisions on priorities for allocating resources and implementing interventions. The concept of any tinnitus does not consider burden, while severe tinnitus does. Differentiating severe tinnitus from any tinnitus seems reasonable. Chronic tinnitus in its debilitating form negatively affects an individual's overall health and emotional well-being and can incur substantial social and financial costs on the individual and the society at large (Hall et al. 2018; Hoekstra et al. 2014; Maes et al. 2013; Stockdale et al. 2017).

A number of epidemiological studies assessing tinnitus prevalence have sought to differentiate severe tinnitus from any tinnitus by seeking information about the bothersomeness of tinnitus symptoms (McCormack et al. 2016). For example, Bhatt et al. (2016) reported a 10% prevalence of any tinnitus and 7.2% prevalence of severe tinnitus (i.e., a big or very big problem) in the US population (Bhatt et al. 2016). For an Italian population, Gallus et al. (2015) reported 6% prevalence of any tinnitus and 1% prevalence of severe tinnitus (Gallus et al. 2015). Both these studies assessed any tinnitus as "ringing or buzzing lasting for five minutes in the past year," while severity was assessed by "how much it "bothered or annoyed" the individual.

Several epidemiological studies have shown a negative relationship between tinnitus severity and quality of life (Nondahl et al. 2007; Zeman et al. 2014). For example, Nondahl et al. (2007) used the Medical Outcomes Study (MOS) Short Form Health Survey (SF-36) to assess the relationship between tinnitus severity and quality of life (QOL) (Nondahl et al. 2007). They analyzed data from EHLS to assess severity. Patients who reported having "ringing or buzzing or noise" in ears over the last year were asked to rate their severity as "Mild/Moderate /Severe/Unknown" (question asked: "How severe is this noise in its worst form?"). The SF-36 assesses eight domains relating to two component scales, namely, the Physical Component Summary Scale (PCS) and the Mental Component Summary Scale (MCS). Increased tinnitus severity was associated with decreased scores for all eight domains and the two component scales (Nondahl et al. 2007). Zeman et al. (2014) also reported a strong negative correlation between tinnitus severity and self-reported well-being measured using the World Health Organization QOL instrument assessed in a large multinational clinical sample (Zeman et al. 2014).

Epidemiological research has also confirmed a positive relationship between tinnitus severity and healthcare costs. In a study conducted in Netherlands, mean annual costs per patient were €767 for people reporting mild tinnitus symptoms, €1,329 for moderate tinnitus, and €2,218 for severe tinnitus (Maes et al. 2013). Therefore, a person with severe tinnitus symptoms accounts for three times the expenditure as a person with mild tinnitus symptoms. A similar pattern was noted by a study in the USA (Goldstein et al. 2015). The increased expenses resulted from a combination of increased doctors' appointments including specialist appointment, cost of therapy, loss of productivity, travel to and from clinics, and other associated expenses like headphones, ear protection, and alternative therapy (Goldstein et al. 2015; Maes et al. 2013; Stockdale et al. 2017).

2 Measures of Effect

Measures of effect are used to quantify the strength of association between a potential risk factor and a health condition. Here the proportion of affected individuals exposed to a potential risk factor is compared with proportion of affected individuals not exposed. This comparison can be made by calculating either the ratio or the difference in the measures of frequency between the two groups. The resultant value gives the increase or decrease in frequency of any given condition in the population exposed to a risk factor when compared with the unexposed population.

Relative risk (RR) is one measure of effect that defines the likelihood of developing a condition or disease of interest linked to an exposure (such as age, sex, smoking, etc.). RR is calculated by the ratio of the probability of the outcome occurring in the exposed group compared to the probability of the outcome occurring in the non-exposed group. For example, if RR is ten for the exposed group, then the exposed individuals are ten times more likely to have the outcome than unexposed individuals. On the other hand, if the RR is 0.5 for the exposed group, then they are 0.5 times as likely to have the outcome as the unexposed. In other words, the exposed are 50% less likely (0.5 = 1.0–0.5) to have the outcome than the unexposed. An RR of 1 indicates that the outcome is no different in the exposed and unexposed groups. If the 95% confidence interval (CI) for the relative risk includes the value of 1.0, then it cannot be concluded that the exposed and unexposed groups are statistically significantly different.

Another measure of effect is the hazard ratio (HR). This is used for dichotomous variables and is defined as the hazard in the exposed groups divided by the hazard in the unexposed groups. Studies commonly use a Cox proportional hazards model for this estimate, so that one estimate of HR summarizes the whole follow-up period experience. In practice, HR is the most common metric of effect reported in many epidemiologic studies, and it is interpreted in the same way as RR.

A third measure of effect is the odds ratios (ORs). This is generated as a statistic to quantify the strength of association between the outcomes of interest (e.g., tinnitus), given a certain exposure. It is the ratio of the odds of an exposure between the cases and controls in a case-control study and is interpreted as the ratio of likelihood in one group with reference to another group. An OR with a value of 1.0 means that exposure is no different between the cases and controls. For meaningful reporting of OR, it is also important to mention the corresponding CI. If the CI spans the value of 1.0, then the association between the exposure and outcome is not statistically significant.

3 Observational Study Designs to Measure Frequency and Effect

Observational study designs form the mainstay of epidemiological research, and well-conducted observational studies can yield qualitatively comparable results to intervention studies such as randomized controlled trials (Benson and Hartz 2000; Concato et al. 2000). In observational study designs, investigators survey a sample population in which individuals are not undergoing any intervention to examine the measures of tinnitus frequency and the effect of risk factors on tinnitus. Cross-sectional, cohort, and case-control studies are all types of observational study designs. Observational studies have an important place in epidemiology since it is not always ethical (or possible) to investigate risk factors in a controlled setting (e.g., assessing the effect of smoking on tinnitus). Nevertheless, observational study designs have been criticized for failing to account for confounding factors whenever an apparently straightforward relationship between a risk factor and an outcome is distorted by the mediating effect of a third variable (the confounder). For tinnitus, one example is the relationship between age and tinnitus which can have hearing loss as a confounding factor because age and hearing loss are strongly correlated with one another.

The following sections give illustrative examples from published research on tinnitus in order to elaborate on the three common observational study designs.

3.1 Cross-Sectional Studies for Estimating Prevalence

Cross-sectional study designs give a snapshot of the exposure status and outcome in the study population, at the same point in time (Fig. 2). Participants are recruited in accordance with exclusion and inclusion criteria specified for the study. The researchers then measure the exposures and outcomes in the participants. Cross-

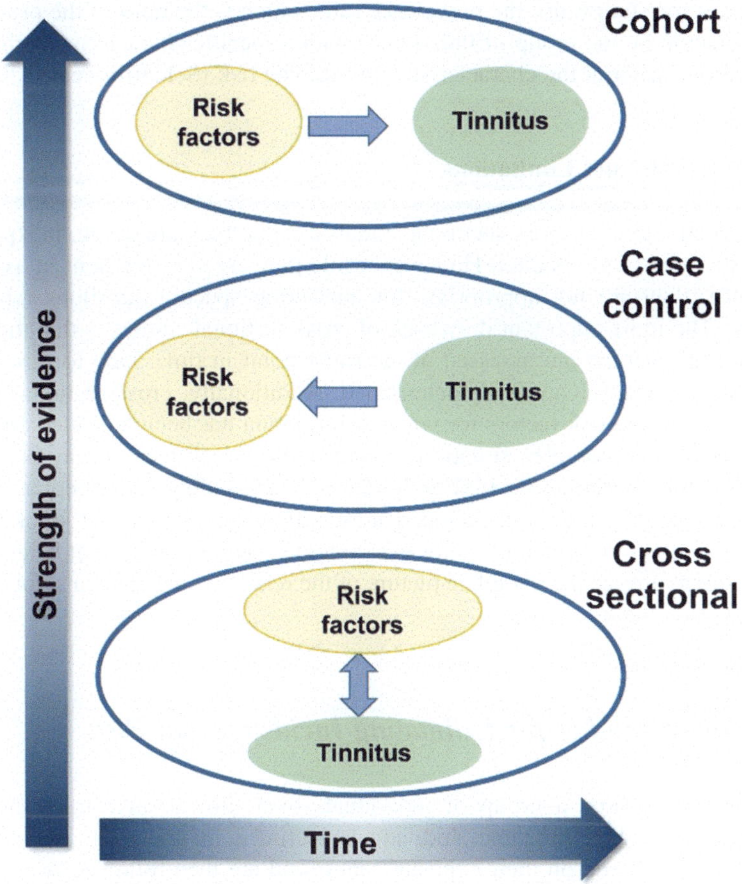

Fig. 2 Study designs, timeline for risk factors and tinnitus, and strength of evidence

sectional studies are particularly important to evaluate the prevalence of a disease or condition. However, while they can help in identifying the relationship between a potential risk factor and tinnitus, they cannot establish causality. In tinnitus research most studies are cross-sectional and usually provide estimates of prevalence. In Sect. 1.1, we provide examples of three prevalence studies that assess point, period, and lifetime prevalence. All three of them are assessed by population-based cross-sectional studies (Bhatt et al. 2016; McCormack et al. 2014; Oiticica and Bittar 2015)

Prevalence odds ratio is the most commonly used measure of effect in cross-sectional studies. Prevalence OR is measured in the same way as conventional OR in case-control studies (Sect. 3.3). For example, McCormack et al. (2014) analyzed the link between neuroticism and tinnitus and found that neuroticism increased odds of having tinnitus (OR = 2.11; 95% CI: 2.00, 2.22) (McCormack et al. 2014).

However, some studies use the prevalence ratio which is the ratio of the prevalence of the outcome in one group of individuals with a specific characteristic relative to another group without the characteristic, to estimate risk (Schiaffino et al. 2003).

3.1.1 Strengths and Limitations

The major strengths of cross-sectional studies are that they are quick, inexpensive, and relatively easy to conduct. However, while they are very efficient in assessing prevalence, they are not appropriate for answering specific questions related to causality. The most important drawback of cross-sectional studies is that the exposure and the outcome are assessed at the same point in time. Due to this lack of temporality, causality cannot be determined. Additionally, cross-sectional studies only assess specific risk factors for which information has been collected and often cannot disentangle complex situations where multiple risk factors may be interrelated. Therefore, observations may be biased by confounding factors not accounted for. For example, if a dataset included information only on current smoking and tinnitus, the effect of past smoking or ever smoking on tinnitus cannot be explored, and one might not see the complete picture of the relationship between smoking and tinnitus.

3.2 Cohort Studies for Estimating Incidence and Risk

Cohort studies follow a group of individuals over time to investigate how the exposures affect their outcome. Individuals recruited in the study are classified into two groups based on their exposure status and are then followed over time to see who develops tinnitus in each group (Fig. 2). Cohort studies can be retrospective or prospective. In prospective studies, condition or disease-free individuals are followed into a future time period to study the development of an outcome from the current time into the future. Retrospective cohort studies are known as historical cohort studies as both exposure and outcome are assessed retrospectively. Here the investigators retrospectively identify a disease-free and "at risk" cohort and use available records to determine exposure status at the beginning of observation period and the subsequent outcome status of the subjects.

In the tinnitus literature, the EHLS is an example of a prospective population-based cohort study that followed individuals who were tinnitus-free in the baseline period (1993 to 1995) for 10 years. In addition to providing the incidence proportion of tinnitus in the study population, this study also reported important associations between tinnitus and potential risk factors. Among 3,753 baseline participants in this study, 3,429 were tinnitus-free. This subset was considered to be at risk of developing tinnitus. Four follow-up surveys were conducted in 1995–1997, 1998–2000, 2000–2002, and 2003–2005. Over this period, 2,922 individuals aged 48 to 92 years provided information on their tinnitus status at least once. Apart from reporting

10-year incidence proportion of 13%, the EHLS data also provided information on various potential risk factors for tinnitus (Nondahl et al. 2010).

In cohort studies, two metrics can be calculated: relative risk and hazard ratio. For example, in EHLS, HR was used to report that the risk of developing tinnitus was higher in subjects having a history of arthritis (Hazards ratio (HR) = 1.37, 95% CI 1.08, 1.73), history of head injury (HR = 1.76, 95% CI 1.40, 2.22), having smoked ever (HR = 1.40, 95% CI 1.10, 1.79), and in women having hearing loss (HR = 2.59, 95% CI 1.79, 3.74). Alcohol consumption (HR = 0.63, 95% CI 0.41, 0.96 for $\geq$141 g/week vs. $<$15 g/week, where 141 g/week is roughly equivalent to 12 light beers per week) was reported to be associated with decreased risk of tinnitus. In men, obesity (HR = 0.55, 95% CI 0.39, 0.78) and increasing age in women (HR = 0.90 for every 5-year increase in age, 95% CI 0.81, 0.99) were also reported to reduce the risk of tinnitus (Nondahl et al. 2010).

3.2.1 Strengths and Limitations

Cohort studies are not always practicable, and in tinnitus research, there are few examples of long-term cohort studies (Martinez et al. 2015; Nondahl et al. 2010, 2011). They come with some inherent advantages and disadvantages. As the individuals recruited are disease-free at baseline, cohort studies are beneficial for establishing the temporal sequence of exposure and outcome and also for assessing incidence. Prospective cohort studies are the best designs to definitely establish causal associations between tinnitus and related risk factors. They are also particularly useful for evaluating the effects of rare exposures, as investigators purposely identify adequate number of individuals with the specific exposure. Moreover, multiple outcomes of the same exposure can be studied simultaneously. Prospective cohort studies are particularly advantageous from the perspective of data quality. The investigator can tailor the data collection methods to have complete information and simultaneously collect data on specific exposures, and data collection is less vulnerable to recall bias. Nonetheless, they are expensive, have long follow-up periods, and, consequently, high loss to follow-up. Benefits of retrospective studies are that investigators use existing data, but this gives limited control over data collection increasing the likelihood of incomplete or inconsistently measured data.

3.3 Case-Control Studies for Estimating Risk

Case-control studies are used to quantify the degree of associations between an exposure (risk factors) and an outcome (e.g., tinnitus). Typically, researchers first identify the cases, that is, individuals having the outcome, and then select an appropriate control group, comprising of individuals without the outcome (Fig. 2). A case-control study on tinnitus, therefore, comprises of two groups: tinnitus cases and non-tinnitus controls. The prevalence of exposures is then compared between the cases and controls. Risk factors can either be beneficial (reduce risk of developing tinnitus) or harmful (increase the risk of developing tinnitus). Case-control studies are retrospective by definition, since the researchers start with the outcome and then retrospectively look in the past for the possible exposures the subjects might have had to a risk factor. Since subjects are selected on the basis of outcome, estimates of incidence and prevalence cannot be measured using case-control studies.

A hallmark of a quality case-control study is one which has a clear and a priori case definition to ensure consistent identification. Selected cases can be either incident or prevalent. Incident cases tend to have a smaller risk of recall bias for exposures because they are newly diagnosed, and it is also easier to assess temporality between exposure and outcome. Selected controls should be representative of the population from which the cases are drawn and well matched to the cases in order to minimize confounding factors affecting the risk of developing the outcome. Common matching parameters include age, sex, and geographical location, but it should be kept in mind that once matched for a factor, the cases and controls can no longer be compared for that factor. In this way, overmatching should be avoided.

As one example, a hospital-based case-control study examined the relationship between tinnitus and mobile phone use (Hutter et al. 2010). The work was conducted at the Department of Ear, Nose and Throat (ENT) at the Medical University of Vienna, Austria. Cases included consecutive patients aged 16 to 80 years with appointments at the outpatient department. Cases were defined as "patients suffering from sound sensations not attributable to external sources and presenting at the ENT outpatient unit after November 2003 (until the projected number of cases was reached in November 2004)." The first date of occurrence of sound sensation was regarded as tinnitus onset. Controls were recruited from the same department matched with cases for age, sex, and ethnic group and within 3 weeks after enrollment of the respective cases. Exposure data, that is, mobile phone use, was censored at the date of tinnitus onset for cases. For controls, the index date was same as that of tinnitus onset in the matched case (Hutter et al. 2010). In case-control studies, ORs are generated as a statistic to quantify risk. In the given example, when comparing mobile phone use in cases and controls, the researchers used the reference category "never use of a mobile phone (prior to the index date) for intensity of use and never use or use for less than 1 year for duration of use." Results showed that mobile phone use had a positive OR of 1.37 but was not statistically significant (95% CI 0.73 to 2.57). Other measures like "average daily duration of use for 10 min or more" (OR = 1.71; 95% CI: 0.85 to 3.45) and "cumulative hours of use" (OR = 1.57; 95% CI: 0.78 to 3.19) showed increased but not statistically significant ORs. Using a mobile phone for 4 years or more did yield a statistically significant increased OR of 1.95 (95% CI 1.00 to 3.80) (Hutter et al. 2010). Therefore, for correct reporting of case-control studies, apart from clearly defining cases and controls, it is important to appropriately state the exposure categories, reference level, and measure of association.

3.3.1 Strengths and Limitations

The retrospective nature of data collection does introduce some limitations to case-control studies. They can be subject to participant recall bias and this will affect risk estimates. In their study, Hutter et al. (2010) pointed out that mobile phone users tended to underestimate the number of calls per month and overestimate duration. In addition, light users were likely to underestimate use, and heavy users were likely to overestimate use (Hutter et al. 2010). Another common bias is observer bias. This describes the situation in which an interviewer records exposure information differently depending on whether the interviewee is a case or control. Observer bias can be avoided by blinding the interviewer to the subject group. These biases can make it quite difficult to determine the temporal sequence of exposure and outcome.

Despite these limitations, case-control studies are cost-effective, efficient, and often less time-consuming. They are particularly advantageous for rare diseases or when little information is currently available on the association between a risk factor and a condition. For tinnitus therefore, case-control studies would be informative. They can establish degrees of association between tinnitus and potential risk factors, which could then inform the design of prospective cohort studies that could more definitely establish causal associations between tinnitus and related risk factors.

4 Some Examples of Risk Factors for Developing Tinnitus

As a symptom, tinnitus is associated with a number of communicable and non-communicable diseases including otological and neurological conditions, especially hearing loss. Epidemiological research is concerned with efforts to identify comorbidities or other factors affecting general health that present a risk for developing tinnitus. Table 1 illustrates conditions which have been reported to co-occur with tinnitus, some only anecdotally (Baguley et al. 2013). This list should therefore be considered a set of potential, not definitive, risk factors.

The following sections describe four of the tinnitus-related risk factors explored in the literature of date.

4.1 Hearing Loss or Hearing Difficulty

Multiple studies have reported hearing impairment to be a major risk factor for tinnitus (Baguley et al. 2013; Davis and Rafaie 2000; Michikawa et al. 2010; Sindhusake et al. 2004), whether it be a self-reported estimate of hearing difficulty or audiometric measure of hearing loss. Globally approximately 7% of the population suffers from disabling hearing loss which is defined as hearing loss of ≥35 dB in the better ear (GBD 2015; Wilson et al. 2017). It is thus evident that hearing loss is a

Table 1 Potential tinnitus-related risk factors (adapted from Baguley et al. 2013)

Otological	Hearing loss, noise exposure, infections, neoplasms, Meniere's disease
Neurological	Meningitis, migraine, epilepsy
Traumatic	Head and neck injury, blast injury
Other medical comorbidities	Hypertension, diabetes, rheumatoid arthritis
Psychological	Anxiety, depression
Lifestyle	Smoking, alcohol consumption, obesity (body mass index), caffeine intake
Ototoxic medications	Antibiotics, antineoplastic drugs, nonsteroidal anti-inflammatory drugs

common otological condition affecting a sizeable part of the global population. Thus, both tinnitus and hearing loss are common medical conditions that are mutually related (Moore et al. 2017; Nondahl et al. 2011; Ratnayake et al. 2009). Therefore, its only reasonable to examine the evidence of increased risk for tinnitus in individuals affected by hearing impairment.

EHLS is a population-based prospective cohort study conducted in the USA to evaluate the characteristics of hearing loss in older adults (48 to 92 years) (Cruickshanks et al. 1998). In this study, audiological tests were used to assess hearing loss, and significant tinnitus was defined as "buzzing, ringing, or noise in the ears in the past year that was at least moderate in severity or that caused problems getting to sleep." In the 5-year follow-up, they found that the likelihood of significant tinnitus increased in participants with hearing loss (odds ratio = 3.90, 95% confidence interval (CI) 2.89, 5.27), after adjusting for age, sex, cardiovascular disease, and head injury (Nondahl et al. 2002). Similar results were seen in the 10-year follow-up but only for women (Hazards ratio = 2.59, 95% CI 1.79, 3.74). In other words, women with hearing loss were found to have 2.59 times increased likelihood of having tinnitus (Nondahl et al. 2010).

In an internet-based study on participants aged 17 to 75 years in the UK and USA, Moore et al. (2017) assessed hearing difficulty using the self-report question – "Do you currently have any difficulty with your hearing?" with five response options ("no difficulty/slight difficulty/moderate difficulty/great difficulty/cannot hear at all"). They assessed tinnitus using the question "How often nowadays do you get tinnitus (noises such as ringing or buzzing in your heard or ears) that lasts for more than 5 min?" again with five response options ("never/rarely/sometimes/usually/constantly"). They found a strong correlation ($r = 0.33$) between increased tinnitus and increasing hearing difficulty, which was present even after adjusting for age and noise exposure (Moore et al. 2017). In a cross-sectional study on German teachers, conducted to assess the prevalence and comorbidity of hearing loss, hyperacusis, and tinnitus, the Mini-Tinnitus Questionnaire (Mini-TQ) was used. Overall, 10.8% subjects were found to experience both tinnitus and hearing difficulty, and there was significant increase in tinnitus-related distress in the presence of self-reported hearing difficulty (Meuer and Hiller 2015).

It is worthwhile recognizing some of the methodological challenges in accurately identifying the relationship between tinnitus and hearing impairment. Hearing loss and hearing difficulty are quite separate concepts that are not mutually inclusive. People with normal audiometric threshold can report hearing difficulty, while other people with audiometric hearing loss can report good hearing ability (Curti et al. 2019; Ramage-Morin et al. 2019). Moreover while self-reported measures of hearing difficulty are subject to reporting bias, even laboratory assessments of hearing do not provide a definitive diagnosis on whether or not there is damage to the hearing system. One obvious limitation is the assessment of high-frequency hearing ability (> 8 kHz) which is not part of any standard clinical audiometric assessment. Another is the recent debate on "hidden hearing loss" which sheds light on the possibility that many people have cochlear damage that impairs hearing but that this is on a subclinical level which cannot be tested using current psychometric tests (Prendergast et al. 2017a, 2017b).

Few population-based surveys have examined tinnitus prevalence in populations with hearing difficulty. But the caveat to these is that, where epidemiological estimates refer to hearing, the observed findings might not necessarily reflect the true nature of the relationship with tinnitus. Moreover, since most evidence is from cross-sectional studies, it is not possible to conclude causality (Meuer and Hiller 2015; Moore et al. 2017). Strong associations are typically reported from available analytical observational data. Therefore, in order to confirm these trends in the relationship between tinnitus and hearing loss, tinnitus researchers could consider study designs that not only provide information on the relationship between variables but also on the direction of association.

4.2 Age

Multiple studies have shown an increase in tinnitus prevalence and in its reported severity as a function of age. Bhatt et al. (2016) conducted their analysis on a nationally representative data in the USA and found that the group that reported tinnitus symptoms was older than the group without symptoms, with mean difference of 8 years. They also reported a direct correlation between increased age and increased tinnitus prevalence ($r = 0.08$) (Bhatt et al. 2016). Gallus et al. (2015) conducted a prevalence study in a nationally representative population in Italy finding that the prevalence of any tinnitus was 3% in individuals younger than 45 years, while it was 9% in individuals older than 45 years of age. They also reported that among study participants who reported severe tinnitus, 97% were of age older than 45 years (Gallus et al. 2015). McCormack et al. (2014) also reported a trend for bothersome tinnitus in both men and women with increasing age, in a study conducted on UK Biobank data (McCormack et al. 2014). Given the age-dependent increase of tinnitus, when pooling data from various studies or when comparing prevalence in different populations, one needs to be careful of the age range. In terms of age at peak prevalence, Shargorodsky et al. (2010) reported peak prevalence of

31.4% in the 60–69 years' age group. In their systematic review, McCormack et al. (2016) also noted a peak prevalence at around 70 years of age (McCormack et al. 2016). In terms of methodology, McCormack et al. (2016) noted inconsistency in the definition of age bands across studies (McCormack et al. 2016). In epidemiological studies, the preferred method of age group distribution is either in 5-year bands (e.g., 20–24, 25–29, 30–34, and so on) or in mid-decade to mid-decade 10-year age groups (25–34, 35–44, 45–54, and so on) (World Health Organization 1999). However, only 2 of the 25 relevant studies have followed this reporting convention (McCormack et al. 2016). The rest used a range of alternatives such as 30–39, 40–49, 50–59 and 18–44, 45–64, and ≥65 years. This disparity can limit the ability to make comparison across studies.

4.3 Sex

The published literature provides no clear consensus on the direction of association between sex and tinnitus. For example, Bhatt et al. (2016) found tinnitus to be 2% more prevalent in men than in women, but with no difference in severity (Bhatt et al. 2016). On the other hand, Gallus et al. (2015) reported no significant difference in prevalence of any tinnitus between men and women but double the prevalence of severe tinnitus in women (1.6%) compared to men (0.8%) (odds ratio = 3.26; 95% CI 1.28,8.21) (Gallus et al. 2015). McCormack et al. (2014) also found current tinnitus to be 4% more common in men than women, and like Gallus and colleagues, they reported women to have slightly higher prevalence of bothersome tinnitus (4.1% in women and 3.5% in men) (McCormack et al. 2014). Park and Moon analyzed data from the Korean National Health and Nutrition Examination Survey (KNHANES) to report women having increased likelihood of any tinnitus compared to men (odds ratio 1.22; 95% CI 1.08, 1.37), although the likelihood of annoying tinnitus did not differ across the two sexes (Park and Moon 2014).

It is worth considering whether study-specific differences in sex could be attributed to differences in the sociocultural and behavioral structure across countries. However, while plausible, this explanation seems unlikely because even studies conducted in same countries report contradictory findings (Bhatt et al. 2016; Nondahl et al. 2002). In conclusion, from currently available information, it is not possible to predict if sex has an effect on tinnitus prevalence. More research needs to be done in the future to understand if sex difference has a role in tinnitus or if confounding factors can explain the apparent effects of sex difference.

4.4 Lifestyle

Smoking, alcohol consumption, obesity (which can be determined by body mass index), and caffeine intake are all lifestyle-related risk factors with a hypothesized relationship to tinnitus (Baguley et al. 2013; Figueiredo et al. 2014). These modifiable risk factors can point toward prevention strategies for reducing tinnitus burden. Nonetheless, results from the current literature are somewhat inconsistent. For example, with respect to smoking, some studies have shown an increased risk of tinnitus in smokers (Kim et al. 2015; Nondahl et al. 2010, 2011), while others have found no such relationship (Gallus et al. 2015; Park and Moon 2014). A recent systematic review and meta-analysis conducted by Veile et al. (2018) on 20 studies related to smoking and tinnitus found an increased tinnitus risk in both current and former smokers (Veile et al. 2018). This study pooled data mostly from cross-sectional studies since there is a lack of data from analytical (case-control and cohort) studies (Veile et al. 2018). Therefore, even though the relationship was statistically significant, the authors could not conclude causality. Obesity, measured as body mass index $\geq$30 kg/m^2, has been reported to increase of risk of tinnitus significantly (Gallus et al. 2015; Shargorodsky et al. 2010). However, one study found no relationship (Kim et al. 2015), and another found a decreased risk (Nondahl et al. 2010).

5 Future Directions

Epidemiological research in tinnitus is plagued by several major gaps in our understanding which identify specific priorities for future research The first calls for standardization in study methods, wherever possible, in order to enhance the confidence in pooling prevalence and incidence estimates across studies. The second calls for investigators to overcome geographical biases and ensure that world regions such as Asia and Africa are better represented in our understanding of tinnitus as a global problem. The third calls for greater efforts to conduct analytical studies in order to better understand the cause-and-effect relationship between potential risk factors and tinnitus.

Standardization of Methods There is a wide range of prevalence estimates, beyond what is expected due to sociodemographic differences. Many authors have highlighted this issue and the need for cross-culturally adapted survey questions instead of verbatim translations. Biswas et al. (2019) addressed this concern and provided standardized questions on tinnitus prevalence translated into 12 European languages (Biswas et al. 2019). Using these standardized questions, a recent study calculated prevalence estimates for tinnitus across Europe. From 12 member states of the European Union (EU), Biswas et al. (2020) found a 15% prevalence for any tinnitus, equivalent to one in seven adults. Despite the standardization in questions and other study methodology, there were some country-specific differences observed, notably with the range in prevalence varying from 9% in Ireland up to 28% in Bulgaria (Biswas et al. 2020). This range is at least broadly consistent with other prevalence estimates (e.g., Bhatt et al. 2016), giving us confidence in accuracy and precision.

Overcoming Biases Most epidemiological research has been limited to Europe and North America, and so our knowledge of tinnitus prevalence is biased toward these world regions (McCormack et al. 2016). As a global community, tinnitus researchers have little insight on the status of tinnitus across continents such as Asia and Africa. Perhaps the rise in interest in global epidemiology in relation to noninfectious chronic conditions, such as cardiovascular disease and diabetes, will open up avenues to developing a more global perspective on the epidemiology for other health conditions such as tinnitus. Certainly the global estimates on disabling hearing loss conducted by the World Health Organization (2018) demonstrate how global epidemiology is possible and highlight how the power of this knowledge can be harnessed to prioritize research and healthcare resources. Supporting and encouraging greater tinnitus research in such countries would also raise priorities for qualitative research necessary to better understand population awareness, attitude, and stigma associated with tinnitus.

Analytical Study Designs The final gap is the lack of analytical studies used to investigate the cause-and-effect relationship between potential risk factors and tinnitus. It is possible to conduct hospital-based case-control studies from ENT clinics or population-based case-control studies from existing healthcare or health insurance databases, by selecting tinnitus cases and tinnitus-free controls matched for index dates. However, to obtain accurate results, in both cases, reliable coding of tinnitus will be essential. In tinnitus research, there are very few examples of such case-control studies (Hutter et al. 2010; Koo and Hwang 2017).

Cohort studies, particularly prospective cohort studies, provide the strongest evidence in risk factor analysis. Martinez et al. (2015) conducted a retrospective cohort study using national healthcare records (Martinez et al. 2015). Additionally, several large population-based prospective cohorts have explored the relationship between tinnitus and various risk factors. Examples include the EHLS, BMHS, Beaver Dam Offspring Study (BOSS), Jackson Heart Study, Nurses' Health Study (I and II), Conservation of Hearing Study (CHEARS), and the UK Biobank (Cruickshanks et al. 1998; Glicksman et al. 2014; House et al. 2018; McCormack et al. 2014; Nondahl et al. 2011; Sindhusake et al. 2004). While conducting new, large, well-designed, prospective cohort studies is needed, perhaps it would also be worthwhile to exhaust the existing cohort datasets to fully examine the interrelations between those risk factors listed in Table 1.

6 Synopsis of Key Points

- We estimate that tinnitus affects about 15% of the adult population, equivalent to one in seven adults.
- We recommend a standardized question for assessing tinnitus prevalence: "Over the past year, have you had noises (such as ringing or buzzing) in your head or in one or both ears that lasts for more than five minutes at a time?," with the

following response options: Yes, most or all of time/Yes, a lot of the time/Yes, some of the time/No, not in the past year/No, and never/Do not know/Prefer not to answer.

- Case-control and retrospective cohort studies can inform our interpretations of associations between tinnitus and potential risk factors, but prospective cohort studies are the best study design to definitely establish causal associations between tinnitus and related risk factors.
- There is strong evidence to support hearing loss as a risk factor for developing tinnitus.

References

Allen N, Sudlow C, Downey P, Peakman T, Danesh J, Elliott P, Gallacher J, Green J, Matthews P, Pell J, Sprosen T, Collins R, on behalf of UK Biobank (2012) UK Biobank: current status and what it means for epidemiology. Health Policy Technol 1(3):123–126. https://doi.org/10.1016/j.hlpt.2012.07.003

Baguley D, McFerran D, Hall D (2013) Tinnitus. Lancet 382(9904):1600–1607. https://doi.org/10.1016/s0140-6736(13)60142-7

Benson K, Hartz AJ (2000) A comparison of observational studies and randomized, controlled trials. N Engl J Med 342(25):1878–1886. https://doi.org/10.1056/nejm200006223422506

Bhatt JM, Lin HW, Bhattacharyya N (2016) Prevalence, severity, exposures, and treatment patterns of tinnitus in the United States. *JAMA Otolaryngol Head Neck Surg* 142(10):959–965. https://doi.org/10.1001/jamaoto.2016.1700

Biswas R, Lugo A, Gallus S, Akeroyd MA, Hall DA (2019) Standardized questions in English for estimating tinnitus prevalence and severity, hearing difficulty and usage of healthcare resources, and their translation into 11 European languages. *Hear Res*. 377:330–338. https://doi.org/10.1016/j.heares.2019.02.008

Biswas R, Lugo A, Hall DA, Akeroyd MA, Liu X, Schlee W, Gallus S (2020) A study of the pan-European prevalence of tinnitus and hearing difficulty using a standardized set of questions. 2020. In: Poster presented at Association of Research in Otolaryngology 43rd Annual MidWinter Meeting. San Jose, California, USA. Abstract retrieved from https://aro.org/wp-content/uploads/2020/02/2020-Abstracts_1-21-20-Web.pdf. Accessed 06 Apr 2020

Concato J, Shah N, Horwitz RI (2000) Randomized, controlled trials, observational studies, and the hierarchy of research designs. N Engl J Med 342(25):1887–1892. https://doi.org/10.1056/nejm200006223422507

Cruickshanks KJ, Wiley TL, Tweed TS, Klein BE, Klein R, Mares-Perlman JA, Nondahl DM (1998) Prevalence of hearing loss in older adults in beaver dam, Wisconsin. The epidemiology of hearing loss study. Am J Epidemiol 148(9):879–886. https://doi.org/10.1093/oxfordjournals.aje.a009713

Curti SA, Taylor EN, Su D, Spankovich C (2019) Prevalence of and characteristics associated with self-reported good hearing in a population with elevated audiometric thresholds. JAMA Otolaryngol Head Neck Surg 145(7):626–633. https://doi.org/10.1001/jamaoto.2019.1020

Davis A, Rafaie EA (2000) Epidemiology of tinnitus. In: Tyler R (ed) Tinnitus handbook, vol 1. Singular, San Diego, p 23

Dawes P, Fortnum H, Moore DR, Emsley R, Norman P, Cruickshanks K et al (2014) Hearing in middle age: a population snapshot of 40- to 69-year olds in the United Kingdom. Ear Hear 35(3):e44–e51. https://doi.org/10.1097/aud.0000000000000010

Degeest S, Keppler H, Corthals P, Clays E (2017) Epidemiology and risk factors for tinnitus after leisure noise exposure in Flemish young adults. Int J Audiol 56(2):121–129. https://doi.org/10.1080/14992027.2016.1236416

Figueiredo RR, Rates MJ, Azevedo AA, Moreira RK, Penido Nde O (2014) Effects of the reduction of caffeine consumption on tinnitus perception. Braz J Otorhinolaryngol 80(5):416–421. https://doi.org/10.1016/j.bjorl.2014.05.033

Folmer RL, McMillan GP, Austin DF, Henry JA (2011) Audiometric thresholds and prevalence of tinnitus among male veterans in the United States: data from the National Health and nutrition examination survey, 1999-2006. J Rehabil Res Dev 48(5):503–516. https://doi.org/10.1682/jrrd.2010.07.0138

Gallus S, Lugo A, Garavello W, Bosetti C, Santoro E, Colombo P et al (2015) Prevalence and determinants of tinnitus in the Italian adult population. Neuroepidemiology 45(1):12–19. https://doi.org/10.1159/000431376

GBD (2015) Disease and Injury Incidence and Prevalence Collaborators. Global, regional, and national incidence, prevalence, and years lived with disability for 310 diseases and injuries, 1990–2015: a systematic analysis for the Global Burden of Disease Study 2015. Lancet 388 (10053):1545–1602. https://doi.org/10.1016/s0140-6736(16)31678-6

Gibrin PC, Melo JJ, Marchiori LL (2013) Prevalence of tinnitus complaints and probable association with hearing loss, diabetes mellitus and hypertension in elderly. Codas 25(2):176–180. https://doi.org/10.1590/s2317-17822013000200014

Glicksman JT, Curhan SG, Curhan GC (2014) A prospective study of caffeine intake and risk of incident tinnitus. Am J Med 127(8):739–743. https://doi.org/10.1016/j.amjmed.2014.02.033

Goldstein E, Ho CX, Hanna R, Elinger C, Yaremchuk KL, Seidman MD, Jesse MT (2015) Cost of care for subjective tinnitus in relation to patient satisfaction. Otolaryngol Head Neck Surg 152 (3):518–523. https://doi.org/10.1177/0194599814566179

Hall DA, Fackrell K, Li AB, Thavayogan R, Smith S, Kennedy V et al (2018) A narrative synthesis of research evidence for tinnitus-related complaints as reported by patients and their significant others. Health Qual Life Outcomes 16(1):61. https://doi.org/10.1186/s12955-018-0888-9

Hannula S, Bloigu R, Majamaa K, Sorri M, Maki-Torkko E (2011) Self-reported hearing problems among older adults: prevalence and comparison to measured hearing impairment. J Am Acad Audiol 22(8):550–559. https://doi.org/10.3766/jaaa.22.8.7

Hasson D, Theorell T, Westerlund H, Canlon B (2010) Prevalence and characteristics of hearing problems in a working and non-working Swedish population. J Epidemiol Community Health 64(5):453–460. https://doi.org/10.1136/jech.2009.095430

Hoekstra CE, Wesdorp FM, van Zanten GA (2014) Socio-demographic, health, and tinnitus related variables affecting tinnitus severity. Ear Hear 35(5):544–554. https://doi.org/10.1097/aud.0000000000000045

House L, Bishop CE, Spankovich C, Su D, Valle K, Schweinfurth J (2018) Tinnitus and its risk factors in African Americans: the Jackson heart study. Laryngoscope 128(7):1668–1675. https://doi.org/10.1002/lary.26964

Hutter HP, Moshammer H, Wallner P, Cartellieri M, Denk-Linnert DM, Katzinger M et al (2010) Tinnitus and mobile phone use. Occup Environ Med 67(12):804–808. https://doi.org/10.1136/oem.2009.048116

Jun HJ, Ok S, Tyler R, Hwang SY, Chae S (2015) Is Hypozincemia related to tinnitus?: a population study using data from the Korea National Health and nutrition examination survey. Clin Exp Otorhinolaryngol 8(4):335–338. https://doi.org/10.3342/ceo.2015.8.4.335

Kim HJ, Lee HJ, An SY, Sim S, Park B, Kim SW et al (2015) Analysis of the prevalence and associated risk factors of tinnitus in adults. PLoS One 10(5):e0127578. https://doi.org/10.1371/journal.pone.0127578

Koo M, Hwang JH (2017) Risk of tinnitus in patients with sleep apnea: a nationwide, population-based, case-control study. Laryngoscope 127(9):2171–2175. https://doi.org/10.1002/lary.26323

Last JM (1988) A dictionary of epidemiology. Oxford University Press, New York

Maes IH, Cima RF, Vlaeyen JW, Anteunis LJ, Joore MA (2013) Tinnitus: a cost study. Ear Hear 34 (4):508–514. https://doi.org/10.1097/AUD.0b013e31827d113a
Martinez C, Wallenhorst C, McFerran D, Hall DA (2015) Incidence rates of clinically significant tinnitus: 10-year trend from a cohort study in England. Ear Hear 36(3):e69–e75. https://doi.org/10.1097/aud.0000000000000121
McCormack A, Edmondson-Jones M, Fortnum H, Dawes P, Middleton H, Munro KJ, Moore DR (2014) The prevalence of tinnitus and the relationship with neuroticism in a middle-aged UK population. J Psychosom Res 76(1):56–60. https://doi.org/10.1016/j.jpsychores.2013.08.018
McCormack A, Edmondson-Jones M, Somerset S, Hall D (2016) A systematic review of the reporting of tinnitus prevalence and severity. Hear Res 337:70–79. https://doi.org/10.1016/j.heares.2016.05.009
Meuer SP, Hiller W (2015) The impact of hyperacusis and hearing loss on tinnitus perception in German teachers. Noise Health 17(77):182–190. https://doi.org/10.4103/1463-1741.160682
Michikawa T, Nishiwaki Y, Kikuchi Y, Saito H, Mizutari K, Okamoto M, Takebayashi T (2010) Prevalence and factors associated with tinnitus: a community-based study of Japanese elders. J Epidemiol 20(4):271–276. https://doi.org/10.2188/jea.JE20090121
Moore DR, Zobay O, Mackinnon RC, Whitmer WM, Akeroyd MA (2017) Lifetime leisure music exposure associated with increased frequency of tinnitus. Hear Res 347:18–27. https://doi.org/10.1016/j.heares.2016.10.030
National Institute of Mental Health (2017) What is prevalence? https://www.nimh.nih.gov/health/statistics/what-is-prevalence.shtml. Accessed 24 Mar 2020
Nondahl DM, Cruickshanks KJ, Wiley TL, Klein R, Klein BE, Tweed TS (2002) Prevalence and 5-year incidence of tinnitus among older adults: the epidemiology of hearing loss study. J Am Acad Audiol 13(6):323–331
Nondahl DM, Cruickshanks KJ, Wiley TL, Klein R, Klein BE, Tweed TS (2004) Comment on tinnitus in older adults from the Blue Mountains study by Sindhusake et al and comparison with tinnitus data from the epidemiology of hearing loss study. Int J Audiol 43(1):60
Nondahl DM, Cruickshanks KJ, Dalton DS, Klein BE, Klein R, Schubert CR et al (2007) The impact of tinnitus on quality of life in older adults. J Am Acad Audiol 18(3):257–266. https://doi.org/10.3766/jaaa.18.3.7
Nondahl DM, Cruickshanks KJ, Wiley TL, Klein BE, Klein R, Chappell R, Tweed TS (2010) The ten-year incidence of tinnitus among older adults. Int J Audiol 49(8):580–585. https://doi.org/10.3109/14992021003753508
Nondahl DM, Cruickshanks KJ, Huang GH, Klein BE, Klein R, Nieto FJ, Tweed TS (2011) Tinnitus and its risk factors in the beaver dam offspring study. Int J Audiol 50(5):313–320. https://doi.org/10.3109/14992027.2010.551220
Oiticica J, Bittar RS (2015) Tinnitus prevalence in the city of Sao Paulo. Braz J Otorhinolaryngol 81 (2):167–176. https://doi.org/10.1016/j.bjorl.2014.12.004
Park RJ, Moon JD (2014) Prevalence and risk factors of tinnitus: the Korean National Health and nutrition examination survey 2010-2011: a cross-sectional study. Clin Otolaryngol 39(2):89–94. https://doi.org/10.1111/coa.12232
Prendergast G, Guest H, Munro KJ, Kluk K, Léger A, Hall DA, Heinz MG, Plack CJ (2017a) Effects of noise exposure on young adults with normal audiograms I: electrophysiology. Hear Res 344:68–81. https://doi.org/10.1016/j.heares.2016.10.028
Prendergast G, Millman RE, Guest H, Munro KJ, Kluk K, Dewey RS, Hall DA, Heinz MG, Plack CJ (2017b) Effects of noise exposure on young adults with normal audiograms II: behavioral measures. Hear Res 356:74–86. https://doi.org/10.1016/j.heares.2017.10.007
Quaranta A, Assennato G, Sallustio V (1996) Epidemiology of hearing problems among adults in Italy. Scand Audiol Suppl 42:9–13
Ramage-Morin PL, Banks R, Pineault D, Atrach M (2019) Unperceived hearing loss among Canadians aged 40 to 79. Health Rep 30(8):11–20. https://doi.org/10.25318/82-003-x201900800002-eng

Ratnayake SA, Jayarajan V, Bartlett J (2009) Could an underlying hearing loss be a significant factor in the handicap caused by tinnitus? Noise Health 11(44):156–160. https://doi.org/10.4103/1463-1741.53362
Schiaffino A, Rodriguez M, Pasarin MI, Regidor E, Borrell C, Fernandez E (2003) Odds ratio or prevalence ratio? Their use in cross-sectional studies. Gac Sanit 17(1):70–74. https://doi.org/10.1016/s0213-9111(03)71694-x
Shargorodsky J, Curhan GC, Farwell WR (2010) Prevalence and characteristics of tinnitus among us adults. Am J Med 123(8):711–718. https://doi.org/10.1016/j.amjmed.2010.02.015
Sindhusake D, Mitchell P, Newall P, Golding M, Rochtchina E, Rubin G (2003) Prevalence and characteristics of tinnitus in older adults: the Blue Mountains hearing study. Int J Audiol 42 (5):289–294
Sindhusake D, Golding M, Wigney D, Newall P, Jakobsen K, Mitchell P (2004) Factors predicting severity of tinnitus: a population-based assessment. J Am Acad Audiol 15(4):269–280. https://doi.org/10.3766/jaaa.15.4.2
Spankovich C, Bishop C, Johnson MF, Elkins A, Su D, Lobarinas E, Le Prell CG (2017) Relationship between dietary quality, tinnitus and hearing level: data from the national health and nutrition examination survey, 1999-2002. Int J Audiol 56(10):716–722. https://doi.org/10.1080/14992027.2017.1331049
Stockdale D, McFerran D, Brazier P, Pritchard C, Kay T, Dowrick C, Hoare DJ (2017) An economic evaluation of the healthcare cost of tinnitus management in the UK. BMC Health Serv Res 17(1):577. https://doi.org/10.1186/s12913-017-2527-2
Veile A, Zimmermann H, Lorenz E, Becher H (2018) Is smoking a risk factor for tinnitus? A systematic review, meta-analysis and estimation of the population attributable risk in Germany. BMJ Open 8(2):*e016589*. https://doi.org/10.1136/bmjopen-2017-016589
Welch D, Dawes PJ (2008) Personality and perception of tinnitus. Ear Hear 29(5):684–692. https://doi.org/10.1097/AUD.0b013e318177d9ac
Wilson BS, Tucci DL, Merson MH, O'Donoghue GM (2017) Global hearing health care: new findings and perspectives. Lancet 390(10111):2503–2515. https://doi.org/10.1016/s0140-6736(17)31073-5
World Health Organization (1999) WHO ear and hearing disorders survey protocol for a population-based survey of prevalence and causes of deafness and hearing impairment and other ear disorders. World Health Organization, Geneva. https://apps.who.int/iris/bitstream/handle/10665/67892/WHO_PBD_PDH_99.8(1).pdf;jsessionid=E4AB918E05F3980A4FF3F10FD5DE3830?sequence=1. Accessed 28 Mar 2020
World Health Organization (2018) Prevention of blindness and deafness. Estimates. World Health Organization, Geneva. https://www.who.int/pbd/deafness/estimates/en/. Accessed 06 Apr 2020
Xu X, Bu X, Zhou L, Xing G, Liu C, Wang D (2011) An epidemiologic study of tinnitus in a population in Jiangsu province, China. J Am Acad Audiol 22(9):578–585. https://doi.org/10.3766/jaaa.22.9.3
Zeman F, Koller M, Langguth B, Landgrebe M (2014) Which tinnitus-related aspects are relevant for quality of life and depression: results from a large international multicentre sample. Health Qual Life Outcomes 12:7. https://doi.org/10.1186/1477-7525-12-7

Genetic Inheritance and Its Contribution to Tinnitus

Sana Amanat, Alvaro Gallego-Martinez, and Jose A. Lopez-Escamez

Contents

S. Amanat and A. Gallego-Martinez
Otology and Neurotology Group CTS495, Department of Genomic Medicine, GENYO – Centre for Genomics and Oncological Research – Pfizer, University of Granada, Junta de Andalucía, PTS, Granada, Spain

J. A. Lopez-Escamez (✉)
Otology and Neurotology Group CTS495, Department of Genomic Medicine, GENYO – Centre for Genomics and Oncological Research – Pfizer, University of Granada, Junta de Andalucía, PTS, Granada, Spain

Department of Otolaryngology, Instituto de Investigación Biosanitaria ibs.GRANADA, Hospital Universitario Virgen de las Nieves, Universidad de Granada, Granada, Spain

Division of Otolaryngology, Department of Surgery, Universidad de Granada, Granada, Spain
e-mail: antonio.lopezescamez@genyo.es

Curr Topics Behav Neurosci (2021) 51: 29–48
https://doi.org/10.1007/7854_2020_155
Published Online: 24 July 2020

Abstract Tinnitus is the abnormal perception of sound that affects more than 15% of adult population around the globe. Severe tinnitus is considered a complex disorder that arises as result of the interaction of genetic and environmental factors, and it is associated with several comorbidities such as hearing loss, anxiety, and insomnia. We begin this review with an introduction to human molecular genetics and the role of genetic variation on the inheritance. There are some genetic reports on tinnitus heritability including concordance studies in twins and adoptees or aggregation in families providing some evidence for familial aggregation in patients with severe tinnitus and high concordance in monozygotic twins with bilateral tinnitus. So, sex differences in familial aggregation and heritability of bilateral tinnitus suggest a potential sexual dimorphism in tinnitus inheritance.

Molecular genetic studies have been demonstrated to be a useful tool to understand the role of genetic variation in rare diseases and complex disorders. The reported associations in common variants in neurotrophic factors such as *GDNF*, *BDNF*, or potassium channels genes were underpowered, and the lack of replication questions these findings. Although candidate gene approaches have failed in replicating these genetic associations, the development of high throughput sequencing technology and the selection of extreme phenotypes are strategies that will allow the clinicians and researchers to combine genetic information with clinical data to implement a personalized diagnosis and therapy in patients with tinnitus.

Keywords Extreme phenotype · Genetics · Heritability · Rare variants · Tinnitus

Abbreviations

A	Adenine
C	Cytosine
DNA	Deoxyribonucleic acid
G	Guanine
GWAS	Genome-wide association study
INDEL	Insertion-deletion
PCR	Polymerase chain reaction
RNA	Ribonucleic acid
SNV	Single-nucleotide variant
T	Thymine
WES	Whole-exome sequencing
WGS	Whole-genome sequencing

1 Introduction to Human Genetics

1.1 Molecular Genetics

Genetic information regarding different human traits is stored across the cells in the form of deoxyribonucleic acid (DNA). DNA molecules are formed by the repetition of four different nucleotide molecules shaping two helical deoxyribose-phosphate backbones (DNA strands). Those four bases are adenine (A) and guanine (G) as purines and cytosine (C) and thymine (T) as pyrimidines. Both pyrimidines and purines are complementary in A-T and G-C pairs. The base pairs build the large scaffold that encodes the entire genetic information of each cell. This information is organized in genes, sequences of nucleotides in DNA that encode the synthesis of a gene product. There are around 19,000 genes in the human genome, and each gene consists of coding (exons) and noncoding sequences (introns), leading to either RNA or proteins (Burgers and Kunkel 2017; Ekundayo and Bleichert 2019).

DNA is packaged inside the nucleus of the cell with proteins called histones forming the nucleosome and chromatin, preventing DNA damage, and regulating gene expression (transcription) and DNA replication. DNA replication is the biological mechanism that ensures the perpetuation of identical DNA molecules across cells. Also, DNA replication is the basis of inheritance. The DNA replication occurs during a phase of cell division known as interphase, where each strand of DNA molecule separates and endorses the production of their complementary strand to produce two identical DNA molecules. This mechanism involves the use of specialized proteins such as DNA polymerases for the insertion of the new bases on the new strand and helicases for the opening of both parental strands (Burgers and Kunkel 2017). This mechanism is also replicated in vitro to produce fragments of DNA than can be sequenced through next-generation sequencing in a process called polymerase chain reaction (PCR).

Transcription is the main process for the expression of different genes. Proteins involved in transcription process lead to the production of certain molecules called acid ribonucleic (RNA) which encode coding segments from the DNA. The main enzyme involved in this process is called RNA polymerase and works similarly to the DNA polymerases. However, this enzyme works in little "bubbles" in the DNA, commonly marked in the DNA by transcription factors, to translate the content of genes encoded in the DNA to the messenger (messenger RNA) that will be translated into a protein. DNA in this fashion acts as a general blueprint that needs to fill a new form in a moment of necessity. The transcription mechanism copies the blueprint on DNA as an easy to understand form called mRNA. This latter form will be translated in the protein through a translation mechanism (Bell and Dutta 2002; Johnson and O'Donnell 2005; Burgers and Kunkel 2017).

1.2 Genetic Variation

Replication machinery is a well-proofed mechanism for the duplication of DNA molecules. Proteins involved in this process usually produce mutations, errors on the insertion of nucleotides. However, DNA polymerases can find those errors, fixing them retroactively, and mark those DNA molecules as failed, so they can be deleted by a depuration mechanism lead by other polymerases (Burgers and Kunkel 2017).

The main mutations in a DNA sequence occur on the first division of the cells, during the germline cell development. Those mutations are carried through the entire development of the individual, leading to a mutation that can arrive to every tissue cell on the developed organism. These variations are usually rare variations, but they can be transmitted to an entire population due to the inheritance of those mutations, making them more common (Bomba et al. 2017).

In terms of population genetics, those differences are called variants. Common variants are usually single base-pair differences or single-nucleotide variants (Fig. 1). Most of the single-nucleotide variants do not have a direct effect on the future phenotype of the individual; however, other structural changes in the DNA sequence (like insertions, deletions, etc.) can result in a permanent damage in the proteins they produce, reducing the probability of its spreading on the population and leading to variants with a very low frequency on the population (also called rare variants).

These variations can occur elsewhere in the DNA molecule, either in genes or intergenic regions; however, most of the genes have an unknown biological function, and the clinical effect on phenotype has not been established.

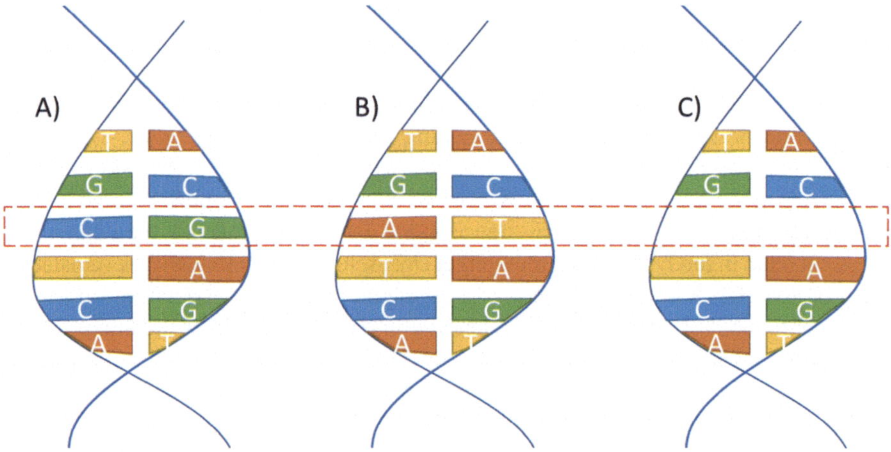

Fig. 1 Types of variations in the DNA. (**a**) Example of reference DNA chain. Color code represents the four nucleotides of the DNA (A,T,G,C). (**b**) Example of single-nucleotide variant. C-G position changed to A-T. (**c**) Example of indel variation. The lack of the C-G position represents a deletion

The contribution of genetic variation is usually modest to small in most diseases (Duzkale et al. 2013). Pathogenicity depends on several factors, and it can be predicted by bioinformatic tools through protein structure stability prediction, conservation of the DNA structure algorithms, or multilevel analysis (Requena et al. 2017; Sun and Yu 2019). Most of the pathogenicity prediction tools usually include different methods for pathogenicity assessment; however, damage in the protein structure or function is only a grain of sand in the explanation of the phenotype in a complex trait such tinnitus. Most of the changes in the later phenotype could occur in other genes affecting the production of a late protein or its expression in a larger network structure. Novel theories in gene network and pathways analysis consider an integration of several variants in different genes regulating expression of different phenotypes. Current standards for the pathogenicity interpretation of genomic variants have been addressed in consensus by the American College of Medical Genetics and Genomics and the Association for Molecular Pathology (ACMG-AMP) (Oza et al. 2018).

1.3 Genetic Inheritance

Most human traits are not strictly determined by genes but rather are influenced by both genes and environment. Genetic inheritance explains the contribution of genetic variants to rare (Mendelian disorders) and complex disorders. Unfortunately, single-nucleotide variation only explains little-to-low well-known Mendelian disorders, such as Duchenne muscular dystrophy and cystic fibrosis. Genetic disorders usually follow certain inheritance patterns that help clinicians to track the disorder on a familial pedigree (Hamosh et al. 2005).

Common Mendelian disorders follow dominant or recessive inheritance patterns, where the variant must be represented in one or both chromosomal copies to affect the protein, respectively. Dominant disorders are usually observed in parents and offspring simultaneously, while recessive disorders are usually observed isolated in one generation with a previous carrying generation without observable effects (they only carry one copy of the variant).

Other inheritance patterns can be observed in other monogenic disorders, such X-linked or mitochondrial inheritance. Mitochondrial diseases are inherited through maternal mitochondrial DNA, and their effects are usually increased due to a high number of mitochondria carrying the affected copy of a variant (homoplasmy) or a mixture of both normal and mutant mitochondrial DNA (heteroplasmy). This situation leads to variable expression or incomplete penetrance of mitochondrial disorders within the same family (Chinnery 2002, Xin and Butow 2005).

More complex inheritance patterns are usually found in polygenic complex disorders and traits, where several genes interact defining a phenotype. Most of the common disorders come under in this definition, including tinnitus disorder, a common condition associated with several diseases with diverse etiologies. There are several examples of complex inheritance, but it usually involves several genes (polygenic disorders) and several alleles per gene (multiallelic disorders) (Mitchell 2012; Boyle et al. 2017). A combination of several alleles and several genes could also drive to syndromic phenotype, such retinitis pigmentosa and hearing impairment.

1.4 Genetic Disorders: Testing and Research

1.4.1 Assessment of Genetic Risk

Patient sequenced datasets usually generate a large quantity of variants with small effects on the phenotype. In complex diseases, the association between single variants and the phenotype is difficult to demonstrate. The common approach for genetic disorders consists in the elaboration of a genetic risk score (Torkamani et al. 2017).

Genetic risk scores are the sum of the possible indirect effects of known sequence variants on each individual and the expected yield those variants suggest to the known phenotype (Khera et al. 2018). Researchers rely on a large amount of genomic data to enable them to calculate which variants are found more frequently in groups of people with a given disease in a known population. There can be hundreds or even thousands of variants for a common disease, and some of them could be population-specific variants. The additive effect of common variants has been used to generate polygenic risk scores, where different variants in different genes yield a unique score for a disease in an individual (Li et al. 2020).

However, genetic risk scores are relative. Polygenic risk scores are made from large-scale genomic studies, where the objective is to find variants in case-control comparison studies. Attending to the studied populations, differences have been observed in the frequencies of the variants in replication cohorts from different populations, and these findings suggest for a more stringent risk score for a disease (Martin et al. 2019).

Another relevant point is that polygenic risk scores are static and determinant of an individual. They may not be causal. Polygenic risk scores are built around correlations. Two individuals with the same polygenic risk scores could have a totally different timeframe for a disorder, according to environmental factors such as diet, aging, and infections. Because of this, polygenic risk scores should not be confused with absolute risk factors of a genetic disorder. Absolute risk factors count the odds of a disease to occur and may be manifested in a specific biomarker. However, this means that both absolute and relative risks can be used together to check the efficacy of drugs and therapeutics (Chatterjee et al. 2016; Torkamani et al. 2018).

1.4.2 Genetic Testing Strategies

Genetic testing has surged toward being a necessity in medical diagnostics. It is the result of the study of genetic variations as genetic markers conferring risk for different disorders. Genetic testing strategies are used to generate genetic risk scores in individuals with suspected genetic conditions and provide health practitioners assistance for early interventions, including genetic counseling. Currently, there are more than 1,000 genetic tests for different genetic disorders in use by clinicians (Phillips et al. 2018).

Genetic testing follows different approaches or methods depending on the genetic target to be assessed:

(a) Molecular genetic tests are common PCR-based tests that study genes of interest or fragments of genes of interest to identify variations that lead to possible genetic disorders.
(b) Chromosomal genetic tests look for long fragments of the DNA with broad changes, such as extra copies of chromosomes or deletions of large parts of the DNA. Common chromosomal studies are fast and visual.
(c) Biochemical genetic tests focus on the expression of proteins of interest, looking for abnormalities in the quantity of protein or derivatives, which could lead to a potential disorder.

The implementation of next-generation sequencing is currently growing in clinical diagnosis, with the development of personalized medicine (Fig. 2). As part of the extensive use, the cost of sequencing is decreasing. However, the different techniques used in sequencing address different issues and target different regions of interest, which makes each of them specially designed for certain situations (Klein and Foroud 2017). The most commonly used approaches for genetic diagnostics are the following:

(a) *Genotyping*: Genotyping is a method that determines alleles in common variants in a set of individuals compared with a control group used as reference. This method is useful to understand which alleles are inherited from the parents of an individual or alleles defining a specific population. This method uses a limited array of targets focusing on some fragments of the DNA. Genotyping usually only need a PCR assay for detection, so it is considered cheap but not so much informative. Genome-wide association studies (GWAS) have been able to

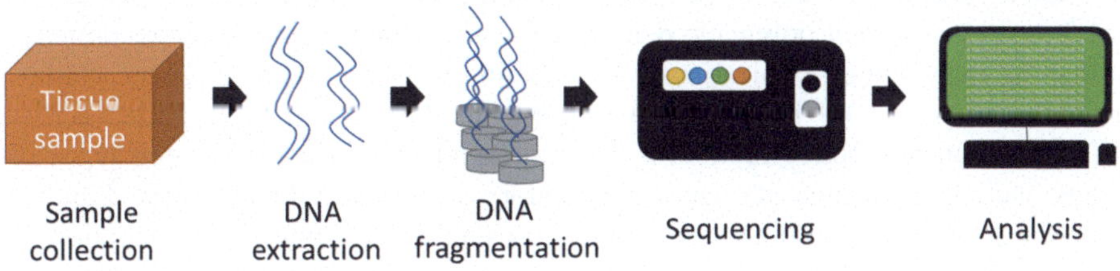

Fig. 2 Genomic sequencing basic workflow

demonstrate the association of common variants with complex traits by using genotyping technology in a case-control design (Tam et al. 2019).

(b) *Gene-targeted panels*: Panels are built around a selection of genes with known markers for a disease. Several gene panels are currently been tested for different disorders, with a large range of target genes. For instance, custom targeted gene panels can be designed to cover entire regions of interest for an affordable diagnosis of complex disorders such as epilepsy or neurodegenerative diseases (Yu et al. 2019).

(c) *Exome sequencing*: Whole-exome sequencing (WES) targets a larger section of the entire genome than gene target panels. Exome sequencing covers coding regions in all human genes. WES has been optimized in recent years and is covering the maximum of the entire coding part of the DNA at a low cost, making this approach affordable for familial studies (Bamshad et al. 2011). However, variants outside protein-coding regions are out of the scope of this technique, so it could represent difficulties in the interpretation for some disorders (Belkadi et al. 2015).

(d) *Genome sequencing*: Whole-genome sequencing (WGS) targets the total DNA sequence of an individual, with a lesser coverage than WES, but with an incredible quantity of information. WGS uniformly covers the entire genome, where WES usually includes the coding regions with different coverage. Compared to than WES, WGS allows sequencing larger fragments for better determination of rearrangements in the DNA or structural variations that are difficult to achieve with WES analysis. WGS is more expensive than WES, although it becomes more and more affordable for clinical testing (Belkadi et al. 2015; Maróti et al. 2018).

1.4.3 Pharmacogenomics

One of the major problems in medical treatments is that many medications do not yield the same results for everyone. While some patients benefit from a medication, others do not respond to the treatment or even show negative side effects for the same disorder (adverse drug reaction). Currently, pharmacogenomics appears as a new field attempting to address the necessity of personalized treatment by studying of how genes affect drug response (Ganesan et al. 2018).

Genetic differences are used to predict the effectiveness of a drug for an individual, the same way that polygenic risk scores work. Although pharmacogenomics is still in its early stages, clinical trials for different disorders such as cancer, cardiovascular diseases, and immune and cerebral disorders have been started across different countries. Since tinnitus is a common adverse effect of several types of drugs, including nonsteroidal anti-inflammatory drugs (NSAIDs) or some diuretics according to American Tinnitus Association (ATA), the genetic contribution to pharmacokinetics during the metabolization of certain drugs could explain the occurrence of tinnitus in some individuals (Elgoyhen et al. 2014).

2 Evidence on Tinnitus Heritability

Tinnitus affects more than 15% of adult population worldwide. It is a complex condition and characterized by a set of several features including perceived sound (ringing, buzzing, or hissing), tinnitus type (objective or subjective), persistent or nonpersistent, unilateral or bilateral, and pulsatile or non-pulsatile (Ueberfuhr et al. 2017; Vona et al. 2017). However, *tinnitus as a distressing disorder*, where tinnitus results in significant disruption of daily life, is a less frequent condition, and it can occur together with common traits such as hypertension (Figueiredo 2016), hearing loss (Shargorodsky et al. 2010), hyperacusis (Ralli et al. 2017), anxiety, sleep disorders, headache (Trevis et al. 2018), or rare diseases such as Meniere disease (Perez-Carpena and Lopez-Escamez 2019), making tinnitus disorder a heterogeneous condition and its characterization more complex.

The evidence supporting a genetic contribution to a tinnitus disorder is limited, and most of the studies have been conducted in Sweden, but it includes a low-to-moderate familial aggregation and high concordance between monozygotic twins with bilateral tinnitus (Fig. 3). These findings indicate that *severe and bilateral tinnitus* may have a genetic inheritance. For this reason, it is important to select an

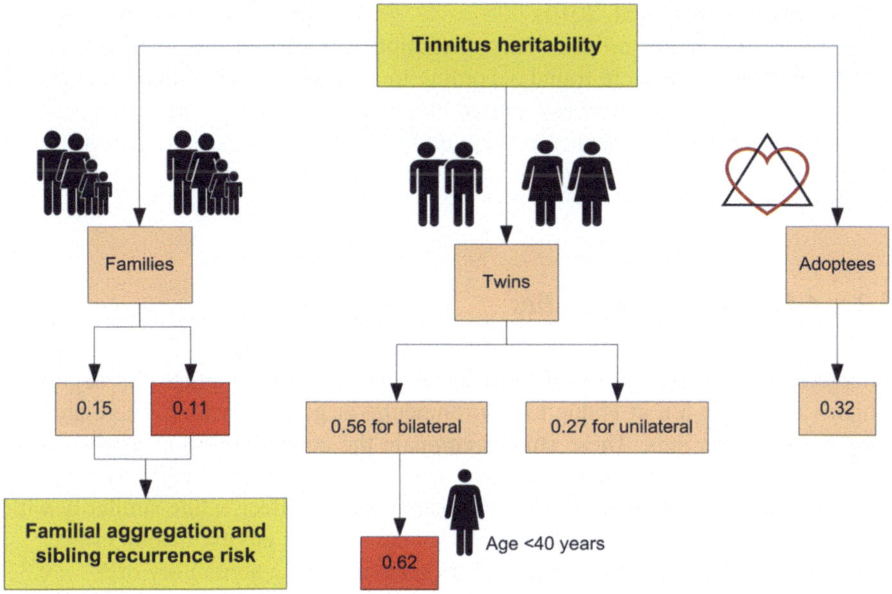

Fig. 3 Tinnitus familial aggregation and sex differences in genetic heritability. Familial aggregation is estimated by calculating the tinnitus recurrence risk ratio between siblings that shared the same familial environment. Heritability has been estimated in twins and adoptees studies by comparing the concordance rate between monozygotic vs. dizygotic twins and between adoptees with biological parents. Red box under families indicates low heritability, and red box under twins indicates increased risk of bilateral tinnitus in young females

appropriate tinnitus phenotype including age of onset, tinnitus severity, family history, or comorbidities linked with tinnitus, but these variables have barely been considered in genetic association studies (Maas et al. 2017).

2.1 *Familial Aggregation*

To investigate the familial aggregation of tinnitus, a large study was conducted in six European countries recruiting 198 families with at least three siblings, resulting in a total of 981 individuals for analysis. To identify tinnitus in these individuals, a question was asked as *Nowadays, do you ever get noises in your head or ear (tinnitus) which usually last longer than 5 min*. This study reported 21.2% of tinnitus prevalence and 0.15 of familial correlation, with a limited information on the role of shared environment and genetic risk factors (Hendrickx et al. 2007). Another study on the heritability of tinnitus recruited 11,498 siblings, 27,607 parent-offspring, and 12,940 spouses from Nord-Trøndelag County. This study reported heritability of tinnitus as $h^2 = 0.06–0.14$ for siblings, for parent-offspring as $h^2 = 0.01–0.07$, and 0.04 for spouse. Although a low heritability of 0.11 was reported without any sex differences, clinical information about the tinnitus profile and severity was missing in the study (Kvestad et al. 2010).

Another large study on familial aggregation was carried out in Sweden, and significant differences were found according to sex; so, a higher sibling recurrence risk ratio for severe tinnitus was reported in women compared to men, suggesting a sexual dimorphism with higher susceptibility in female cases (Trpchevska et al. 2020).

2.2 *Adoptees-Based Studies*

Adoptees-based studies are used in behavioral genetics to estimate the degree to which variation in a trait is due to environmental or genetic factors. There are two adoption study designs. These studies compare the concordance in the phenotype between the adoptee and their biological and adoptive parents. Similarity with the biological parents is associated with a heritable genetic effect, while similarity with the adoptive parent is associated with the shared familial, environmental effect.

A Swedish study using national registry data reported a heritability of $h^2 = 0.32$ with an association between tinnitus and adoptees in relation to their biological parents, but not to their adoptive parents. So, tinnitus in adoptive parents does not increase the odds of tinnitus among adoptees, suggesting a limited association of familial environment with tinnitus heritability (Cederroth et al. 2019).

2.3 *Twins-Based Studies*

Twin-based epidemiological studies are used to estimate heritability by comparing disease concordance in monozygotic (MZ) vs. dizygotic (DZ) twins. So, a high concordance in genetically identical twins, who share on average half of their alleles, suggests a genetic inheritance. It is assumed both MZ and DZ twins share the same family environment, thus yielding relevant information about the genetic contribution to disease etiology. Two twin studies have been published, and both of them independently concluded that genetic factors contribute to tinnitus.

Bogo et al. (2017) evaluated the genetic contribution of self-reported tinnitus in male twins aged 52–96 years with age-related hearing loss. Twins assessments were performed at baseline ($n = 1{,}084$ individuals) and after 18 years of follow-up ($n = 576$ individuals), and they included audiometry and self-reported tinnitus perception. The hypothesis was that individuals with faster hearing deterioration had the greatest tinnitus risk and that genetic factors influenced tinnitus (Bogo et al. 2017). However, no difference in tinnitus prevalence between MZ and DZ twins at either time point was found. As expected, the hearing thresholds among MZ twins discordant for tinnitus were more similar than for discordant DZ twins, which may be explained by the shared genetic background. An overall heritability of $h^2 = 0.4$ was estimated for tinnitus, which support a moderate genetic influence.

A large Swedish cohort of 10,464 twin pairs was selected to evaluate the concordance for tinnitus based on a question *"Do you have buzzing in the ears?"*. The investigation showed a higher concordance of bilateral tinnitus in MZ twins as compared to DZ twin pairs. Further investigations revealed significant sex differences in the heritability tinnitus: for unilateral tinnitus, the observed heritability was 0.29 for male and $h^2 = 0.25$ for female; but for bilateral tinnitus, it was reported as $h^2 = 0.68$ for male and $h^2 = 0.41$ for female. Furthermore, some sex differences in the heritability of bilateral tinnitus with different age groups were also observed. The selection of young female pairs with age < 40 years for analysis found an increase in the heritability ($h^2 = 0.62$), providing additional evidence of sex-mediated effect on tinnitus heritability (Maas et al. 2017).

3 Genetic Contribution to Tinnitus

3.1 *Neurotrophic Factors BDNF, GDNF*

Several genetic association studies have been conducted in patients with chronic tinnitus across different ethnic backgrounds and age groups, but most of them have not been replicated in an independent cohort (Table 1). For example, common variants in some candidate genes have been genotyped under the assumption that they could be involved in tinnitus mechanisms. A genetic study was performed on 240 Caucasian patients, using the Tinnitus Questionnaire (TQ) to assess the severity

Table 1 Genetic studies on chronic tinnitus and suggested candidate genes

Reference	Population	Sample	Sex		Sequencing method	Candidate genes
			F	M		
Deniz et al. (2010)	Turkish	54	33	21	Genotyping	*SLC6A4*
Sand et al. (2010)	Caucasian	201	49	152	Genotyping	*KCNE1*
Sand et al. (2011)	Caucasian	288	86	202	Genotyping	*KCNE3*
Sand et al. (2012b)[a]	Caucasian	240	69	171	Genotyping	*GDNF, BDNF*
Sand et al. (2012a)[b]	Caucasian	95	28	67	Genotyping	*KCTD12*
Orenay-Boyacioglu et al. (2016)	Turkish	52	19	33	Genotyping	*GDNF*
Gilles et al. (2017)	Belgium	167	67	100	GWAS	Metabolic pathway
Orenay-Boyacioglu et al. (2019)	Turkish	60	21	39	Methylation of 12 CpG sites in the promoter	*GDNF, BDNF*

level (Goebel and Hiller 1994). Genotyping was performed for *BDNF* (rs2049046, rs6265) and *GDNF* (rs1110149, rs884344, rs3812047). After correcting by multiple testing, no significant association was found when compared with controls. However, it was found that in the female group, variants in *BDNF* and *GDNF* were associated with tinnitus severity (Sand et al. 2012b).

Further investigations on the contribution of *GDNF* gene (rs884344, rs3812047, rs1110149) were carried out in a group of 52 Turkish individuals diagnosed with chronic tinnitus. Although, no association was found between tinnitus and these three variants of allelic frequency, the heterozygosity (C: G) was lower for *GDNF* rs1110149 in tinnitus patients compared to controls ($p = 0.02$). However, no definite conclusions were reached, and the authors suggested the need of a large sample size and detailed investigation of different expression patterns of *GDNF* gene (Orenay-Boyacioglu et al. 2016).

Recently, the methylation status of the *GDNF/BDNF* CpG promoter was examined in a group of 60 Turkish cases. A significant difference of methylation ratio was observed between patients and controls for *GDNF* CpG3–5-6 (CpG3 ($p = 0.0005$), CpG5 ($p = 0.00003$), CpG6 ($p = 0.0029$)), and CpG6 BDNF ($p = 0.002$); however, a larger sample size to replicate this findings is needed (Orenay-Boyacioglu et al. 2019).

3.2 Regulation of Serotonin Transporter

Patients with severe tinnitus show overlap with other comorbidities and a particularly strong association among patients with depressive disorders that affect

~5–10% of the general population. There is an estimated 30% concordant overlap between comorbid depressive disorder and tinnitus, and probably common variants with pleiotropic effects contribute to molecular to both phenotypes (Tyler et al. 2006). As such, genes involved in serotonin regulation, a critical process associated with depressive disorders, have been proposed as tinnitus candidate genes.

The gene *SLC6A4* regulates serotonin neurotransmission, and it was evaluated for tinnitus association in a genotyping study performed on 54 Turkish patients to identify the role of a 44 bp Indel and 17 bp tandem repeat in the *SLC6A4* gene in tinnitus development. Additionally, Tinnitus Handicap Inventory (THI) and Beck Depression Inventory (BDI) were used to assess tinnitus severity and depression, respectively. However, this study did not find any significant association of these variants in *SLC6A4* with tinnitus development (Deniz et al. 2010).

3.3 Potassium Channels

Another study recruited 288 Caucasian individuals with chronic tinnitus, and TQ was used to assess the severity level of tinnitus. This study did not report any novel variant, but rs17215437G > A: R83H was observed only in three carriers, and the impact of this variant on tinnitus severity cannot be fully assessed due to limited power of the study. The detailed exploration of common variants in *KCNE3* was suggested since the effects of potassium channel on tinnitus severity cannot be excluded (Sand et al. 2011).

A study on 201 Caucasian patients with chronic tinnitus was performed to examine the association of *KCNE1* with chronic tinnitus. A novel missense mutation ch12: 35821794G > A: V47I with MAF of 0.002 in tinnitus cases was found, and the detailed examination of *KCNE1* subunits was suggested (Sand et al. 2010). Considering $GABA_B$ receptor subunit of *KCTD12* as a candidate gene for tinnitus development, a study was performed on 95 Caucasian individuals with chronic tinnitus. Although rs34544607 was associated with tinnitus ($p = 0.04$) when compared with European controls, the association was not replicated when 50 additional patients were screened ($p = 0.07$). The authors hypothesized that the interaction of variants in regulatory regions of $GABA_{B1}$ and $GABA_{B2}$ receptors could be related with tinnitus (Sand et al. 2012a). However, all these genotyping studies were underpowered, and their lack of significant association was expected.

A pilot GWAS was performed on 167 individuals with tinnitus episodes lasting more than 5 min. After performing gene enrichment analysis, no significant SNPs were identified that could reach the threshold of $p < 5.0e\text{–}8$. However, several metabolic pathways were suggested to be involved in tinnitus development, which include oxidative stress or serotonin receptor-mediated signaling. This study was underpowered, and the authors indicated the necessity of a large sample size for any GWAS, and more specific tinnitus subtypes were suggested for future studies (Gilles et al. 2017).

4 Strategies for Designing Tinnitus Genetic Studies

Genetic research in tinnitus disorders is still in its early stages. The involvement of auditory healthcare practitioners (audiologists, otolaryngology physicians, psychologists, and general practitioners) in multicenter studies to recruit well-phenotyped patients and prospectively collect clinical data and DNA samples has been promoted by the EU in the last few years, including TINNET,[1] ESIT (Schlee et al. 2018), TIN-ACT,[2] or UNITI.[3] Moreover, a guideline including recommendations on collecting and storing samples for genetic studies on tinnitus for clinicians has been released (Szczepek et al. 2019).

4.1 Well-Defined Tinnitus Phenotype with a Homogenous Profile

Tinnitus is a common symptom, but *tinnitus as a disorder* has a prevalence around 1%, and this phenotype can be narrowed by selecting specific comorbid conditions. Genetic studies with a broad tinnitus phenotype could produce false-negative results. For a well-defined clinical phenotype, the precise and right selection of the following variables is very important for the design of genetic studies (Lopez-Escamez et al. 2016):

- Early age of onset
- Major comorbidities: hearing loss, headache, insomnia, hypertension
- Emotional and psychological factors: anxiety, depression
- Characterization of the hearing profile
- Tinnitus questionnaires to assess functional impact (THI, TFI)
- Psychoacoustic evaluation of tinnitus

The precise and homogenous profile could provide the essential grounds required to identify the allelic variants associated with a particular tinnitus subtype or endophenotype. An example of tinnitus profiling in well-phenotyped patients is Meniere disease, a rare disorder which shows different tinnitus subtypes (Perez-Carpena et al. 2019). This phenotype-based strategy can help in the reduction of background noise in the genetic structure and can circumvent the heterogeneity nature of clinical profile.

---

[1] http://tinnet.tinnitusresearch.net/

[2] https://tinact.eu/

[3] https://uniti.tinnitusresearch.net/

4.2 *Extreme Phenotype Studies*

Let us consider tinnitus as a quantitative trait showing a normal distribution, where patients may show a mild perception of the associated distress according to tinnitus questionnaires (low expressivity) or severe perception of tinnitus distress (high expressivity). A tinnitus extreme phenotype will include a subgroup of patients with severe tinnitus that would be located at the right tail of the normal distribution (Fig. 4). So, by selecting patients with severe tinnitus for genetic studies (i.e., THI score > 90 percentile), we can expect an enrichment of rare variants in their genomic data that would be associated with extreme phenotype.

However, the prevalence of tinnitus and hearing loss increases with aging (Ciorba et al. 2015), and we can also expect that elderly individuals would also report more severe tinnitus. Aging is associated with random mutations and epigenetic changes on DNA including structural changes and accumulation of copy number variants; these mutations associated with environmental factors exposure along life (ultraviolet radiation, chemical agents, pollutants) are accumulated in elderly individuals, and they may cause de novo somatic mutations in elderly patients. These novel mutations may generate "environmental noise" in the interpretation of sequencing data from elderly patients. So, it is preferred to select young individuals (ideally children) with "low exposure" to environmental factors for genetic studies. Also, the combined effect of environmental noise and epigenetic modifications endorses the onset and progression of hearing loss and tinnitus in the elderly population (Forsberg et al. 2012; Guo et al. 2017).

Therefore, the high heritability expected among young patients with bilateral tinnitus supports the selection of extreme phenotype cases for genetic studies (Lopez-Escamez et al. 2016; Maas et al. 2017). Since chronic tinnitus is not common among young individuals, they will be considered a rare disorder with an expected enrichment of rare variants. The study design to investigate extreme individuals in a

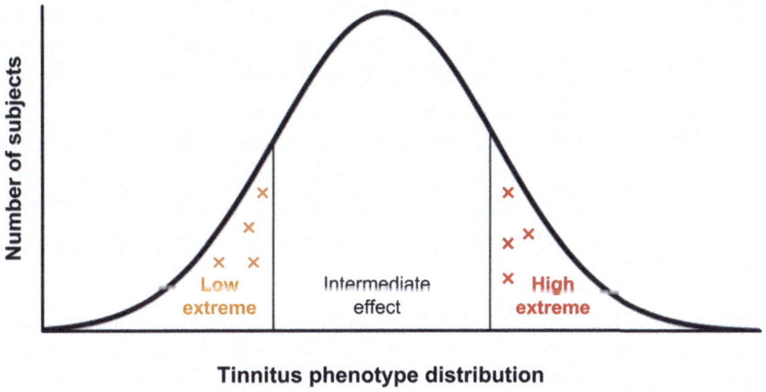

Fig. 4 Tinnitus phenotype with a normal distribution. Individuals at both extremes will show very mild and very severe tinnitus perception

quantitative trait is known as an "*extreme phenotype*" study which could include patients at both extremes of the normal distribution (i.e., patients with low and high expressivity). The rare variants that contribute to a particular disease are probably enriched in the extremes of a quantitative trait distribution, and the sequencing of a small subset of extreme individuals has enough power to identify potential candidate genes (Bamshad et al. 2011).

4.3 Tinnitus Extreme Phenotype in Meniere Disease

We have defined tinnitus extreme phenotype in Meniere disease (MD) according to the following features: (a) persistent and constant tinnitus considered as severe when with high scores on the Tinnitus Handicap Inventory (THI) (Newman et al. 1996), (b) early onset of the symptoms, (c) positive family history, and (d) fast progression of MD symptoms (Perez-Carpena et al. 2019). By using this approach, we are conducting a gene burden analysis in WES dataset to search for an enrichment of rare variations in patients with MD and tinnitus extreme phenotype.

4.4 Limitations and Future Directions

There are several limitations observed in the existing genetic studies and study designs. Some of the major limitations are:

1. *Lack of replication in most case-control studies with candidate genes:* there is a need of validating the few reported associations in independent replication cohorts to bring up consistent association of candidate genes with tinnitus and to reduce the false association of genes with tinnitus.
2. *Selection bias with inclusion criteria for patients with a broad tinnitus phenotype:* a precise and homogenous clinical profile should be considered for the selection of tinnitus phenotype among cases. It would be fruitful to identify rare variants associated with a particular tinnitus subgroup.
3. *Use of standardized tinnitus questionnaire*: there are many tools to evaluate and assess tinnitus severity; the utilization of the same questionnaire should be preferred to reduce the heterogeneity of phenotype and to facilitate the comparison between different studies (Lopez-Escamez et al. 2016; Vona et al. 2017).

Acknowledgments This project is a part of European School for Interdisciplinary Tinnitus (ESIT) research, and Sana Amanat is a PhD student in the Biomedicine Program at the University of Granada. This study has been funded by H2020 MSC-ITN-2017–722046 Grant. The project leading to these results has received funding by H2020 ERA-NET Cofund GENDER-NET Plus (GNP-182 Grant) from "la Caixa" Foundation *(ID 100010434)*, under agreement LCF/PR/DE18/52010002.

Conflict of Interest The authors declare no conflict of interest.

References

Bamshad MJ, Ng SB, Bigham AW et al (2011) Exome sequencing as a tool for Mendelian disease gene discovery. Nat Rev Genet 12:745–755. https://doi.org/10.1038/nrg3031

Belkadi A, Bolze A, Itan Y et al (2015) Whole-genome sequencing is more powerful than whole-exome sequencing for detecting exome variants. Proc Natl Acad Sci U S A 112:5473–5478. https://doi.org/10.1073/pnas.1418631112

Bell SP, Dutta A (2002) DNA replication in eukaryotic cells. Annu Rev Biochem 71:333–374. https://doi.org/10.1146/annurev.biochem.71.110601.135425

Bogo R, Farah A, Karlsson KK et al (2017) Prevalence, incidence proportion, and heritability for tinnitus: a longitudinal twin study. Ear Hear 38:292–300. https://doi.org/10.1097/AUD.0000000000000397

Bomba L, Walter K, Soranzo N (2017) The impact of rare and low-frequency genetic variants in common disease. Genome Biol 18:77

Boyle EA, Li YI, Pritchard JK (2017) An expanded view of complex traits: from polygenic to omnigenic. Cell 169:1177–1186

Burgers PMJ, Kunkel TA (2017) Eukaryotic DNA replication fork. Annu Rev Biochem 86:417–438. https://doi.org/10.1146/annurev-biochem-061516-044709

Cederroth CR, Pirouzifard M, Trpchevska N et al (2019) Association of genetic vs environmental factors in Swedish adoptees with clinically significant tinnitus. JAMA Otolaryngol Head Neck Surg 145:222–229. https://doi.org/10.1001/jamaoto.2018.3852

Chatterjee N, Shi J, García-Closas M (2016) Developing and evaluating polygenic risk prediction models for stratified disease prevention. Nat Rev Genet 17:392–406

Chinnery PF (2002) Inheritance of mitochondrial disorders. Mitochondrion 2:149–155. https://doi.org/10.1016/S1567-7249(02)00046-6

Ciorba A, Hatzopoulos S, Bianchini C et al (2015) Genetics of presbycusis and presbystasis. Int J Immunopathol Pharmacol 28:29–35

Deniz M, Bayazit YA, Celenk F et al (2010) Significance of serotonin transporter gene polymorphism in tinnitus. Otol Neurotol 31:19–24. https://doi.org/10.1097/MAO.0b013e3181c2dcbc

Duzkale H, Shen J, Mclaughlin H et al (2013) A systematic approach to assessing the clinical significance of genetic variants. Clin Genet 84:453–463. https://doi.org/10.1111/cge.12257

Ekundayo B, Bleichert F (2019) Origins of DNA replication. PLoS Genet 15:e1008320

Elgoyhen AB, Langguth B, Nowak W et al (2014) Identifying tinnitus-related genes based on a side-effect network analysis. CPT Pharmacometrics Syst Pharmacol 3:3. https://doi.org/10.1038/psp.2013.75

Figueiredo RR (2016) Positive association between Tinnitus and arterial hypertension. Front Neurol 7:3–8. https://doi.org/10.3389/fneur.2016.00171

Forsberg LA, Rasi C, Razzaghian HR et al (2012) Age-related somatic structural changes in the nuclear genome of human blood cells. Am J Hum Genet 90:217–228. https://doi.org/10.1016/j.ajhg.2011.12.009

Ganesan P, Schmiedge J, Manchaiah V et al (2018) Ototoxicity: a challenge in diagnosis and treatment. J Audiol Otol 22:59–68

Gilles A, Camp G, Van de Heyning P, Fransen E (2017) A pilot genome-wide association study identifies potential metabolic pathways involved in tinnitus. Front Neurosci 11:1–10. https://doi.org/10.3389/fnins.2017.00071

Goebel G, Hiller W (1994) The tinnitus questionnaire. A standard instrument for grading the degree of tinnitus results of a multicenter study with the tinnitus questionnaire. HNO 42:166–172

Guo L, Li P, Li H et al (2017) Effects of environmental noise exposure on DNA methylation in the brain and metabolic health. Environ Res 153:73–82. https://doi.org/10.1016/j.envres.2016.11.017

Hamosh A, Scott AF, Amberger JS et al (2005) Online Mendelian inheritance in man (OMIM), a knowledgebase of human genes and genetic disorders. Nucleic Acids Res 33:D514. https://doi.org/10.1093/nar/gki033

Hendrickx JJ, Huyghe JR, Demeester K et al (2007) Familial aggregation of tinnitus: a European multicentre study. B-ENT 3:51–60

Johnson A, O'Donnell M (2005) Cellular DNA replicases: components and dynamics at the replication fork. Annu Rev Biochem 74:283–315. https://doi.org/10.1146/annurev.biochem.73.011303.073859

Khera AV, Chaffin M, Aragam KG et al (2018) Genome-wide polygenic scores for common diseases identify individuals with risk equivalent to monogenic mutations. Nat Genet 50:1219–1224

Klein CJ, Foroud TM (2017) Neurology individualized medicine: when to use next-generation sequencing panels. Mayo Clin Proc 92:292–305

Kvestad E, Czajkowski N, Engdahl B et al (2010) Low heritability of tinnitus. Arch Otolaryngol Neck Surg 136:178. https://doi.org/10.1001/archoto.2009.220

Li R, Chen Y, Ritchie MD, Moore JH (2020) Electronic health records and polygenic risk scores for predicting disease risk. Nat Rev Genet. https://doi.org/10.1038/s41576-020-0224-1

Lopez-Escamez JA, Bibas T, Cima RFF et al (2016) Genetics of tinnitus: an emerging area for molecular diagnosis and drug development. Front Neurosci 10:1–13. https://doi.org/10.3389/fnins.2016.00377

Maas IL, Brüggemann P, Requena T et al (2017) Genetic susceptibility to bilateral tinnitus in a Swedish twin cohort. Genet Med 19:1007–1012. https://doi.org/10.1038/gim.2017.4

Maróti Z, Boldogkoi Z, Tombácz D et al (2018) Evaluation of whole exome sequencing as an alternative to BeadChip and whole genome sequencing in human population genetic analysis 06 biological sciences 0604 genetics. BMC Genomics 19:778. https://doi.org/10.1186/s12864-018-5168-x

Martin AR, Kanai M, Kamatani Y et al (2019) Clinical use of current polygenic risk scores may exacerbate health disparities. Nat Genet 51:584–591. https://doi.org/10.1038/s41588-019-0379-x

Mitchell KJ (2012) What is complex about complex disorders? Genome Biol 13:237

Newman CW, Jacobson GP, Spitzer JB et al (1996) Development of the tinnitus handicap inventory. Arch Otolaryngol Head Neck Surg 122:143–148

Orenay-Boyacioglu S, Coskunoglu A, Caki Z, Cam FS (2016) Relationship between chronic tinnitus and glial cell line-derived Neurotrophic factor gene rs3812047, rs1110149, and rs884344 polymorphisms in a Turkish population. Biochem Genet 54:552–563. https://doi.org/10.1007/s10528-016-9741-1

Orenay-Boyacioglu S, Caliskan M, Boyacioglu O et al (2019) Chronic tinnitus and BDNF/GDNF CpG promoter methylations: a case–control study. Mol Biol Rep 46:3929–3936. https://doi.org/10.1007/s11033-019-04837-0

Oza AM, DiStefano MT, Hemphill SE et al (2018) Expert specification of the ACMG/AMP variant interpretation guidelines for genetic hearing loss. Hum Mutat 39:1593–1613. https://doi.org/10.1002/humu.23630

Perez-Carpena P, Lopez-Escamez JA (2019) Current understanding and clinical Management of Meniere's disease: a systematic review. Semin Neurol 40:138–150. https://doi.org/10.1055/s-0039-3402065

Perez-Carpena P, Martinez-Martinez M, Martínez Carranza RA et al (2019) A tinnitus symphony in 100 patients with Meniere's disease. Clin Otolaryngol 44:1176–1180

Phillips KA, Deverka PA, Hooker GW, Douglas MP (2018) Genetic test availability and spending: where are we now? Where are we going? Health Aff 37:710–716. https://doi.org/10.1377/hlthaff.2017.1427

Ralli M, Salvi RJ, Greco A et al (2017) Characteristics of somatic tinnitus patients with and without hyperacusis. PLoS One 12:1–15. https://doi.org/10.1371/journal.pone.0188255

Requena T, Gallego-Martinez A, Lopez-Escamez JA (2017) A pipeline combining multiple strategies for prioritizing heterozygous variants for the identification of candidate genes in exome datasets. Hum Genomics 11:11. https://doi.org/10.1186/s40246-017-0107-5

Sand PG, Luettich A, Kleinjung T et al (2010) An examination of KCNE1 mutations and common variants in chronic tinnitus. Genes (Basel) 1:23–37. https://doi.org/10.3390/genes1010023

Sand PG, Langguth B, Kleinjung T (2011) Deep resequencing of the voltage-gated potassium channel subunit KCNE3 gene in chronic tinnitus. Behav Brain Funct 7:39. https://doi.org/10.1186/1744-9081-7-39

Sand PG, Langguth B, Itzhacki J et al (2012a) Resequencing of the auxiliary GABA B receptor subunit gene KCTD12 in chronic tinnitus. Front Syst Neurosci 6:1–5. https://doi.org/10.3389/fnsys.2012.00041

Sand PG, Langguth B, Schecklmann M, Kleinjung T (2012b) GDNF and BDNF gene interplay in chronic tinnitus. Int J Mol Epidemiol Genet 3:245–251

Schlee W, Hall DA, Canlon B et al (2018) Innovations in doctoral training and research on Tinnitus : the European School on Interdisciplinary Tinnitus Research (ESIT) perspective. Front Aging Neurosci 9:1–7. https://doi.org/10.3389/fnagi.2017.00447

Shargorodsky J, Curhan GC, Farwell WR (2010) Prevalence and characteristics of tinnitus among US adults. Am J Med 123:711–718. https://doi.org/10.1016/j.amjmed.2010.02.015

Sun H, Yu G (2019) New insights into the pathogenicity of non-synonymous variants through multi-level analysis. Sci Rep 9:1–11. https://doi.org/10.1038/s41598-018-38189-9

Szczepek AJ, Frejo L, Vona B et al (2019) Recommendations on collecting and storing samples for genetic studies in hearing and tinnitus research. Ear Hearing 40:219–226

Tam V, Patel N, Turcotte M et al (2019) Benefits and limitations of genome-wide association studies. Nat Rev Genet 20:467–484

Torkamani A, Andersen KG, Steinhubl SR, Topol EJ (2017) High-definition medicine. Cell 170:828–843

Torkamani A, Wineinger NE, Topol EJ (2018) The personal and clinical utility of polygenic risk scores. Nat Rev Genet 19:581–590

Trevis KJ, McLachlan NM, Wilson SJ (2018) A systematic review and meta-analysis of psychological functioning in chronic tinnitus. Clin Psychol Rev 60:62–86. https://doi.org/10.1016/j.cpr.2017.12.006

Trpchevska N, Bulla J, Hellberg MP et al (2020) Familial aggregation and sexual dimorphism of tinnitus in the Swedish population. Under review

Tyler RS, Coelho C, Noble W (2006) Tinnitus: standard of care, personality differences, genetic factors. Orl 68:14–19. https://doi.org/10.1159/000090486

Ueberfuhr MA, Wiegrebe L, Krause E et al (2017) Tinnitus in normal-hearing participants after exposure to intense low-frequency sound and in Ménière's disease patients. Front Neurol 7:1–11. https://doi.org/10.3389/fneur.2016.00239

Vona B, Nanda I, Shehata-Dieler W, Haaf T (2017) Genetics of tinnitus: still in its infancy. Front Neurosci 11

Xin JC, Butow RA (2005) The organization and inheritance of the mitochondrial genome. Nat Rev Genet 6:815–825

Yu ACS, Yim AKY, Chan AYY et al (2019) A targeted gene panel that covers coding, non-coding and short tandem repeat regions improves the diagnosis of patients with neurodegenerative diseases. Front Neurosci 13:1324. https://doi.org/10.3389/fnins.2019.01324

Part II
Preclinical and Clinical Models of Tinnitus

Pharmacological Evaluation of Drugs in Animal Models of Tinnitus

Yiwen Zheng, Jessica McTavish, and Paul F. Smith

Contents

Abstract Despite the pressing need for effective drug treatments for tinnitus, currently, there is no single drug that is approved by the FDA for this purpose. Instead, a wide range of unproven over-the-counter tinnitus remedies are available on the market with little or no benefit for tinnitus but with potential harm and adverse effects. Animal models of tinnitus have played a critical role in exploring the pathophysiology of tinnitus, identifying therapeutic targets and evaluating novel and existing drugs for tinnitus treatment. This review summarises and compares the studies on pharmacological evaluation of tinnitus treatment in different animal models based on the pharmacological properties of the drug and provides insights into future directions for tinnitus drug discovery.

Y. Zheng (✉), J. McTavish, and P. F. Smith
Department of Pharmacology and Toxicology, School of Biomedical Sciences, University of Otago, Dunedin, New Zealand

Brain Research New Zealand, Auckland, New Zealand

Brain Health Research Centre, University of Otago, Dunedin, New Zealand

Eisdell Moore Centre for Hearing and Balance Research, University of Auckland, Auckland, New Zealand
e-mail: yiwen.zheng@otago.ac.nz

Curr Topics Behav Neurosci (2021) 51: 51–82
https://doi.org/10.1007/7854_2020_212
Published Online: 16 February 2021

Keywords Acoustic trauma · Animal model · Behaviour · Pharmacology · Salicylate · Tinnitus

Abbreviations

5-HT	Serotonin
AC	Auditory cortex
ACEA	Arachidonyl-2′-chloroethylamide
BDNF	Brain-derived neurotrophic factor
CBD	Cannabidiol
ChAT	Choline acetyltransferase
CN	Cochlear nucleus
C-PC	C-phycocyanin
dTT	3,6′-Dithiothalidomide
FFA	Flufenamic acid
GABA	γ-Aminobutyric acid
GAD_{65}	Glutamic acid decarboxylase-65
GAD_{67}	Glutamic acid decarboxylase-67
GPIAS	Gap-prepulse inhibition of the acoustic startle reflex
IC	Inferior colliculus
IL-1β	Interleukin 1 beta
IL-6	Interleukin 6
LMM	Linear mixed model
NAcc	Nucleus accumbens
NMDA	N-methyl-D-aspartate
ROC	Receiver-operating characteristic
SIPAC	Schedule-induced polydipsia avoidance conditioning
TNF-α	Tumour necrosis factor-α
VAChT	Vesicular acetylcholine transporter
vmPFC	Ventromedial prefrontal cortex
Δ-9-THC	Delta-9-tetrahydrocannabinol

1 Animal Models of Tinnitus

The use of laboratory animals to evaluate the safety and efficacy of drugs is part of the regulatory process for drug research and development (European Commission 2004; US Congress 2018). Animal models also play a critical role in characterising the pathophysiology of disease, identifying therapeutic targets and biomarkers, and understanding the mechanisms of action of drugs. However, animal models of human disease are often criticised for their validity in reproducing the clinical symptoms and manifestations of a disease, replicating the underlying biological causes in the human condition and predicting therapeutic effects in humans

(McGonigle and Ruggeri 2014). This is even more challenging in the case of tinnitus, the perception of sound(s) with no corresponding external sound source, as it can be caused by many factors.

Although the main factors that can cause tinnitus include noise exposure, head and neck injury, drug toxicity, ear infection, Meniere's disease, ageing, and even affective disorders, such as depression (Baguley et al. 2013), noise exposure is believed to be the most common cause of tinnitus in humans (Cooper 1994). Therefore, in most recent animal models, tinnitus is induced by exposing the animals to noise trauma (Brozoski et al. 2002, 2007; Dehmel et al. 2012; Heffner 2011; Heffner and Koay 2005; Nowotny et al. 2011; Turner et al. 2012; Zheng et al. 2011b; Zuo et al. 2017). Noise-induced tinnitus can last for a very long time in animals, which represents the chronic nature of tinnitus in humans, albeit the tinnitus induction rate in animals can vary from 30% to 80% (von der Behrens 2014). Another frequently used tinnitus animal model is induced by a high dose of salicylate (Jastreboff et al. 1988b; Lobarinas et al. 2004; Stolzberg et al. 2013). Salicylate is the active ingredient in aspirin, which is widely known to induce hearing loss and tinnitus in humans at high doses (Cazals 2000; Cianfrone et al. 2011). In animals, a single high dose of salicylate (200–350 mg/kg, i.p.) has been shown to induce acute tinnitus that occurs within 2–3 h and lasts for about 72 h (Jastreboff et al. 1988b; Lobarinas et al. 2004; Stolzberg et al. 2013), although administration of salicylate in drinking water over a long period has been used to induce chronic tinnitus in animals (Bauer et al. 1999).

In addition to inducing tinnitus, a more challenging question is how to measure the subjective tinnitus perception in animals. In 1988, Jastreboff and colleagues developed the first behavioural method enabling the objective measurement of tinnitus in rats (Jastreboff et al. 1988a, b), which made it possible to not only correlate pathophysiological changes with tinnitus itself, but also conduct preclinical evaluation of different treatments for tinnitus. Since then, various behavioural paradigms have been developed for the assessment of acute or chronic tinnitus in rats (Bauer et al. 1999; Heffner 2011; Lobarinas et al. 2004; Rüttiger et al. 2003; Stolzberg et al. 2013; Turner et al. 2006; Zheng et al. 2011b), mice (Longenecker and Galazyuk 2011; Zuo et al. 2017), guinea pigs (Berger et al. 2013; Dehmel et al. 2012), chinchillas (Brozoski et al. 2002), gerbils (Nowotny et al. 2011), and hamsters (Heffner and Koay 2005). The establishment of animal models of tinnitus has provided great tools for preclinical evaluation of drugs that are potentially useful in treating tinnitus.

2 Current Drug Evaluation in Animal Models of Tinnitus

Although a number of tinnitus management options are available, medication is rated by patients as the most preferred form of treatment over external devices, cochlear implants, brain surface implants, or intra-brain implants (Tyler 2012). Despite the strong desire, there is no FDA-approved drug for tinnitus. In comparison, a wide range of unproven over-the-counter tinnitus remedies are available on the market with little or no benefit for tinnitus but with potential harm and adverse effects (Vendra et al. 2019). Therefore, the need to develop an effective drug treatment for tinnitus is pressing. A review of the clinical trials using pharmacological treatments for tinnitus showed that most of the clinical trials have been hampered by the heterogeneity of tinnitus, the methodological limitations, and the lack of reliable and repeatable objective measures of tinnitus, which leads to the high variability in the treatment outcomes (Langguth and Elgoyhen 2012; McFerran et al. 2019). The use of animal models in tinnitus drug development, on the other hand, offers a number of advantages. These include the homogenous genetic background of the testing subjects, the well-controlled aetiology of tinnitus, and balanced experimental design and powerful statistical analysis. In addition, it also allows the use of invasive methods to correlate drug effects on tinnitus perception with molecular, neurochemical, and electrophysiological changes, to inform underlying mechanisms. This review summarises and compares the studies on the pharmacological evaluation of tinnitus treatment in animal models based on the pharmacological properties of the drug.

2.1 Drugs Enhancing GABA-Mediated Inhibitory Neurotransmission

Animal studies using molecular, neurochemical, or electrophysiological approaches have suggested that an imbalance between inhibitory and excitatory neurotransmission may play a role in tinnitus development (Eggermont 2005; Roberts et al. 2010) (Eggermont and Roberts 2004; Wang et al. 2011). In particular, a down-regulation of GABAergic (γ-aminobutyric acid) neurotransmission, the main inhibitory neurotransmission in the brain, is evident at multiple levels along the auditory pathways. For example, significantly reduced $GABA_A$ receptor expression as well as the expression of the enzyme responsible for the synthesis of GABA, glutamic acid decarboxylase-67 (GAD_{67}), was found in the auditory brain regions following acoustic trauma (Browne et al. 2012; Dong et al. 2010b). Similarly, salicylate has been reported to cause a decrease in glutamic acid decarboxylase-65 (GAD_{65}) levels as well as the levels of $GABA_A$ (Zou and Shang 2012) and $GABA_B$ (Butt et al. 2016) receptors in the inferior colliculus (IC). A decrease in the number of $GABA_A$ receptor binding sites has also been found in the IC following chronic salicylate treatment; however, there was a significant increase in GAD_{65} protein expression and $GABA_A$ receptor affinity (Bauer et al. 2000). Therefore, a number of drugs with GABAergic enhancing properties have been tested in animal models of tinnitus.

Gabapentin and tiagabine were the first two GABAergic enhancing drugs that were tested in an animal model of tinnitus (Bauer and Brozoski 2001). Gabapentin is an anticonvulsant, anti-hyperanalgesic and antinociceptive drug. Some studies suggested that gabapentin might act as a $GABA_B$ receptor agonist in the brain (Bertrand et al. 2001; Ng et al. 2001), while others suggested otherwise (Lanneau et al. 2001). Tiagabine, by contrast, enhances GABAergic inhibition at synapses where it acts as a potent, selective, and competitive inhibitor of GABA reuptake (Krogsgaard-Larsen et al. 1987). Both drugs were tested for their effects on chronic tinnitus induced by unilateral acoustic trauma in rats (Bauer and Brozoski 2001). In this study, male Long-Evans rats were exposed to acoustic trauma (a narrowband noise centred at 16 kHz, at 105 dB for 1 or 2 h) or a control condition and behavioural evidence of tinnitus was determined using a lever-pressing procedure that required the animal's auditory discrimination ability. It was confirmed that the animals exhibited tinnitus-like behaviour when tested at 2 months and 17 months after noise exposure, which suggested the persistent nature of noise-induced tinnitus. In addition, various behavioural control tests were conducted to ensure that tinnitus-like behaviour was frequency-specific and not due to unilateral hearing loss. After the confirmation of tinnitus, gabapentin or tiagabine were delivered in the drinking water for 10 days for each dose and the perception of tinnitus was tested. Gabapentin, but not tiagabine, significantly reduced acoustic trauma-induced tinnitus. Since the main aim of this study was to validate acoustic trauma-induced tinnitus in rats using a psychophysical procedure, the pharmacological mechanisms underlying the effects of gabapentin on tinnitus and the difference between gabapentin and tiagabine were not discussed. Compared with the relatively well-defined mechanisms of action for tiagabine as a GABAergic drug, gabapentin has also been shown to inhibit glutamatergic neurotransmission in the CA1 area of the hippocampus through presynaptic N-methyl-D-aspartate (NMDA) autoreceptors (Suarez et al. 2005), which suggests that enhancing GABAergic neurotransmission alone may not be sufficient to reduce tinnitus. However, the same research group tested another GABAergic drug, vigabatrin, which is an irreversible inhibitor of the mitochondrial enzyme GABA-transaminase (Jung et al. 1977), in the same animal model of tinnitus, and showed that vigabatrin completely eliminated the behavioural evidence of tinnitus. The study also showed that tinnitus returned in those animals after 7 weeks of the drug washout period, which suggests that maintaining a constant level of GABAergic neurotransmission is crucial for the treatment of tinnitus. This was supported by another study that tested L-baclofen, a $GABA_B$ receptor agonist, in acoustic trauma-induced tinnitus (Zheng et al. 2012b). In this study, tinnitus was induced by exposing the rats to a 16 kHz pure tone at 110 dB in one of the ears and the perception of tinnitus was measured using a conditioned lick suppression paradigm. Following the confirmation of tinnitus, a vehicle and 3 different doses of L-baclofen (1, 3, and 5 mg/kg) were administered to all the animals with a washout period inserted between the 3 and 5 mg/kg L-baclofen treatments. It was found that L-baclofen dose-dependently and reversibly reduced the behavioural evidence of tinnitus in the animals.

Benzodiazepines are a class of psychoactive drugs that enhance inhibitory GABAergic neurotransmission by acting as positive allosteric modulators at the $GABA_A$ receptors and produce anxiolytic effects. Since anxiety is one of the most common comorbidities of tinnitus (Bhatt et al. 2017), benzodiazepines have been used as an option for its management. However, a systematic review found that there was a large variation in the effects of different benzodiazepines on tinnitus in clinical trials, with the overall conclusion that there was no robust evidence supporting the use of benzodiazepines for tinnitus (Jufas and Wood 2015). Using an animal model of tinnitus, the only benzodiazepine that has been tested on the perception of tinnitus is midazolam (Panford-Walsh et al. 2008). In this study, tinnitus was induced using two different methods, a local application of salicylate through the round window (a 5, 10, or 20 μl of 70 mg/ml solution) or a systemic injection of a single high dose of salicylate (350 mg/kg, i.p.). Both methods induced comparable hearing loss and tinnitus behaviour using a rewarding based conditioned auditory discrimination paradigm (Rüttiger et al. 2003), as well as comparable levels of salicylate in the cochlear fluid. Salicylate treatment using both methods also caused a significant increase in brain-derived neurotrophic factor (BDNF) expression in the cochlear spiral ganglion neurons and a significant decrease in Arg3.1 expression in the auditory cortex (AC). Midazolam (0.5 mg/kg, i.p.) injected 2.5 h after systemic treatment with salicylate reversed the changes in BDNF and Arg3.1. In addition, pre-treatment with midazolam (1 mg/ml, 1 μl/h) for 1 week through a cannula directed at the round window attenuated tinnitus behaviour and reversed the changes in BDNF and Arg3.1. These results suggest that peripheral loss of $GABA_A$ receptor-mediated inhibition in the cochlea plays an important role in the salicylate-induced imbalance of central auditory neuronal activity and tinnitus. However, the direct link between GABA inhibition and tinnitus was weakened by the very small sample size used in the tinnitus reversal experiment. There were only two animals in each group and the results could not be statistically analysed. Nevertheless, this research provided additional evidence on the role of peripheral GABAergic inhibition in salicylate-induced central neuronal plasticity.

Based on the anxiolytic property of benzodiazepines and the involvement of the limbic system in tinnitus generation and tinnitus-related distress (Rauschecker et al. 2010), the effects of diazepam, another benzodiazepine, on neuronal excitability in both the AC and lateral amygdala, were investigated in the salicylate model of tinnitus (de la Iglesia-Larrad et al. 2020; Song et al. 2016). Using single unit recording, it was found that the spontaneous firing rate of neurons in both the AC and lateral amygdala increased significantly at 2 h after salicylate injection (350 mg/kg, i.p.). Systemic injection of diazepam significantly reversed the increased firing rate in the lateral amygdala and microinjection of diazepam into the lateral amygdala significantly reversed the increased firing rate in the AC. These results suggested that benzodiazepines could not only suppress salicylate-induced neuronal hyperactivity in the lateral amygdala, but also suppress the neuronal hyperactivity in the AC by enhancing the inhibitory control from the lateral amygdala. Unfortunately, the perception of tinnitus was not measured in these animals. Therefore, the suppression of neuronal excitability by diazepam could not be directly related to the suppression of tinnitus perception. Taken together, animal research into the effects of benzodiazepines on tinnitus is sparse and has only been conducted on salicylate-induced tinnitus. Therefore, more studies are needed to test benzodiazepines in acoustic trauma-induced tinnitus and to explore the underlying mechanisms. Since the use of benzodiazepines is also associated with some significant side effects, including withdrawal syndromes, cognitive decline, respiratory distress, sedation, falls, as well as the potential for misuse and abuse (Panes et al. 2020), the development of novel benzodiazepines (e.g. partial agonists) with a better safety profile, favourable pharmacokinetic properties, and a wide therapeutic index is also needed.

Another chemical that interacts with GABAergic neurotransmission is taurine, which is an amino acid involved in a range of physiological processes. Taurine has been shown to inhibit neuronal activity by activating $GABA_A$ receptors. Specifically, taurine displays a low affinity at $GABA_A$ receptors with α1 or α2 subunits (Kletke et al. 2013) and a high affinity at $GABA_A$ receptors containing α4, β2, and δ or α6, β2, and δ subunits, which are located extrasynaptically (Ahring et al. 2016). Therefore, the effects of taurine on tinnitus were evaluated in the acoustic trauma-induced tinnitus model (Brozoski et al. 2010). Tinnitus was induced by exposing the rats to a band limited noise (centred at 16 kHz, 116 dB) unilaterally for 90 min and the perception of tinnitus was confirmed using an operant conditioned lever-pressing paradigm at 22 weeks after acoustic trauma. Then, all of the rats received taurine treatment delivered in drinking water using an ascending dose series (0, 1, and 4 mg/ml) followed by two washout periods. It was found that a dose (4 mg/ml) of taurine significantly reduced tinnitus and then the therapeutic effects of taurine slowly declined, with tinnitus returning 70–90 days after the taurine was discontinued. The expression of α4 and δ-subunit- containing $GABA_A$ receptors has been found in the medial geniculate body (MGB), part of the auditory thalamus, and functionally mediates slow tonic inhibitory currents (Richardson et al. 2011). Unlike the down-regulation of GABAergic neurotransmission associated with tinnitus in other auditory brain areas, MGB neurons showed an increased number of spikes per burst, increased tonic inhibition mediated by $GABA_A$ receptors containing α4 and δ-subunits, and increased $GABA_A$ receptor δ-subunit gene expression in animals with chronic tinnitus induced by acoustic trauma (Sametsky et al. 2015). Taurine has been shown to reduce the excitability of thalamocortical relay neurons by activating extrasynaptic $GABA_A$ receptors in the mouse ventrobasal thalamus (Jia et al. 2008), which may account for its effects on tinnitus. In addition to its interaction with GABAergic neurotransmission, taurine has also been shown to activate glycine receptors and reduce neuronal excitability in the rat IC (Xu et al. 2006) and AC (Tang et al. 2008), which may also contribute to the tinnitus suppression by taurine.

2.2 *Drugs Reducing Glutamate-Mediated Excitatory Neurotransmission*

In addition to reduced inhibition, increased excitatory neurotransmission has also been proposed as one of the underlying mechanisms of tinnitus. For example, there is some evidence that salicylate-induced tinnitus may involve an increase in glutamatergic neurotransmission at the NMDA subtype of glutamate receptor in the cochlea (Guitton et al. 2003) as well as in the cochlear nucleus (CN) (Hu et al. 2015), IC (Hwang et al. 2013) and AC (Jang et al. 2019). Surprisingly, evidence of increased glutamatergic neurotransmission in the acoustic trauma-induced tinnitus model is very limited. Godfrey et al. (2012) reported a long-term increase in tissue concentrations of glutamate in the IC of the hamster after intense tone exposure (Godfrey et al. 2012). In another study, glutamate receptor AMPA subunit α2 and glutamate receptor NMDA subunit 1 gene expression were up-regulated in the ipsilateral CN and contralateral IC of guinea pigs at 2–4 weeks after acoustic trauma (Dong et al. 2010a). However, behavioural evidence of tinnitus was not confirmed in these animals in either study. Nevertheless, given that glutamate is the major excitatory neurotransmitter, drugs that reduce glutamate-mediated neurotransmission have be tested in animal models of tinnitus.

NMDA receptor antagonists have been the most frequently tested drugs in tinnitus animal models. Since NMDA receptors are expressed in the cochlea (Niedzielski and Wenthold 1995), the first attempt to use NMDA receptor antagonists to treat tinnitus was performed by delivering the drugs locally into the cochlea. In the salicylate-induced tinnitus model, the effects of three different NMDA receptor antagonists, MK-801 (a channel blocker), 7-chlorokynurenate (a glycine-site antagonist), and gacyclidine (a phencyclidine-site antagonist), were tested (Guitton et al. 2003). In this study, the NMDA antagonists were applied to the round window using gelfoam soaked with the respective drug, 1 day before the salicylate injection. Salicylate (300 mg/kg, i.p.) was administered once a day for 4 days and behavioural evidence of tinnitus was measured every day at 2 h after salicylate injection and for 4 more days after the last dose of salicylate, using an active avoidance paradigm. It was found that all three NMDA receptor antagonists significantly reduced tinnitus-like behaviour induced by salicylate. Furthermore, given that salicylate is a cyclooxygenase inhibitor (Wu 2003), the study also investigated the involvement of the cyclooxygenase pathway in salicylate-induced tinnitus by testing another cyclooxygenase inhibitor, mefenamate. The results showed that mefenamate treatment induced tinnitus-like behaviour similar to that following salicylate treatment, which was prevented by the NMDA receptor antagonist, 7-chlorokynurenate. Taken together, this study suggests that salicylate may cause tinnitus through the activation of cochlear NMDA receptors, that is likely to be due to accumulation of arachidonic acid by the inhibition of cyclooxygenase (Guitton et al. 2003). However, the role of cochlear NMDA receptors in acoustic trauma-induced chronic tinnitus was questioned by a subsequent study published by the same group (Guitton and Dudai 2007). In this study, the behavioural evidence of tinnitus was measured either at 2 h after a 4-day treatment of salicylate or at 2 weeks after acoustic trauma (16 kHz, 130 dB for 15 min) using a place-tone conditioning paradigm. Ifenprodil, an NMDA receptor antagonist at the polyamine site, was applied to the round window either on the same day of the first salicylate injection, or at day 0 immediately before acoustic trauma, or day 4, 8, or 12 after acoustic trauma. As expected, ifenprodil supressed salicylate-induced tinnitus. However, for acoustic trauma-induced chronic tinnitus, ifenprodil was effective only when it was administered either immediately before acoustic trauma or at day 4 after acoustic trauma. This suggests that activation of NMDA receptors in the cochlea may serve as a trigger for acoustic trauma-induced tinnitus and then chronic tinnitus is established beyond the pathological changes in the cochlea. More importantly, this transition from a peripheral mechanism to a central mechanism occurs in the first few days after the acoustic trauma, which suggests a potential therapeutic window for preventing the development of chronic tinnitus by targeting the cochlea.

The effects of systemic administration of NMDA receptor antagonists on tinnitus have also been investigated in animal models. There were 3 studies that tested the effects of systemic injection of memantine on salicylate-induced tinnitus. Memantine is an uncompetitive NMDA receptor antagonist which blocks the Ca^{2+} channel, similar to MK-801. The first study was carried out by Lobarinas et al. (2006), where memantine (1.5 or 3 mg/kg, i.p.) was co-administered with salicylate (150 mg/kg, i.p.) and tinnitus in animals was measured using a schedule-induced polydipsia avoidance conditioning paradigm. Sound-evoked neuronal activity was also recorded in the AC. Memantine did not affect salicylate-induced tinnitus, nor did it reduce salicylate-induced neuronal hyperactivity in the AC, although no statistical analysis was performed on the local field potential data (Lobarinas et al. 2006). By contrast, Ralli et al. (2014) reported that memantine (5 mg/kg, i.p.) significantly reduced tinnitus-like behaviour induced by salicylate (300 mg/kg), measured by the gap-prepulse inhibition of the acoustic startle reflex (GPIAS) paradigm (Ralli et al. 2014). This was supported by a more recent study showing that co-administration of memantine (5 mg/kg, i.p.) and salicylate (400 mg/kg, i.p.) significantly attenuated tinnitus-like behaviour measured with the GPIAS (Jang et al. 2019). This study also demonstrated that memantine could decrease salicylate-induced upregulation of activity-regulated cytoskeleton-associated protein and tumour necrosis factor-α (TNF-α) genes in cell culture, as well as the NMDA receptor subtype NR2B gene and protein expression in both the cell culture and the AC of rats receiving the combination of salicylate and memantine. Comparing these three studies, the dose of memantine used was much lower in Lobarinas et al. (2006), which may have contributed to the lack of positive effects on tinnitus in this study.

Taken together, studies have shown that both cochlear and systemic application of NMDA receptor antagonists can attenuate salicylate induced-tinnitus and that cochlear application of NMDA receptor antagonists within a certain time window after an insult can prevent the development of acoustic trauma-induced chronic tinnitus. However, one important question that remains to be answered is whether NMDA receptor antagonists are able to reverse chronic tinnitus that has already developed. This possibility was tested in a study using the acoustic trauma-induced tinnitus model (Zheng et al. 2012a). Animals were exposed to unilateral acoustic trauma (16 kHz, 110 dB for 1 h) or a sham procedure and tested for behavioural evidence of tinnitus using a conditioned lick suppression paradigm at 2 weeks after acoustic trauma. Following the confirmation of tinnitus, each animal received the same order of treatment while being tested for the presence of tinnitus: vehicle, memantine (5 mg/kg, s.c.), and a drug washout period. Following acoustic trauma, 5 out of 8 rats exhibited tinnitus-like behaviour, which was reduced to 2 out of 8 rats during memantine treatment. When memantine treatment was stopped during the washout period, the number of tinnitus rats increased to 3. This study suggests that

NMDA receptor blockade in the central nervous system is necessary for the treatment of chronic tinnitus. However, since memantine was administered systemically, it is not clear which brain area(s) are responsible for tinnitus maintenance and the tinnitus attenuation effects of memantine. Using local infusion of an NMDA antagonist, D(2)-2-amino-5-phosphonopentanoic acid (D-AP5), into the cerebellar paraflocculus unilateral to the acoustic trauma side, chronic tinnitus was significantly reduced within 3 days of treatment and the effect lasted for 23 days after the treatment was discontinued (Brozoski et al. 2013). D-AP5 also significantly reduced the neuronal hyperactivity associated with acoustic trauma and tinnitus in the bilateral paraflocculus and dorsal CN, as well as the contralateral ventral CN. This suggests that reducing glutamatergic neurotransmission in the paraflocculus is important for the treatment of chronic tinnitus. The role of the paraflocculus in acoustic trauma-induced tinnitus was further analysed by surgical ablation of the paraflocculus before tinnitus induction, after the establishment of tinnitus, or by reversible inactivation using lidocaine (Bauer et al. 2013). It was found that both paraflocculus ablation and inactivation eliminated established tinnitus, while paraflocculus ablation before tinnitus induction only partially attenuated tinnitus. The question is whether and how neurotransmission in the paraflocculus changes over time during tinnitus development. It was reported that mRNA levels of glutamate receptor NMDA subunit 1 did not change at 2 weeks after either unilateral acoustic trauma or mechanical trauma. Instead, there was a significant increase in mRNA levels of glutamate decarboxylase 1 in the ipsilateral paraflocculus of guinea pigs (Mulders et al. 2014b). However, it must be borne in mind that gene expression was measured at 2 weeks after acoustic trauma in Mulders et al. (2014b), while D-AP5 treatment and neuronal hyperactivity measurements were conducted several months after the confirmation of tinnitus in the study of Brozoski et al. (2013). In a more recent study, neurotransmitter levels were measured in the paraflocculus of rats after the injection of salicylate using microdialysis. It was found that extracellular levels of glutamate in the paraflocculus were significantly increased from 2 h after salicylate injection, which was accompanied by a significant increase of spontaneous firing rate of the excitatory interneurons in the same area (Du et al. 2017). Therefore, future studies need to correlate behavioural evidence of tinnitus, extracellular levels of excitatory and inhibitory neurotransmitters, and neuronal activity in the paraflocculus in acoustic trauma animals in order to provide a better understanding of the pharmacological target in the paraflocculus for the treatment of tinnitus.

2.3 Drugs Modulating Other Neurotransmitter Systems

In addition to GABAergic and glutamatergic neurotransmission, other neurotransmitters, such as serotonin, acetylcholine, noradrenaline, dopamine, and endocannabinoids, have also been implicated in tinnitus generation. It has been shown that serotoninergic, cholinergic, adrenergic, and dopaminergic signalling play important roles in auditory information processing (Lustig 2006; Papesh and

Hurley 2016; Schicknick et al. 2012, 2019; Weinberger 2007). In addition, these neurotransmitters are widely distributed in multiple neural networks associated with attention, stress, emotion, learning, memory, and motivational behaviour (Chudasama and Robbins 2004; Fisher and Hariri 2013; Jay 2003; Lee and Han 2019), which are similar to the network changes observed in tinnitus patients (Henry et al. 2014; Leaver et al. 2016b; Roberts et al. 2010; Simonetti and Oiticica 2015). Therefore, they represent potential pharmacological targets for tinnitus treatment. In fact, some selective serotonin reuptake inhibitors, serotonin–norepinephrine reuptake inhibitors, and dopamine receptor agonists have been trialled in tinnitus patients, with variable outcomes in their effectiveness in reducing tinnitus loudness and tinnitus-related distress (Chang and Wu 2012; Cil et al. 2020 In press; Lopez-Gonzalez et al. 2007a, b; Oishi et al. 2010; Sziklai et al. 2011). Hence, there is a need for more preclinical studies using animal models of tinnitus to objectively evaluate their efficacy on tinnitus perception and to understand the pharmacological mechanisms of these drugs. However, to date, only a very few drugs that modulate these neurotransmitter systems have been tested in animals.

Cyclobenzaprine is a tricyclic antidepressant and an FDA-approved drug for the treatment of muscle spasm. Its pharmacological mechanisms are not very clear. Its muscle relaxing effect is thought to be mediated by action on the locus coeruleus through noradrenergic pathways and by the inhibition of serotoninergic pathways at 5-HT_2 receptors in the spinal cord (Cimolai 2009). However, none of this explains the CNS effects of cyclobenzaprine, such as sedation and mood altering properties. Cyclobenzaprine was tested for the treatment of tinnitus in animals due to its effects on the locus coeruleus and the involvement of the locus coeruleus in attention, awareness, and memory (Lobarinas et al. 2015). In this study, rats were exposed to unilateral acoustic trauma (narrow band noise centred at 16 kHz, 126 dB SPL) for 1 h and tinnitus behaviour was assessed using the GPIAS paradigm. Forty days after acoustic trauma and the confirmation of tinnitus, cyclobenzaprine (0.5 mg/kg, i.p) was administered. When animals were tested for behavioural evidence of tinnitus 1 h after the drug treatment, tinnitus was significantly suppressed by cyclobenzaprine. However, when tested 48 h later, tinnitus returned. Although the authors attributed the tinnitus suppression effect to cyclobenzaprine's action in the locus coeruleus, further studies are needed to link cyclobenzaprine's tinnitus suppression effect to its effects on neurotransmission and neuronal activity in the locus coeruleus as well as in other tinnitus-related brain areas.

Anxiety and depression are the most common comorbidities of tinnitus and there is a strong relationship between tinnitus severity and the prevalence of anxiety and depression (Bhatt et al. 2017). However, the causative relationship between tinnitus and mood disorders is not clear. In animals with acoustic trauma-induced tinnitus, there was no evidence of increased anxiety (Pace and Zhang 2013; Zheng et al. 2011a). Salicylate has been shown to induce anxiety-like behaviour and classic anxiety is associated with slow theta oscillation in the ventral hippocampus (Winne et al. 2019). To investigate the influence of anxiety on tinnitus perception, meta-chlorophenylpiperazine (mCPP), a non-selective serotonin receptor agonist known to produce anxiety-like behaviour, was tested in animals with salicylate-induced tinnitus (Guitton et al. 2005). In this study, tinnitus was induced by a high dose of salicylate (300 mg/kg, i.p.) and tinnitus-like behaviour was measured using an active avoidance paradigm, 2 h after salicylate injection. mCPP (0.1 mg/kg, i.p.) was administered on the second day of salicylate injection at 45 min before the tinnitus testing session. It was found that mCPP significantly enhanced the perception of tinnitus in salicylate-treated animals. Furthermore, local administration of an NMDA receptor antagonist into the round window blocked tinnitus-like behaviour in animals that received both salicylate and mCPP injections. The results suggest that the presence of anxiety may increase the perception of tinnitus. It would be interesting to investigate whether mCPP would also increase tinnitus perception in acoustic trauma-induced tinnitus.

Changes in the cholinergic system have also been implicated in tinnitus pathophysiology. Cholinergic innervations are widely distributed in the auditory system, from the cochlea to the AC (Schofield et al. 2011). It has been shown that choline acetyltransferase (ChAT) activity was significantly increased at 5 months after acoustic trauma in the anteroventral CN and the lateral superior olivary nucleus, but not in the IC or AC (Godfrey et al. 2013). Zhang et al. (2019) examined hippocampal vesicular acetylcholine transporter (VAChT) expression in relation to acoustic trauma and tinnitus in guinea pigs (Zhang et al. 2019). It was found that the VAChT level was significantly decreased at 2 weeks after acoustic trauma in the bilateral hippocampus and more interestingly, at 12 weeks after acoustic trauma, VAChT recovered to baseline levels in acoustic trauma animals that did not develop tinnitus, but remained decreased in animals that developed tinnitus. However, the effects of pharmacological modulation of the cholinergic system on tinnitus and tinnitus-related pathophysiology were only tested in the salicylate model, but not in the acoustic trauma model. Using Arg3.1 and c-Fos as markers for neuroplasticity, it was found that salicylate (350 mg/kg, i.p.) induced plastic changes in the AC in the Mongolian gerbil amygdala and scopolamine, a non-selective muscarinic acetylcholine receptor antagonist, dose-dependently (0.5–3 mg/kg, i.p.) reduced salicylate-related plasticity in the AC, but not in the amygdala (Wallhausser-Franke et al. 2006). However, scopolamine (1 mg/kg, i.p.) did not affect salicylate (150 mg/kg, i.p.)-induced tinnitus in rats when tested using a schedule-induced polydipsia avoidance conditioning (SIPAC) paradigm (Lobarinas et al. 2008). Taken together, the involvement of the cholinergic system in tinnitus perception is rather complicated. Further pharmacological investigation, such as directly correlating the activation and inactivation of the cholinergic system with tinnitus perception and tinnitus-related neuronal activity in different brain regions in both the salicylate and acoustic trauma models of tinnitus, would contribute to a better understanding of the causal relationship.

Another neurotransmitter system that has a long relationship with tinnitus, but is under-investigated, is the endocannabinoid system. Endocannabinoids and their receptors are widely distributed in the central nervous system and contribute to virtually all brain functions through short- and long-term modulation of synaptic transmission (Fride 2005; Pazos et al. 2005). In the auditory system, the cannabinoid receptor subtype 1 (CB_1) was found in the dorsal CN (Tzounopoulos 2008; Zheng et al. 2007) and activation of CB_1 receptors has been shown to inhibit the release of both glutamate and glycine and modulate the balance of excitation and inhibition in auditory circuits (Tzounopoulos 2008; Zhao et al. 2009). A number of natural or synthetic cannabinoid CB_1 receptor agonists have been tested for their effects in animal models of tinnitus. Zheng et al. (2010a) investigated the effects of two CB_1 receptor agonists, WIN55,212–2 and CP-55940, on tinnitus induced by salicylate in rats. The animals were injected with salicylate (350 mg/kg, i.p.) and tinnitus-like behaviour was measured 2 h after salicylate injection using a conditioned lick suppression paradigm (Zheng et al. 2010a). WIN55,212–2 (3 mg/kg, s.c.) or CP55,940 (0.1 or 0.3 mg/kg, s.c.) was administered 30 min before salicylate injection. Neither CB_1 agonist reduced tinnitus-like behaviour induced by salicylate. Instead, 3 mg/kg WIN55,212–2 and 0.3 mg/kg CP55,940 induced tinnitus-like behaviour in normal control animals, suggesting that these cannabinoids could not prevent salicylate-induced tinnitus and might actually induce tinnitus themselves. The same research group also tested the combination of two natural cannabinoids, delta-9-tetrahydrocannabinol (Δ-9-THC) and cannabidiol (CBD), on acoustic trauma-induced tinnitus in rats (Zheng et al. 2015b). Tinnitus was induced by exposing the animals to unilateral acoustic trauma and the perception of tinnitus was confirmed using a conditioned lick suppression paradigm 1 month after acoustic trauma. Following the confirmation of tinnitus, the effects of the cannabinoids on tinnitus were investigated by administering either a vehicle or a mixture of Δ-9-THC (1.5 mg/kg, s.c) and CBD (1.5 mg/kg, s.c) every day, 30 min before tinnitus testing, throughout the tinnitus testing period, for a total of 27 days. It was found that cannabinoids significantly increased the number of tinnitus animals in the acoustic trauma group, which suggests that cannabinoids may promote the development of tinnitus following acoustic trauma. However, there are some questions regarding the selectivity of these cannabinoids for the CB_1 receptors (Pertwee et al. 2010). Therefore, a highly selective CB_1 receptor agonist, arachidonyl-2′-chloroethylamide (ACEA), was tested in both the salicylate- and acoustic trauma-induced tinnitus models in guinea pigs (Berger et al. 2017). Tinnitus was induced by either an injection of salicylate (350 mg/kg, i.p.) or exposure to unilateral acoustic trauma (narrow band noise centred at 10 kHz, 116 dB for 1 h) and behavioural evidence of tinnitus was measured using the GPIAS paradigm adapted for use with the pinna reflex. The results showed that ACEA (1 mg/kg, i.p.) was not able to reduce tinnitus induced by either salicylate or acoustic trauma. In agreement with the animal studies, a cross-sectional analysis of nationally representative data collected from 2,705 adults, revealed that regular marijuana use was associated with an increased prevalence of tinnitus (Qian and Alyono 2019). A study of medicinal *Cannabis* use in Tuscany reported that ‘ear and labyrinth disorders’ were among the most commonly reported adverse effects (Crescioli et al. 2020). Therefore, studies in both animals and humans suggest that activation of the endocannabinoid system may promote the development of tinnitus. However, the link between the endocannabinoid system and tinnitus perception remains to be fully explored. One possibility could be the involvement of the endocannabinoid system in auditory gating, which is an inhibitory process, reducing the attention that the brain directs to repeated irrelevant sounds. Auditory gating is believed to involve the frontal-striatal and limbic networks, including brain structures such as the ventromedial prefrontal cortex (vmPFC), hippocampus, and nucleus accumbens (NAcc) (Rauschecker et al. 2010, 2015), and the auditory gating hypothesis is that the perception of tinnitus may arise from the failure of auditory gating (Rauschecker et al. 2015). It was reported that the CB_1 receptor antagonist, CP55,940, disrupted auditory gating and neural oscillations in the hippocampus and entorhinal cortex of rats (Hajos et al. 2008). This is not surprising, given the role of endocannabinoids in modulating corticostriatal functional connectivity. Studies have shown that long-term exposure to Δ-9-THC reduced input from the mPFC to the NAcc while increasing the input from the ventral hippocampus and basolateral amygdala to the NAcc (Hwang and Lupica 2020). The reduced functional connectivity between the NAcc and cortical regions was also observed in humans following the administration of Δ-9-THC (Mason et al. 2019). Therefore, further studies in this area will contribute to a better understanding of tinnitus and its treatment.

2.4 *Drugs Modulating Ion Channels*

Ion channels are membrane-bound proteins that allow the movement of specific ions across cell membranes and play an important role in modulating neuronal excitability. Therefore, drugs shown to reduce neuronal activity through various ion channels have been tested for their effects on tinnitus in animal models. Carbamazepine is an anti-epileptic drug which reduces neuronal hyperactivity by inhibiting repetitive firing, by increasing the number of voltage-gated Na^+ channels in the inactivated state (Yang et al. 2010). There are some clinical studies showing that carbamazepine either reduced tinnitus in some of patients or had no effect on tinnitus (Hulshof and Vermeij 1985; Melding and Goodey 1979; Sanchez et al. 1999). Therefore, more preclinical investigations are needed. Zheng et al. (2008) evaluated the effects of carbamazepine on tinnitus in the salicylate-induced tinnitus model in rats. Tinnitus was induced by salicylate (350 mg/kg, i.p.) and the perception of tinnitus was measured using a conditioned lick suppression paradigm. Carbamazepine was administered (i.p.) 30 min before salicylate injection and 3 doses of carbamazepine (5, 15 and 30 mg/kg) were tested to generate a dose-response curve. The results showed that 15 mg/kg carbamazepine, but not 5 or 30 mg/kg, significantly reduced salicylate-induced tinnitus, which suggests that carbamazepine may have the potential to be used to treat salicylate-induced tinnitus (Zheng et al. 2008).

Another ion channel-modulating drug that has been tested in the salicylate-induced tinnitus model is Maxipost and its enantiomer, R-Maxipost. Maxipost is a positive Kv7.2–Kv7.5 channel modulator, which increases the depolarisation threshold for cells and produces anxiolytic effects; R-Maxipost is a negative modulator for Kv7.2–Kv7.5 channels without anxiolytic effects (Korsgaard et al. 2005). In this study, tinnitus was induced by salicylate (150 mg/kg, i.p.) and confirmed using the SIPAC paradigm 1 h after salicylate injection. Maxipost (5 or 10 mg/kg, i.p.) or

R-Maxipost (1, 3, 5 or 10 mg/kg, i.p.) was administered 30 min before the tinnitus testing. It was found that both Maxipost and R-Maxipost dose-dependently reduced salicylate-induced tinnitus (Lobarinas et al. 2011). This was unexpected as Maxipost and R-Maxipost have opposite effects on Kv7.2–Kv7.5 channels and R-Maxipost does not have anxiolytic effects. Therefore, their effects on tinnitus could not be explained by the effects on these channels and on anxiety. The authors suggested a possible mechanism through other ion channels, such as Kv7.1 or BK channels; however, further investigations are necessary to understand the involvement of ion channels in tinnitus and its treatment.

Flufenamic acid (FFA) is an anti-inflammatory and analgesic drug used since the 1960s, but its use was limited due to its harmful side effect profile and weak beneficial effects. However, it was discovered later to be an ion channel modulator. FFA not only affects non-selective cation channels and chloride channels, but also modulates potassium, calcium, and sodium channels (Guinamard et al. 2013). The rationale for using FFA to treat tinnitus was based on its blocking effect of TRPM2 (Transient receptor potential melastatin type-2) channels, which is believed to decrease neuronal excitability (Bal et al. 2016). Salicylate (400 mg/kg, s.c.) was injected to induce tinnitus, which was behaviourally confirmed using a conditioned lick suppression paradigm at 2 h after salicylate injection. FFA (66 mg/kg, i.p.) administered 5 h before the tinnitus testing significantly reduced salicylate-induced tinnitus. However, due to the broad spectrum of FFA's effects on ion channels, the tinnitus suppression effects cannot be attributed solely to TRPM2 channels.

Furosemide is a loop diuretic that has been shown to reduce the endocochlear potential and the spontaneous discharge rates of auditory nerve fibres by affecting ion transporters in the stria vascularis and causing a subsequent reduction of neurotransmitter release from inner hair cells (Rybak and Morizono 1982; Sewell 1984). Therefore, furosemide was evaluated as a potential treatment for tinnitus in a guinea pig model of acoustic trauma-induced tinnitus (Mulders et al. 2014a). Animals were exposed unilaterally to a loud tone (10 kHz, 124 dB SPL) for 2 h and behavioural evidence of tinnitus was tested weekly using the GPIAS paradigm. Immediately after the confirmation of tinnitus, furosemide was injected (80 mg/kg, i.p.) and tinnitus tested 1 h later. In some of the animals, the spontaneous activity of auditory nerve fibres and spontaneous firing rates of neurons in the central nucleus of the IC were also recorded. The results showed that intraperitoneal injection of furosemide significantly reduced tinnitus-like behaviour and also decreased spontaneous neuronal firing rate in both the primary auditory nerves and IC. In order to determine whether the effects of furosemide on IC neurons could be caused by its effects in the cochlea, furosemide was infused into the cochlea and this manipulation also suppressed hyperactivity in the IC. However, furosemide may also reduce tinnitus through a direct central effect by acting on the ion transporters in the brain. Further animal studies are needed to investigate the effects of furosemide on chronic tinnitus. In addition, high doses of furosemide have been associated with an increased risk of temporary deafness and tinnitus in humans (Ho and Sheridan 2006), possibly by abolishing blood flow in the vessels supplying the lateral wall and causing transient ischaemia (Ding et al. 2016). It is necessary to find out whether prolonged use of furosemide can also cause ototoxicity.

It is noteworthy that most of these drugs were tested only in the salicylate model of tinnitus. However, salicylate-induced tinnitus does not happen very often in humans and will normally resolve by itself, so does not require treatment. More studies need to be conducted in the acoustic trauma-induced tinnitus model, which represents the most common form of tinnitus in humans.

2.5 Drugs Modulating Neuroinflammatory Pathways

Although it is controversial, steroids have been used in treating various forms of tinnitus, including tinnitus resulting from acoustic trauma (Cesarani et al. 2002; Markou et al. 2004; Shulman and Goldstein 2000; Silverstein et al. 1996), suggesting a possible involvement of inflammation in tinnitus. It has been shown that inflammatory cytokines, such as interleukin 1 beta (IL-1β), TNF-α, and interleukin 6 (IL-6) have neuromodulatory properties and are involved in a range of pathophysiological functions, including the modulation of synaptic transmission and neuronal excitability by modulating both voltage-gated and receptor-coupled ion channels (Vezzani and Viviani 2015). In salicylate-treated animals, changes in a range of inflammatory gene expression were found in the cochlea (Hwang et al. 2011), CN (Hu et al. 2014), IC (Hwang et al. 2011) and AC (Chen and Zheng 2017), and there was a positive association between tinnitus scores and the expression levels of TNF-α, IL-1β, and NR2B (Hwang et al. 2011). In addition, peripheral cochlear inflammation has also been reported following acoustic trauma (Adams et al. 2009; Fujioka et al. 2006; Hirose et al. 2005; Miyao et al. 2008; Tornabene et al. 2006). The relationship between acoustic trauma, tinnitus, neuroinflammation, and synaptic function was further investigated by Wang et al. (2019) in mice. The study firstly showed that unilateral acoustic trauma (narrow band noise centred at 8 kHz, 112–114 dB SPL) for 2 h induced a long-lasting (up to 10 days post-exposure) hearing threshold shift in the exposed ear and a significant upregulation of TNF-α in the primary AC (AI) at 12 h after acoustic trauma, which lasted for up to 10 days. Proinflammatory cytokines IL-1β and IL-18 were also up-regulated in the AI at 10 days following acoustic trauma. In addition, microglial activation was evident in the AI at 5 days after acoustic trauma. However, noise-induced microglial activation was absent in TNF-α knockout mice. Furthermore, the authors also showed acoustic trauma-induced tinnitus-like behaviour in wild-type (WT) mice, but not in the TNF-α knockout mice, when measured using a GPIAS paradigm (Wang et al. 2019). However, a close inspection of the results raised some questions regarding the interpretation of the lack of tinnitus-like behaviour in TNF-α knockout mice. The authors showed that during gap-induced PPI, the startle response ratio was 0.4–0.6 for the WT mice, which increased to 0.6–0.8 after acoustic trauma. This increase in the startle response ratio after acoustic trauma has been used as an indication of the perception of a sound during the gap period, i.e. tinnitus. However, in TNF-α knockout mice, even before the acoustic trauma, the startle response ratio for gap detection in response to 14 and 20 kHz carrier tones (i.e. 0.6–0.8) was higher than that in WT mice and similar to that in WT mice after the acoustic trauma. This might suggest that TNF-α knockout mice actually experience tinnitus at high frequencies or exhibit a temporal processing deficit. However, the startle response ratio in another group of TNF-α knockout mice was very similar to that in WT mice in the next experiment, which suggests that there might be large variations in the animals' startle reflex response between different experiments. Testing the TNF-α knockout mice in a conditioned lick suppression paradigm would help to clarify the issue. Nevertheless, the role of TNF-α in tinnitus was further analysed by infusing TNF-α into the AI in WT and TNF-α knockout mice. It was found that the TNF-α infusion caused tinnitus-like behaviour at 20 kHz in both the WT and TNF-α knockout mice, evidenced by an impaired gap detection. Since increased TNF-α and tinnitus could be a result of microglial activation, microglia were depleted with i.p. injections of PLX3397 and animals were examined for acoustic trauma-induced TNF-α expression and tinnitus. The administration of PLX3397 significantly reduced the TNF-α expression in control mice and prevented the increased TNF-α expression and tinnitus induced by acoustic trauma. Interestingly, PLX3397 also significantly improved gap detection in mice before and after acoustic trauma, which returned to baseline level after drug washout. This leads to an important question. If the reduced TNF-α expression induced by PLX3397 is responsible for the improved gap detection observed, the impaired gap detection following TNF-α infusion may reflect an inability to detect the gap rather than the gap being masked by the tinnitus sound. Along the same lines, an improved gap detection might also contribute to the lack of acoustic trauma-induced tinnitus-like behaviour in TNF-α knockout mice. Therefore, if TNF-α is able to modulate the animal's gap detection ability, caution needs to be taken when interpreting the results. In order to investigate the role of neuroinflammation in tinnitus, a TNF-α inhibitor that has been shown to reduce neuroinflammation, 3,6′-dithiothalidomide (dTT), was tested. It was found that dTT prevented acoustic trauma-induced microglial activation and proinflammatory cytokine expression, as well as acoustic trauma-induced tinnitus. However, there was no control group to show that the vehicle treatment did not prevent acoustic trauma-induced tinnitus in this experiment. Nevertheless, the authors did validate the effects of dTT on tinnitus using a different behavioural paradigm – a conditioned lick suppression paradigm – and the results showed that dTT significantly reduced acoustic trauma-induced tinnitus. Finally, the functional effects of TNF-α on acoustic trauma-induced excitatory and inhibitory imbalance were investigated by recording the excitatory and inhibitory synaptic currents in AI pyramidal neurons of mice following acoustic trauma, with or without dTT treatment. The results showed that acoustic trauma induced a reduction of inhibitory currents and an increase of excitatory currents and that these effects were reversed by the administration of dTT, which suggests that the acoustic trauma-induced excitatory and inhibitory imbalance is largely mediated by neuroinflammation. Using a combination of molecular, genetic, pharmacological, electrophysiological, and behavioural approaches, this study demonstrated the link between acoustic trauma, neuroinflammation, synaptic plasticity, and tinnitus generation, which suggests a potential pharmacological target for tinnitus treatment.

Another anti-inflammatory drug that has been tested in the acoustic trauma-induced tinnitus model is RO27–3225, which is a selective melanocortin receptor 4 agonist. In recent years, the anti-inflammatory properties of melanocortins, a class of peptides known as α-, β-, and γ-melanocyte-stimulating hormones and adrenocorticotrophic hormone, have gained increasing interest in pharmacology due to their promising therapeutic potential in the treatment of inflammatory-mediated diseases (Catania 2008). RO27–3225 has been shown to improve neuronal function following cerebral ischaemia by counteracting inflammatory and apoptotic responses, such as the changes in TNF-α, BAX, ERK1/2, JNK1/2, caspase-3, and Bcl-2 protein expression (Spaccapelo et al. 2011). Therefore, the effects of RO27–3225 on tinnitus development were investigated in rats by administering the drug 30 min before acoustic trauma and then every 12 h for 10 days and behavioural evidence of tinnitus was measured at 5 months after acoustic trauma using a conditioned lick suppression paradigm (Zheng et al. 2015a). Two doses of RO27–3225 (90 and 180 μg/kg, s.c.) were used in the study and tinnitus was induced by exposing the animals unilaterally to a 16 kHz tone at 115 dB for 1 h. It was found that neither dose of RO27–3225 prevented the development of tinnitus following acoustic trauma. Unfortunately, changes in inflammatory cytokines in the brain were not measured in these animals; therefore, it is impossible to determine whether the lack of effect on tinnitus development was due to insufficient anti-inflammatory effects of the RO27–3225 doses or the anti-inflammatory effects occurring at inappropriate time points. Further studies need to address these issues.

C-phycocyanin (C-PC) is the active component of *Spirulina* (a microscopic blue-green algae) and exhibits anti-oxidant and anti-inflammatory properties in pathological conditions (McCarty et al. 2010; Romay et al. 1999). Therefore, C-PC and *Spirulina platensis* water extract (15–25% phycobiliproteins containing 15–25% C-PC and allophycocyanin, 35–45% polysaccharides, 10–20% proteins other than phycobiliproteins, 5–8% water, and 10–12% ash) were tested in salicylate-induced tinnitus in mice (Hwang et al. 2013). C-PC (130 mg/kg/day) and *Spirulina* (1,000 mg/kg/day) were administered as daily dietary supplements starting on the first day of behavioural training, i.e. 5 days before salicylate injection. Salicylate (300 mg/kg, i.p.) was injected on day 6 and tinnitus behaviour was measured 2 h later using an active avoidance paradigm. The results showed that both C-PC and *Spirulina* significantly reduced salicylate-induced tinnitus to a similar extent. In addition, both C-PC and *Spirulina* also significantly reduced NR2B, TNF-α, IL-1β, and COX-2 gene expression. Therefore, C-PC and *Spirulina* may reduce tinnitus through anti-inflammatory effects.

Given the role of inflammatory cytokines in neuromodulation, anti-inflammatory drugs hold potential as an effective pharmacological treatment for tinnitus. More studies need to be conducted to understand the time course of their effects on tinnitus as well as the molecular mechanisms in different animal models of tinnitus.

2.6 Herbal Medicines

Due to the dissatisfaction with the conventional pharmacological treatment for tinnitus and the desperate need to control it, complementary or alternative medicines, including herbal medicines, are often used by tinnitus patients (Enrico et al. 2007). These include *Ginkgo biloba* extracts, Yoku-kan-san, *Cimicifuga racemosa, Cornus officinalis, Verbascum densiflorum, Rhodiola rosea, Hydrastis canadensis, Sesamum indicum* (seeds), *Helianthus annuus* (seeds), and many more. The publications on the use of herbal medicines for tinnitus treatment are either case reports or poorly controlled clinical trials; therefore, information regarding their efficacy and safety is unreliable and controversial (Darlington et al. 2009; Enrico et al. 2007).

One of the herbal medicines that has been tested in multiple clinical trials for its effects on tinnitus is *Ginkgo biloba* extract, which has been used for hundreds of years in Chinese traditional medicine. It is also commonly used in Western countries, being one of the top ten best-selling herbal medicines in Europe and the USA (Drew and Davies 2001; Sierpina et al. 2003). Despite the numerous clinical trials that have been conducted in tinnitus patients, the results have been controversial, with the two most systematic, double-blind and placebo controlled clinical trials showing negative results and suggesting that *Ginkgo biloba* extracts are no more effective in the treatment of tinnitus than a placebo (Smith et al. 2005). By contrast, the effects of *Ginkgo biloba* extract on tinnitus in animals were tested in only one study, where the salicylate-induced tinnitus model was used (Jastreboff et al. 1997). In this study, EGb-761, a standardised *Ginkgo biloba* extract containing 24% flavonoids, 7% proanthocyanidins, and 6% terpenoids was administered orally (10–100 mg/kg) every day for 2 weeks before salicylate was injected (275 mg/kg, s.c.) to induce tinnitus. Tinnitus-like behaviour was measured using a conditioned lick suppression paradigm. It was found that EGb 761 dose-dependently (25, 50 and 100 mg/kg) decreased salicylate-induced tinnitus. As shown in a systematic review and meta-analysis of 90 randomised controlled trials of EGb 761 in patients with cognitive impairment and dementia, it may improve cognition with no important safety concerns (Tan et al. 2015). However, other studies suggest that it has no significant effect on cognitive performance in humans without dementia and even little effect on those with dementia (Dongen 2003; Snitz et al. 2009; Solomon et al. 2002). Nonetheless, it might be worth evaluating the effects of EGb 761 in the acoustic trauma-induced tinnitus model and to further explore its mechanisms of action.

A mixture of a number of herbal medicines, Er Ming Fang (EMF01), was also tested in the salicylate-induced tinnitus model (Zheng et al. 2010b). Er Ming Fang (EMF01) is modified from the traditional Chinese medicinal prescription, LiuWei Dihuang Wan, which has long been used in the treatment of diabetes mellitus. It is formed by retaining the main components of LiuWei Dihuang Wan, namely Shu Dihuang (Rhizoma of *Rehmannia glutinosa* LIBOSCH), Shanzhuyu (Fructus of *Cornus officinalis* SIEB. Et ZUCC.), and Fuling (*Poria cocos* (SCHW.) WOLF) and adding Danshen (Radix of *Salvia Miltiorrhiza* BGE), Gegen (Radix of *Pueraria lobata* (WILD.) OHWI), Wuweizi (Fructus of *Schisandra Chinensis* (TURCX.) BAILL), and Jiegeng (Radix of *Platycodon Grandiflorum* (JACQ.) A. DC). It has been shown to relieve tinnitus in a small number of tinnitus patients with kidney essence deficiency – a traditional Chinese Medicine diagnosis. Twenty out of twenty-four patients reported that their tinnitus either disappeared or was reduced following the treatment (Unpublished observation, Xuan-xuan Zhu, Research Centre of JiangSu Provincial Hospital of Traditional Chinese Medicine, Nanjing, P.R. China). Therefore, the effects of EMF01 on tinnitus were tested in a salicylate model in rats (Zheng et al. 2010b). Acute tinnitus was induced by a single injection of sodium salicylate (350 mg/kg, s.c.) and tinnitus-like behaviour was measured using a conditioned lick suppression paradigm. EMF01 (8.75 or 17.5 g/kg, oral gavage) or the vehicle (tap water) was administered for 20 days before the induction of tinnitus. On the day of the tinnitus induction, EMF01 or vehicle was administered 30 min prior to salicylate or saline administration. It was found that EMF01 did not affect salicylate-induced tinnitus (Zheng et al. 2010b).

Based on clinical and limited animal studies, it seems that there is a lack of scientific evidence supporting the use of herbal medicines for tinnitus treatment. However, given the vast number of herbal medicines available, this may be an unexplored avenue for developing pharmacological treatment for tinnitus. In particular, the idea that tinnitus is a result of abnormal interactions in multiple networks/systems is consistent with Traditional Chinese Medicine's view of considering the human body as a holistic system where different organs/systems are interrelated (Castaneda et al. 2019; Yap et al. 2009). Based on Huangdi Neijing (The Yellow Emperor's Inner Canon), the 'ear' has a close relationship with the 'kidney'. It is well documented in Traditional Chinese Medicine that 'kidneys open into ears' and the function of the ear reflects the function of the kidney. The concept of 'kidney function' in Chinese medicine consists of growth, development, reproduction, urine secretion, endocrine function, as well as water and salt metabolism. Although there is yet to be a solid scientific basis for the relationship between the kidney and the ears, hearing loss and tinnitus are the most common symptoms accompanying kidney dysfunction (Govender et al. 2013; Ikeda et al. 1987; Renda et al. 2015) and drugs, such as aminoglycoside antibiotics, that produce ototoxicity, almost always cause nephrotoxicity (Pagkalis et al. 2011). Therefore, Traditional Chinese Medicine takes a network approach in treating hearing loss and tinnitus by restoring kidney function and there are hundreds of herbal medicines that have been used either alone or in a mixture, for the treatment of hearing loss and tinnitus in Chinese medical practice (Castaneda et al. 2019; Yap et al. 2009). Therefore, studies to evaluate more herbal medicines in different animal models of tinnitus and to explore their underlying mechanisms, may contribute to potential drug discovery for tinnitus.

3 Advantages and Limitations of Drug Evaluation Using Animal Models

Animal studies are necessary in order to be able to fully explore the beneficial and adverse effects of drugs that may be useful in the treatment of tinnitus, and to be able to investigate the neural bases of any drug effect. Despite the utility of human studies employing fMRI and other imaging techniques, animal studies will always be necessary to fully comprehend the effects that drugs used to treat tinnitus, have on the brain. Nonetheless, subjective tinnitus can only be inferred from animal models using methods such as gap detection and conditioned responses.

The ability to test the effects of drugs on tinnitus-related behaviour in animals is predicated upon the ability to reliably induce tinnitus, and all of the available methods have their limitations. The salicylate method is limited by the fact that the tinnitus-related behaviour that is induced is short-lived and that salicylate can have direct effects on the brain itself. The noise-induced tinnitus model is limited by the fact that only 30–80% of animals exhibit tinnitus-like behaviour (von der Behrens 2014), so there are always animals that appear to be resistant to developing tinnitus, or at least resistant to expressing tinnitus-related behaviour. It is difficult to assess the success of a test for tinnitus when it is impossible to know whether animals really have the condition. Usually, the sensitivity and specificity of a test could be evaluated using receiver-operating characteristic (ROC) curves, in which the correct identification ('hit') rate is plotted against 1 – specificity ('false alarm') rate (Hsieh and Turnball 1996). However, without a 'gold standard' to compare against, one model, such the gap detection method, can only be compared against another, such as the conditioned lick response model. This has never been done in the same study because, without a gold standard test, it would be impossible to know which test was more accurate.

Drug studies have varied enormously in the time points that have been used following the induction of tinnitus, from several weeks to months. Given that tinnitus in humans is often a chronic disorder when it is troublesome, long-term studies in animals may be particularly useful. However, as animals such as rats and mice age, they become more prone to diseases such as cancer, and because their weight escalates, they become more difficult to use for electrophysiological and neurochemical experiments, not least because they respond to anaesthetics in a less predictable way. A non-human primate model may be developed in the future.

The way in which the data for tinnitus-related behaviour have been analysed has varied substantially across the available drug studies. Since the statistical analysis is the final arbiter of whether an animal is deemed to have tinnitus or not, as well as any drug effect on it, it is a crucial step in determining the outcome of any experiment. Analyses have ranged from the use of t tests and repeated measures ANOVAs, to linear mixed model (LMM) analyses. An important consideration here is that most behavioural models of tinnitus necessarily involve repeated measures, and repeated measures ANOVAs are notoriously susceptible to the problem of violating the assumption of 'sphericity', i.e. that the data are correlated over time (Smith 2017).

In biology, data from the same subjects are usually correlated across time; therefore, it is likely that this ANOVA assumption is violated in many of these analyses. Corrections such as the Geisser–Greenhouse and Huynh–Feldt corrections can be used, but they simply make significant differences more difficult to achieve by reducing the type I error rate, thereby potentially increasing the type II error rate. LMM analyses, by contrast, control for potential correlation in the repeated measures data by modelling it using various covariance matrix structure models (Smith 2017).

There are also pharmacological issues that impact on the use of animal models for testing drugs. Many studies use intraperitoneal (i.p) injections to deliver drugs, as an approximation to oral administration in humans. However, the exact location of i.p injections into the peritoneal cavity can vary considerably, altering absorption and therefore distribution of the drug to the brain (Steward et al. 1968). Differences in blood protein binding, penetration of the blood–brain barrier, and elimination (i.e. half-life and clearance) can cause large differences in drug action compared to humans. Probably the method that best represents oral administration in humans is oral gavage, but this is difficult to do in animals and can raise ethical issues. When comparing doses used in animals and humans, relatively few studies have employed the kind of dose-translation equations employed by the FDA (Reagan-Shaw et al. 2008). Animals obviously have different body surface areas and metabolic rates compared to humans; therefore, converting animal drug doses to human doses by simply multiplying the animal dose by the difference in body weight is incorrect (Reagan-Shaw et al. 2008). The FDA requires calculation of a human equivalent dose (HED) that is the animal dose in mg/kg multiplied by the animal Km divided by the human Km, where Km controls for differences in body surface area (Reagan-Shaw et al. 2008). In order to calculate the dose in a rat, for example, for baclofen that corresponds to the usual dose used in adult humans, the usual HED must be multiplied by the human Km/rat Km, which equals 6.17. Using this equation, many drug doses used in rats often appear very large compared to the HED. The use of the HED equation increases the chances of research using animal models of tinnitus being relevant to human therapeutics. One way of enhancing the validity of animal studies of drug effects on tinnitus is to have follow-up studies of investigational drugs, in human patients, so that the preclinical and clinical studies are coordinated.

4 Future Directions in Tinnitus Drug Discovery

Traditionally, the gold standard, modern drug discovery approach seeks to design more selective drugs with ideally one specific target in order to reduce side effects. Despite excessive efforts and investment routinely made in discovering individual molecular targets over the last two decades, the rate of new drug candidates being translated into clinical use is decreasing (Kola and Landis 2004). Hopkins argued that the 'one gene, one drug, one disease' drug design philosophy might be the fundamental problem (Hopkins 2008). This is because biological functions, or dysfunctions in disease states, are more likely to be a consequence of complex biochemical regulation processes driven by interactive networks within the genome (Chen et al. 2008), transcriptome (Iancu et al. 2014), proteome (Ebhardt et al. 2015), and metabolome (Shah et al. 2015). Targeting such dynamic network biology by identifying disease-causing networks rather than disease-causing genes is likely to be a more effective approach for drug discovery (Hopkins 2007, 2008; Kell and Goodacre 2014; Roth et al. 2004). Since tinnitus has been shown to be associated with changes in neuronal activity and connectivity involving multiple neural networks, both in human patients (Boyen et al. 2014; Husain and Schmidt 2014; Kraus and Canlon 2012; Leaver et al. 2011; Maudoux et al. 2012; Song et al. 2012; Vanneste et al. 2011; Vanneste and De Ridder 2012) and an animal model (Chen et al. 2014), it has been increasingly recognised that tinnitus is unlikely to be generated by a single pathological source, but rather complex network changes involving not only the auditory system but also systems related to memory, emotion, and stress (Henry et al. 2014; Leaver et al. 2016a; Roberts et al. 2010; Simonetti and Oiticica 2015). Therefore, identifying pharmacological targets at the network level using animal models of tinnitus may open up new avenues for the development of effective treatments for tinnitus.

References

Adams JC, Seed B, Lu N, Landry A, Xavier RJ (2009) Selective activation of nuclear factor kappa B in the cochlea by sensory and inflammatory stress. Neuroscience 160(2):530–539

Ahring PK, Bang LH, Jensen ML, Strobaek D, Hartiadi LY, Chebib M, Absalom N (2016) A pharmacological assessment of agonists and modulators at alpha4beta2gamma2 and alpha4beta2delta GABAA receptors: the challenge in comparing apples with oranges. Pharmacol Res 111:563–576

Baguley D, McFerran D, Hall D (2013) Tinnitus. Lancet 382(9904):1600–1607

Bal R, Ustundag Y, Bulut F, Demir CF, Bal A (2016) Flufenamic acid prevents behavioral manifestations of salicylate-induced tinnitus in the rat. Arch Med Sci 12(1):208–215

Bauer CA, Brozoski TJ (2001) Assessing tinnitus and prospective tinnitus therapeutics using a psychophysical animal model. J Assoc Res Otolaryngol 2(1):54–64

Bauer CA, Brozoski TJ, Rojas R, Boley J, Wyder M (1999) Behavioral model of chronic tinnitus in rats. Otolaryngol Head Neck Surg 121(4):457–462

Bauer CA, Brozoski TJ, Holder TM, Caspary DM (2000) Effects of chronic salicylate on GABAergic activity in rat inferior colliculus. Hear Res 147(1–2):175–182

Bauer CA, Kurt W, Sybert LT, Brozoski TJ (2013) The cerebellum as a novel tinnitus generator. Hear Res 295:130–139

Berger JI, Coomber B, Shackleton TM, Palmer AR, Wallace MN (2013) A novel behavioural approach to detecting tinnitus in the guinea pig. J Neurosci Methods 213(2):188–195

Berger JI, Coomber B, Hill S, Alexander SPH, Owen W, Palmer AR, Wallace MN (2017) Effects of the cannabinoid CB1 agonist ACEA on salicylate ototoxicity, hyperacusis and tinnitus in guinea pigs. Hear Res 356:51–62

Bertrand S, Ng GY, Purisai MG, Wolfe SE, Severidt MW, Nouel D, Robitaille R, Low MJ, O'Neill GP, Metters K, Lacaille JC, Chronwall BM, Morris SJ (2001) The anticonvulsant, antihyperalgesic agent gabapentin is an agonist at brain gamma-aminobutyric acid type B

receptors negatively coupled to voltage-dependent calcium channels. J Pharmacol Exp Ther 298 (1):15–24
Bhatt JM, Bhattacharyya N, Lin HW (2017) Relationships between tinnitus and the prevalence of anxiety and depression. Laryngoscope 127(2):466–469
Boyen K, de Kleine E, van Dijk P, Langers DRM (2014) Tinnitus-related dissociation between cortical and subcortical neural activity in humans with mild to moderate sensorineural hearing loss. Hear Res 312:48–59
Browne CJ, Morley JW, Parsons CH (2012) Tracking the expression of excitatory and inhibitory neurotransmission-related proteins and neuroplasticity markers after noise induced hearing loss. PLoS One 7(3):e33272
Brozoski TJ, Bauer CA, Caspary DM (2002) Elevated fusiform cell activity in the dorsal cochlear nucleus of chinchillas with psychophysical evidence of tinnitus. J Neurosci 22:2383–2390
Brozoski TJ, Spires TJD, Bauer CA (2007) Vigabatrin, a GABA transaminase inhibitor, reversibly eliminates tinnitus in an animal model. J Assoc Res Otolarnygol 8(1):105–118
Brozoski TJ, Caspary DM, Bauer CA, Richardson BD (2010) The effect of supplemental dietary taurine on tinnitus and auditory discrimination in an animal model. Hear Res 270(1–2):71–80
Brozoski TJ, Wisner KW, Odintsov B, Bauer CA (2013) Local NMDA receptor blockade attenuates chronic tinnitus and associated brain activity in an animal model. PLoS One 8(10):e77674
Butt S, Ashraf F, Porter LA, Zhang H (2016) Sodium salicylate reduces the level of GABAB receptors in the rat's inferior colliculus. Neuroscience 316:41–52
Castaneda R, Natarajan S, Jeong SY, Hong BN, Kang TH (2019) Traditional oriental medicine for sensorineural hearing loss: can ethnopharmacology contribute to potential drug discovery? J Ethnopharmacol 231:409–428
Catania A (2008) Neuroprotective actions of melanocortins: a therapeutic opportunity. Trends Neurosci 31(7):353–360
Cazals Y (2000) Auditory sensori-neural alterations induced by salicylate. Prog Neurobiol 62 (6):583–631
Cesarani A, Capobianco S, Soi D, Giuliano DA, Alpini D (2002) Intratympanic dexamethasone treatment for control of subjective idiopathic tinnitus: our clinical experience. Int Tinnitus J 8 (2):111–114
Chang JP, Wu CC (2012) Serotonin-norepinephrine reuptake inhibitor treatment for tinnitus and depression. J Clin Psychopharmacol 32(5):729
Chen XH, Zheng LL (2017) Expression of pro-inflammatory cytokines in the auditory cortex of rats with salicylate-induced tinnitus. Mol Med Rep 16(4):5643–5648
Chen YQ, Zhu J, Lum PY, Yang X, Pinto S, MacNeil DJ, Zhang CS, Lamb J, Edwards S, Sieberts SK, Leonardson A, Castellini LW, Wang S, Champy MF, Zhang B, Emilsson V, Doss S, Ghazalpour A, Horvath S, Drake TA, Lusis AJ, Schadt EE (2008) Variations in DNA elucidate molecular networks that cause disease. Nature 452(7186):429–435
Chen Y-C, Wang J, Jiao Y, Zang F-C, Yang M, Tong J-X, Salvi R, Teng G-J (2014) Changes in resting-state fMRI activity during salicylate-induced tinnitus and sound stimulation. In: 8th international TRI tinnitus conference abstract, p 106
Chudasama Y, Robbins TW (2004) Psychopharmacological approaches to modulating attention in the five-choice serial reaction time task: implications for schizophrenia. Psychopharmacology 174(1):86–98
Cianfrone G, Pentangelo D, Cianfrone F, Mazzei F, Turchetta R, Orlando MP, Altissimi G (2011) Pharmacological drugs inducing ototoxicity, vestibular symptoms and tinnitus: a reasoned and updated guide. Eur Rev Med Pharmacol Sci 15(6):601–636
Cil OC, Zateri C, Guclu O, Oymak S, Tezcan E (2020) The effect of fibromyalgia treatment on tinnitus. Am J Otolaryngol 41(3):102390
Cimolai N (2009) Cyclobenzaprine: a new look at an old pharmacological agent. Expert Rev Clin Pharmacol 2(3):255–263
Cooper JC (1994) Health and nutrition examination survey of 1971-75: part II. Tinnitus subjective hearing loss and well-being. J Am Acad Audiol 5:37–43

Crescioli G, Lombardi N, Bettiol A, Menniti-Ippolito F, Da Cas R, Parrilli M, Del Lungo M, Gallo E, Mugelli A, Maggini V, Firenzuoli F, Vannacci A (2020) Adverse events following cannabis for medical use in Tuscany: an analysis of the Italian Phytovigilance database. Br J Clin Pharmacol 86(1):106–120

Darlington CL, Zheng Y, Smith PF (2009) Herbal medicinal products for tinnitus: perspectives and evaluation. In: Houghton P, Mukherjee PK (eds) Evaluation of herbal medicinal products. Pharmaceutical Press, London, pp 314–319

de la Iglesia-Larrad JI, Barral C, Casado-Espada NM, de Alarcon R, Macia-Casas A, Vicente Hernandez B, Roncero C (2020) Benzodiazepine abuse, misuse, dependence, and withdrawal among schizophrenic patients: a review of the literature. Psychiatry Res 284:112660

Dehmel S, Eisinger D, Shore SE (2012) Gap prepulse inhibition and auditory brainstem-evoked potentials as objective measures for tinnitus in guinea pigs. Front Syst Neurosci 6:42

Ding D, Liu H, Qi W, Jiang H, Li Y, Wu X, Sun H, Gross K, Salvi R (2016) Ototoxic effects and mechanisms of loop diuretics. J Otol 11(4):145–156

Dong S, Mulders WH, Rodger J, Woo S, Robertson D (2010a) Acoustic trauma evokes hyperactivity and changes in gene expression in guinea-pig auditory brainstem. Eur J Neurosci 31 (9):1616–1628

Dong S, Rodger J, Mulders WH, Robertson D (2010b) Tonotopic changes in GABA receptor expression in guinea pig inferior colliculus after partial unilateral hearing loss. Brain Res 1342:24–32

Dongen M (2003) Ginkgo for elderly people with dementia and age-associated memory impairment: a randomized clinical trial. J Clin Epidemiol 56(4):367–376

Drew S, Davies E (2001) Effectiveness of *Ginkgo biloba* in treating tinnitus: double blind, placebo controlled trial. BMJ 322(7278):73

Du Y, Liu J, Jiang Q, Duan Q, Mao L, Ma F (2017) Paraflocculus plays a role in salicylate-induced tinnitus. Hear Res 353:176–184

Ebhardt HA, Root A, Sander C, Aebersold R (2015) Applications of targeted proteomics in systems biology and translational medicine. Proteomics 15(18):3193–3208

Eggermont JJ (2005) Tinnitus: neurobiological substrates. Drug Discov Today 10(19):1283–1290

Eggermont JJ, Roberts LE (2004) The neuroscience of tinnitus. Trends Neurosci 27(11):676–682

Enrico P, Sirca D, Mereu M (2007) Antioxidants, minerals, vitamins, and herbal remedies in tinnitus therapy. Prog Brain Res 166:323–330

European Commission (2004) Directive 2004/27/EC of the European Parliament and of the Council of 31 March 2004 Amending Directive 2001/83/EC on the Community Code Relating to Medicinal Products for Human Use. https://ec.europa.eu/health/sites/health/files/files/eudralex/vol-1/dir_2004_27/dir_2004_27_en.pdf

Fisher PM, Hariri AR (2013) Identifying serotonergic mechanisms underlying the corticolimbic response to threat in humans. Philos Trans R Soc Lond B Biol Sci 368(1615):20120192

Fride E (2005) Endocannabinoids in the central nervous system: from neuronal networks to behavior. Curr Drug Targets CNS Neurol Disord 4(6):633–642

Fujioka M, Kanzaki S, Okano HJ, Masuda M, Ogawa K, Okano H (2006) Proinflammatory cytokines expression in noise-induced damaged cochlea. J Neurosci Res 83(4):575–583

Godfrey DA, Kaltenbach JA, Chen K, Ilyas O, Liu X, Licari F, Sacks J, McKnight D (2012) Amino acid concentrations in the hamster central auditory system and long-term effects of intense tone exposure. J Neurosci Res 90(11):2214–2224

Godfrey DA, Kaltenbach JA, Chen K, Ilyas O (2013) Choline acetyltransferase activity in the hamster central auditory system and long-term effects of intense tone exposure. J Neurosci Res 91(7):987–996

Govender SM, Govender CD, Matthews G (2013) Cochlear function in patients with chronic kidney disease. S Afr J Commun Disord 60:44–49

Guinamard R, Simard C, Del Negro C (2013) Flufenamic acid as an ion channel modulator. Pharmacol Ther 138(2):272–284

Guitton MJ, Dudai Y (2007) Blockade of cochlear NMDA receptors prevents long-term tinnitus during a brief consolidation window after acoustic trauma. Neural Plast 2007:80904

Guitton MJ, Caston J, Ruel J, Johnson RM, Pujol R, Puel J-L (2003) Salicylate induces tinnitus through activation of cochlear NMDA receptors. J Neurosci 23(9):3944–3952

Guitton MJ, Pujol R, Puel JL (2005) m-Chlorophenylpiperazine exacerbates perception of salicylate-induced tinnitus in rats. Eur J Neurosci 22(10):2675–2678

Hajos M, Hoffmann WE, Kocsis B (2008) Activation of cannabinoid-1 receptors disrupts sensory gating and neuronal oscillation: relevance to schizophrenia. Biol Psychiatry 63(11):1075–1083

Heffner HE (2011) A two-choice sound localization procedure for detecting lateralized tinnitus in animals. Behav Res Methods 43(2):577–589

Heffner HE, Koay G (2005) Tinnitus and hearing loss in hamsters (*Mesocricetus auratus*) exposed to loud sound. Behav Neurosci 119(3):734–742

Henry JA, Roberts LE, Caspary DM, Theodoroff SM, Salvi RJ (2014) Underlying mechanisms of tinnitus: review and clinical implications. J Am Acad Audiol 25(1):5–22

Hirose K, Discolo CM, Keasler JR, Ransohoff R (2005) Mononuclear phagocytes migrate into the murine cochlea after acoustic trauma. J Comp Neurol 489(2):180–194

Ho KM, Sheridan DJ (2006) Meta-analysis of frusemide to prevent or treat acute renal failure. BMJ 333(7565):420

Hopkins AL (2007) Network pharmacology. Nat Biotechnol 25(10):1110–1111

Hopkins AL (2008) Network pharmacology: the next paradigm in drug discovery. Nat Chem Biol 4 (11):682–690

Hsieh F, Turnball BW (1996) Non-parametric and semi-parametric estimating of the receiver-operating characteristic curve. Ann Stats 24(1):25–40

Hu SS, Mei L, Chen JY, Huang ZW, Wu H (2014) Effects of salicylate on the inflammatory genes expression and synaptic ultrastructure in the cochlear nucleus of rats. Inflammation 37 (2):365–373

Hu SS, Mei L, Chen JY, Huang ZW, Wu H (2015) Expression of immediate-early genes in the dorsal cochlear nucleus in salicylate-induced tinnitus. Eur Arch Otorhinolaryngol 273 (2):325–332

Hulshof JH, Vermeij P (1985) The value of carbamazepine in the treatment of tinnitus. ORL J Otorhinolaryngol Relat Spec 47(5):262–266

Husain FT, Schmidt SA (2014) Using resting state functional connectivity to unravel networks of tinnitus. Hear Res 307:153–162

Hwang EK, Lupica CR (2020) Altered corticolimbic control of the nucleus accumbens by long-term delta(9)-tetrahydrocannabinol exposure. Biol Psychiatry 87(7):619–631

Hwang JH, Chen JC, Yang SY, Wang MF, Chan YC (2011) Expression of tumor necrosis factor-alpha and interleukin-1beta genes in the cochlea and inferior colliculus in salicylate-induced tinnitus. J Neuroinflammation 8:30

Hwang JH, Chen JC, Chan YC (2013) Effects of c-phycocyanin and spirulina on salicylate-induced tinnitus, expression of NMDA receptor and inflammatory genes. PLoS One 8(3):e58215

Iancu OD, Colville A, Darakjian P, Hitzemann R (2014) Coexpression and cosplicing network approaches for the study of mammalian brain transcriptomes. Int Rev Neurobiol 116:73–93

Ikeda K, Kusakari J, Arakawa E, Ohyama K, Inamura N, Kawamoto K (1987) Cochlear potentials of guinea pigs with experimentally induced renal failure. Acta Otolaryngol Suppl 435:40–45

Jang CH, Lee S, Park IY, Song A, Moon C, Cho GW (2019) Memantine attenuates salicylate-induced tinnitus possibly by reducing NR2B expression in auditory cortex of rat. Exp Neurobiol 28(4):495–503

Jastreboff PJ, Brennan JF, Coleman JK, Sasaki CT (1988a) Phantom auditory sensation in rats: an animal model for tinnitus. Behav Neurosci 102(6):811–822

Jastreboff PJ, Brennan JF, Sasaki CT (1988b) An animal model for tinnitus. Laryngoscope 98 (3):280–286

Jastreboff PJ, Zhou S, Jastreboff MM, Kwapisz U, Gryczynska U (1997) Attenuation of salicylate-induced tinnitus by *Ginkgo biloba* extract in rats. Audiol Neurootol 2(4):197–212

Jay TM (2003) Dopamine: a potential substrate for synaptic plasticity and memory mechanisms. Prog Neurobiol 69(6):375–390

Jia F, Yue M, Chandra D, Keramidas A, Goldstein PA, Homanics GE, Harrison NL (2008) Taurine is a potent activator of extrasynaptic GABA(A) receptors in the thalamus. J Neurosci 28 (1):106–115

Jufas NE, Wood R (2015) The use of benzodiazepines for tinnitus: systematic review. J Laryngol Otol 129(Suppl 3):S14–S22

Jung MJ, Lippert B, Metcalf BW, Bohlen P, Schechter PJ (1977) Gamma-vinyl GABA (4-amino-hex-5-enoic acid), a new selective irreversible inhibitor of GABA-T: effects on brain GABA metabolism in mice. J Neurochem 29(5):797–802

Kell DB, Goodacre R (2014) Metabolomics and systems pharmacology: why and how to model the human metabolic network for drug discovery. Drug Discov Today 19(2):171–182

Kletke O, Gisselmann G, May A, Hatt H, Sergeeva OA (2013) Partial agonism of taurine at gamma-containing native and recombinant GABAA receptors. PLoS One 8(4):e61733

Kola I, Landis J (2004) Can the pharmaceutical industry reduce attrition rates? Nat Rev Drug Discov 3(8):711–715

Korsgaard MP, Hartz BP, Brown WD, Ahring PK, Strobaek D, Mirza NR (2005) Anxiolytic effects of Maxipost (BMS-204352) and retigabine via activation of neuronal Kv7 channels. J Pharmacol Exp Ther 314(1):282–292

Kraus KS, Canlon B (2012) Neuronal connectivity and interactions between the auditory and limbic systems. Effects of noise and tinnitus. Hear Res 288(1–2):34–46

Krogsgaard-Larsen P, Falch E, Larsson OM, Schousboe A (1987) GABA uptake inhibitors: relevance to antiepileptic drug research. Epilepsy Res 1(2):77–93

Langguth B, Elgoyhen AB (2012) Current pharmacological treatments for tinnitus. Expert Opin Pharmacother 13(17):2495–2509

Lanneau C, Green A, Hirst WD, Wise A, Brown JT, Donnier E, Charles KJ, Wood M, Davies CH, Pangalos MN (2001) Gabapentin is not a GABAB receptor agonist. Neuropharmacology 41 (8):965–975

Leaver AM, Renier L, Chevillet MA, Morgan S, Kim HJ, Rauschecker JP (2011) Dysregulation of limbic and auditory networks in tinnitus. Neuron 69(1):33–43

Leaver AM, Seydell-Greenwald A, Rauschecker JP (2016a) Auditory-limbic interactions in chronic tinnitus: challenges for neuroimaging research. Hear Res 334:49–57

Leaver AM, Turesky TK, Seydell-Greenwald A, Morgan S, Kim HJ, Rauschecker JP (2016b) Intrinsic network activity in tinnitus investigated using functional MRI. Hum Brain Mapp 37 (8):2717–2735

Lee EH, Han PL (2019) Reciprocal interactions across and within multiple levels of monoamine and cortico-limbic systems in stress-induced depression: a systematic review. Neurosci Biobehav Rev 101:13–31

Lobarinas E, Sun W, Cushing R, Salvi R (2004) A novel behavioral paradigm for assessing tinnitus using schedule-induced polydipsia avoidance conditioning (SIP-AC). Hear Res 190 (1–2):109–114

Lobarinas E, Yang G, Sun W, Ding D, Mirza N, Dalby-Brown W, Hilczmayer E, Fitzgerald S, Zhang L, Salvi R (2006) Salicylate- and quinine-induced tinnitus and effects of memantine. Acta Otolaryngol Suppl 126(sup556):13–19

Lobarinas E, Sun W, Stolzberg D, Lu J, Salvi R (2008) Human brain imaging of tinnitus and animal models. Semin Hear 29(4):333–349

Lobarinas E, Dalby-Brown W, Stolzberg D, Mirza NR, Allman BL, Salvi R (2011) Effects of the potassium ion channel modulators BMS-204352 Maxipost and its R-enantiomer on salicylate-induced tinnitus in rats. Physiol Behav 104(5):873–879

Lobarinas E, Blair C, Spankovich C, Le Prell C (2015) Partial to complete suppression of unilateral noise-induced tinnitus in rats after cyclobenzaprine treatment. J Assoc Res Otolaryngol 16 (2):263–272

Longenecker RJ, Galazyuk AV (2011) Development of tinnitus in CBA/CaJ mice following sound exposure. J Assoc Res Otolaryngol 12(5):647–658
Lopez-Gonzalez MA, Moliner-Peiro F, Alfaro-Garcia J, Esteban-Ortega F (2007a) Sulpiride plus hydroxyzine decrease tinnitus perception. Auris Nasus Larynx 34(1):23–27
Lopez-Gonzalez MA, Santiago AM, Esteban-Ortega F (2007b) Sulpiride and melatonin decrease tinnitus perception modulating the auditolimbic dopaminergic pathway. J Otolaryngol 36 (4):213–219
Lustig LR (2006) Nicotinic acetylcholine receptor structure and function in the efferent auditory system. Anat Rec A Discov Mol Cell Evol Biol 288(4):424–434
Markou K, Nikolaou A, Petridis DG, Vlachtsis KC, Kouloulas A, Daniilidis IC (2004) Evaluation of various therapeutic schemes in the treatment of tinnitus due to acute acoustic trauma. Kulak Burun Bogaz Ihtis Derg 12(5–6):107–114
Mason NL, Theunissen EL, Hutten NRPW, Tse DHY, Toennes SW, Stiers P, Ramaekers JG (2019) Cannabis induced increase in striatal glutamate associated with loss of functional corticostriatal connectivity. Eur Neuropsychopharmacol 29(2):247–256
Maudoux A, Lefebvre P, Cabay JE, Demertzi A, Vanhaudenhuyse A, Laureys S, Soddu A (2012) Connectivity graph analysis of the auditory resting state network in tinnitus. Brain Res 1485:10–21
McCarty MF, Barroso-Aranda J, Contreras F (2010) Oral phycocyanobilin may diminish the pathogenicity of activated brain microglia in neurodegenerative disorders. Med Hypotheses 74(3):601–605
McFerran DJ, Stockdale D, Holme R, Large CH, Baguley DM (2019) Why is there no cure for tinnitus? Front Neurosci 13:802
McGonigle P, Ruggeri B (2014) Animal models of human disease: challenges in enabling translation. Biochem Pharmacol 87(1):162–171
Melding PS, Goodey RJ (1979) The treatment of tinnitus with oral anticonvulsants. J Laryngol Otol 93(2):111–122
Miyao M, Firestein GS, Keithley EM (2008) Acoustic trauma augments the cochlear immune response to antigen. Laryngoscope 118(10):1801–1808
Mulders WH, Barry KM, Robertson D (2014a) Effects of furosemide on cochlear neural activity, central hyperactivity and behavioural tinnitus after cochlear trauma in guinea pig. PLoS One 9 (5):e97948
Mulders WH, Rodger J, Yates CG, Robertson D (2014b) Modulation of gene expression in guinea pig paraflocculus after induction of hearing loss. F1000Res 3:63
Ng GY, Bertrand S, Sullivan R, Ethier N, Wang J, Yergey J, Belley M, Trimble L, Bateman K, Alder L, Smith A, McKernan R, Metters K, O'Neill GP, Lacaille JC, Hebert TE (2001) Gamma-aminobutyric acid type B receptors with specific heterodimer composition and postsynaptic actions in hippocampal neurons are targets of anticonvulsant gabapentin action. Mol Pharmacol 59(1):144–152
Niedzielski AS, Wenthold RJ (1995) Expression of AMPA, kainate, and NMDA receptor subunits in cochlear and vestibular ganglia. J Neurosci 15(3):2338–2353
Nowotny M, Remus M, Kossl M, Gaese BH (2011) Characterization of the perceived sound of trauma-induced tinnitus in gerbils. J Acoust Soc Am 130(5):2827–2834
Oishi N, Kanzaki S, Shinden S, Saito H, Inoue Y, Ogawa K (2010) Effects of selective serotonin reuptake inhibitor on treating tinnitus in patients stratified for presence of depression or anxiety. Audiol Neurootol 15(3):187–193
Pace E, Zhang JS (2013) Noise-induced tinnitus using individualized gap detection analysis and its relationship with hyperacusis, anxiety, and spatial cognition. PLoS One 8(9):e75011. https://doi.org/10.71371/journal.pone.0075011
Pagkalis S, Mantadakis E, Mavros MN, Ammari C, Falagas ME (2011) Pharmacological considerations for the proper clinical use of aminoglycosides. Drugs 71(17):2277–2294

Panes A, Verdoux H, Fourrier-Reglat A, Berdai D, Pariente A, Tournier M (2020) Use of benzodiazepines non-compliant with guidelines in patients with psychiatric and non-psychiatric chronic disorders. Gen Hosp Psychiatry 65:21–27

Panford-Walsh R, Singer W, Ruttiger L, Hadjab S, Tan J, Geisler HS, Zimmermann U, Kopschall I, Rohbock K, Vieljans A, Oestreicher E, Knipper M (2008) Midazolam reverses salicylate-induced changes in brain-derived neurotrophic factor and arg3.1 expression: implications for tinnitus perception and auditory plasticity. Mol Pharmacol 74(3):595–604

Papesh MA, Hurley LM (2016) Modulation of auditory brainstem responses by serotonin and specific serotonin receptors. Hear Res 332:121–136

Pazos MR, Nunez E, Benito C, Tolon RM, Romero J (2005) Functional neuroanatomy of the endocannabinoid system. Pharmacol Biochem Behav 81(2):239–247

Pertwee RG, Howlett AC, Abood ME, Alexander SP, Di Marzo V, Elphick MR, Greasley PJ, Hansen HS, Kunos G, Mackie K, Mechoulam R, Ross RA (2010) International Union of Basic and Clinical Pharmacology. LXXIX. Cannabinoid receptors and their ligands: beyond CB (1) and CB(2). Pharmacol Rev 62(4):588–631

Qian ZJ, Alyono JC (2019) An association between marijuana use and tinnitus. Am J Otolaryngol 41(1):102314

Ralli M, Troiani D, Podda MV, Paciello F, Eramo SL, de Corso E, Salvi R, Paludetti G, Fetoni AR (2014) The effect of the NMDA channel blocker memantine on salicylate-induced tinnitus in rats. Acta Otorhinolaryngol Ital 34(3):198–204

Rauschecker JP, Leaver AM, Muhlau M (2010) Tuning out the noise: limbic-auditory interactions in tinnitus. Neuron 66(6):819–826

Rauschecker JP, May ES, Maudoux A, Ploner M (2015) Frontostriatal gating of tinnitus and chronic pain. Trends Cogn Sci 19(10):567–578

Reagan-Shaw S, Nihal M, Ahmad N (2008) Dose translation from animal to human studies revisited. FASEB J 22(3):659–661

Renda R, Renda L, Selcuk OT, Eyigor H, Yilmaz MD, Osma U (2015) Cochlear sensitivity in children with chronic kidney disease and end-stage renal disease undergoing hemodialysis. Int J Pediatr Otorhinolaryngol 79(12):2378–2383

Richardson BD, Ling LL, Uteshev VV, Caspary DM (2011) Extrasynaptic GABA(A) receptors and tonic inhibition in rat auditory thalamus. PLoS One 6(1):e16508

Roberts LE, Eggermont JJ, Caspary DM, Shore SE, Melcher JR, Kaltenbach JA (2010) Ringing ears: the neuroscience of tinnitus. J Neurosci 30(45):14972–14979

Romay C, Ledon N, Gonzalez R (1999) Phycocyanin extract reduces leukotriene B4 levels in arachidonic acid-induced mouse-ear inflammation test. J Pharm Pharmacol 51(5):641–642

Roth BL, Sheffler DJ, Kroeze WK (2004) Magic shotguns versus magic bullets: selectively non-selective drugs for mood disorders and schizophrenia. Nat Rev Drug Discov 3(4):353–359

Rüttiger L, Ciuffani J, Zenner H-P, Knipper M (2003) A behavioral paradigm to judge acute sodium salicylate-induced sound experience in rats: a new approach for an animal model on tinnitus. Hear Res 180(1–2):39–50

Rybak LP, Morizono T (1982) Effect of furosemide upon endolymph potassium concentration. Hear Res 7(2):223–231

Sametsky EA, Turner JG, Larsen D, Ling L, Caspary DM (2015) Enhanced GABAA-mediated tonic inhibition in auditory thalamus of rats with behavioral evidence of tinnitus. J Neurosci 35 (25):9369–9380

Sanchez TG, Balbani AP, Bittar RS, Bento RF, Camara J (1999) Lidocaine test in patients with tinnitus: rationale of accomplishment and relation to the treatment with carbamazepine. Auris Nasus Larynx 26(4):411–417

Schicknick H, Reichenbach N, Smalla KH, Scheich H, Gundelfinger ED, Tischmeyer W (2012) Dopamine modulates memory consolidation of discrimination learning in the auditory cortex. Eur J Neurosci 35(5):763–774

Schicknick H, Henschke JU, Budinger E, Ohl FW, Gundelfinger ED, Tischmeyer W (2019) Beta-adrenergic modulation of discrimination learning and memory in the auditory cortex. Eur J Neurosci 50(7):3141–3163

Schofield BR, Motts SD, Mellott JG (2011) Cholinergic cells of the pontomesencephalic tegmentum: connections with auditory structures from cochlear nucleus to cortex. Hear Res 279 (1–2):85–95

Sewell WF (1984) The relation between the endocochlear potential and spontaneous activity in auditory nerve fibres of the cat. J Physiol 347:685–696

Shah NJ, Sureshkumar S, Shewade DG (2015) Metabolomics: a tool ahead for understanding molecular mechanisms of drugs and diseases. Indian J Clin Biochem 30(3):247–254

Shulman A, Goldstein B (2000) Intratympanic drug therapy with steroids for tinnitus control: a preliminary report. Int Tinnitus J 6(1):10–20

Sierpina VS, Wollschlaeger B, Blumenthal M (2003) *Ginkgo biloba*. Am Fam Physician 68 (5):923–926

Silverstein H, Choo D, Rosenberg SI, Kuhn J, Seidman M, Stein I (1996) Intratympanic steroid treatment of inner ear disease and tinnitus (preliminary report). Ear Nose Throat J 75 (8):468–471

Simonetti P, Oiticica J (2015) Tinnitus neural mechanisms and structural changes in the brain: the contribution of neuroimaging research. Int Arch Otorhinolaryngol 19(3):259–265

Smith PF (2017) A guerrilla guide to common problems in 'Neurostatistics': essential statistical topics in neuroscience. J Undergrad Neurosci Educ 16(1):R1–R12

Smith PF, Zheng Y, Darlington CL (2005) *Ginkgo biloba* extracts for tinnitus: more hype than hope? J Ethnopharmacol 100(1–2):95–99

Snitz BE, O'Meara ES, Carlson MC, Arnold AM, Ives DG, Rapp SR, Saxton J, Lopez OL, Dunn LO, Sink KM, DeKosky ST, Ginkgo Evaluation of Memory Study I (2009) *Ginkgo biloba* for preventing cognitive decline in older adults: a randomized trial. JAMA 302(24):2663–2670

Solomon PR, Adams F, Silver A, Zimmer J, DeVeaux R (2002) Ginkgo for memory enhancement: a randomized controlled trial. JAMA 288(7):835–840

Song JJ, De Ridder D, Van de Heyning P, Vanneste S (2012) Mapping tinnitus-related brain activation: an activation-likelihood estimation metaanalysis of PET studies. J Nucl Med 53 (10):1550–1557

Song Y, Liu J, Ma F, Mao L (2016) Diazepam reduces excitability of amygdala and further influences auditory cortex following sodium salicylate treatment in rats. Acta Otolaryngol 136 (12):1220–1224

Spaccapelo L, Bitto A, Galantucci M, Ottani A, Irrera N, Minutoli L, Altavilla D, Novellino E, Grieco P, Zaffe D, Squadrito F, Giuliani D, Guarini S (2011) Melanocortin MC(4) receptor agonists counteract late inflammatory and apoptotic responses and improve neuronal functionality after cerebral ischemia. Eur J Pharmacol 670(2–3):479–486

Steward JP, Ornellas EP, Beernink KD, Northway WH (1968) Errors in the technique of intraperitoneal injection in mice. Appl Microbiol 16(9):1418–1419

Stolzberg D, Hayes SH, Kashanian N, Radziwon K, Salvi RJ, Allman BL (2013) A novel behavioral assay for the assessment of acute tinnitus in rats optimized for simultaneous recording of oscillatory neural activity. J Neurosci Methods 219(2):224–232

Suarez LM, Suarez F, Del Olmo N, Ruiz M, Gonzalez-Escalada JR, Solis JM (2005) Presynaptic NMDA autoreceptors facilitate axon excitability: a new molecular target for the anticonvulsant gabapentin. Eur J Neurosci 21(1):197–209

Sziklai I, Szilvassy J, Szilvassy Z (2011) Tinnitus control by dopamine agonist pramipexole in presbycusis patients: a randomized, placebo-controlled, double-blind study. Laryngoscope 121 (4):888–893

Tan MS, Yu JT, Tan CC, Wang HF, Meng XF, Wang C, Jiang T, Zhu XC, Tan L (2015) Efficacy and adverse effects of *Ginkgo biloba* for cognitive impairment and dementia: a systematic review and meta-analysis. J Alzheimers Dis 43(2):589–603

Tang ZQ, Lu YG, Chen L (2008) Developmental stability of taurine's activation on glycine receptors in cultured neurons of rat auditory cortex. Neurosci Lett 430(1):54–59

Tornabene SV, Sato K, Pham L, Billings P, Keithley EM (2006) Immune cell recruitment following acoustic trauma. Hear Res 222(1–2):115–124

Turner JG, Brozoski TJ, Bauer CA, Parrish JL, Myers K, Hughes LF, Caspary DM (2006) Gap detection deficits in rats with tinnitus: a potential novel screening tool. Behav Neurosci 120 (1):188–195

Turner J, Larsen D, Hughes L, Moechars D, Shore S (2012) Time course of tinnitus development following noise exposure in mice. J Neurosci Res 90(7):1480–1488

Tyler RS (2012) Patient preferences and willingness to pay for tinnitus treatments. J Am Acad Audiol 23(2):115–125

Tzounopoulos T (2008) Mechanisms of synaptic plasticity in the dorsal cochlear nucleus: plasticity-induced changes that could underlie tinnitus. Am J Audiol 17(2):S170–S175

US Congress (2018) Federal Food, Drug, and Cosmetic Act (FD&C Act). https://www.fda.gov/regulatoryinformation/lawsenforcedbyfda/federalfooddrugandcosmeticactfdcact/default.htm

Vanneste S, De Ridder D (2012) The auditory and non-auditory brain areas involved in tinnitus. An emergent property of multiple parallel overlapping subnetworks. Front Syst Neurosci 6:31. https://doi.org/10.3389/fnsys.2012.00031

Vanneste S, de Heyning PV, De Ridder D (2011) The neural network of phantom sound changes over time: a comparison between recent-onset and chronic tinnitus patients. Eur J Neurosci 34 (5):718–731

Vendra V, Vaisbuch Y, Mudry AC, Jackler RK (2019) Over-the-counter tinnitus "cures": marketers' promises do not ring true. Laryngoscope 129(8):1898–1906

Vezzani A, Viviani B (2015) Neuromodulatory properties of inflammatory cytokines and their impact on neuronal excitability. Neuropharmacology 96(Pt A):70–82

von der Behrens W (2014) Animal models of subjective tinnitus. Neural Plast 2014:741452

Wallhausser-Franke E, Cuautle-Heck B, Wenz G, Langner G, Mahlke C (2006) Scopolamine attenuates tinnitus-related plasticity in the auditory cortex. Neuroreport 17(14):1487–1491

Wang H, Brozoski TJ, Caspary DM (2011) Inhibitory neurotransmission in animal models of tinnitus: maladaptive plasticity. Hear Res 279(1–2):111–117

Wang W, Zhang LS, Zinsmaier AK, Patterson G, Leptich EJ, Shoemaker SL, Yatskievych TA, Gibboni R, Pace E, Luo H, Zhang J, Yang S, Bao S (2019) Neuroinflammation mediates noise-induced synaptic imbalance and tinnitus in rodent models. PLoS Biol 17(6):e3000307

Weinberger NM (2007) Auditory associative memory and representational plasticity in the primary auditory cortex. Hear Res 229(1–2):54–68

Winne J, Franzon R, de Miranda A, Malfatti T, Patriota J, Mikulovic S, Leao KE, Leao RN (2019) Salicylate induces anxiety-like behavior and slow theta oscillation and abolishes the relationship between running speed and fast theta oscillation frequency. Hippocampus 29(1):15–25

Wu KK (2003) Aspirin and other cyclooxygenase inhibitors: new therapeutic insights. Semin Vasc Med 3(2):107–112

Xu H, Wang W, Tang ZQ, Xu TL, Chen L (2006) Taurine acts as a glycine receptor agonist in slices of rat inferior colliculus. Hear Res 220(1–2):95–105

Yang YC, Huang CS, Kuo CC (2010) Lidocaine, carbamazepine, and imipramine have partially overlapping binding sites and additive inhibitory effect on neuronal Na+ channels. Anesthesiology 113(1):160–174

Yap L, Pothula VB, Warner J, Akhtar S, Yates E (2009) The root and development of otorhinolaryngology in traditional Chinese medicine. Eur Arch Otorhinolaryngol 266(9):1353–1359

Zhang L, Wu C, Martel DT, West M, Sutton MA, Shore SE (2019) Remodeling of cholinergic input to the hippocampus after noise exposure and tinnitus induction in Guinea pigs. Hippocampus 29 (8):669–682

Zhao Y, Rubio ME, Tzounopoulos T (2009) Distinct functional and anatomical architecture of the endocannabinoid system in the auditory brainstem. J Neurophysiol 101(5):2434–2446

Zheng Y, Baek JH, Smith PF, Darlington CL (2007) Cannabinoid receptor down-regulation in the ventral cochlear nucleus in a salicylate model of tinnitus. Hear Res 228(1–2):105–111

Zheng Y, Hooton K, Smith PF, Darlington CL (2008) Carbamazepine reduces the behavioural manifestations of tinnitus following salicylate treatment in rats. Acta Otolaryngol 128(1):48–52

Zheng Y, Stiles L, Hamilton E, Smith PF, Darlington CL (2010a) The effects of the synthetic cannabinoid receptor agonists, WIN55,212-2 and CP55,940, on salicylate-induced tinnitus in rats. Hear Res 268(1–2):145–150

Zheng Y, Vagal S, Zhu XX, de Waele C, Smith PF, Wang G, Zhang M, Darlington CL (2010b) The effects of the Chinese herbal medicine EMF01 on salicylate-induced tinnitus in rats. J Ethnopharmacol 128(2):545–548

Zheng Y, Hamilton E, McNamara E, Smith PF, Darlington CL (2011a) The effects of chronic tinnitus caused by acoustic trauma on social behaviour and anxiety in rats. Neuroscience 193:143–153

Zheng Y, Hamilton E, Stiles L, McNamara E, de Waele C, Smith PF, Darlington CL (2011b) Acoustic trauma that can cause tinnitus impairs impulsive control but not performance accuracy in the 5-choice serial reaction time task in rats. Neuroscience 180:75–84

Zheng Y, McNamara Y, Stiles L, Darlington CL, Smith PF (2012a) Evidence that memantine reduces chronic tinnitus caused by acoustic trauma in rats. Front Neurol 3:127

Zheng Y, Vagal S, Hamilton E, Darlington CL, Smith PF (2012b) A dose-response analysis of the effects of L-baclofen on chronic tinnitus caused by acoustic trauma in rats. Neuropharmacology 62(2):940–946

Zheng Y, McPherson K, Reid P, Smith PF (2015a) The anti-inflammatory selective melanocortin receptor subtype 4 agonist, RO27-3225, fails to prevent acoustic trauma-induced tinnitus in rats. Eur J Pharmacol 761:206–210

Zheng Y, Reid P, Smith PF (2015b) Cannabinoid CB1 receptor agonists do not decrease, but may increase acoustic trauma-induced tinnitus in rats. Front Neurol 6:60

Zou QZ, Shang XL (2012) Effect of salicylate on the large GABAergic neurons in the inferior colliculus of rats. Acta Neurol Belg 112(4):367–374

Zuo H, Lei D, Sivaramakrishnan S, Howie B, Mulvany J, Bao J (2017) An operant-based detection method for inferring tinnitus in mice. J Neurosci Methods 291:227–237

Animal Models of Tinnitus Treatment: Cochlear and Brain Stimulation

Jinsheng Zhang, Ethan Firestone, and Ahmed Elattma

Contents

Abstract Neuromodulation, via stimulation of a variety of peripheral and central structures, is used to suppress tinnitus. However, investigative limitations in humans due to ethical reasons have made it difficult to decipher the mechanisms underlying treatment-induced tinnitus relief, so a number of animal models have arisen to address these unknowns. This chapter reviews animal models of cochlear and brain stimulation and assesses their modulatory effects on behavioral evidence of

J. Zhang (✉)
Department of Otolaryngology-Head and Neck Surgery, Wayne State University School of Medicine, Detroit, MI, USA

Department of Communication Sciences and Disorders, Wayne State University College of Liberal Arts and Sciences, Detroit, MI, USA
e-mail: jinzhang@wayne.edu

E. Firestone and A. Elattma
Department of Otolaryngology-Head and Neck Surgery, Wayne State University School of Medicine, Detroit, MI, USA

Curr Topics Behav Neurosci (2021) 51: 83–130
https://doi.org/10.1007/7854_2021_227
Published Online: 20 July 2021

tinnitus and its related neural correlates. When a structure is stimulated, localized modulation, often presenting as downregulation of spontaneous neuronal spike firing rate, bursting and neurosynchrony, occurs within the brain area. Through anatomical projections and transmitter pathways, the interventions activate both auditory- and non-auditory structures by taking bottom-up ascending and top-down descending modes to influence their target brain structures. Furthermore, it is the brain oscillations that cochlear or brain stimulation evoke and connect the prefrontal cortex, striatal systems, and other limbic structures to refresh neural networks and relieve auditory, attentive, conscious, as well as emotional reactive aspects of tinnitus. This oscillatory neural network connectivity is achieved via the thalamocorticothalamic circuitry including the lemniscal and non-lemniscal auditory brain structures. Beyond existing technologies, the review also reveals opportunities for developing advanced animal models using new modalities to achieve precision neuromodulation and tinnitus abatement, such as optogenetic cochlear and/or brain stimulation.

Keywords Animal model · Auditory and limbic brain structures · Behavioral assays · Brain stimulation · Cochlear stimulation · Tinnitus

Abbreviations

ABI	Auditory brainstem implant
AC	Auditory cortex
ACC	Anterior cingulate cortex
ACES	Auditory cortex electrical stimulation
AMI	Auditory midbrain implant
BDNF	Brain derived neurotrophic factor
CBT	Cognitive behavioral therapy
CES	Cochlear electrical stimulation
CFC	Cross-frequency coupling
CI	Cochlear implant
CIC	Central nucleus of the inferior colliculus
DBS	Deep brain stimulation
DCIC	Dorsal cortex of the inferior colliculus
DCN	Dorsal cochlear nucleus
dlPFC	Dorsolateral prefrontal cortex
dMGB	Dorsal subdivision of the medial geniculate body
ECIC	External cortex of the inferior colliculus
ECS	Epi-cranial electrical stimulation
EEG	Electroencephalogram
FLS	Fronto-limbic-striatal
fMRI	Functional magnetic resonance imaging
GAD	Glutamate decarboxylase
GPIAS	Gap-prepulse inhibition of the acoustic startle reflex

HPC	Hippocampus
IC	Inferior colliculus
LFP	Local field potential
MDT	Mediodorsal thalamus
MGB	Medial geniculate body
mMGB	Medial subdivision of the medial geniculate body
MRI	Magnetic resonance imaging
NAc	Nucleus accumbens
NF2	Neurofibromatosis type II
PET	Positron emission tomography
PFC	Prefrontal cortex
rTMS	Repetitive transcranial magnetic stimulation
SFR	Spontaneous firing rate
SGN	Spiral ganglion neuron
SR	Stochastic resonance
STN	Subthalamic nucleus
tACS	Transcranial alternate current stimulation
TBI	Traumatic brain injury
TCD	Thalamocortical dysrhythmias
TCT	Thalamocorticothalamic
tDCS	Transcranial direct current stimulation
TES	Transcranial electrical stimulation
TMS	Transcranial magnetic stimulation
TRN	Thalamic reticular nucleus
tRNS	Random-noise stimulation
vMGB	Ventral subdivision of the medial geniculate body

1 Significance of Developing Animal Models

Tinnitus is a prevalent health condition that affects millions of people. Nearly 10–15% of the adult population is stricken with the condition, and 1–2% of these patients experience an unremitting form that significantly impacts their quality of life and daily functioning, sometimes even leading to suicide (Axelsson and Ringdahl 1989; Crocetti et al. 2009; Hasson et al. 2011; Stevens et al. 2007). In addition to civilian populations, tinnitus is a top comorbidity of war among military personnel and Veterans. Service-related loud noises from equipment operations and explosion-derived blast shockwaves damage cochlear hair cells and trigger maladaptive neural plasticity that leads to tinnitus, hearing loss, and other related neurological disorders (Zhang 2019). Economically, tinnitus-related compensation for healthcare management reaches billions of dollars annually (Zhang 2019). Given this large caseload and economic burden, there is significant incentive to develop effective treatment regimens and even a cure for tinnitus.

Thus far, numerous therapeutic strategies have been studied, which include noise-masking (Roberts et al. 2006), Tinnitus Retraining Therapy (Jastreboff and Jastreboff 2000), Neuromonics (Davis et al. 2007), acoustic coordinated reset neuromodulation (Wegger et al. 2017), pharmaceuticals (Langguth and Elgoyhen 2012), and electrical, transcranial magnetic or direct current stimulation of numerous sites, including the cochlea (Rubinstein et al. 2003; Zeng et al. 2015), cochlear nucleus (Luo et al. 2012; Soussi and Otto 1994), inferior colliculus (Smit et al. 2016b), auditory cortex (Claes et al. 2014; De Ridder et al. 2010; Zhang et al. 2011; Zhang 2013), somatosensory structures (De Ridder et al. 2007; Folmer et al. 2006; Marks et al. 2018), vagal nerve (De Ridder et al. 2014a; Engineer et al. 2011), caudate nucleus (Ahsan et al. 2018; Cheung and Larson 2010), and dorsolateral prefrontal cortex (De Ridder et al. 2012; Shekhawat et al. 2016) or frontal cortex (De Ridder and Vanneste 2012). Despite these efforts, establishment of a universal, reliable treatment that warrants clinical standardization has yet to occur. According to the clinical practice guidelines recommended by the American Academy of Otolaryngology – Head and Neck Surgery, the standard-of-care remains sound-based modulation and cognitive behavioral therapy (CBT) (Tunkel et al. 2014) (https://journals.sagepub.com/doi/pdf/10.1177/0194599814545325). However, CBT focuses on managing limbic comorbidities such as anxiety, depression, disturbed sleep, and impaired memory (Axelsson and Ringdahl 1989; Crocetti et al. 2009; Hasson et al. 2011; Stevens et al. 2007), rather than targeting the underlying tinnitus pathophysiology (McFerran et al. 2019).

Although the exact mechanisms are unknown, a considerable amount of research supports the generally agreed-upon hypothesis that disruption of the peripheral auditory apparatus triggers a cascade of maladaptive plasticity in the central nervous system, resulting in the generation of phantom auditory perception and related, reactive limbic comorbidities. Supporting a central origin of tinnitus, studies have shown that pathologic sound sensation persists even after auditory nerve transection (House and Brackmann 1981; Jackson 1985). Furthermore, animal work detailing neural spike information suggests that the trauma-induced peripheral deafferentation of the auditory system sparks hyperactivity, hypersynchrony, increased bursting, and/or tonotopic map reorganization (Eggermont 2015) in brain structures such as the dorsal cochlear nucleus (DCN) (Wu et al. 2016; Zhang and Kaltenbach 1998), inferior colliculus (IC) (Bauer et al. 2008), medial geniculate body (MGB) (Kalappa et al. 2014), auditory cortex (AC) (Eggermont and Roberts 2004; Seki and Eggermont 2003), and a host of fronto-limbic-striatal centers (Rauschecker et al. 2010, 2015; Vanneste et al. 2011b). Additionally, human experiments investigating large-scale brain oscillations revealed that tinnitus sufferers display aberrant rhythms, especially theta- and gamma-frequency, in many of the same structures (De Ridder et al. 2011; Llinas et al. 2005; Weisz et al. 2007). While transient gamma activity underlies sensation, cognition, and a number of other physiological functions (Fries 2009; Jacobs and Kahana 2010; Nozaradan et al. 2017; Singer 1999; Vianney-Rodrigues et al. 2011), it is believed that peripheral deafferentation leads to pathologic entrainment of the theta-gamma signals, as thalamocortical dysrhythmias (TCD), which are implicated in conditions including tinnitus, schizophrenia, neurogenic pain, Parkinson's, and major depressive disorder (De Ridder et al. 2015; Llinas et al. 1999, 2005). Attempting to alter this abnormal, brain activity landscape has been the focus of clinical therapies such as intra-cochlear electrical stimulation (Arts et al. 2015; Kim et al. 2015), extra-cochlear electrical stimulation (CES) (Mulders et al. 2016a; Wenzel et al. 2014), transcranial magnetic stimulation (TMS) (Langguth 2020), transcranial direct current stimulation (tDCS) (Vanneste and De Ridder 2011), and auditory cortex electrical stimulation (ACES) (Seidman et al. 2008; Zhang et al. 2011). However, these strategies yield variable success rates and, in some cases, even exacerbate clinical symptoms. The vast range of responses highlights the critical need for better understanding of both the underlying tinnitus pathophysiology and how the candidate treatment modalities manipulate it.

Given the ethical considerations in human investigations, it is imperative to develop animal models that allow much greater freedom for investigating the underlying mechanisms, testing new therapeutics, and exploring their effects on tinnitus and other compounded auditory disorders. Unlike treatment development in many other fields where animal models precede translation, many candidate tinnitus therapies started as preliminary clinical evidence that prompted further investigation and refinement in sub-human species. Decades of research for this purpose has produced a host of animal systems targeting numerous structures along the lemniscal and extra-lemniscal auditory pathways that have been fundamental in pushing the field closer to standardized clinical treatments. This chapter will focus on the current state of the animal models of cochlear and brain modulation-based tinnitus treatments, relate their findings to the underlying pathophysiology, and offer insight on possible avenues for future improvement.

2 Cochlear Stimulation to Suppress Tinnitus

Cochlear implants (CIs) have been successfully used to restore hearing in patients who cannot benefit from acoustic amplification, via hearing aids. In addition to benefiting audition, numerous clinical trials and case studies have demonstrated that this intra-cochlear electrical stimulation (CES) may simultaneously relieve patients' tinnitus (Arts et al. 2012; Baguley and Atlas 2007; Mertens et al. 2016; Poncet-Wallet et al. 2020). These results spurred strong interest in conducting extra-cochlear electrical stimulation experiments, avoiding the need for surgical implantation of electrodes, and it was also found to have a positive outcome on tinnitus (Daneshi et al. 2005; Di et al. 2009; Ito and Sakakihara 1994; Rubinstein et al. 2003; Wenzel et al. 2014). However, the CES-induced therapeutic efficacy is highly variable across subjects, and standardized guidelines for clinical application are lacking, as a result of limited understanding of the mechanisms underlying CES-induced tinnitus modulation (Mertens et al. 2016; Perreau et al. 2020; Song et al. 2017; Zeng et al. 2015). This prompts great need to establish animal models to gain knowledge regarding how exactly CES-related treatment relieves tinnitus. Thus far, limited cochlear electrical stimulation in animal studies, as described below, has demonstrated promising results in re-establishing normal peripheral (cochlear) neural activity and modulating of tinnitus-related activity in the central auditory system.

2.1 Cochlear Electrical Stimulation

2.1.1 Intra-cochlear Electrical Stimulation

The majority of clinical studies that demonstrate CES-induced tinnitus suppression are typically geared toward hearing restoration, as the cochlear implantees routinely receive benefits to their tinnitus as well. Prior to studies of the underlying mechanisms of CES-induced tinnitus in human subjects (Song et al. 2017), very limited research had been conducted in animals, but the results do provide hints about how CES suppresses tinnitus. In a rat model of intra-cochlear electrical stimulation, Argence and colleagues performed cochleectomies, a surgical procedure that destroys the cochlea to deafen the animals, and witnessed a decrease of the inhibitory GlyRα1 and glutamate decarboxylase 67 (GAD67) mRNA expressions, leading to increased neural excitability in the IC; they then implanted electrodes in the cochlea and demonstrated that chronic CES reversed the cochleectomy-induced downregulation of collicular GlyRα1 and GAD67 mRNA expression (Fig. 1) (Argence et al. 2008). The findings from this study support the hypothesis that CES of deafferented auditory nerves downregulates pathologic neural excitability in the auditory brain, mediating CES-induced suppression of tinnitus. Additionally, Basta and colleagues implanted a standard HiFocus1j electrode array in the guinea pig cochlea and recorded from several auditory brain structures following CES with a HiRes strategy (Basta et al. 2015). The authors found that low-rate electrical stimulation significantly decreased spontaneous activity in the DCN and MGB, and high-rate electrical stimulation reduced spontaneous activity in the DCN and AC. This demonstrates that CES facilitates the homeostasis of neural networks by downregulating peripheral deafferentation-induced hyperactivity in the central auditory system.

2.1.2 Extra-cochlear Electrical Stimulation

Following the reported clinical benefits of intra-cochlear electrical stimulation on relieving tinnitus, there is a strong desire to develop a less invasive and specialized device to stimulate different parts of the cochlea such as the round window, promontory, and cochlear wall, without penetrating electrodes. To investigate the mechanisms underlying extra-cochlear electrical stimulation-induced tinnitus, a limited number of animal studies have also been conducted. For example, using a guinea pig model, Norena and colleagues demonstrated that positive current stimulation of IC neurons with high characteristic frequencies decreased spontaneous

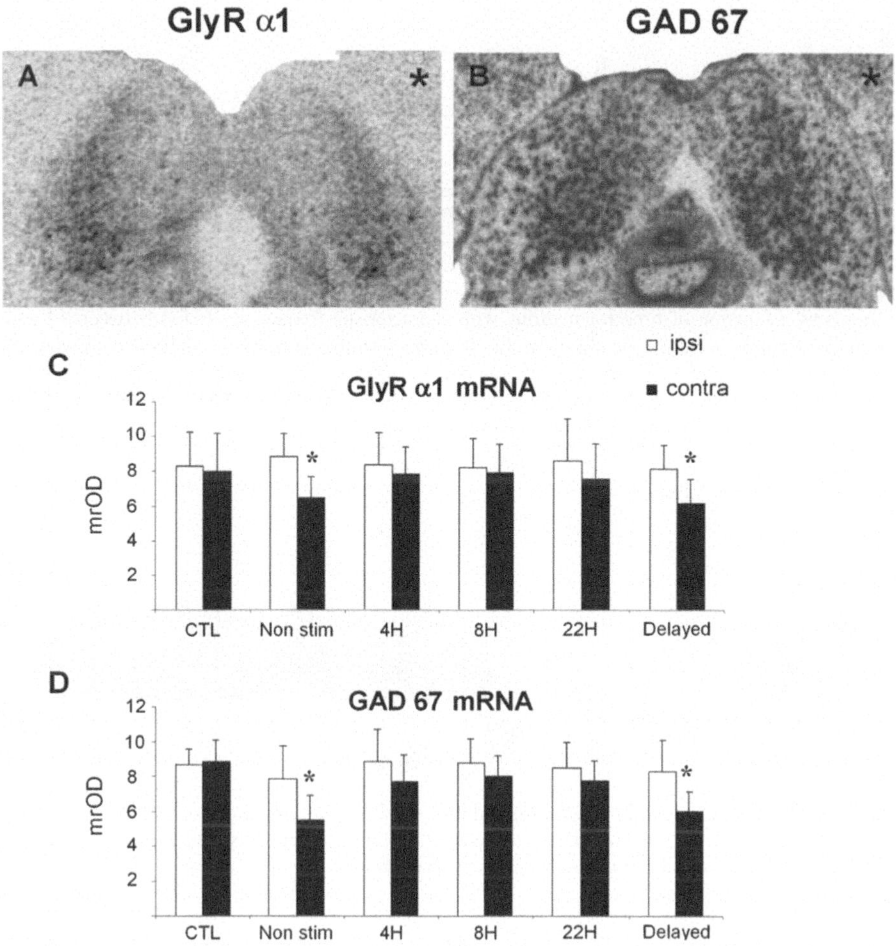

Fig. 1 GlyRα1 (glycine receptor) and glutamine acid decarboxylase (GAD)67 mRNA in stimulated animals. (**a**, **b**) In situ hybridization autoradiographs showing the distribution of 35S-labeled GlyRα1 (**a**) and GAD67 (**b**) antisense probes in sections of implanted animals after 8 h of daily stimulation for 5 days. The star indicates the side ipsilateral to the lesion. (**c**, **d**) Graphs showing the GlyRα1 mRNA level (**c**) and GAD67 mRNA level (**d**) in nonstimulated and stimulated animals, in the ipsilateral (white bar) and contralateral (black bar) central nucleus of the inferior colliculus. A large contralateral decrease, relative to control values, was observed in nonstimulated animals and delayed animals. In contrast, there was no difference between stimulated and control animals. rOD, relative optical density. $*P < 0.01$ compared with the control. Adapted from Argence et al. (2008)

firing rates, whereas negative current increased their neural activity, suggesting that the monophasic, positive current relieves tinnitus via quelling hyperactivity (Fig. 2) (Norena et al. 2015). Along the same line, Mulders and colleagues used biphasic train stimulation of the guinea pig round window. They found that the noise trauma-induced, elevated spontaneous firing rates of IC neurons were suppressed for up to

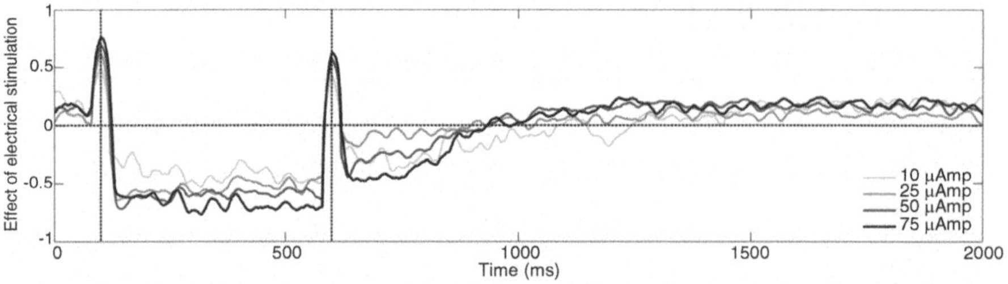

Fig. 2 Example showing averaged effects of extra-cochlear electrical stimulation with positive currents as a function of time for neurons with characteristic frequencies (CFs) between 14 and 28 kHz. The time windows of electrical stimulation (500-ms duration) are delimited by vertical dotted lines. Inserts above panels indicate the pattern of the electric stimulation. The peaks at the onset and offset of the electrical stimulation were due to the electrical artifacts generated when the stimulation is turned on and off. Adapted from Norena et al. (2015)

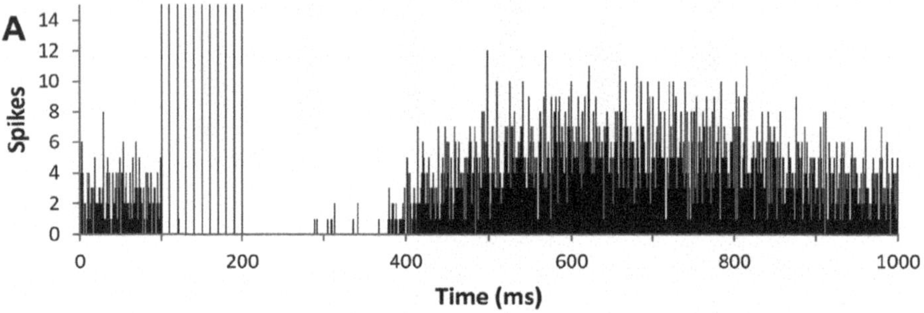

Fig. 3 A histogram example illustrating inhibitory effects of round window (extra-cochlear) electrical stimulation at 100 Hz on spontaneous firing rate of hyperactive neurons of the guinea pig inferior colliculus. Adapted from Mulders et al. (2016a)

hundreds of milliseconds after the cessation of electrical stimulation (Fig. 3), but during stimulation both inhibition and excitation were noted (Mulders et al. 2016a). Such extended post-CES suppression of hyperactivity in the auditory midbrain may be the critical mechanistic factor contributing to tinnitus relief. It was postulated that this inhibition was the result of activating GABA-ergic neuronal circuitry in the IC (Mulders et al. 2016a), which is in line with the commonly held notion that tinnitus is associated with decreased inhibitory drive (Argence et al. 2008). Apart from the CES-induced suppression of tentative neural correlates of tinnitus, the authors concurrently demonstrated excitation of other neurons in the IC, which may be attributed to diffusive and non-specific activation of certain auditory fibers from the CES (Mulders et al. 2016a). Further pursuit of this line of research is needed to investigate how tinnitus-related activity is modulated by CES and how the neuromodulation contributes to any corresponding changes in the behavioral evidence of tinnitus.

2.1.3 CES to Normalize Spontaneous Firing in the Auditory Nerve

Tinnitus may be due to a lack of normal spontaneous neural firing in the cochlear nerve following acoustic trauma (Kiang et al. 1976). Taking this into consideration, it was hypothesized that high-rate electrical stimulation of the cochlea – 5,000-pps – could produce endogenous-like patterns of spike activity in the auditory nerve, which would presumably have therapeutic effects on tinnitus (Rubinstein et al. 2003). Rubinstein and colleagues then used transtympanic stimulation of the round window by delivering charge-balanced, biphasic pulses of various widths, at 4,800-pps, on an electrode that was pitch-matched to the tinnitus percept. Their results showed that this stimulation paradigm generated substantial or complete tinnitus suppression among 45% of their participants (Rubinstein et al. 2003). Despite the promising outcomes, there was a substantial range of efficacy across the cohort, and thus it is not currently used as a routine clinical tool. However, there are remaining questions to be answered, especially why some participants experienced positive tinnitus relief outcomes, while others did not. In addition, one wonders how a brain with acoustic trauma-deprived peripheral input would respond when stimulating the auditory nerve at such a high rate. Experiments with high-rate electrical stimulation of the cochlea in animal models with noise trauma should be conducted to answer the question.

2.1.4 CES and Modulation of Brain Plasticity

Tinnitus is a type of maladaptive, plastic process in response to peripheral and/or central injury that causes neuronal loss, disruptions of ion channel functions, alterations to synaptic transmission and neurotransmitter pathways, as well as sensory map reorganizations (Zhang 2019). It would make sense that targeting and effectively mitigating and/or resetting these processes would ideally have potent therapeutic effects on tinnitus and should serve as a basis when designing new treatment modalities. Over the last four decades, a number of studies have attempted to decipher how long-term use of CIs improves speech performance, mainly in animal models. For example, using congenitally deaf cats with cochlear implants, chronic CES produced high-amplitude field potentials, expanded response map areas, and increased synaptic efficacy, in the auditory cortex (Klinke et al. 1999). Another team using a cat CI model for hearing restoration further demonstrated that chronic CES had a remarkable capacity to alter brain plasticity (Fig. 4) (Fallon et al. 2008, 2009). More specifically, they found that in some animals with chronic deafness, CES was able to restore cochleotopy, as cochelotopic map formation of the primary auditory cortex determines the functional organization and effectiveness of sound perception (Fallon et al. 2014). The restored cochleotopy following CES may have re-activated functional afferent activity in previously damaged tonotopic regions. Such CES-induced re-establishment of cochleotopy and re-afferentation may help reset homeostasis and reverse the maladaptive plasticity, providing tinnitus relief. In a

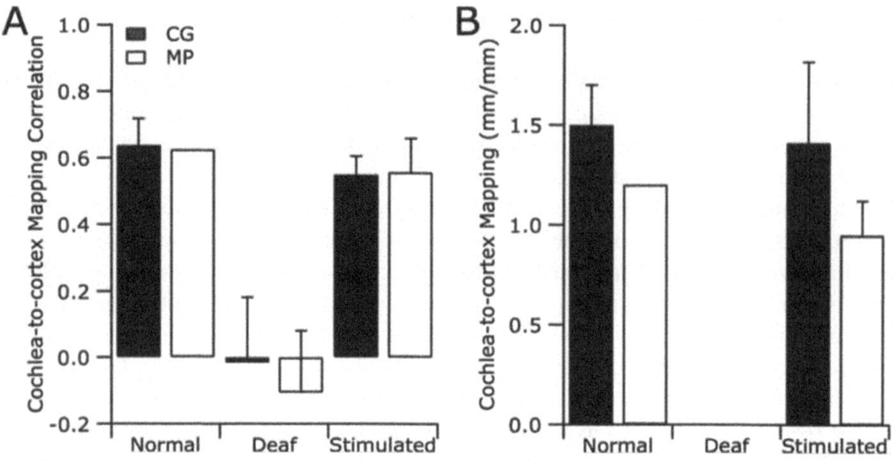

Fig. 4 Cochlea-to-cortex mapping showing the relationship between the location of cochlea stimulation and the location of neurons tuned preferentially to the stimulating electrode in the primary auditory cortex. (**a**) There were significant cochlea-to-cortex mapping correlations (Pearson correlations) in normal hearing and chronically stimulated animals (CG: common-ground; MP: monopolar), but not in long-term deaf animals. (**b**) Mean cochlea-to-cortex mapping in chronically stimulated animals was not significantly different from that in normal hearing animals. Error bar = SEM. Adapted from Fallon et al. (2009)

more detailed study, it was shown that CES in ototoxically deafened cats preserved postsynaptic densities in the endbulbs of Held, the large endings of the auditory nerve which hold a key position in the timing pathway for sound localization (Ryugo et al. 2010). Along the same line, the CES-restored postsynaptic densities facilitate peripheral re-afferentation that helps restore the excitatory–inhibitory balance in the auditory brain, relieving tinnitus. Finally, it has recently been demonstrated in a rat model that CES induces robust local field potentials (LFPs) in the hippocampus, implicating auditory spatial cognition (auditory perception to a location in space) as a potential area-of-interest (Hitier et al. 2020). Considering the fact that synaptic processing in the hippocampus may play an important role in the pathophysiology of tinnitus (Zhang et al. 2018), the CES-induced activity there (Hitier et al. 2020) may modulate the neural networks underlying tinnitus. This helps us understand that CES-induced tinnitus relief involves modulation of neural plasticity at synaptic to systemic levels and throughout both lemniscal and non-lemniscal pathways.

2.1.5 Directions of Future Studies with Cochlear Electrical Stimulation

The above information concerning the underlying mechanisms of CES-induced tinnitus relief illustrates a promising outlook for further development of animal models and more extensive investigations into the molecular, ion channel, synaptic transmission, neuronal circuitry, systemic, and tinnitus behavioral changes.

Additionally, the pressing questions that remain to be answered are: (1) what are the best stimulation strategies to simultaneously achieve long-lasting suppression of both tinnitus-related neural signals and behavioral manifestations? and (2) how to differentially achieve suppression of noise tinnitus and tonal tinnitus?

As for some potential areas to address and improve, firstly, the above animal studies lack behavioral assays in their experiment designs. Thus, during experimentation, there is a need to include behavioral data and associate it with the pathophysiological data at different levels, in order to understand the causal relationships. Secondly, there is a need to achieve controlled suppression of tinnitus at different pitches, effectively suppressing noise tinnitus and/or tonal tinnitus. For instance, it has been reported that low-rate stimulation of the apical turn of the cochlea yields large suppression of high-pitch tinnitus (Zeng et al. 2011). This suggests that stimulation of specific regions of the cochlea affect specific neural pathways, possibly modulating their hyperactive neurons and ultimately yielding desirable suppressive effects on tinnitus and its pitch. Technically, stimulation-specific peripheral loci in the cochlea may be practically achievable. However, to specifically modulate certain neural pathways and their corresponding networks to achieve modulation of tinnitus of specific pitches by specific stimulation loci in the cochlea remains challenging, even though neurons in the AC are known to tune to the stimulated loci/electrodes in the cochlea (Fallon et al. 2009). Finally, although hearing is restored by CIs, especially when accompanied by cochleotopic map reorganization following chronic stimulation, one would expect that the hearing loss- and/or peripheral deafferentation-derived tinnitus should permanently benefit from this desirable reset of maladaptive neural plasticity. Such a reset may be equivalent to chronic CES-induced cochleotopic map reorganization for hearing restoration. However, the clinical reality is that tinnitus relief may not synchronize with hearing restored by chronic CES. Thus, there is a need to investigate why CES that restores hearing may not always suppress tinnitus, which possibly involves complex non-auditory limbic structures. These CES ideas should be explored using existing and new methodologies such as optogenetic cochlear stimulation.

2.2 *Optogenetic Cochlear Stimulation*

The CES method has provided a wealth of information regarding hearing restoration and tinnitus suppression. However, the current cochlear prothesis provides limited spatial and frequency resolutions, due to the spread of electrical current across areas neighboring the active electrodes (Kral et al. 1998). Optogenetics may overcome these CES issues. Optogenetics was developed to optically control neurons via photo-sensitive proteins expressed on the cell surface, effectively controlling excitatory and inhibitory synaptic transmission with high spatial and temporal precision (Boyden et al. 2005). While optogenetics has vastly progressed methodologically in a variety of fields (de Mena et al. 2018; DiGuiseppi and Zuo 2019; Shaaya et al. 2021), it has only been recently used in attempts to improve cochlear stimulation, and with increased numbers of independent stimulation channels in the cochlea, it promises spatially more confined activation of spiral ganglion neurons (SGNs) and, hence, higher frequency resolution of coding (Weiss et al. 2016) and better spectral selectivity of artificial sound encoding, in the gerbil IC (Dieter et al. 2019). More recently, efforts have been made to increase temporal fidelity of optogenetic stimulation and achieve efficient, non-traumatic and neuron-specific expression of fast-switching opsins. For example, a visible red-light-activated f-Chrimson with helix 6 mutations has been used in mouse SGNs by way of postnatal AAV-injection into the scala tympani through the round window, which resulted in restoration of central auditory neural activity in deaf mice (Mager et al. 2018). In addition, f-Chrimson was introduced to overcome the shortcomings from using ChR-based cochlear optogenetics (Keppeler et al. 2018), such as low-temporal bandwidth of optical coding. The fast-evolving research in this area over the last decade has laid important foundations allowing development of a reliable and safe optogenetic cochlear implant for hearing restoration in humans (Dieter et al. 2020). Although research in this field is moving forward speedily, there is no research applying this newer technology to the treatment of tinnitus in an animal model. Based on the etiology of tinnitus and its modulation via CES, it is intuitive that the available tools developed for optogenetic cochlear implant stimulation may be utilized for neuromodulation and treating tinnitus.

3 Intraparenchymal Brain Electrical Stimulation

3.1 Targets in the Brain Network and Modalities of Stimulation

3.1.1 Brain Network Dysfunction and Informatics

Tinnitus and its associated limbic dysfunctions are known to result from a cascade of maladaptive plasticity and pathologic network informatics that occur in the brain following loud noise exposure or traumatic brain injury (TBI) (De Ridder et al. 2014b; Zhang 2019). Tinnitus perception is driven by these anatomical anomalies since they sub-serve the higher-order information processing, and its dysfunction within and between default-mode, auditory, limbic, frontal, and striatal networks that produce clinical symptoms (Chen et al. 2018; De Ridder et al. 2014c; Leaver et al. 2011; Rauschecker et al. 2010, 2015; Seeley et al. 2007). Therefore, electrical stimulation of the brain parenchyma – both invasively and non-invasively can alter the electrophysiological output of specified neural ensembles and ultimately manipulate neuropsychiatric phenomena, such as phantom auditory perception in tinnitus. Below is a review of animal models that aim to address methodological and mechanistical issues that remain from clinical investigations.

3.1.2 Lemniscal vs. Extra-lemniscal and Their Associated Neural Network

Nearly every station within the auditory pathway and associated-networks has been implicated in the etiology of tinnitus (De Ridder et al. 2015; Leaver et al. 2011; Rauschecker et al. 2015). This raises the question – what areas should be stimulated? The afferent, lemniscal or classical arm of the auditory pathway carries coded neural signals from the cochlea to the cochlear nucleus: they then decussate to the central nucleus of the inferior colliculus (CIC), followed by a relay at the ventral subdivision of the medial geniculate body (vMGB) of the thalamus, and finally arrive at the primary auditory cortex (AI) – and surrounding areas (Møller 2011; Nozaradan et al. 2017; Vianney-Rodrigues et al. 2011). At the same time, the information is carried via massive collaterals and descending loops throughout the central auditory system (Coomes et al. 2005; Meltzer and Ryugo 2006; Pinault 2004). Among numerous descending pathways, there is heavy feedback stemming from pyramidal neurons in cortical layer VI that exist among all the modality-specific thalamic nuclei. When all corticothalamic cells are considered, they outnumber their thalamocortical counterparts by a factor of ten (Deschenes et al. 1998). It is the neural impulses through these thalamocorticothalamic (TCT) loops at various frequencies that generate dynamic oscillatory patterns that uniquely describe real-time, transduced sensory and phantom information (Llinas et al. 1999, 2005; Vianney-Rodrigues et al. 2019). If these signals resonate with the greater perceptual network of the brain (the sensory, limbic, frontal, and striatal regions), conscious sensation of tinnitus results (De Ridder et al. 2015; Leaver et al. 2011; Rauschecker et al. 2015).

Aside from the classic, lemniscal system, integration of sensory pathways with executive, memory, and emotional processing affords the brain the capability to assign salience to various stimuli, filter unwanted information, and guide real-time input–output functions through attention and will, allowing us to exist in-, make predictions about-, and thrive in an ever changing environment. Anatomically, this is accomplished through the extra-lemniscal –"non-classical" – pathways that carry a parallel stream of afferent sensory information from the auditory nerve, through the belt structures of the central auditory system, such as the granule region of the cochlear nucleus, external (ECIC) and dorsal cortices of the inferior colliculus (DCIC), the medial (mMGB) and dorsal thalamic areas (dMGB), and secondary auditory cortex (AII) (Malmierca et al. 2002; Møller 2011). From there, instead of terminating mostly in AI like their "classical" cousins, these extra-lemniscal neurons send axons to structures such as AII and other areas of association cortex for higher-level processing (Møller 2011). By doing so, the extra-lemniscal pathway infuses auditory processing with connections to memory centers in the hippocampus and parahippocampus, emotional processing areas in the amygdala, prefrontal executive attentional circuits involving the dorsolateral and ventromedial prefrontal cortices, striatal systems such as the nucleus accumbens and caudate, and other regions such as the insular and anterior cingulate cortices (De Ridder et al. 2014c, 2015; Leaver et al. 2011; Ledoux et al. 1990; Rauschecker et al. 2010, 2015; Seeley et al. 2007).

The brain takes the real-time, sensory stream, splits the information into two identical quanta, each carrying the entire, transduced environmental representation, and sends the first copy to primary sensory cortices via the lemniscal system, creating a candidate virtual-model of the external environment that is represented as electrically oscillating TCT signatures (Llinas et al. 1999; Steriade et al. 1991); the other copy of this same information is simultaneously carried down the extra-lemniscal pathway where the fronto-limbic-striatal (FLS) systems (Rauschecker et al. 2010, 2015) compares the sensory engram with contextualized, experience-derived "virtual states of reality" to gate, filter, refine, and assign emotions and salience to the various components of the sensory scene. Through connections back to the classical pathway, the extra-lemniscal system can manipulate the thalamocortical broadcast signal to ensure that the perceptual event and motor output being experienced by an individual reflect the will and experiences of the executive control centers.

Given that the neural language is expressed through varied patterns of different frequency waves, one could imagine an interpretation where the thalamocortical oscillations encoding sensory information are viewed as literal and interacting waves that form dynamic representations of sensory scenes. These sensory scenes are carried by the classical pathway, and via manipulation of the thalamocortical discharge frequencies, possibly through the thalamic reticular nucleus (TRN) (Rauschecker et al. 2010), the "live" version is edited, creating a final signature perceived by the individual that reflects the corrections and outputs intended by the FLS. Dysfunction in either the bottom-up lemniscal-TC signal generator or the top-down extra-lemniscal-FLS contextualization, gating, and salience-machinery can cause phantom auditory perception and other TCDs such as schizophrenia and chronic pain. Many stations along both pathways have been the focus of brain-based, electrical treatment paradigms (Ahsan et al. 2018; Cheung et al. 2019; De Ridder et al. 2011; Jakobs and Lozano 2019; Luo et al. 2012; Seidman et al. 2008; Zhang et al. 2011). Both human and animal studies yield a mixed-bag of responders, non-responders, tinnitus-exacerbators, and new-onset tinnitus generators, and below is a review of animal models of brain stimulation for treating tinnitus by targeting the brain stem, midbrain, thalamus, and the cortex in the central auditory system and basal ganglion.

3.2 Auditory Brainstem Implant (ABI)

Auditory brainstem implants (ABIs) were introduced to stimulate the cochlear nucleus (especially the DCN) for hearing restoration among neurofibromatosis type II (NF2) patients who undergo translabyrinthine removal of vestibular schwannomas and cannot benefit from CIs (Fernandes et al. 2020; Gilles et al. 2020; Roberts et al. 2017; Soussi and Otto 1994). It has been reported from tracking 112 patients who were implanted with ABIs between 1994 and 2015 that NF2 patients who have undergone removal of vestibular schwannomas experience tinnitus symptoms that are alleviated when their ABIs are switched on (Deklerck et al. 2020; Roberts et al. 2017). One explanation of such tinnitus relief is that NF2 patients may develop a tinnitus handicap due to sudden, diminished input following surgery, which in turn causes an ascending information deficit that triggers increased neural hypersensitivity, spontaneous spike firing rates, bursting, and synchrony in the IC and AC (Gerken et al. 1984; Seki and Eggermont 2003; Wu et al. 2016). In addition, the topographic nature of the deafferentation induces a surrounding brain region to expand its characteristic-frequency receptive field into damaged areas, known as "map reorganization." This "map reorganization" has been noted throughout the auditory axis in both humans and animals (Komiya and Eggermont 2000; Shore et al. 2016; Wienbruch et al. 2006; Yang et al. 2011), which may have been mitigated by ABIs in tinnitus relief. At the very least, the promising clinical results prompted animal studies to elucidate the underlying mechanisms of ABI-induced tinnitus suppression.

The early animal models were mainly designed to improve hearing outcomes from ABIs, so as such, rodent (Fig. 5) (Vachicouras et al. 2019; Zhang and Zhang 2010), feline (McCreery et al. 2010, 2013, 2018), and non-human primate models (Wang et al. 2015) have been developed for ABI implantation in the DCN and ventral cochlear nucleus. Efforts to improve the technology include electrically evoked auditory brainstem responses (eABR)-guided fitting to better align with the tonotopic axis (Lachowska et al. 2020; O'Driscoll et al. 2011), flexible implants to better fit the curvature of the neuro-organ surface (Vachicouras et al. 2019), and varied stimulation paradigms to better mimic physiologic signals (Mauger et al. 2012; McCreery et al. 2013, 2018).

To address the mechanisms underlying ABI-induced tinnitus relief, our group chronically implanted platinum/iridium microwire arrays into the DCN of rats, conducted electrical stimulation with charged-balanced, biphasic electric pulses of 50 μA at 10 pps, and demonstrated suppression of behavioral evidence of noise-induced tinnitus, as measured by the gap-prepulse inhibition of the acoustic startle reflex (GPIAS) behavioral paradigm (Luo et al. 2012). Taking a similar approach, another group recently targeted the DCN with high-rate stimulation (bipolar, monophasic pulses of 100 μA at 1,000 Hz), and their results substantiate the earlier findings that delivering electrical current to this lower brainstem region reduces tinnitus-like behavior in rats (van Zwieten et al. 2019). To explain the suppressive effects of ABI on behavioral evidence of tinnitus, it is possible that ABIs may have corrected aberrant neural activity that is classically associated with tinnitus. More specifically, at the lower brainstem level, ABI stimulation may exert its therapeutic effects on tinnitus behavior by interfering with DCN fusiform cells, because their increased spontaneous firing rates (SFR), spike-synchrony, and bursting along with altered plasticity are correlated with tinnitus-like behavior in animals (Marks et al. 2018; Martel et al. 2019; Wu et al. 2016). At the midbrain level, Mauger et al. (2012) demonstrated that ABIs suppress spontaneous activity in a subset of IC neurons (Mauger et al. 2012), possibly increasing GABA and glutamate levels in target tissue and restoring synaptic transmission (Ghafouri et al. 2019). At the cortical level, the hyperactive signals are intimately linked to the tonotopic map reorganization. Thus,

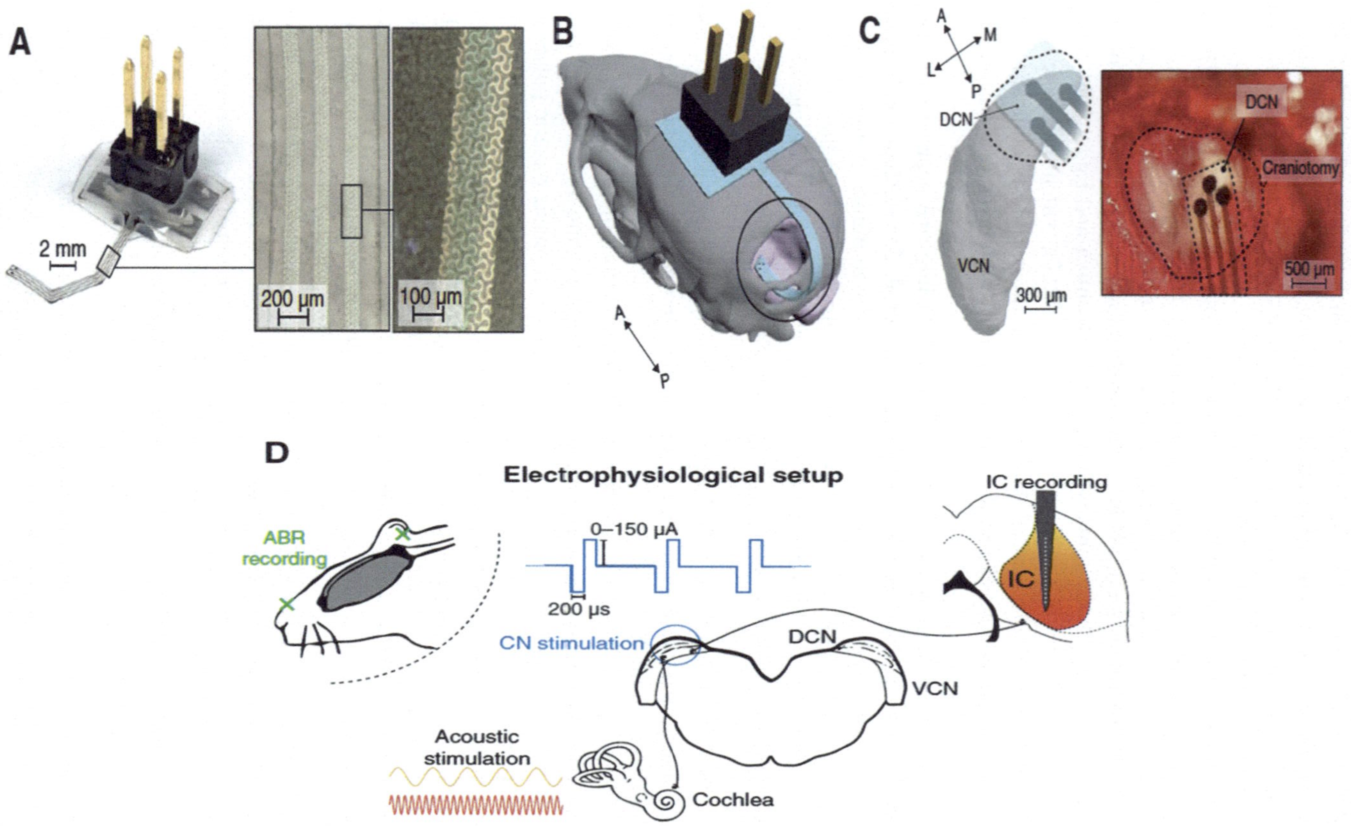

Fig. 5 Rodent model of an ABI (Adapted from Vachicouras et al. 2019). (**a**) Photograph of ABI device showing the external interface (black box with metal prongs) and intracranial portion (small, flexible pad) with inset zooming in on electrode contacts. (**b**) Cartoon depicting macroscopic position of device on rodent skull. (**c**) Anatomical location of electrode pad draped on the surface on the DCN. (**d**) Cartoon describing the electrophysiological setup whereby the ABI in the cochlea delivers electrical stimulation and both ABR and recording probes in the IC detect the consequent electrophysiological changes

by correcting the GABAergic deficit and suppressing hyper neural activity, as well as modulation map reorganization, ABI could also be alleviating tinnitus via patching the topographic insult; to the point, ABI stimulation is capable of restoring plasticity to normal homeostatic levels (Ghafouri et al. 2019) and reversing deafferentation-induced topographic deficits.

Another theory to explain how this hyperactivity translates into conscious perception of tinnitus is through the phenomenon of stochastic resonance (SR). In this framework, subthreshold information can be pushed into awareness by introducing a critical amount of statistical noise that, by chance, resonates with the suspect signal, increasing its amplitude just enough to be detected (Krauss et al. 2016, 2017). In the case of tinnitus, damage to cochlear hair cells reduces ascending input which is sensed by the brain, and it responds by increasing internal noise to activate SR, raising the diminished auditory nerve input above threshold. In this context, the trauma-induced hyperactivity believed to underlie tinnitus could be the physical analogue for the "noise" that's increased to restore afferent signaling (Krauss et al. 2016). Krauss et al. (2016) tested this idea using computer simulations showing that SR was capable of enhancing detection of auditory stimuli following hearing loss. The effects of SR presumably explain why multiple studies analyzing audiometric data from tinnitus patients found they actually had better low-frequency hearing detection and steeper audiograms, compared to controls (Konig et al. 2006; Krauss et al. 2016). Within this framework, one could imagine that stimulation of the cochlear nucleus using an ABI reduces the hyperactivity or "noise" which sinks the pathologic signal back under the SR-detector threshold, or alternatively, since DCN stimulation directly activates connections between both the VCN and DCN and the IC (Mauger et al. 2012; McCreery et al. 2010, 2018), it could be restoring physiologic information flow along the auditory pathway. Realistically, it is probably a combination of both acting on the disparate, tonotopic sub-circuits because too-little and too-much noise disrupt SR (Krauss et al. 2016). Additionally, ABI stimulation has been shown to simultaneously increase and decrease different neuronal populations in the IC, which may be a demonstration of "turning-down" tinnitus and "turning-up" physiologic auditory information flow (Mauger et al. 2012). Putting it all together, the available information suggests that ABI-induced tinnitus suppression may operate through a combination of modulating the salience network and restoring physiologic information flow along the ascending auditory pathway to draw pathological attention away from the aberrant signal and onto a "normal" input. This may be accomplished by temporarily reversing excitatory–inhibitory imbalance to both reduce and mask the aberration via mitigation of trauma-induced elevated spontaneous firing, hypersynchrony, and gain and re-energizing endogenous afferent pathways, respectively. Unfortunately, tinnitus symptoms return when devices are powered off, suggesting only a transient therapeutic effect. This motivates further investigation as to what constitutes ABI-induced tinnitus suppression and how to extend its therapeutic time window.

3.3 Auditory Midbrain Implant (AMI)

While ABI is useful for hearing restoration in NF2 patients, the labyrinthine craniotomy and surgical exposure of the cochlear nucleus is rather challenging and risky, as the cochlear nucleus is a brainstem structure, which is near the respiratory and cardiac centers. Furthermore, following surgical removal of vestibular schwannomas in cases where tumor significantly invades the cochlear nucleus complex, the language performance from the ABI is often limited. From this context, auditory midbrain implants (AMIs) were introduced to bypass the lower auditory brainstem to implant and electrically stimulate the IC for hearing restoration (Calixto et al. 2013; Lim and Lenarz 2015; Neuheiser et al. 2010; Pages et al. 2016; Quass et al. 2018; Schierholz et al. 2017). In addition to the beneficial effects on deafness, a portion of these AMI clinical trial patients also experienced disruption to their coexisting tinnitus, which spurred the possibility that IC stimulation via an AMI could be a potential treatment for tinnitus (Offutt et al. 2014). This is because the IC is the main auditory processing center in the midbrain responsible for functions such as encoding amplitude modulation and neural envelope synchrony and identification and categorization of natural sounds (Henry et al. 2016; Rode et al. 2013; Sadeghi et al. 2019), and hyperactivity in the IC is believed to contribute to tinnitus (Bauer et al. 2008). Anatomically, the IC is mainly divided into the lemniscal central nucleus (CIC) or tonotopically organized, core auditory processing center, non-lemniscal external cortex (ECIC), involved in multimodal integration, and the dorsal cortex (DCIC), whose role is unclear but may be modulatory (Offutt et al. 2014).

Delving deeper into the underlying mechanisms of AMI-induced tinnitus relief, Smit et al. (2016b) developed a rat AMI model to specifically deliver bilateral, high-frequency electrical stimulation (100 Hz at 100 μA) of the ECIC – a subregion with tinnitus-related hyperactivity – and demonstrated AMI's capability of reducing tinnitus-like behavior in rats, as measured with the GPIAS paradigm (Smit et al. 2016b) (Fig. 6). This group's follow-up study using the same rat model confirmed that the IC-AMI stimulation did not interfere with physiologic auditory information processing, making it an even more attractive approach for treating tinnitus (Smit et al. 2017). Although these studies lack a comprehensive mechanistic analysis to probe how IC-AMI relieves tinnitus pathology, high-frequency stimulation is thought to induce a lesion-like effect (Smit et al. 2016b) to quell the hyperactivity that has been implicated in tinnitus etiology (Bauer et al. 2008) and suppress IC neuronal firing (Offutt et al. 2014). Considering that Smit et al. (2016b) targeted the ECIC, which connects with extra-lemniscal, medial, and dorsal thalamic areas, the AMI-induced tinnitus reduction may be attributed to downstream disruption of the implicated fronto-limbic-striatal pathways as previously discussed (Smit et al. 2016b). Pertinently, salicylate – a common tinnitus inducer – causes aberrant hippocampus (HPC) theta rhythms in mice (Winne et al. 2019) and IC discharge is temporally correlated with these rhythms, in a stimulus-dependent manner (Liberman et al. 2009). By inducing a lesion effect, it's plausible that AMI

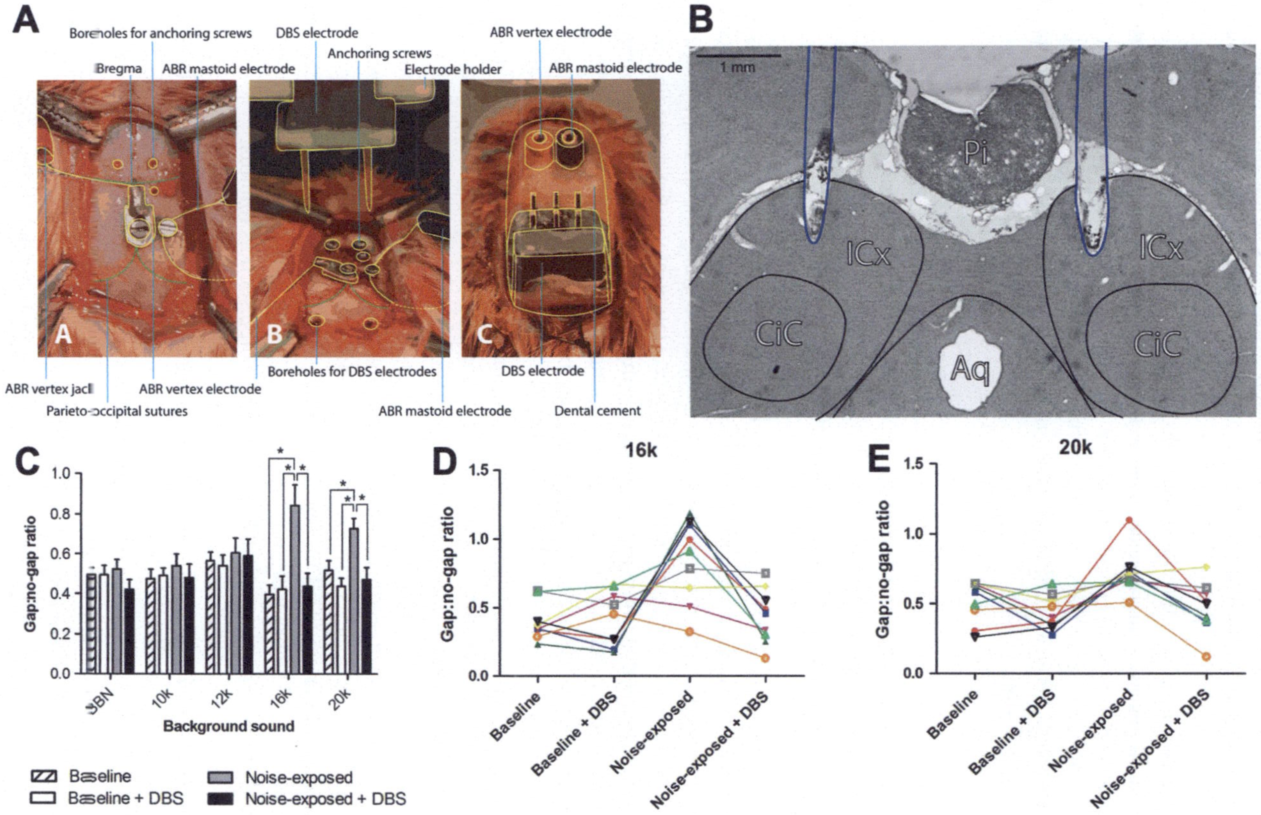

Fig. 6 The auditory midbrain implant (ABI) suppresses tinnitus behavior in rodents (Adapted from Smit et al. 2016b, 2017). (**a**) Photograph of surgical field and overlying schematic of auditory midbrain implant and ABR recording setup. (**b**) Brain slice verifying AMI electrode tracts penetrating the ICx. (**c**) Gap ratios expressed as an average of all animals showing tinnitus behavior and DBS-induced suppression at 16 kHz and 20 kHz frequencies. (**d**) Gap ratios in the 16 kHz frequency band for the individual animals represented in "C". (**e**) Gap ratios in the 20 kHz frequency band for the individual animals represented in "C"

stimulation provides tinnitus relief by interfering with this IC-HPC theta correlation to decouple the two regions: whose connection seems to be an important auditory process. Neurophysiologically, AMI stimulation of the DCIC is known to cause both suppression and facilitation of CIC neuronal firing that carries auditory information along the lemniscal pathway (Offutt et al. 2014). Thus, it is possible that AMI-induced tinnitus suppression may also be due to disruption of this lemniscal signal within the IC and beyond. Taken together, AMI stimulation represents a potentially viable treatment for intractable tinnitus in a subset of patients that are indicated for such an invasive procedure, but existing animal models are still at the early stage and require further expansion to fully understand IC-AMI-induced tinnitus suppression.

3.4 Thalamocortical Modulation

3.4.1 Thalamic Modulation and Deep Brain Stimulation (DBS)

While ABI and AMI do offer a semblance of hearing restoration for patients who are not eligible for CIs, their therapeutic efficacy in tinnitus treatment varies greatly. Efforts have been made to target higher-order stations along the afferent pathways for more successful tinnitus treatments. The next station along the ascending auditory pathway is the thalamus: the final converging center for all peripheral, sensory modalities, before reaching the cortex. The auditory portion consists of the lemniscal, ventral subdivision of the medial geniculate body (vMGB) and extra-lemniscal, medial, and dorsal regions (mMGB and dMGB), which connect to the primary AC and secondary AC, association cortex, limbic, prefrontal, and other higher processing areas, respectively (Hoover and Vertes 2007; Jang and Yeo 2014; Mitchell 2015; Møller 2011). However, the thalamus is not just a simple relay, but rather a central hub that allows for integration and rapid communication across the brain. It forms intrinsic feedback loops with connected areas of the cortex, such that when signals reach this level of the brain, they resonate through the thalamocorticothalamic (TCT) loops in the form of different frequency oscillations for nearby and remote brain regions, and this level of activity is believed to represent the neural correlates of consciously perceived events, such as tinnitus.

Like the lower stations along the ascending pathway, the MGB may serve as a DBS target for modulation of tinnitus, as hyperactivity and bursting there following auditory trauma is associated with tinnitus-like behavior in rats (Barry et al. 2019; Kalappa et al. 2014). Additionally, hyperactivity – due to reduced GABA – in the mediodorsal thalamus (MDT) has been shown to correlate with anxiety and other features of limbic dysfunction that are frequently comorbid with compounded auditory disorders, as reported in a non-human primate model (Rotge et al. 2012). Importantly, the auditory thalamus is also a key site in the generation of the aberrant theta-gamma oscillations found in tinnitus and a host of other TCDs (De Ridder et al. 2015; Llinas et al. 1999). Clinical results corroborate the animal studies in implicating the thalamus as a key factor in tinnitus etiology, as resting state fMRI experiments show tinnitus associated with abnormal thalamic activity and connectivity to the hippocampus (Chen et al. 2017; Ueyama et al. 2013), and human surface-based vertex analysis showed thalamic anatomical expansion in those suffering from tinnitus (Tae et al. 2018). Furthermore, Lv et al. (2020) demonstrated that the thalamus had abnormal fMRI functional connectivity with the inferior frontal gyrus and ACC, and when sound therapy was applied, the abnormal functional connectivity along with tinnitus symptoms disappeared (Lv et al. 2020).

Considering direct electrical modulation, a clinical case series and retrospective questionnaire showed that patients implanted with DBS in the ventralis intermedius nucleus of the thalamus for other conditions also experienced favorable effects on coexisting tinnitus (Deklerck et al. 2020; Shi et al. 2009; Smit et al. 2016a). The above clinical evidence prompted the need for animal systems to validate thalamic-DBS as a suitable treatment and explore its underlying therapeutic mechanisms. Recently Van Zwieten et al. (2019) developed a rat paradigm with DBS electrodes implanted into the ventral MGB for tinnitus relief (van Zwieten et al. 2019). Using loud noise exposure for tinnitus induction and GPIAS for behavioral evaluation, the authors reported amelioration of tinnitus symptoms in their murine model by delivering bipolar and monophasic high-rate stimulation at 100 Hz and 100 μA, directly to the auditory thalamus; note that low-rate stimulation at 10 Hz did not have a significant impact on tinnitus behavior (van Zwieten et al. 2019). In addition to modulating tinnitus behavior, Van Zwieten and colleagues subjected their rats to zero maze and open field behavioral tests to evaluate anxiety and other limbic conditions that are key features of compounded auditory disorders, but MGB-DBS failed to produce any meaningful changes in these behavioral outputs (van Zwieten et al. 2019). This makes sense given that the MGB is a lemniscal region, as opposed to the MDT that gives rise to FLS connections and is more involved with the affective and cognitive components of tinnitus and hearing loss.

In the study by Van Zwieten et al. (2019), the authors did not delve into therapeutic mechanisms (van Zwieten et al. 2019). However, other groups using thalamic DBS for a variety of conditions offer hints at how this modality may be impacting tinnitus. First, one has to consider that electrical currents tend to have non-specific current spread; thus, if the MDT is involved, though not directly enough to impact limbic symptoms but just enough to alter sensory percepts, it's possible that modulation of the FLS circuit may be involved in tinnitus relief. This is because MDT-DBS influences activity possibly in other brain regions such as the NAc, cingulate cortex, striatum, and thalamus (Casquero-Veiga et al. 2016). Secondly, high-frequency DBS is believed to create a lesion-like effect on target brain tissue (Lee et al. 2019). If pathologic gamma signals are oscillating in the TCT loops, then DBS could be inhibiting the thalamocortical relay neurons that are responsible for these aberrant signals. Alternatively, DBS partially operates by restoring physiologic information flow (Lee et al. 2019), and to the point, electrical stimulation of the vMGB is able to elicit responses in tonotopically-linked areas of the AC (Atencio et al. 2014), which in turn mask any tinnitus-related signals. Since acoustic-trauma has been shown to increase tonic GABAergic inhibition in the MGB (Sametsky et al.

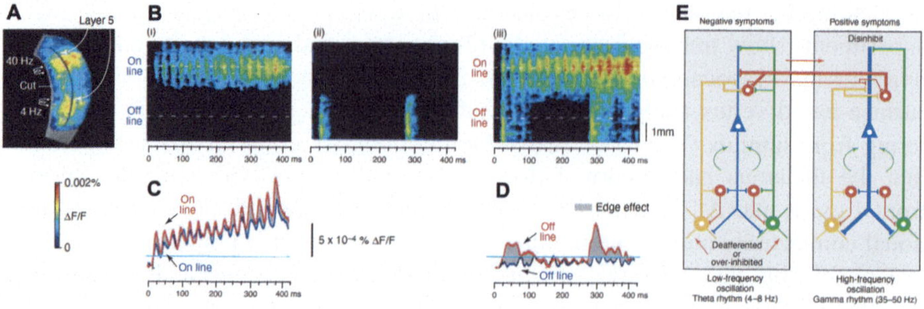

Fig. 7 Thalamocortical dysrhythmias and the edge-effect (Adapted from Llinas et al. 2005). (**a**) Schematic to show the experimental setup, where an arbitrary line is drawn across the cortical layer V field and following stimulation via two electrodes 2 mm apart, the fluorescence – activation – along that line is quantified. (**b**) Activation profiles with 40 Hz stimulation (i), 4 Hz (ii), or both (iii). "On line" refers to values taken from the arbitrary line and "Off line" refers to values from a random spot not on the line. (**c**) The difference in fluorescence between the 40 & 4 Hz-together (red) and 40 Hz-only (blue) at the "on-line." (**d**) The difference in fluorescence between the 40 & 4 Hz-together (red) and 40 Hz-only (blue) at the "off-line." Note more activation when 40 Hz and 4 Hz are co-applied, with this "edge-effect" denoted as the gray shaded area between the red and blue curves in (**c**, **d**). (**e**) A cartoon representation of TCD where low- frequency cortical oscillations – responsible for negative symptoms – impair lateral inhibition, disinhibiting an "edge-effect" of gamma waves believed to drive positive symptoms

2015), which drives hyperpolarization-induced burst mode of TC relays and consequent slow-wave cortical activity that are implicated in tinnitus, perhaps MGB-DBS is concurrently interfering with these pathologic oscillations. This mechanism has particular relevance to the thalamocortical dysrythmia (TCD) model of tinnitus and other related neurological conditions. The TCD theory posits that deafferentation of sensory inputs to the thalamus hyperpolarizes thalamocortical relays, and in doing so disinhibits their T-type calcium channels that then cause the neurons to enter a bursting mode (Jahnsen and Llinas 1984a, b). In turn, these bursts resonating through the TCT loops cause slowing of alpha oscillations to theta waves in the cortex, which impairs lateral inhibition and ignites a surrounding region of pathologic gamma rhythms – dubbed the "edge-effect" – believed to drive positive symptomolgy in a host of disorders such as tinnitus, schizophrenia, chronic pain, and Parkinson's (De Ridder et al. 2015; Gault et al. 2018; Llinas et al. 1999, 2005) (Fig. 7). Considering TCD in regard to tinnitus, one would expect the thalamus to simultaneously show both excessive bursting in certain areas and gamma activity in others. Thus, to alleviate phantom auditory perception, MGB-DBS may have interfered with the coexisting slow-wave and high-frequency components of the pathologic signal, which may help explain the various observed effects. Aside from its invasive nature, thalamic-DBS is an attractive candidate tinnitus therapy. Animal models should be expanded to cement this claim and understand its therapeutic mechanisms, in depth.

3.4.2 Auditory Cortex Electrical Stimulation (ACES)

The AC has been the focus of multiple TCD treatment modalities because a slew of pathologic features in the AC correlate with tinnitus, such as altered tonotopic organization, gain adjustment, elevated neural spiking activity, and intractable theta and gamma oscillations (De Ridder et al. 2011; Eggermont 2015; Komiya and Eggermont 2000; Seidman et al. 2008; van der Loo et al. 2009; Weisz et al. 2005, 2007; Zhang et al. 2011). One technique that has shown promise is auditory cortex electrical stimulation (ACES), as multiple human studies have demonstrated that direct stimulation of the AC via an implanted electrode relieves tinnitus (De Ridder et al. 2006, 2011; Engelhardt et al. 2014; Friedland et al. 2007; Seidman et al. 2008). However, the therapeutic efficacy varies greatly across patients (Engelhardt et al. 2014) and the underlying modulatory mechanisms remain elusive. To circumvent this issue, an animal model of ACES was established by our group to help shed light on the mechanism driving ACES-induced tinnitus suppression. Zhang et al. (2011) noise-exposed rats for induction of behavioral evidence of tinnitus and then implanted intra-parenchymal, microwire electrode arrays into the primary AC (Zhang et al. 2011) (Fig. 8).

Delivering continuous, biphasic, square-wave, electrical pulses at 50 μA and 10 pps, directly to the primary AC, relieved tinnitus-like behavior and hearing deficits in a GPIAS paradigm (Zhang et al. 2011). A follow-up study in rats delved deeper into the electrophysiological mechanics underlying the ACES-induced

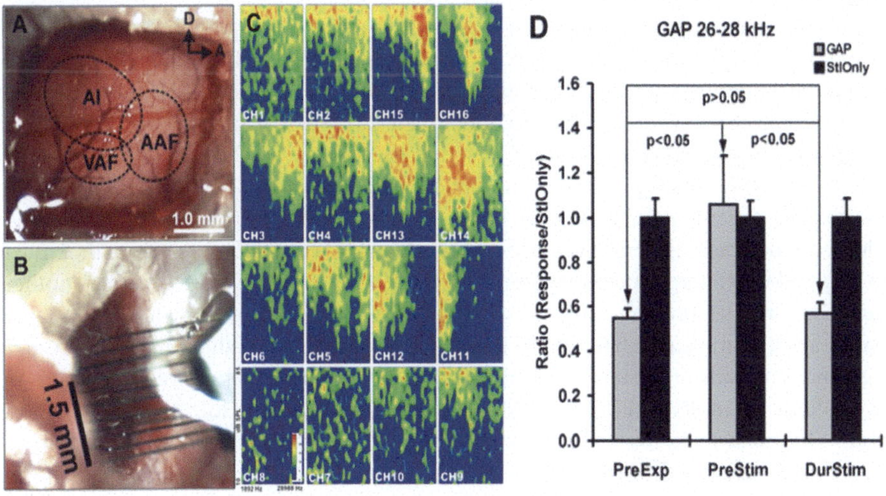

Fig. 8 Auditory cortex electrical stimulation (adapted from Zhang et al. 2011). (**a**) Peri-implantation craniotomy to expose the AC, detailing the primary (AI), anterior auditory field (AAF), and ventral auditory field (VAF). (**b**) Electrode arrays inserted into the AI. (**c**) Representative tonotopic frequency tuning curves recorded from AI neurons. (**d**) GAP ratios at 26–28 kHz of tinnitus (+) animals displaying tinnitus-like behavior after noise exposure (PreStim) that is relieved with ACES (DurStim)

reduction of tinnitus behavior by combining the therapeutic design with multi-structure electrode recordings (Firestone et al. 2019). Although preliminary, it suggested that tinnitus correlated with aberrant oscillatory neural activity – especially theta and gamma-band – within and between the DCN, IC and AC, and ACES-driven alleviation of tinnitus behavior was paralleled by reduction of AC theta-gamma coherence and restoration of physiologic information flow throughout the auditory axis (Firestone et al. 2019). Another recent study by Vianney-Rodrigues et al. (2019) also found that an animal model of tinnitus displayed hypercoherent gamma signatures within and between the AC and MGB, strengthening the possiblity that the pathologic signals are key features of tinnitus pathophysiology and likely targets of ACES (Vianney-Rodrigues et al. 2019). Many clinical studies further corroborate these rodent results, as they too show enhanced gamma oscillations correlating with tinnitus sensation and disappearing during treatment (De Ridder et al. 2011; Llinas et al. 1999; Vanneste et al. 2017; Weisz et al. 2007). However, it's not simply the presence of these TCT fast rhythms that generate consciously perceived sensation, but rather it's critical they are synchronized/coherent, since that is how the brain temporally binds spatially separate gamma islands into single, complex sensory experiences (Gray et al. 1989; Gray and Singer 1989). In fact, research has shown that, of the many gamma waves triggered by external stimuli, phase-synchrony increases only for those oscillations that represent information translated into perception (Melloni et al. 2007), and Vanneste et al. (2017) demonstrated that tinnitus suppression correlates with de-synchronization of gamma rhythms in the AC, which is supported by similar results in the aforementioned pilot study in rats (Firestone et al. 2019).

However, the role of these fast rhythms in perceptual pathologies is not so straightforward because a number of clinical imaging studies failed to identify gamma oscillations in tinnitus patients (Adjamian et al. 2012; Kahlbrock and Weisz 2008; Sedley et al. 2015a), and their presence has both positively and negatively correlated with tinnitus severity (Sedley et al. 2012). This is in line with other work which posits that the pathological gamma oscillations identified in tinnitus, neurogenic pain, and other TCSs (De Ridder et al. 2015; Llinas et al. 1999, 2005), are not causing sensation, per se, but are rather a critical prerequisite that must be plugged into a larger network for actual perception. That is, there must be another factor in tinnitus broadcasting these signatures into conscious awareness. Theta waves are a prime candidate due to their presumed role in modulating, synchronizing, and connecting distant gamma signatures in remote brain regions both in physiologic (Canolty et al. 2006; Canolty and Knight 2010; Holz et al. 2010; Tort et al. 2008) and pathologic (De Ridder et al. 2015; Weisz et al. 2005) sensation, as slow waves are capable of traversing extremely large cortical distances (Canolty et al. 2006; Canolty and Knight 2010; Sedley et al. 2015a). For example, elevated slow-waves have been found to serve as the main factor distinguishing patients with compounded tinnitus and hearing loss versus those with only impaired audition (Adjamian et al. 2012), abnormal slow waves (delta–theta) have been identified in a host of animal and human tinnitus sufferers (Adjamian et al. 2012; De Ridder et al. 2015; Firestone et al. 2019; Vianney-Rodrigues et al. 2019; Weisz et al. 2005), and they tend to be present in those tinnitus patients who failed to yield gamma rhythms (Adjamian et al. 2012; Sedley et al. 2015a). Interestingly, the animals tested by Vianney-Rodrigues et al. (2019) actually experienced a depression of theta activity following salicylate application to induce tinnitus. Another attrative role for theta waves is facilitating the phenomenon of cross-frequency coupling (CFC), whereby the amplitude and phase of numerous high-frequency oscillations are phase-locked to and modulated by slow waves that can span the entire cerebral cortex (Canolty et al. 2006; Canolty and Knight 2010). CFC has been proposed as a central tenant of the TCD theory and is believed to play a major role in tinnitus etiology (De Ridder et al. 2011, 2015). While abnormal theta-gamma CFC has been identified in animal models and human tinnitus patients (Adamchic et al. 2014; Vianney-Rodrigues et al. 2019), Ahn and colleagues noted the lack of any aberrant CFC in human tinnitus sufferers (Ahn et al. 2017). Alternatively, a neurosurgical case study detailed a patient with an electrode implanted over the AC, and it recorded theta-gamma sigantures peaking with tinnitus sensation, which were concomiantly abolished following ACES (De Ridder et al. 2011). Clearly, the exact role of theta-gamma oscillations in regard to tinnitus is still murky, but, at the very least, they do seem to be involved in the pathophysiology.

On a cellular level, the rapid brain rhythms likely originate from parvalbumin neurons, as optical excitation of this subset of interneurons drives gamma-range LFP waves and gates sensory perception (Cardin et al. 2009). In addition, animal and human studies suggest that imbalance of excitatory/inhibitory activity due to decreased GABA is implicated in tinnitus etiology, and restoring cortical, GABAergic balance can alleviate tinnitus perception (Sedley et al. 2015b; Wang et al. 2018b; Yang et al. 2011). In fact, a recent study by Miyakawa and colleagues was able to parse apart the various pathologic components following acoustic trauma and showed that tinnitus was specifically associated with decreased glutamate decarboxylase 65 expression, while hearing loss, exclusively, was correlated with tonotopic map reorganization (Miyakawa et al. 2019). Apart from fast rhythms, optogenetic manipulation of pyramidal neurons is capable of generating slow-frequency oscillations (Cardin et al. 2009). Since both fast and slow waves are implicated in tinnitus (De Ridder et al. 2015; Llinas et al. 1999; Weisz et al. 2005, 2007), it's plausible that ACES directly alters the electrophysiologic responses of multiple neuronal sub-types to influence both theta and gamma waves. Considering the AC has extensive connections with both the auditory lemniscal and extra-lemniscal pathways, it is well positioned to simultaneously turn off the pathologic sensory signal resonating in the ascending pathways and associated neural loops, via cochlear – AC stimulation, and also mitigate its counterpart on the FLS side.

Taken together, while there is clearly more work needed to fully understand the exact role of the above-mentioned theta, gamma, and their coupling, as well as the causal-linkage of TCD in tinnitus, the overall evidence implicates them as key neural components driving tinnitus perception. More studies are needed to determine exactly how the dynamic interplay between these neural populations leads to complex tinnitus perception. In addition, although these experiments support the notion that this abberant oscillatory activity could be the long-coveted neural correlate of tinnitus and prime target of ACES, more experimentation and expanded animal models are needed to: determine the causal nature of these phenomenon, decide whether or not they are necessary and/or sufficient to drive tinnitus sensation, and refine ACES so that it can become a more effective tinnitus treatment.

3.5 The Basal Ganglion and Deep Brain Stimulation (DBS)

3.5.1 Subthalamic Nucleus (STN)-DBS

Numerous basal ganglia structures are attractive candidate regions for DBS-driven therapeutics. For example, DBS directed to the STN and globus pallidus pars interna are FDA approved methods for treating various Parkinson's symptoms (Lee et al. 2019). Extending this paradigm to tinnitus treatment, retrospective survey questionnaires highlighted that some humans with STN-electrodes implanted for other conditions experienced tinnitus repression with device usage (Smit et al. 2016a). The ability of STN-DBS to interfere with a pathologic, perceptual signal can be anatomically explained by direct connections between the STN and the TRN, sensory cortices, and other nodes of the tinnitus network such as the amygdala, insula, caudate, and NAc (Cavdar et al. 2018). To delve into the mechanisms underlying basal ganglia-DBS-induced tinnitus relief in patients, our research group created the first pilot rat model implanting DBS electrode arrays bilaterally in the anterior caudate nucleus (Ahsan et al. 2018). Following surgical recovery, the rats were noise-exposed to induce tinnitus, as measured by the GIPAS behavioral paradigm. After establishing behavioral evidence of tinnitus, single charge-balanced unipolar pulses were administered at 50/75/150 μA and 10/20/40 pps, for a duration of 30 min. We demonstrated that DBS suppressed behavioral evidence of tinnitus, especially at high-frequency bands (26–28 kHz), and the suppression lasted up to 5 days. The behavioral changes were accompanied by increased spontaneous and bursting activity in the caudate nucleus, as well as decreased correlation between the AC and caudate nucleus, especially in the lower frequency bands (Fig. 9) (Ahsan et al. 2018). STN-DBS has also been known to increase both striatal glutamate and GABA (Lee et al. 2019). We postulated that the activation of the caudate nucleus amplifies inhibition of the globus pallidus via GABAergic neurons, which in turn reduces thalamocortical input to the AC (Ahsan et al. 2018). In addition, stimulating different STN regions differentially alters striatal dopamine levels (Min et al. 2016). Since tinnitus is associated with abnormal levels of different neurotransmitters (Wang et al. 2018b), DBS might be relieving perceptual pathology by correcting the local chemical imbalances to restore physiologic output of the basal ganglia. Yet, until a tinnitus-focused STN-DBS animal model is further established, its detailed effect on the pathophysiology of tinnitus remains to be addressed.

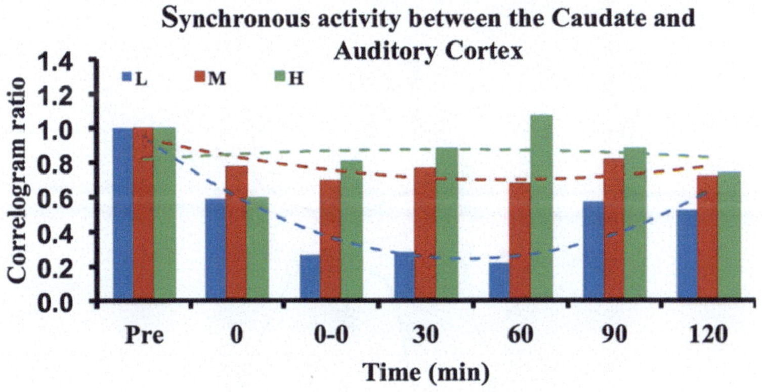

Fig. 9 Decrease in correlation between the AC and the caudate nucleus after DBS of the caudate nucleus. The decrease was noted in the lower frequency bands. This suggests that DBS of the caudate nucleus leads to a change in how the caudate nucleus interacts with the AC. Dashed lines illustrate the change in correlation over time after stimulation of the caudate nucleus. Adapted from Ahsan et al. (2018)

3.5.2 Nucleus Accumbens (NAc)-DBS

The fronto-limbic-striatal (FLS) pathway has been proposed to function as a noise-canceling and salience-assigning mechanism that normally filters unwanted signals from sensory perception (Fig. 10), and its dysfunction allows aberrant neural rhythms to enter real-time conscious awareness, producing tinnitus symptoms (De Ridder et al. 2011; Rauschecker et al. 2010, 2015; Seeley et al. 2007; Vanneste et al. 2010). In this pathway, the NAc – a major center in the ventral striatum that is critical for reward, avoidance, and addiction behaviors (Blood and Zatorre 2001; McCullough et al. 1993; Rauschecker et al. 2010) – has been found to be hyperactive in tinnitus patients (Hullfish et al. 2019; Leaver et al. 2011). Mechanistically, the NAc has extensive connections with subcallosal-related cortical areas, TRN, and via the ventral pallidum (Lavin and Grace 1996), the mediodorsal thalamus (Johansen-Berg et al. 2008; O'Donnell and Grace 1995; O'Donnell et al. 1997; Ongur and Price 2000), serving as gatekeeper with the capability to turn on or off sections of the thalamocortical sensory stream based on attention and salience (Leaver et al. 2011; Rauschecker et al. 2010). More specifically, between NAc connections to the ventral pallidum – exerting an inhibitory effect on the mediodorsal thalamus (Mogenson et al. 1987; O'Donnell et al. 1997; Rauschecker et al. 2010) – and TRN, which drives GABAergic tone onto thalamocortical neurons (Pinault 2004), the NAc is able to toggle the TC relays between spike and burst mode through depolarization or hyperpolarization (Jahnsen and Llinas 1984a, b; Steriade et al. 1991), respectively. This provides a direct mechanism for gating and modulating the electrophysiological signatures resonating in the TCT loops that are the neural correlates of physiologic and pathologic sensory percepts such as tinnitus.

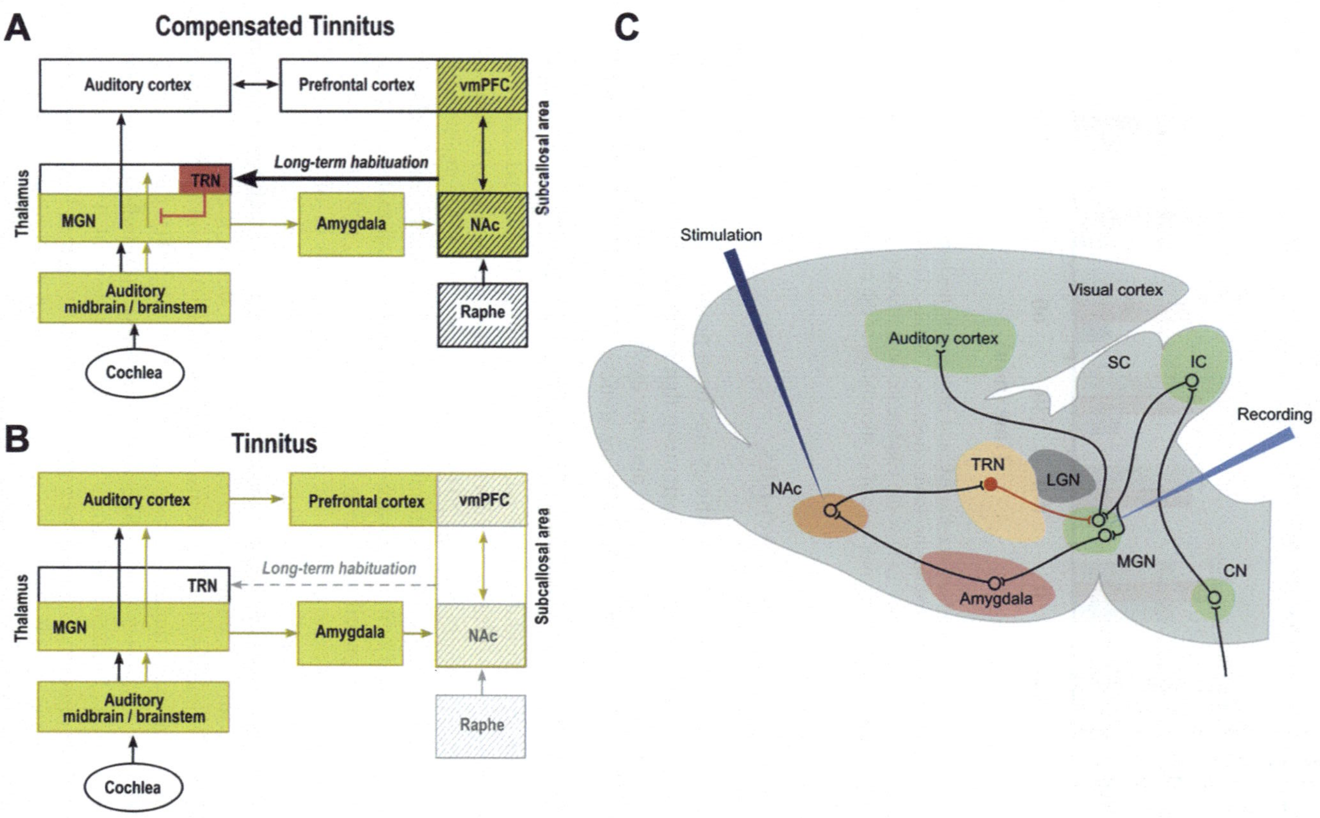

Fig. 10 Proposed fronto-limbic-striatal signal canceling mechanism (Adapted from Rauschecker et al. 2010). (**a**) Afferent auditory signals enter the FLS system via the MGB and AC, and in compensated tinnitus, the vmPFC–NAc circuit is canceling out abberant signals at the level of the MGB, via the TRN. (**b**) In tinnitus, the vmPFC–NAc signaling is defunct and the pathologic signal is not filtered and allowed to propagate throughout the system. (**c**) A schematic of the noise canceling mechanism, complete with the recording (MGB) and stimulation (NAc) paradigm to test the theory

Using a rat model, Barry and colleagues carried out electrical stimulation of the NAc and measured NAc-DBS-induced effects on neural activity in the auditory thalamus (Barry et al. 2015). Briefly, the authors implanted stimulating electrodes in the NAc and another set of recording probes in the MGB. Following recovery from surgery, they delivered bipolar electrical current, either single pulses or pulse trains at 50 μA to 1 mA and 125–300 pps, and found reduced spontaneous firing rates in the majority of thalamic neurons, although a minority displayed excitation (Barry et al. 2015). Since increased SFRs, altered bursting patterns, and increased rate-level function slopes in the MGB and the topographically linked AC are considered tinnitus correlates (Kalappa et al. 2014), the NAc-DBS may ameliorate tinnitus by directly reducing this aberrant activity in the MGB. Alternatively, NAc-DBS-induced downregulation of MGB activity may be achieved by activation of GABAergic neurons in the TRN (O'Donnell et al. 1997) exerting predominant inhibitory innervation onto MGB neurons. In addition, it has been hypothesized that aberrant activity originates from a lesion-induced plasticity of the auditory pathways and the lesion-comprised NAc fails to "noise-cancel," which gives rise to tinnitus percept (Rauschecker et al. 2010). It is possible that NAc-DBS may have restored the "noise-cancelation" mechanism. Furthermore, the NAc and parahippocampus – also MGB-IC and HPC – are hyperconnected in tinnitus (Hullfish et al. 2019). It is possible that by enhancing local GABAergic tonic drive, DBS disrupts these abnormal connections, preventing the hippocampal-induced activation and by extension blocking the dysfunctional prefrontal influence believed to contribute to tinnitus etiology. However, the authors of this animal study (Barry et al. 2015) did not conduct behavioral assays to document tinnitus status, thus, the observed neural activity changes could not be directly associated with tinnitus manifestation. Further animal studies should incorporate behavioral measures of tinnitus into the electrophysiological work.

4 Transcranial Brain Stimulation

Although the above brain stimulation treatments have shown promising results, their caveat is the invasive nature of the surgical procedures. Thus, several non- or semi-invasive techniques, such as repetitive transcranial magnetic (rTMS), direct current (tDCS), alternating current (tACS), and random-noise stimulation (tRNS), have been developed to treat tinnitus (see details in the chapter by De Ridder). Below is a review of the opportunities for developing animal models using these and related modalities.

4.1 Transcranial Magnetic Stimulation (TMS)

TMS has great potential to ameliorate tinnitus and, to this it effect, it has been used to target a number of brain regions including the temporal cortex, temporoparietal junction, dlPFC, and the ACC (De Ridder et al. 2016; Kreuzer et al. 2017; Langguth 2020; Sahlsten et al. 2019; Schwippel et al. 2019). Mechanistically, it has been proposed that TMS-induced tinnitus relief results from suppression of steady-state auditory evoked potentials (Li et al. 2019); looking at specific areas, TMS stimulation of the temporoparietal junction is capable of relieving tinnitus and altering PET signals in the right parahippocampal gyrus and superior frontal gyrus (De Ridder et al. 2016). TMS stimulation of the ACC also relieves tinnitus and increases functional connectivity between the parahippocampus and subgenual ACC (De Ridder et al. 2016), and TMS-induced tinnitus suppression – via stimulation of either temporal or frontal cortices – was paralleled by decreased EEG theta and delta and increased beta, in the frontal region, and decreased beta and gamma, in the temporal region, respectively (Schecklmann et al. 2016). Taken together, TMS mainly interferes with the tinnitus network through reduction of the TCT, theta-gamma lemniscal signal and/or its analogue in FLS circuits, decoupling areas that are pathologically linked, and simultaneously reducing the underlying inflammation and cell death, following trauma.

Despite being used in over 200 instances as an investigative tinnitus treatment, every cohort has a sizeable portion of non-responders or sometimes even exacerbation, the suppression is typically transient and relatively meager, and a number of studies even fail to find a clinically significant impact (Godbehere et al. 2019; Plewnia et al. 2012; Plewnia 2018; Schecklmann et al. 2016). Not only do the results of numerous, randomized sham-controlled clinical trials contradict one another, but there are even conflicting systematic reviews and meta-analyses that advocate for and against the fidelity of TMS as an effective tinnitus treatment (Dong et al. 2020; Folmer et al. 2015; Godbehere et al. 2019; Marcondes et al. 2010; Plewnia et al. 2012; Plewnia 2018; Schecklmann et al. 2016; Soleimani et al. 2016). However, since TMS is truly non-invasive and does represent a potentially useful treatment modality, more research is needed to determine the extent of efficacy for TMS in treating tinnitus and to conduct further investigations to uncover the underlying mechanisms of exactly how it suppresses phantom auditory perception. Critically, an expansion of animal models will provide an arena for testing new parameters and uncovering these unknowns.

Thus far, a number of groups have developed various animal models of TMS for tinnitus treatment (Fig. 11). For example, Mulders and colleagues unilaterally noise-exposed guinea pigs to induce tinnitus, as measured by GPIAS, and then treated the animals with 10 min of TMS – 1 Hz, base coil intensity 90 mT – over the contralateral AC, daily, for ten sessions over 2 weeks. The intervention was able to demonstrate a therapeutic effect on tinnitus (Mulders et al. 2019). Mechanistically, while behavior changed, spontaneous firing rates in the IC slightly decreased, albeit insignificantly compared to controls, and BDNF remained unaffected in the IC and AC (Mulders et al. 2019).

Conversely, a follow-up study testing TMS stimulation of the prefrontal cortex in guinea pigs (either 1 Hz or 10 Hz for 14 days) was unable to suppress behavioral manifestation of tinnitus, although it did increase spontaneous firing rates in the MGB and alter the levels of calbindin and parvalbumin positive neurons (Mulders

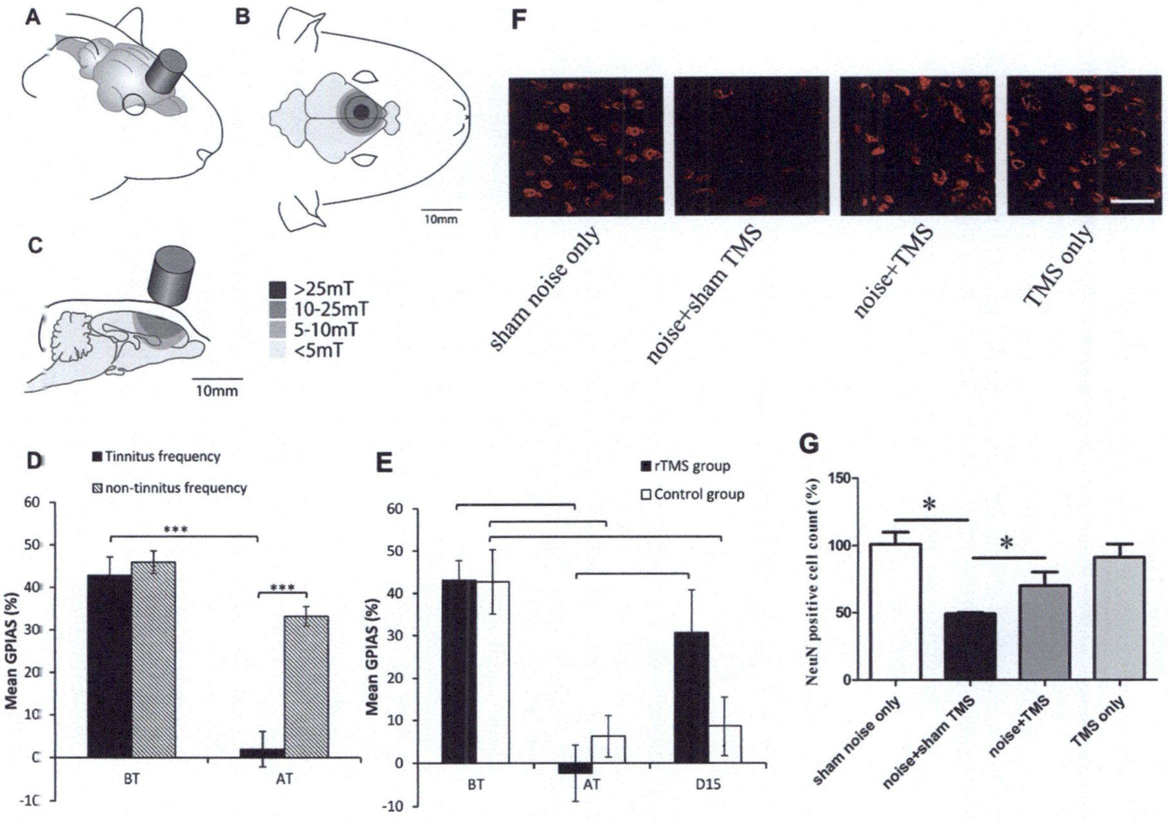

Fig. 11 Therapeutic effects of transcranial magnetic stimulation (TMS) on tinnitus (Adapted from Mulders et al. 2016a, b, 2019; Yang et al. 2016). (**a**) Schematic of a prefrontal cortex-targeted TMS paradigm in guinea pigs, 3D view. (**b**) Top-down view of schematic in "A". (**c**) Sagittal view of schematic in "A". (**d**) Tinnitus behavior measured by the GPIAS test before (BT) and after noise trauma (AT), with the deficit denoting pathologic behavior. (**e**) Tinnitus behavior measured by the GPIAS test before trauma (BT), after trauma (AT), and 15 days following trauma (D15), with a TMS-induced recovery of GPIAS values. (**f**) Neural cell density in the auditory cortex in the presence and absence of trauma and TMS. (**g**) Quantification of cell counts in "F". (Mulders et al. 2016a, b)

et al. 2019). Another group created a rat model of acoustic trauma and was able to demonstrate that TMS stimulation over the temporal cortex (daily, 1,800 pulses per session, at 1 Hz, for 14 days) showed a therapeutic effect on BDNF and neuronal loss in the AC, with no effect on hearing thresholds: though tinnitus behavior was not measured (Yang et al. 2016). These results are in agreement with TMS animal models of other disorders which also show that TMS has an anti-apoptotic and anti-inflammatory effect (Sasso et al. 2016; Yoon et al. 2011), and these features could directly contribute to alleviating tinnitus, considering they are known etiologic factors (Wang et al. 2018b, 2019). Beyond cytologic alterations, TMS stimulation of the AC and PFC in animals is able to influence neural activity in the IC and MGB, respectively (Mulders et al. 2016b, 2019), demonstrating the plausibility that TMS suppresses tinnitus perception by interfering with the underlying neural correlates through both auditory and non-auditory limbic systems.

4.2 Transcranial (TES) and Epi-Cranial Electrical Stimulation (ECS)

TES, in the common forms of tDCS, tACS, and tRNS, is achieved by placing the device on the scalp to deliver a weak electrical current through the skull and into certain brain regions of interest, ideally modulating neuronal membrane potentials and network activity (Faber et al. 2012; Vanneste et al. 2013a, b; Zaehle et al. 2011). In tDCS, two surface electrodes are placed such that a current of electricity continually flows from one electrode to another to depolarize or hyperpolarize the underlying tissue, whether under anode or cathode, respectively (Vanneste et al. 2013b). Differently, tACS relies on alternating currents via a single electrode, creating a sinusoidal form with a specific frequency that is theoretically better suited to manipulate neural oscillations, given its inherent wave structure (Vanneste et al. 2013b). The tRNS approach excites underlying cortex and is a special variant of tACS that alternates randomly across a normally distributed frequency range, mimicking "white-noise": low-frequency (1–100 Hz), high-frequency (101–650 Hz), or whole frequency (1–650 Hz) (Kreuzer et al. 2019; Vanneste et al. 2013a). It has been reported that tRNS had superior tinnitus-suppressive performance, compared to tDCS and tACS (Vanneste et al. 2013a). Mechanistically, tDCS has been shown to reduce tinnitus and the associated hyperactivity in the AC, as measured with functional near-infrared spectroscopy (Verma et al. 2019); furthermore, it has been shown to reduce fMRI functional connectivity between the AC and somatosensory, motor, and visual areas (Minami et al. 2015), reduce theta and beta oscillations in frontal, temporoparietal, and limbic areas (Souza et al. 2020), and alter gamma oscillations depending on the stimulation polarity (Vanneste et al. 2011a, 2013b). By targeting the AC and dlPFC, tRNS-induced tinnitus relief was accompanied by increased EEG alpha rhythms in the AC and PFC, decreased delta and beta waves in the dlPFC, OFC, ACC, and parahippocampus, and reduced functional connectivity between the prefrontal and auditory cortices (Mohsen et al. 2019). However, the role of TES for treating tinnitus is debated due to a large variety of stimulus parameters used across studies, meager or conflicting results, and even exacerbation in some cases (Chen et al. 2020; Kreuzer et al. 2019; Souza et al. 2020; Wang et al. 2018a; Yuan et al. 2018). Thus, there are plenty of opportunities for establishing animal models of TES to elucidate the underlying mechanisms of TES-induced tinnitus modulation and improve clinical outcomes. Due to the challenges of limited spatial resolution and the much smaller skull dimension of the commonly used rodent animal models, there is a need to develop micro-TES devices by collaborating with engineers and conducting animal studies that have been carried out as described in the above sections. Alternatively, once a behavioral paradigm is established for measuring tinnitus in non-human primates, it will ideally be applied to the TES devices used for humans.

To enhance more localized stimulation, ECS is a semi-invasive technique whereby electrodes are implanted subcutaneously, overlaying the skull, and electricity passes right through the bone onto the target brain region (Khatoun et al. 2019). Although this method is semi-invasive, in that it does require opening the scalp, the cranium remains closed, and it has the advantage of implantability so subjects can be mobile and receive chronic stimulation. For example, recent work demonstrated encouraging results for its capability of stimulating the rat motor cortex (Khatoun et al. 2019). At the same time, if a multi-channel device is adopted, EEG data may be recorded, which may be used to bridge the knowledge between animal and human studies. Thus, ECS is an attractive avenue for future application to the tinnitus field.

Taken together, TES may relieve tinnitus suffering by interfering with aberrant gamma auditory percepts and their slow – delta and theta – carrier waves, decreasing beta oscillations in key tinnitus network nodes which can be thought of as knocking the system out of its pathologic "status quo" (Spitzer and Haegens 2017), and reducing the hyper-connectivity keeping PFC attention on the tinnitus signal. By manipulating activity in both the lemniscal AC and extra-lemniscal prefrontal and limbic areas, TES interferes with both the TCT signal generator and the executive, affective FLS component. Given the information gleaned and potential clinical benefit gained from the clinical studies over recent years, it is imperative to develop animal model systems to gain better understanding of the underlying TES-induced mechanisms, which will help further shed light on tinnitus etiology and better shape these technologies for more effective clinical translation.

5 Bottom-Up and Top-Down Modulations

This chapter focuses on reviews and discussions of cochlear and brain stimulation to modulate tinnitus. The cochlear stimulation, via a number of different manipulations of the peripheral structures – cochlea and cochlear nerve, takes a bottom-up approach by exerting modulatory effects through the afferent ascending pathway.

The artificially generated neural signals replenish the lost peripheral input to the central nervous system, downregulate the central aberrant hyperactivity, and restore the lost central excitatory and inhibitory balance, ultimately relieving tinnitus symptoms.

For brain stimulation, it may be implemented via modulations of the central auditory system using ABI, AMI, MGB-DBS, and ACES or modulations of non-auditory systems such as DBS of the basal ganglion and related brain structures. For those modalities modulating the central auditory system, both ascending and descending pathways may have been impacted by non-specific electrical activation. More specifically, it is expected that ABI, AMI, MGB-DBS, and ACES would first exert local modulation, while ABI and AMI may take bottom-up approaches to activate more brain stations along the ascending pathways and MGB-DBS and ACES may take a top-down approach to activate more brain stations along the descending pathways. Indeed, abundant evidence has demonstrated that the top-down auditory corticofugal fibers project to almost all the sub- and brainstem structures, modulating neural activity in an "egocentric" manner (Suga et al. 2000; Zhang and Suga 2000) and maintaining a dynamic homeostasis in the brain. In addition, both bottom-up and top-down treatment modalities inevitably activate the MGB, which plays a pivotal role in the modulation of tinnitus, as the MGB carries sensory information to the cortex in the form of gamma rhythms, integrates the signals with the phase of slower frequency waves, binds these spatially separate gamma islands scattered throughout the brain into single perceived moments, and plugs them into the consciousness broadcasting system (Canolty et al. 2006; De Ridder et al. 2015; Steriade et al. 1991). The primary AC is the pinnacle of the afferent, lemniscal auditory pathway where sensory – physiologic and phantom – information is coded into gamma oscillations. This area is connected to the prefrontal, limbic, and striatal systems and projects the gamma oscillations into this parallel processing stream in order to make executive decisions using the available stimuli based off emotional responses and memory, focus large swaths of first-order cortex using attention, and enact willed outputs to execute these decisions. To determine which sensory streams are allowed to enter conscious awareness, the TRN receives both reciprocal connections from topographically linked areas of thalamus and cortex, and it sends GABAergic projections to inhibit the TCT loops (Pinault 2004), keeping those particular neural channels on and plugged into the larger consciousness broadcasting system (Crandall et al. 2015; Halassa et al. 2014) for tinnitus perception. Thus, these auditory modulation modalities not only regulate neural activity within the auditory brain stations and the corresponding auditory sensory structures, but also regulate neural activity in many non-auditory brain structures.

For those modulation modalities of the non-auditory structures such as the basal ganglion, STN-DBS and NAc-DBS, neural activity in both auditory and non-auditory brain structures has been altered to eventually change the perception of tinnitus. For instance, STN-DBS interferes with tinnitus-related pathological signals by affecting the anatomical and functional connections between the STN and the TRN, amygdala, insula, caudate, NAc, and sensory cortices (Cavdar et al.

2018). NAc-DBS can serve as gatekeeper to turn on or off sections of the thalamocortical sensory stream, canceling out tinnitus-noise signals, based on attention and salience (Leaver et al. 2011; Rauschecker et al. 2010), as the NAc has broad connections such as TRN, subcallosal-related cortical areas, and mediodorsal thalamus (Johansen-Berg et al. 2008; Lavin and Grace 1996; O'Donnell and Grace 1995; O'Donnell et al. 1997; Ongur and Price 2000). Depending on the cortical regions stimulated, TMS and TES may broadly interfere with the tinnitus network through reduction of the TCT, theta-gamma lemniscal signal and/or its analogue in FLS circuits, decoupling areas that are pathologically linked, and simultaneously reducing the underlying inflammation and cell death, following trauma. Among the cortices modulated, the dlPFC is an important center that is heavily involved with contextualizing the primary sensory signals and directing attention to guide which will be plugged into the larger tinnitus network, and these areas are believed to be hyperconnected in tinnitus brains (Araneda et al. 2018; Lin et al. 2020). Apparently, modulation of these non-auditory brain structures may activate a variety of cortices via a bottom-up approach and the thalamo-TRN network that projects to lower brain structures via a top-down approach.

Thus far, there exist no animal models to specifically address how the bottom-up and top-down auditory or non-auditory modulatory machinery, either independently or jointly, contributes to the modulation of neural correlates of tinnitus. It demands well-designed, future investigations using animal models and multidisciplinary techniques to quantify the contributions of bottom-up ascending, top-down descending and in situ localized modulation in both auditory and non-auditory systems.

6 Summary, Limitations, and Future Directions

Pulling it together, if the ultimate goal of tinnitus research is unfettered understanding of the underlying etiology, for the purpose of developing a universal, highly efficacious therapeutic, then considerable progress has been made. However, consistent and lasting tinnitus suppression has not yet been achieved. This chapter explored the many modulatory interventions that have been tested and/or are in a developmental stage for treating tinnitus, particularly their animal models, and they target nearly every structure along the auditory pathway including but not limited to the cochlea, DCN, IC, MGB, Vim, AC, caudate, NAc, ACC, STN, and dlPFC (Fig. 12). It is interesting that perturbing such a range of locations all results in similar functional outcome: tinnitus relief, considering each region is believed to have a unique function. In general, the treatments seem to be broadly effecting two main facets of the tinnitus network: the lemniscal TCT signal, itself, and the extralemniscal FLS salience and attention focused upon it. Although every location modulates tinnitus in its own special manner, seeing the forest through the trees is not focusing as much on any given point along the pathway but rather understanding that the broader underlying mechanism is disrupting the larger tinnitus network

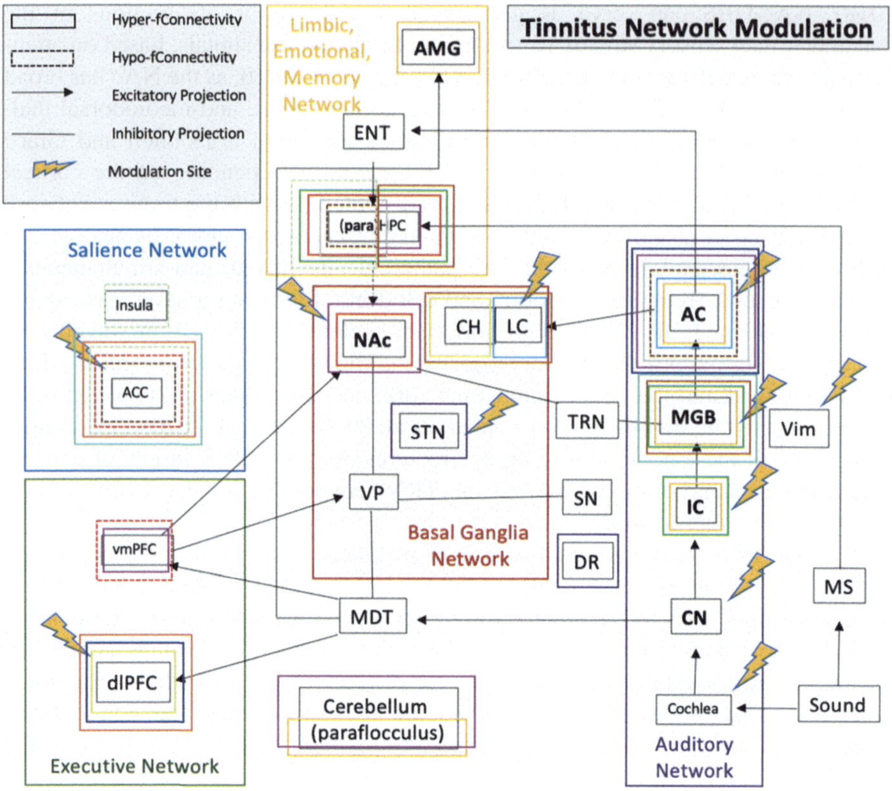

Fig. 12 Electrical interventions in the tinnitus network. This schematic outlines the key structures and networks that are implicated in tinnitus etiology, and it also shows which sites have been targeted for therapeutic intervention. The lines represent anatomical connectivity, the colored boxes represent functional connectivity, and the lightning bolt symbol denotes which areas have been targeted for tinnitus treatment. *AC* auditory cortex, *ACC* anterior cingulate cortex, *AMG* amygdala, *CH* caudate head, *CN* cochlear nucleus, *dlPFC* dorsolateral prefrontal cortex, *DR* dorsal raphe, *ENT* entorhinal cortex, *HPC* hippocampus, *IC* inferior colliculus, *LC* locus of caudate, *MDT* mediodorsal thalamus, *MGB* medial geniculate body of thalamus, *MS* medial septum, *NAc* nucleus accumbens, *para* paraHPC, *SN* substantia nigra, *STN* subthalamic nucleus, *TRN* thalamic reticular nucleus, *Vim* ventral intermediate nucleus of thalamus, *vmPFC* ventromedial prefrontal cortex, *VP* ventral pallidum

that's broadcasting these signals and their affective components into conscious awareness.

Whether interfering with the lemniscal TCT signal generators or the extra-lemniscal FLS directed attention, the treatments reviewed in this chapter were only successful for a portion of patients and animals, underscoring the need for more studies. One major issue that needs to be addressed moving forward is that the cochlea and brain are unique, and to that same point, each patient's and animal's tinnitus has its own unique features; however, these bottom-up and/or top-down interventions use blanket parameters for everyone, in each given study. Clearly, the exact location for intervention is not necessarily critical, given it is plugged strongly enough into the larger tinnitus network, so the more important factor may be determining the unique electrophysiologic features that comprise any given subject's tinnitus and tailoring the stimulation location and stimulus paradigms accordingly. In fact, personalizing stimuli has been recently explored using TMS for tinnitus (Kreuzer et al. 2017) and application of deep learning algorithms to identify unique aberrant neural signals and output customized, dynamic stimulus trains is already being investigated in the auditory system (Lee et al. 2019). Additionally, many of the animal models focused mainly on behavior, without diving deep into the underlying mechanisms; therefore, it is critical to combine behavioral measures of tinnitus with more detailed electrophysiologic and immunohistochemical experiments to probe how these modulatory devices, including the newly emerging optogenetic cochlear stimulation (Dieter et al. 2020) for enhanced spatial and temporal specificity and fidelity as well as optogenetic brain stimulation (Darrow et al. 2015), impact the neural circuits driving symptom generation and maintenance. Although incomplete, the hard work of so many different groups have provided a plethora of potential new therapeutic options that have given hope and relief to many patients, and as the animal models provide better understanding of how these technologies operate, they will continue becoming more efficacious until a universal, definitive treatment modality is established for tinnitus.

Acknowledgements This work was partially supported by the NIH/NIDCD R21DC014335-01, 1R21 DC010059-01.

References

Adamchic I et al (2014) Abnormal cross-frequency coupling in the tinnitus network. Front Neurosci 8:284

Adjamian P et al (2012) Neuromagnetic indicators of tinnitus and tinnitus masking in patients with and without hearing loss. J Assoc Res Otolaryngol 13:715–731

Ahn MH, Hong SK, Min BK (2017) The absence of resting-state high-gamma cross-frequency coupling in patients with tinnitus. Hear Res 356:63–73

Ahsan SF et al (2018) An animal model of deep brain stimulation for treating tinnitus: a proof of concept study. Laryngoscope 128:1213–1222

Araneda R et al (2018) A key role of the prefrontal cortex in the maintenance of chronic tinnitus: an fMRI study using a Stroop task. Neuroimage Clin 17:325–334

Argence M et al (2008) Stimulation by cochlear implant in unilaterally deaf rats reverses the decrease of inhibitory transmission in the inferior colliculus. Eur J Neurosci 28:1589–1602

Arts RA et al (2012) Review: cochlear implants as a treatment of tinnitus in single-sided deafness. Curr Opin Otolaryngol Head Neck Surg 20:398–403

Arts RA et al (2015) Tinnitus suppression by Intracochlear electrical stimulation in single-sided deafness: a prospective clinical trial – part I. Audiol Neurootol 20:294–313

Atencio CA et al (2014) Primary auditory cortical responses to electrical stimulation of the thalamus. J Neurophysiol 111:1077–1087

Axelsson A, Ringdahl A (1989) Tinnitus – a study of its prevalence and characteristics. Br J Audiol 23:53–62

Baguley DM, Atlas MD (2007) Cochlear implants and tinnitus. Prog Brain Res 166:347–355

Barry KM et al (2015) Modulation of medial geniculate nucleus neuronal activity by electrical stimulation of the nucleus accumbens. Neuroscience 308:1–10

Barry KM, Robertson D, Mulders W (2019) Changes in auditory thalamus neural firing patterns after acoustic trauma in rats. Hear Res 379:89–97

Basta D et al (2015) Bilateral changes of spontaneous activity within the central auditory pathway upon chronic unilateral intracochlear electrical stimulation. Otol Neurotol 36:1759–1765

Bauer CA et al (2008) Tinnitus and inferior colliculus activity in chinchillas related to three distinct patterns of cochlear trauma. J Neurosci Res 86:2564–2578

Blood AJ, Zatorre RJ (2001) Intensely pleasurable responses to music correlate with activity in brain regions implicated in reward and emotion. Proc Natl Acad Sci U S A 98:11818–11823

Boyden ES et al (2005) Millisecond-timescale, genetically targeted optical control of neural activity. Nat Neurosci 8:1263–1268

Calixto R et al (2013) Investigation of a new electrode array technology for a central auditory prosthesis. PLoS One 8:e82148

Canolty RT, Knight RT (2010) The functional role of cross-frequency coupling. Trends Cogn Sci 14:506–515

Canolty RT et al (2006) High gamma power is phase-locked to theta oscillations in human neocortex. Science 313:1626–1628

Cardin JA et al (2009) Driving fast-spiking cells induces gamma rhythm and controls sensory responses. Nature 459:663–667

Casquero-Veiga M et al (2016) Response to deep brain stimulation in three brain targets with implications in mental disorders: a PET study in rats. PLoS One 11:e0168689

Cavdar S et al (2018) Afferent projections of the subthalamic nucleus in the rat: emphasis on bilateral and interhemispheric connections. Acta Neurobiol Exp (Wars) 78:251–263

Chen YC et al (2017) Resting-state brain abnormalities in chronic subjective tinnitus: a meta-analysis. Front Hum Neurosci 11:22

Chen YC et al (2018) Alterations of the default mode network and cognitive impairment in patients with unilateral chronic tinnitus. Quant Imaging Med Surg 8:1020–1029

Chen JJ et al (2020) Association of central noninvasive brain stimulation interventions with efficacy and safety in tinnitus management: a meta-analysis. JAMA Otolaryngol Head Neck Surg 146:801–809

Cheung SW, Larson PS (2010) Tinnitus modulation by deep brain stimulation in locus of caudate neurons (area LC). Neuroscience 169:1768–1778

Cheung SW et al (2019) Phase I trial of caudate deep brain stimulation for treatment-resistant tinnitus. J Neurosurg:1–10

Claes L et al (2014) Auditory cortex tACS and tRNS for tinnitus: single versus multiple sessions. Neural Plast 2014:436713

Coomes DL, Schofield RM, Schofield BR (2005) Unilateral and bilateral projections from cortical cells to the inferior colliculus in guinea pigs. Brain Res 1042:62–72

Crandall SR, Cruikshank SJ, Connors BW (2015) A corticothalamic switch: controlling the thalamus with dynamic synapses. Neuron 86:768–782

Crocetti A et al (2009) Questionnaires to evaluate anxiety and depressive levels in tinnitus patients. Otolaryngol Head Neck Surg 140:403–405

Daneshi A et al (2005) Auditory electrical tinnitus suppression in patients with and without implants. Int Tinnitus J 11:85–91

Darrow KN et al (2015) Optogenetic stimulation of the cochlear nucleus using channelrhodopsin-2 evokes activity in the central auditory pathways. Brain Res 1599:44–56

Davis PB, Paki B, Hanley PJ (2007) Neuromonics tinnitus treatment: third clinical trial. Ear Hear 28:242–259

de Mena L, Rizk P, Rincon-Limas DE (2018) Bringing light to transcription: the optogenetics repertoire. Front Genet 9:518

De Ridder D, Vanneste S (2012) EEG driven tDCS versus bifrontal tDCS for tinnitus. Front Psych 3:84
De Ridder D et al (2006) Primary and secondary auditory cortex stimulation for intractable tinnitus. ORL J Otorhinolaryngol Relat Spec 68:48–54
De Ridder D et al (2007) Electrical stimulation of auditory and somatosensory cortices for treatment of tinnitus and pain. Prog Brain Res 166:377–388
De Ridder D et al (2010) Burst stimulation of the auditory cortex: a new form of neurostimulation for noise-like tinnitus suppression. J Neurosurg 112:1289–1294
De Ridder D et al (2011) Theta-gamma dysrhythmia and auditory phantom perception. J Neurosurg 114:912–921
De Ridder D et al (2012) Dorsolateral prefrontal cortex transcranial magnetic stimulation and electrode implant for intractable tinnitus. World Neurosurg 77:778–784
De Ridder D et al (2014a) Safety and efficacy of vagus nerve stimulation paired with tones for the treatment of tinnitus: a case series. Neuromodulation 17:170–179
De Ridder D, Vanneste S, Freeman W (2014b) The Bayesian brain: phantom percepts resolve sensory uncertainty. Neurosci Biobehav Rev 44:4–15
De Ridder D et al (2014c) An integrative model of auditory phantom perception: tinnitus as a unified percept of interacting separable subnetworks. Neurosci Biobehav Rev 44:16–32
De Ridder D et al (2015) Thalamocortical dysrhythmia: a theoretical update in tinnitus. Front Neurol 6:124
De Ridder D, Joos K, Vanneste S (2016) Anterior cingulate implants for tinnitus: report of 2 cases. J Neurosurg 124:893–901
Deklerck AN et al (2020) Invasive neuromodulation as a treatment for tinnitus: a systematic review. Neuromodulation 23:451–462
Deschenes M, Veinante P, Zhang ZW (1998) The organization of corticothalamic projections: reciprocity versus parity. Brain Res Brain Res Rev 28:286–308
Di NW et al (2009) Transtympanic electrical stimulation for immediate and long-term tinnitus suppression. Int Tinnitus J 15:100–106
Dieter A et al (2019) Near physiological spectral selectivity of cochlear optogenetics. Nat Commun 10:1962
Dieter A, Keppeler D, Moser T (2020) Towards the optical cochlear implant: optogenetic approaches for hearing restoration. EMBO Mol Med 12:e11618
DiGuiseppi J, Zuo J (2019) The awesome power of optogenetics in hearing research. Neurosci Lett 701:175–179
Dong C et al (2020) Low-frequency repetitive transcranial magnetic stimulation for the treatment of chronic tinnitus: a systematic review and meta-analysis of randomized controlled trials. Biomed Res Int 2020:3141278
Eggermont JJ (2015) The auditory cortex and tinnitus – a review of animal and human studies. Eur J Neurosci 41:665–676
Eggermont JJ, Roberts LE (2004) The neuroscience of tinnitus. Trends Neurosci 27:676–682
Engelhardt J et al (2014) Effect of chronic cortical stimulation on chronic severe tinnitus: a prospective randomized double-blind cross-over trial and long-term follow up. Brain Stimul 7:694–700
Engineer ND et al (2011) Reversing pathological neural activity using targeted plasticity. Nature 470:101–104
Faber M et al (2012) Top down prefrontal affective modulation of tinnitus with multiple sessions of tDCS of dorsolateral prefrontal cortex. Brain Stimul 5:492–498
Fallon JB, Irvine DR, Shepherd RK (2008) Cochlear implants and brain plasticity. Hear Res 238:110–117
Fallon JB, Irvine DR, Shepherd RK (2009) Neural prostheses and brain plasticity. J Neural Eng 6:065008

Fallon JB, Shepherd RK, Irvine DR (2014) Effects of chronic cochlear electrical stimulation after an extended period of profound deafness on primary auditory cortex organization in cats. Eur J Neurosci 39:811–820

Fernandes NF et al (2020) Auditory and language skills in children with auditory brainstem implants. Int J Pediatr Otorhinolaryngol 132:110010

Firestone E, Luo H, Pace E, Liu B, Agabigum B, Zhu L, Zhang XG, Moran J, Zhang JS (2019) Pathology-specific brain oscillations serve as underpinnings of tinnitus and its modulation. Assoc Res Otolaryngol Abs 804:501

Folmer RL et al (2006) Effects of repetitive transcranial magnetic stimulation (rTMS) on chronic tinnitus. Acta Otolaryngol Suppl:96–101

Folmer RL et al (2015) Repetitive transcranial magnetic stimulation treatment for chronic tinnitus: a randomized clinical trial. JAMA Otolaryngol Head Neck Surg 141:716–722

Friedland DR et al (2007) Feasibility of auditory cortical stimulation for the treatment of tinnitus. Otol Neurotol 28:1005–1012

Fries P (2009) Neuronal gamma-band synchronization as a fundamental process in cortical computation. Annu Rev Neurosci 32:209–224

Gault R, McGinnity TM, Coleman S (2018) A computational model of thalamocortical dysrhythmia in people with tinnitus. IEEE Trans Neural Syst Rehabil Eng 26:1845–1857

Gerken GM, Saunders SS, Paul RE (1984) Hypersensitivity to electrical stimulation of auditory nuclei follows hearing loss in cats. Hear Res 13:249–259

Ghafouri S et al (2019) Deep brain stimulation restores the glutamatergic and GABAergic synaptic transmission and plasticity to normal levels in kindled rats. PLoS One 14:e0224834

Gilles A et al (2020) Neural substrates of tinnitus in an auditory brainstem implant patient: a preliminary molecular imaging study using H2 15 O-PET including a 5-year follow-up of auditory performance and tinnitus perception. Otol Neurotol 41:e15–e20

Godbehere J et al (2019) Treatment of tinnitus using theta burst based repetitive transcranial magnetic stimulation-a single blinded randomized control trial. Otol Neurotol 40:S38–S42

Gray CM, Singer W (1989) Stimulus-specific neuronal oscillations in orientation columns of cat visual cortex. Proc Natl Acad Sci U S A 86:1698–1702

Gray CM et al (1989) Oscillatory responses in cat visual cortex exhibit inter-columnar synchronization which reflects global stimulus properties. Nature 338:334–337

Halassa MM et al (2014) State-dependent architecture of thalamic reticular subnetworks. Cell 158:808–821

Hasson D et al (2011) Stress and prevalence of hearing problems in the Swedish working population. BMC Public Health 11:130

Henry KS et al (2016) Neural correlates of behavioral amplitude modulation sensitivity in the budgerigar midbrain. J Neurophysiol 115:1905–1916

Hitier M et al (2020) The effects of selective electrical stimulation of the rat cochlea on hippocampal field potentials. Hear Res 395:108023

Holz EM et al (2010) Theta-gamma phase synchronization during memory matching in visual working memory. Neuroimage 52:326–335

Hoover WB, Vertes RP (2007) Anatomical analysis of afferent projections to the medial prefrontal cortex in the rat. Brain Struct Funct 212:149–179

House JW, Brackmann DE (1981) Tinnitus: surgical treatment. Ciba Found Symp 85:204–216

Hullfish J et al (2019) Frontostriatal network dysfunction as a domain-general mechanism underlying phantom perception. Hum Brain Mapp 40:2241–2251

Ito J, Sakakihara J (1994) Tinnitus suppression by electrical stimulation of the cochlear wall and by cochlear implantation. Laryngoscope 104:752–754

Jackson P (1985) A comparison of the effects of eighth nerve section with lidocaine on tinnitus. J Laryngol Otol 99:663–666

Jacobs J, Kahana MJ (2010) Direct brain recordings fuel advances in cognitive electrophysiology. Trends Cogn Sci 14:162–171

Jahnsen H, Llinas R (1984a) Ionic basis for the electro-responsiveness and oscillatory properties of Guinea-pig thalamic neurones in vitro. J Physiol 349:227–247
Jahnsen H, Llinas R (1984b) Electrophysiological properties of guinea-pig thalamic neurones: an in vitro study. J Physiol 349:205–226
Jakobs M, Lozano AM (2019) Editorial. Deep brain stimulation for tinnitus: exploring the frontier between sensory perception and awareness. J Neurosurg:1–4
Jang SH, Yeo SS (2014) Thalamocortical connections between the mediodorsal nucleus of the thalamus and prefrontal cortex in the human brain: a diffusion tensor tractographic study. Yonsei Med J 55:709–714
Jastreboff PJ, Jastreboff MM (2000) Tinnitus retraining therapy (TRT) as a method for treatment of tinnitus and hyperacusis patients. J Am Acad Audiol 11:162–177
Johansen-Berg H et al (2008) Anatomical connectivity of the subgenual cingulate region targeted with deep brain stimulation for treatment-resistant depression. Cereb Cortex 18:1374–1383
Kahlbrock N, Weisz N (2008) Transient reduction of tinnitus intensity is marked by concomitant reductions of delta band power. BMC Biol 6:4
Kalappa BI et al (2014) Single unit hyperactivity and bursting in the auditory thalamus of awake rats directly correlates with behavioural evidence of tinnitus. J Physiol 592:5065–5078
Keppeler D et al (2018) Ultrafast optogenetic stimulation of the auditory pathway by targeting-optimized chronos. EMBO J:37
Khatoun A, Asamoah B, Mc Laughlin M (2019) Investigating the feasibility of epicranial cortical stimulation using concentric-ring electrodes: a novel minimally invasive neuromodulation method. Front Neurosci 13:773
Kiang NY, Liberman MC, Levine RA (1976) Auditory-nerve activity in cats exposed to ototoxic drugs and high-intensity sounds. Ann Otol Rhinol Laryngol 85:752–768
Kim K et al (2015) A novel method for device-related electroencephalography artifact suppression to explore cochlear implant-related cortical changes in single-sided deafness. J Neurosci Methods 255:22–28
Klinke R et al (1999) Recruitment of the auditory cortex in congenitally deaf cats by long-term cochlear electrostimulation. Science 285:1729–1733
Komiya H, Eggermont JJ (2000) Spontaneous firing activity of cortical neurons in adult cats with reorganized tonotopic map following pure-tone trauma. Acta Otolaryngol 120:750–756
Konig O et al (2006) Course of hearing loss and occurrence of tinnitus. Hear Res 221:59–64
Kral A et al (1998) Spatial resolution of cochlear implants: the electrical field and excitation of auditory afferents. Hear Res 121:11–28
Krauss P et al (2016) Stochastic resonance controlled upregulation of internal noise after hearing loss as a putative cause of tinnitus-related neuronal hyperactivity. Front Neurosci 10:597
Krauss P et al (2017) Adaptive stochastic resonance for unknown and variable input signals. Sci Rep 7:2450
Kreuzer PM et al (2017) Individualized repetitive transcranial magnetic stimulation treatment in chronic tinnitus? Front Neurol 8:126
Kreuzer PM et al (2019) Daily high-frequency transcranial random noise stimulation of bilateral temporal cortex in chronic tinnitus – a pilot study. Sci Rep 9:12274
Lachowska M et al (2020) Detailed insight in intraoperative eABR measurements to assist auditory brainstem implantation in a patient with neurofibromatosis type 2. Acta Neurol Belg 120:1371–1378
Langguth B (2020) Non-invasive neuromodulation for tinnitus. J Audiol Otol 24:113–118
Langguth B, Elgoyhen AB (2012) Current pharmacological treatments for tinnitus. Expert Opin Pharmacother 13:2495–2509
Lavin A, Grace AA (1996) Physiological properties of rat ventral pallidal neurons recorded intracellularly in vivo. J Neurophysiol 75:1432–1443
Leaver AM et al (2011) Dysregulation of limbic and auditory networks in tinnitus. Neuron 69:33–43

Ledoux JE, Farb C, Ruggiero DA (1990) Topographic organization of neurons in the acoustic thalamus that project to the amygdala. Hear Res 10:1043–1054
Lee DJ et al (2019) Current and future directions of deep brain stimulation for neurological and psychiatric disorders. J Neurosurg 131:333–342
Li LP et al (2019) Steady-state auditory evoked fields reflect long-term effects of repetitive transcranial magnetic stimulation in tinnitus. Clin Neurophysiol 130:1665–1672
Liberman T, Velluti RA, Pedemonte M (2009) Temporal correlation between auditory neurons and the hippocampal theta rhythm induced by novel stimulations in awake guinea pigs. Brain Res 1298:70–77
Lim HH, Lenarz T (2015) Auditory midbrain implant: research and development towards a second clinical trial. Hear Res 322:212–223
Lin X et al (2020) Altered topological patterns of Gray matter networks in tinnitus: a graph-theoretical-based study. Front Neurosci 14:541
Llinas RR et al (1999) Thalamocortical dysrhythmia: a neurological and neuropsychiatric syndrome characterized by magnetoencephalography. Proc Natl Acad Sci U S A 96:15222–15227
Llinas R et al (2005) Rhythmic and dysrhythmic thalamocortical dynamics: GABA systems and the edge effect. Trends Neurosci 28:325–333
Luo H et al (2012) Tinnitus suppression by electrical stimulation of the rat dorsal cochlear nucleus. Neurosci Lett 522:16–20
Lv H et al (2020) Altered functional connectivity of the thalamus in tinnitus patients is correlated with symptom alleviation after sound therapy. Brain Imaging Behav 14:2668–2678
Mager T et al (2018) High frequency neural spiking and auditory signaling by ultrafast red-shifted optogenetics. Nat Commun 9:1750
Malmierca MS et al (2002) Direct projections from cochlear nuclear complex to auditory thalamus in the rat. J Neurosci 22:10891–10897
Marcondes RA et al (2010) Repetitive transcranial magnetic stimulation improve tinnitus in normal hearing patients: a double-blind controlled, clinical and neuroimaging outcome study. Eur J Neurol 17:38–44
Marks KL et al (2018) Auditory-somatosensory bimodal stimulation desynchronizes brain circuitry to reduce tinnitus in guinea pigs and humans. Sci Transl Med 10
Martel DT, Pardo-Garcia TR, Shore SE (2019) Dorsal Cochlear nucleus fusiform-cell plasticity is altered in salicylate-induced tinnitus. Neuroscience 407:170–181
Mauger SJ et al (2012) An in vivo investigation of inferior colliculus single neuron responses to cochlear nucleus pulse train stimulation. J Neurophysiol 108:2999–3008
McCreery D, Han M, Pikov V (2010) Neuronal activity evoked in the inferior colliculus of the cat by surface macroelectrodes and penetrating microelectrodes implanted in the cochlear nucleus. IEEE Trans Biomed Eng 57:1765–1773
McCreery D et al (2013) Encoding of the amplitude modulation of pulsatile electrical stimulation in the feline cochlear nucleus by neurons in the inferior colliculus; effects of stimulus pulse rate. J Neural Eng 10:056010
McCreery D, Yadev K, Han M (2018) Responses of neurons in the feline inferior colliculus to modulated electrical stimuli applied on and within the ventral cochlear nucleus; implications for an advanced auditory brainstem implant. Hear Res 363:85–97
McCullough LD, Sokolowski JD, Salamone JD (1993) A neurochemical and behavioral investigation of the involvement of nucleus accumbens dopamine in instrumental avoidance. Neuroscience 52:919–925
McFerran DJ et al (2019) Why is there no cure for tinnitus? Front Neurosci 13:802
Melloni L et al (2007) Synchronization of neural activity across cortical areas correlates with conscious perception. J Neurosci 27:2858–2865
Meltzer NE, Ryugo DK (2006) Projections from auditory cortex to cochlear nucleus: a comparative analysis of rat and mouse. Anat Rec A Discov Mol Cell Evol Biol 288:397–408

Mertens G, De Bodt M, Van de Heyning P (2016) Cochlear implantation as a long-term treatment for ipsilateral incapacitating tinnitus in subjects with unilateral hearing loss up to 10 years. Hear Res 331:1–6
Min HK et al (2016) Dopamine release in the nonhuman primate caudate and putamen depends upon site of stimulation in the subthalamic nucleus. J Neurosci 36:6022–6029
Minami SB et al (2015) Auditory resting-state functional connectivity in tinnitus and modulation with transcranial direct current stimulation. Acta Otolaryngol 135:1286–1292
Mitchell AS (2015) The mediodorsal thalamus as a higher order thalamic relay nucleus important for learning and decision-making. Neurosci Biobehav Rev 54:76–88
Miyakawa A et al (2019) Tinnitus correlates with downregulation of cortical glutamate decarboxylase 65 expression but not auditory cortical map reorganization. J Neurosci 39:9989–10001
Mogenson GJ et al (1987) Ventral pallidum projections to mediodorsal nucleus of the thalamus: an anatomical and electrophysiological investigation in the rat. Brain Res 404:221–230
Mohsen S et al (2019) Multisite transcranial random noise stimulation (tRNS) modulates the distress network activity and oscillatory powers in subjects with chronic tinnitus. J Clin Neurosci 67:178–184
Møller AR (2011) Anatomy and physiology of the auditory system. Springer, New York
Mulders WH, Spencer TC, Robertson D (2016a) Effects of pulsatile electrical stimulation of the round window on central hyperactivity after cochlear trauma in guinea pig. Hear Res 335:128–137
Mulders WH et al (2016b) The effects of repetitive transcranial magnetic stimulation in an animal model of tinnitus. Sci Rep 6:38234
Mulders W et al (2019) Low-intensity repetitive transcranial magnetic stimulation over prefrontal cortex in an animal model alters activity in the auditory thalamus but does not affect behavioural measures of tinnitus. Exp Brain Res 237:883–896
Neuheiser A et al (2010) Effects of pulse phase duration and location of stimulation within the inferior colliculus on auditory cortical evoked potentials in a guinea pig model. J Assoc Res Otolaryngol 11:689–708
Norena AJ, Mulders WH, Robertson D (2015) Suppression of putative tinnitus-related activity by extra-cochlear electrical stimulation. J Neurophysiol 113:132–143
Nozaradan S et al (2017) Intracerebral evidence of rhythm transform in the human auditory cortex. Brain Struct Funct 222:2389–2404
O'Donnell P, Grace AA (1995) Synaptic interactions among excitatory afferents to nucleus accumbens neurons: hippocampal gating of prefrontal cortical input. J Neurosci 15:3622–3639
O'Donnell P et al (1997) Interconnected parallel circuits between rat nucleus accumbens and thalamus revealed by retrograde transynaptic transport of pseudorabies virus. J Neurosci 17:2143–2167
O'Driscoll M, El-Deredy W, Ramsden RT (2011) Brain stem responses evoked by stimulation of the mature cochlear nucleus with an auditory brain stem implant. Ear Hear 32:286–299
Offutt SJ et al (2014) Suppression and facilitation of auditory neurons through coordinated acoustic and midbrain stimulation: investigating a deep brain stimulator for tinnitus. J Neural Eng 11:066001
Ongur D, Price JL (2000) The organization of networks within the orbital and medial prefrontal cortex of rats, monkeys and humans. Cereb Cortex 10:206–219
Pages DS et al (2016) Effects of electrical stimulation in the inferior colliculus on frequency discrimination by rhesus monkeys and implications for the auditory midbrain implant. J Neurosci 36:5071–5083
Perreau A, Tyler R, Mancini PC (2020) Programming a Cochlear Implant for tinnitus suppression. J Am Acad Audiol 31:302–308
Pinault D (2004) The thalamic reticular nucleus: structure, function and concept. Brain Res Brain Res Rev 46:1–31
Plewnia C (2018) Transcranial brain stimulation for the treatment of tinnitus: positive lessons from a negative trial. Brain Stimul 11:1–2

Plewnia C et al (2012) Treatment of chronic tinnitus with theta burst stimulation: a randomized controlled trial. Neurology 78:1628–1634
Poncet-Wallet C et al (2020) Prospective multicentric follow-up study of Cochlear implantation in adults with single-sided deafness: tinnitus and audiological outcomes. Otol Neurotol 41:458–466
Quass GL et al (2018) Electrical stimulation of the midbrain excites the auditory cortex asymmetrically. Brain Stimul 11:1161–1174
Rauschecker JP, Leaver AM, Muhlau M (2010) Tuning out the noise: limbic-auditory interactions in tinnitus. Neuron 66:819–826
Rauschecker JP et al (2015) Frontostriatal gating of tinnitus and chronic pain. Trends Cogn Sci 19:567–578
Roberts LE, Moffat G, Bosnyak DJ (2006) Residual inhibition functions in relation to tinnitus spectra and auditory threshold shift. Acta Otolaryngol Suppl:27–33
Roberts DS et al (2017) Tinnitus suppression after auditory brainstem implantation in patients with neurofibromatosis type-2. Otol Neurotol 38:118–122
Rode T et al (2013) Neural representation in the auditory midbrain of the envelope of vocalizations based on a peripheral ear model. Front Neural Circuits 7:166
Rotge JY et al (2012) The associative and limbic thalamus in the pathophysiology of obsessive-compulsive disorder: an experimental study in the monkey. Transl Psychiatry 2:e161
Rubinstein JT et al (2003) Electrical suppression of tinnitus with high-rate pulse trains. Otol Neurotol 24:478–485
Ryugo DK et al (2010) Synaptic plasticity after chemical deafening and electrical stimulation of the auditory nerve in cats. J Comp Neurol 518:1046–1063
Sadeghi M et al (2019) A neural ensemble correlation code for sound category identification. PLoS Biol 17:e3000449
Sahlsten H et al (2019) Neuronavigated versus non-navigated repetitive transcranial magnetic stimulation for chronic tinnitus: a randomized study. Trends Hear 23:2331216518822198
Sametsky EA et al (2015) Enhanced GABAA-mediated tonic inhibition in auditory thalamus of rats with behavioral evidence of tinnitus. J Neurosci 35:9369–9380
Sasso V et al (2016) Repetitive transcranial magnetic stimulation reduces remote apoptotic cell death and inflammation after focal brain injury. J Neuroinflammation 13:150
Schecklmann M et al (2016) Neuronavigated left temporal continuous theta burst stimulation in chronic tinnitus. Restor Neurol Neurosci 34:165–175
Schierholz I et al (2017) Auditory and audio-visual processing in patients with cochlear, auditory brainstem, and auditory midbrain implants: an EEG study. Hum Brain Mapp 38:2206–2225
Schwippel T et al (2019) Clinical review: the therapeutic use of theta-burst stimulation in mental disorders and tinnitus. Prog Neuropsychopharmacol Biol Psychiatry 92:285–300
Sedley W et al (2012) Single-subject oscillatory gamma responses in tinnitus. Brain 135:3089–3100
Sedley W et al (2015a) Intracranial mapping of a cortical tinnitus system using residual inhibition. Curr Biol 25:1208–1214
Sedley W et al (2015b) Human auditory cortex neurochemistry reflects the presence and severity of tinnitus. J Neurosci 35:14822–14828
Seeley WW et al (2007) Dissociable intrinsic connectivity networks for salience processing and executive control. J Neurosci 27:2349–2356
Seidman MD et al (2008) Direct electrical stimulation of Heschl's gyrus for tinnitus treatment. Laryngoscope 118:491–500
Seki S, Eggermont JJ (2003) Changes in spontaneous firing rate and neural synchrony in cat primary auditory cortex after localized tone-induced hearing loss. Hear.Res. 180:28–38
Shaaya M, Fauser J, Karginov AV (2021) Optogenetics: the art of illuminating complex signaling pathways. Physiology (Bethesda) 36:52–60
Shekhawat GS et al (2016) Intensity, duration, and location of high-definition transcranial direct current stimulation for tinnitus relief. Neurorehabil Neural Repair 30:349–359

Shi Y et al (2009) Deep brain stimulation effects in patients with tinnitus. Otolaryngol Head Neck Surg 141:285–287
Shore SE, Roberts LE, Langguth B (2016) Maladaptive plasticity in tinnitus – triggers, mechanisms and treatment. Nat Rev Neurol 12:150–160
Singer W (1999) Neuronal synchrony: a versatile code for the definition of relations? Neuron 24 (49–65):111–125
Smit JV et al (2016a) The impact of deep brain stimulation on tinnitus. Surg Neurol Int 7:S848–S854
Smit JV et al (2016b) Deep brain stimulation of the inferior colliculus in the rodent suppresses tinnitus. Brain Res 1650:118–124
Smit JV et al (2017) Hearing assessment during deep brain stimulation of the central nucleus of the inferior colliculus and dentate cerebellar nucleus in rat. PeerJ 5:e3892
Soleimani R, Jalali MM, Hasandokht T (2016) Therapeutic impact of repetitive transcranial magnetic stimulation (rTMS) on tinnitus: a systematic review and meta-analysis. Eur Arch Otorhinolaryngol 273:1663–1675
Song JJ et al (2017) A quantitative electroencephalography study on Cochlear implant-induced cortical changes in single-sided deafness with tinnitus. Front Hum Neurosci 11:210
Soussi T, Otto SR (1994) Effects of electrical brainstem stimulation on tinnitus. Acta Otolaryngol 114:135–140
Souza DDS et al (2020) Transcranial direct current stimulation improves tinnitus perception and modulates cortical electrical activity in patients with tinnitus: a randomized clinical trial. Neurophysiol Clin 50:289–300
Spitzer B, Haegens S (2017) Beyond the status quo: a role for beta oscillations in endogenous content (re)activation. eNeuro 4
Steriade M et al (1991) Fast oscillations (20-40 Hz) in thalamocortical systems and their potentiation by mesopontine cholinergic nuclei in the cat. Proc Natl Acad Sci U S A 88:4396–4400
Stevens C et al (2007) Severe tinnitus and its effect on selective and divided attention. Int J Audiol 46:208–216
Suga N et al (2000) The corticofugal system for hearing: recent progress. Proc Natl Acad Sci U S A 97:11807–11814
Tae WS et al (2018) Changes in the regional shape and volume of subcortical nuclei in patients with tinnitus comorbid with mild hearing loss. Neuroradiology 60:1203–1211
Tort AB et al (2008) Dynamic cross-frequency couplings of local field potential oscillations in rat striatum and hippocampus during performance of a T-maze task. Proc Natl Acad Sci U S A 105:20517–20522
Tunkel DE et al (2014) Clinical practice guideline: tinnitus executive summary. Otolaryngol Head Neck Surg 151:533–541
Ueyama T et al (2013) Brain regions responsible for tinnitus distress and loudness: a resting-state FMRI study. PLoS One 8:e67778
Vachicouras N et al (2019) Microstructured thin-film electrode technology enables proof of concept of scalable, soft auditory brainstem implants. Sci Transl Med 11(514):eaax9487
van der Loo E et al (2009) Tinnitus intensity dependent gamma oscillations of the contralateral auditory cortex. PLoS One 4:e7396
van Zwieten G et al (2019) Inhibition of experimental tinnitus with high frequency stimulation of the rat medial geniculate body. Neuromodulation 22:416–424
Vanneste S, De Ridder D (2011) Bifrontal transcranial direct current stimulation modulates tinnitus intensity and tinnitus-distress-related brain activity. Eur J Neurosci 34:605–614
Vanneste S et al (2010) The neural correlates of tinnitus-related distress. Neuroimage 52:470–480
Vanneste S, Langguth B, De Ridder D (2011a) Do tDCS and TMS influence tinnitus transiently via a direct cortical and indirect somatosensory modulating effect? A combined TMS-tDCS and TENS study. Brain Stimul 4:242–252

Vanneste S, van de Heyning P, De Ridder D (2011b) The neural network of phantom sound changes over time: a comparison between recent-onset and chronic tinnitus patients. Eur J Neurosci 34:718–731
Vanneste S, Fregni F, De Ridder D (2013a) Head-to-head comparison of transcranial random noise stimulation, transcranial AC stimulation, and transcranial DC stimulation for tinnitus. Front Psych 4:158
Vanneste S et al (2013b) Comparing immediate transient tinnitus suppression using tACS and tDCS: a placebo-controlled study. Exp Brain Res 226:25–31
Vanneste S et al (2017) Pairing sound with vagus nerve stimulation modulates cortical synchrony and phase coherence in tinnitus: an exploratory retrospective study. Sci Rep 7:17345
Verma R, Jha A, Singh S (2019) Functional near-infrared spectroscopy to probe tDCS-induced cortical functioning changes in tinnitus. J Int Adv Otol 15:321–325
Vianney-Rodrigues P, Iancu OD, Welsh JP (2011) Gamma oscillations in the auditory cortex of awake rats. Eur J Neurosci 33(1):119–129
Vianney-Rodrigues P, Auerbach BD, Salvi R (2019) Aberrant thalamocortical coherence in an animal model of tinnitus. J Neurophysiol 121:893–907
Wang ZM et al (2015) Auditory rehabilitation in rhesus macaque monkeys (Macaca mulatta) with auditory brainstem implants. Chin Med J (Engl) 128:1363–1369
Wang TC et al (2018a) Effect of transcranial direct current stimulation in patients with tinnitus: a meta-analysis and systematic review. Ann Otol Rhinol Laryngol 127:79–88
Wang W et al (2018b) Blocking tumor necrosis factor-alpha expression prevents blast-induced excitatory/inhibitory synaptic imbalance and parvalbumin-positive interneuron loss in the hippocampus. J Neurotrauma 35:2306–2316
Wang W et al (2019) Neuroinflammation mediates noise-induced synaptic imbalance and tinnitus in rodent models. PLoS Biol 17:e3000307
Wegger M, Ovesen T, Larsen DG (2017) Acoustic coordinated reset neuromodulation: a systematic review of a novel therapy for tinnitus. Front Neurol 8:36
Weiss RS, Voss A, Hemmert W (2016) Optogenetic stimulation of the cochlea-a review of mechanisms, measurements, and first models. Network 27:212–236
Weisz N et al (2005) Tinnitus perception and distress is related to abnormal spontaneous brain activity as measured by magnetoencephalography. PLoS Med 2:e153
Weisz N et al (2007) The neural code of auditory phantom perception. J Neurosci 27:1479–1484
Wenzel GI et al (2014) Non-penetrating round window electrode stimulation for tinnitus therapy followed by cochlear implantation. Eur Arch Otorhinolaryngol 272 (11):3283–3293
Wienbruch C et al (2006) Frequency organization of the 40-Hz auditory steady-state response in normal hearing and in tinnitus. Neuroimage 33:180–194
Winne J et al (2019) Salicylate induces anxiety-like behavior and slow theta oscillation and abolishes the relationship between running speed and fast theta oscillation frequency. Hippocampus 29:15–25
Wu C, Martel DT, Shore SE (2016) Increased synchrony and bursting of dorsal Cochlear nucleus fusiform cells correlate with tinnitus. J Neurosci 36:2068–2073
Yang S et al (2011) Homeostatic plasticity drives tinnitus perception in an animal model. Proc Natl Acad Sci U S A 108:14974–14979
Yang H et al (2016) Effect of repetitive transcranial magnetic stimulation on auditory function following acoustic trauma. Neurol Sci 37:1511–1516
Yoon KJ, Lee YT, Han TR (2011) Mechanism of functional recovery after repetitive transcranial magnetic stimulation (rTMS) in the subacute cerebral ischemic rat model: neural plasticity or anti-apoptosis? Exp Brain Res 214:549–556
Yuan T et al (2018) Transcranial direct current stimulation for the treatment of tinnitus: a review of clinical trials and mechanisms of action. BMC Neurosci 19:66
Zaehle T et al (2011) Excitability changes induced in the human auditory cortex by transcranial direct current stimulation: direct electrophysiological evidence. Exp Brain Res 215:135–140

Zeng FG et al (2011) Tinnitus suppression by low-rate electric stimulation and its electrophysiological mechanisms. Hear Res 277:61–66

Zeng FG, Djalilian H, Lin H (2015) Tinnitus treatment with precise and optimal electric stimulation: opportunities and challenges. Curr Opin Otolaryngol Head Neck Surg 23:382–387

Zhang J (2013) Auditory cortex stimulation to suppress tinnitus: mechanisms and strategies. Hear Res 295:38–57

Zhang J (2019) Blast-induced tinnitus: animal models. J Acoust Soc Am 146:3811

Zhang JS, Kaltenbach JA (1998) Increases in spontaneous activity in the dorsal cochlear nucleus of the rat following exposure to high-intensity sound. Neurosci Lett 250:197–200

Zhang Y, Suga N (2000) Modulation of responses and frequency tuning of thalamic and collicular neurons by cortical activation in mustached bats. J Neurophysiol 84:325–333

Zhang J, Zhang X (2010) Electrical stimulation of the dorsal cochlear nucleus induces hearing in rats. Brain Res 1311:37–50

Zhang J, Zhang Y, Zhang X (2011) Auditory cortex electrical stimulation suppresses tinnitus in rats. J Assoc Res Otolaryngol 12:185–201

Zhang L et al (2018) Remodeling of cholinergic input to the hippocampus after noise exposure and tinnitus induction in guinea pigs. Hippocampus 29 (8):669–682

Part III
Developing Biomarkers for Tinnitus

Functional Neuroanatomy of Salicylate- and Noise-Induced Tinnitus and Hyperacusis

Richard Salvi, Benjamin D. Auerbach, Condon Lau, Yu-Chen Chen, Senthilvelan Manohar, Xiaopeng Liu, Dalian Ding, and Guang-Di Chen

Contents

Abstract Tinnitus and hyperacusis are debilitating conditions often associated with aging or exposure to intense noise or ototoxic drugs. One of the most reliable methods of inducing tinnitus is with high doses of sodium salicylate, the active ingredient in aspirin. High doses of salicylate have been widely used to investigate the functional neuroanatomy of tinnitus and hyperacusis. High doses of salicylate have been used to develop novel behavioral methods to detect the presence of

R. Salvi (✉), B. D. Auerbach, S. Manohar, X. Liu, D. Ding, and G.-D. Chen
Center for Hearing and Deafness, University at Buffalo, Buffalo, NY, USA
e-mail: salvi@buffalo.edu

C. Lau
Department of Physics, City University of Hong Kong, Hong Kong, China

Y.-C. Chen
Department of Radiology, Nanjing First Hospital, Nanjing Medical University, Nanjing, China

Curr Topics Behav Neurosci (2021) 51: 133–160
https://doi.org/10.1007/7854_2020_156
Published Online: 12 July 2020

tinnitus and hyperacusis in animal models. Salicylate typically induces a hearing loss of approximately 20 dB which greatly reduces the neural output of the cochlea. As this weak neural signal emerging from the cochlea is sequentially relayed to the cochlear nucleus, inferior colliculus, medial geniculate, and auditory cortex, the neural response to suprathreshold sounds is progressively amplified by a factor of 2–3 by the time the signal reaches the auditory cortex, a phenomenon referred to as enhanced central gain. Sound-evoked hyperactivity also occurred in the amygdala, a region that assigns emotional significance to sensory stimuli. Resting state functional magnetic imaging of the BOLD signal revealed salicylate-induced increases in spontaneous neural activity in the inferior colliculus, medial geniculate body, and auditory cortex as well as in non-auditory areas such as the amygdala, reticular formation, cerebellum, and other sensory areas. Functional connectivity of the BOLD signal revealed increased neural coupling between several auditory areas and non-auditory areas such as the amygdala, cerebellum, reticular formation, hippocampus, and caudate/putamen; these strengthened connections likely contribute to the multifaceted dimensions of tinnitus. Taken together, these results suggest that salicylate-induced tinnitus disrupts a complex neural network involving many auditory centers as well as brain regions involved with emotion, arousal, memory, and motor planning. These extra-auditory centers embellish the basic auditory percepts that results in tinnitus and which may also contribute to hyperacusis.

Keywords Amygdala · Arousal · Auditory cortex · Auditory nerve · Central gain · Cerebellum · Cochlear nucleus · Emotion · fMRI · Functional connectivity · Inferior colliculus · Medial geniculate body · Reticular formation · Salicylate · Tinnitus · Hyperacusis

Abbreviations

AC	Auditory cortex
ALFF	Amplitude of low-frequency fluctuation
AMY	Amygdala
BOLD	Blood-oxygen-level-dependent
CAP	Compound action potential
CB	Cerebellum
CF	Characteristic frequency
CN	Cochlear nucleus
CPU	Caudate/putamen
FC	Functional connectivity
fMRI	Functional magnetic resonance imaging
HIP	Hippocampus
IC	Inferior colliculus
MEM	Memantine
MGB	Medial geniculate body
RF	Reticular formation

RMP	R Maxipost
ROI	Region of interest
SS	Sodium salicylate
SSC	Somatosensory cortex
VC	Visual cortex

1 Introduction

The phantom sound of tinnitus is often triggered by hearing loss caused by aging or exposure to noise or ototoxic drugs. Approximately 12% of adults have tinnitus, but only 1% have tinnitus severe enough that they seek medical treatment (Baguley et al. 2013; Lockwood et al. 2002). Among combat personnel, the prevalence of tinnitus is especially high (Helfer et al. 2011). This has resulted in enormous tinnitus disability payments to veterans paid for by the United States Veterans Administration (Yankaskas 2013). To add insult to injury, tinnitus is often accompanied by hyperacusis, a condition in which normal everyday sounds are perceived as extremely loud, annoying, or even painful (Baguley 2003; Aazh and Moore 2018; Pienkowski et al. 2014; Tyler et al. 2014). Among patients whose primary concern is hyperacusis, up to 86% also suffer from tinnitus. Among those whose primary complaint is tinnitus, approximately 40% also have tinnitus. The prevalence of hyperacusis is estimated to be around 9%, although the percentage is likely much higher because many individuals are unaware that they have a loudness intolerance problem (Gu et al. 2010). Hyperacusis is associated with many other neurological disorders such as autism, Williams syndrome, fibromyalgia, and chronic pain syndrome (Khalfa et al. 2004; Suhnan et al. 2017; Johnson et al. 2001) suggesting that it may result from dysfunctions in the central nervous system. While most patients with tinnitus and hyperacusis have hearing loss, the majority of those with hearing loss do not suffer from tinnitus or hyperacusis. This suggests that other factors in addition to hearing loss may be necessary to induce tinnitus and/or hyperacusis.

1.1 Consistent Induction of Tinnitus and Hyperacusis

Scientific advances are most often made when the experimental manipulation establishes a strong cause and effect relationship. Clinical observations long ago revealed that patients would develop roaring tinnitus when they ingested high doses of aspirin to treat their rheumatoid arthritis. When they stopped taking the drug, their tinnitus disappeared in a day or 2 (Bernstein and Weiss 1967; Mongan et al. 1973; Myers and Bernstein 1965). These clinical observations provided auditory researchers with a critical clue on how to reliably turn tinnitus on and off in humans in order to investigate perceptual consequences (McFadden and Plattsmier 1983;

Wier et al. 1988; Cazals 2000). A critical step in using aspirin to study the biological basis of tinnitus in animal models is verity that animals treated with high doses of salicylate, the active ingredient in aspirin, would develop tinnitus. This was accomplished by the pioneering studies of Jastreboff who showed that large doses of sodium salicylate would induce tinnitus-like behaviors in rats (Jastreboff et al. 1988). Animal models of hyperacusis initially focused on use of startle reflex amplitude to assess abnormal loudness growth, but more sophisticated and compelling methods were developed by Radziwon using reaction time to assess abnormal loudness growth (Radziwon et al. 2017). Since then, a variety of novel behavioral models have been developed to assess tinnitus and hyperacusis (Hayes et al. 2014).

1.2 Tinnitus Evaluated with Schedule-Induced Polydipsia Avoidance Conditioning (SIPAC)

Polydipsia refers to excessive drinking even when a subject is not thirsty. Polydipsia can be experimentally induced in animals by scheduled delivery of a food pellet to a food-restricted rat at a fixed time interval (e.g., every 30 s) (Falk 1966). Even though the animal is not thirsty, it will nevertheless begin to lick for water while waiting for the next food pellet to be delivered. We then employed the polydipsia phenomenon to develop the SIPAC method to test for tinnitus (Lobarinas et al. 2004). To accomplish this, we put the licking behavior under stimulus control (i.e., the animal responded correctly on most sound trials and quiet trials) by allowing the rat to drink for water only during 30-s quiet intervals interspersed in an ongoing noise; the quiet intervals occurred at random times within the ongoing noise. However, if the rat attempted to lick for water in the presence of the background noise (30-s blocks of noise, roved in level and frequency), foot shock was delivered. Rats were trained to suppress licking in the presence of any types of background noises and to only lick for water during 30-s quiet intervals. After about 4 days of baseline training, the rats licked extensively on 30-s quiet trials, but seldom licked during the 30-s blocks of noise. Figure 1 is a schematic showing the number of licks per daily session. During the first 4 days of baseline testing, roughly 4,000 licks/session occurred during the 30-s quiet trials, whereas very few licks occurred on quiet trials during each daily session (~1.5 h). When the rats were treated with 50 mg/kg or 100 mg/kg of sodium salicylate, the licking behavior remained largely unchanged from baseline. However, when either 150 or 250 mg/kg of salicylate was administered, the number of licks on quiet trials decreased dramatically to levels comparable to sound trials. When the 150- or 250-mg/kg dose of salicylate was discontinued, the number of licks/session on quiet trials gradually recovered to baseline levels, whereas the number of licks on sound trials remained similar to baseline. The most parsimonious explanation for these results is that on the days when either 150 or 250 mg/kg of sodium salicylate was administered, the rats perceived the phantom sound of tinnitus on quiet trials and

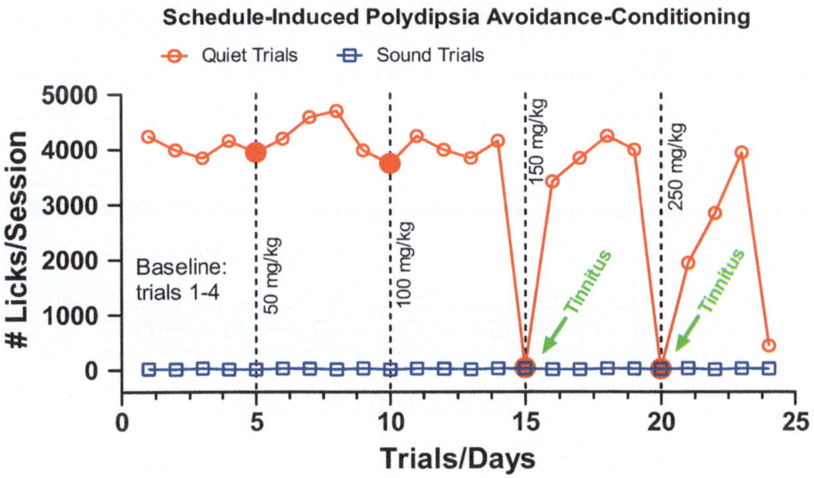

Fig. 1 Schematic of schedule-induced polydipsia avoidance conditioning (SIPAC) paradigm illustrating the # of licks/daily session. Rats trained to avoid foot shock by not licking for water on 30-s intervals in which sounds are presented. Rats allowed to lick for water on quiet trials (sound absent). Vertical lines are days on which sodium salicylate was administered at the dose indicated. Rats seldom licked for water on sound trials, but often licked for water on quiet trials. However licks during quiet trials were almost completely suppressed on the day when 150 or 250 mg/kg of salicylate was administered. Licking rate gradually increased 1–3 days after these high doses of salicylate

therefore suppressed their licking to levels comparable to sound trials. When salicylate was withdrawn, licks in quiet returned to baseline in a day or 2.

1.3 How Much Salicylate Is Required to Induce Tinnitus?

Using the SIPAC paradigm, a dose-response study was conducted to identify the dose of salicylate needed to reliably induce tinnitus (Lobarinas et al. 2004). As schematized in Fig. 2, licks on quiet trials remained equal to baseline (no salicylate, 0 mg/kg) for the 50 and 100 mg/kg doses of salicylate. However, licks on quiet trials declined dramatically for the doses of salicylate from 150 to 350 mg/kg. Licks on sound trials for salicylate doses from 50 to 350 mg/kg were similar to those at baseline, results indicating that the behavior of the rats was under stimulus control. Because the rats stopped licking on quiet trials when the dose of salicylate was 150 mg/kg or higher, it is reasonable to conclude that the rats were hearing the phantom sound of tinnitus which would explain why they stopped licking.

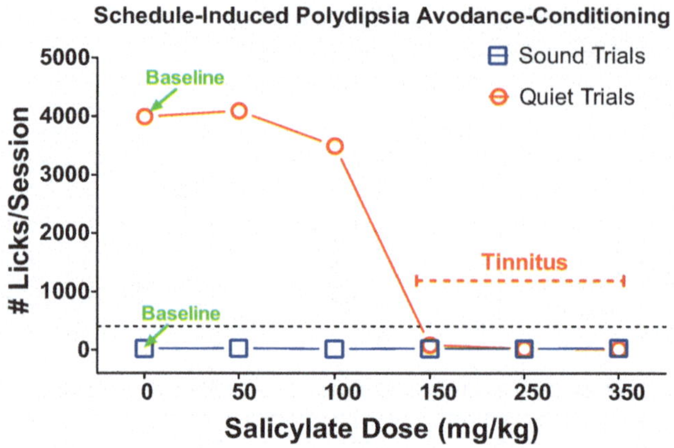

Fig. 2 Schematic of sodium salicylate dose-response function obtained with the schedule-induced polydipsia avoidance conditioning paradigm. Baseline licks/session during sound trials and quiet trials measured in the absence of salicylate (0 mg/kg). Licks on sound trials were low, but were high on quiet trials, behaviors indicating that the rats were able to distinguish quiet trials from sound trials. Licks/session after treatment with 50- and 100-mg/kg salicylate were similar to baseline indicating that the rats were not experiencing tinnitus, but licks on quiet trials decreased significantly when 150, 250, or 350 mg/kg of salicylate was administered, behaviors indicative of tinnitus

1.4 Using SIPAC to Screen Drugs to Suppress Tinnitus

It has been suggested that tinnitus could arise from increased excitatory neurotransmission at various sites along the auditory pathway such as the synapse between the inner hair cell and auditory nerve fiber (Oestreicher et al. 1998; Ruel et al. 2008). Memantine, which suppresses NMDA-mediated glutamatergic neurotransmission between the inner hair cell and auditory nerve, has been suggested as a treatment for tinnitus. Support of this hypothesis comes from a study in which cochlear infusion of an NMDA antagonist suppressed behavioral evidence of salicylate-induced tinnitus (Guitton et al. 2003). Based on these encouraging results with local drug delivery, we hypothesized that memantine, a clinically approved NMDA antagonist, would provide a more practical method to treat tinnitus. To test this hypothesis, we treated a group of rats with 150 mg/kg of sodium salicylate, a dose that significantly (#) reduced the number of licks in quiet compared to baseline licks in quiet (Fig. 3a) (Lobarinas et al. 2006). Afterward, the 150-mg/kg dose of salicylate was administered along with 1.5 mg/kg of memantine. This dose of memantine increased the number of licks in quiet, but the number of licks in quiet was still significantly less than baseline. To determine if a higher dose of memantine would be more effective in alleviating the tinnitus-like behavior, 3 mg/kg of memantine was administered together with the 150-mg/kg dose of salicylate. Instead of increasing the number of licks in quiet, the 3-mg/kg dose of memantine decreased the number of licks, ostensibly making the tinnitus-like behavior worse. These results suggest that systemic administration of memantine can at best only partially suppress tinnitus.

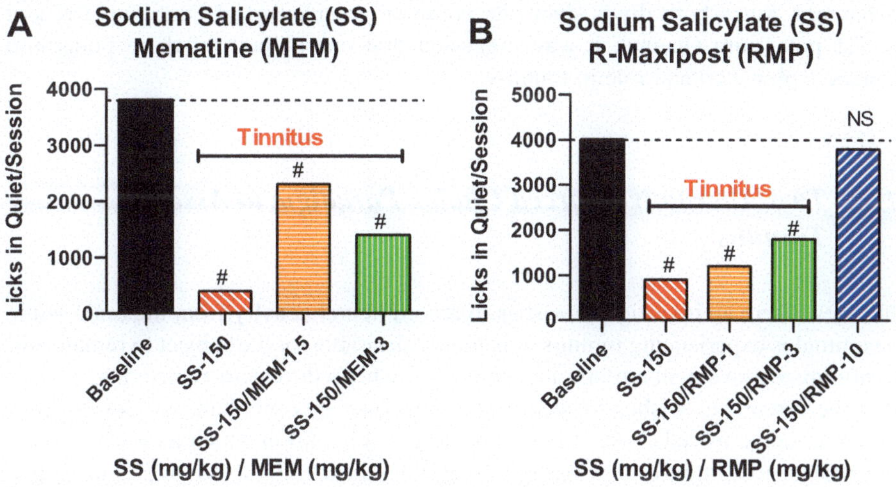

Fig. 3 Use of SIPAC to test drugs to suppress tinnitus. (**a**) Schematic illustrating the effects of memantine (MEM), an NMDA antagonist, on sodium salicylate-induced tinnitus. Results illustrate the number of licks in quiet obtained at baseline and then after treatment with 150-mg/kg sodium salicylate (SS), SS plus 1.5 mg/kg of memantine (MEM), and SS + 3-mg/kg MEM. The 150-mg/kg dose of SS suppressed licks in quite, behaviors indicative of tinnitus. The 1.5- and 3-mg/kg dose of MEM increased the number of licks in quiet, behaviors indicative of partial but incomplete reduction of tinnitus (# = significant difference between baseline and other conditions). (**b**) Schematic illustrating the effect of R-Maxipost (RMP) on salicylate-induced tinnitus. The 150-mg/kg dose of SS suppressed licks in quiet (#). Treatment with 1- and 3-mg/kg RMP increased the number of licks in quiet, but did not completely suppress salicylate-induced behavior. The 10-mg/kg dose of RMP completely suppressed tinnitus-like behavior

Similarly, other investigators have reported that memantine partially suppressed tinnitus-like behavior in two of five rats with noise-induced tinnitus (Zheng et al. 2012). On the other hand, a human clinical trial found that memantine was ineffective in suppressing tinnitus symptoms and that it caused significant side effects in 10% of cases (Figueiredo et al. 2008).

Some clinical studies have reported that anxiolytic drugs were sometimes effective in reducing the perception of tinnitus (Dobie 1999). Maxipost, an experimental drug that modulates potassium channels, was found to be effective in reducing anxiety-like behaviors in rodents (Korsgaard et al. 2005). These clinical and experimental studies, along with other studies showing that Kv7 potassium channels are expressed in the peripheral and central auditory pathway, raised the possibility that Maxipost might be effective in suppressing tinnitus. To test this hypothesis, rats were treated with 150-mg/kg sodium salicylate which reduced licking in quiet, behaviors consistent with tinnitus (Fig. 3b). When rats were treated with salicylate in combination with either Maxipost or R-Maxipost, the drugs dose-dependently increased the number of licks in quiet. The highest dose completely abolished the tinnitus-like behavior (Lobarinas et al. 2011). Interestingly, R-Maxipost, which has no anxiolytic properties, was as effective as Maxipost in suppressing tinnitus-like behaviors. Since both drugs share the common mechanism of inhibiting BK and Kv7.1 potassium channel, it was suggested that one or both of these potassium channels play a critical role in tinnitus.

1.5 Two-Alternative Forced Choice Paradigm to Assess Tinnitus

In cases where an investigator wishes to record neural activity from the brain while an animal is experiencing tinnitus, it is highly desirable for the subject to remain still to minimize movement artifacts that could overwhelm the electrical neural responses that the researcher wishes to record. To overcome this problem, we developed a two-alternative forced choice (2AFC) paradigm (Fig. 4a) in which a rat was trained to insert its nose into a nose-poke hole and hold it there without moving for up to 10 s while listening to the presence or absence of sounds in the environment (amplitude-modulated noise = AM, steady narrow band noise = NBN (various types), or quiet = no stimulus) (Stolzberg et al. 2013). At the end of the hold-still period, a light was turned on to inform the rat to make a decision about the acoustic

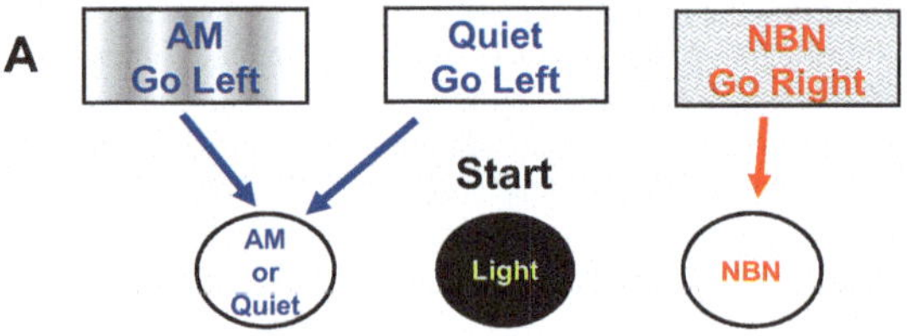

B Normal

Stimulus	% Left	% Right
NBN	7	93
AM	94	6
Quiet	91	9

C Salicylate/Tinnitus

Stimulus	% Left	% Right
NBN	7	93
AM	94	6
Quiet	20	80

Tinnitus

Green = correct; Red = incorrect

Fig. 4 Schematic of positive-reinforcement, two-alternative forced choice paradigm to assess acute tinnitus. (**a**) Apparatus consists of three nose-poke holes. The nose-poke hole in the middle is used to start at trial. When the light comes on, the rat must decide whether to respond to the nose-poke hole on the left or on the right. During training, the nose-poke hole on the left records a correct response and delivers a food reward if the rat places its nose on the left hole on trials when amplitude-modulated (AM) noise or quite is presented. The nose-poke hole on the right records a correct response and delivers a food reward if the rat places its nose on the right hole on trials when a steady narrow band noise (NBN) is presented. (**b**) When a normal rat is trained to criterion, a high percentage (>90%) of correct responses occur on NBN, AM, and quiet trials as schematized in the response matrix. (**c**) A few hours after a rat is treated with a high dose of salicylate sufficient to induce tinnitus (e.g., 200 mg/kg), the rat continues to have a high percentage of correct responses when the NBN and AM sounds are presented. However, if the rat perceives the phantom sound of tinnitus on quiet trials, it incorrectly responds to the right because it perceives the phantom sound of tinnitus instead of quiet environment (Fig. 4a, i.e., AM, NBN, or quiet). If an AM noise was presented during the hold-still period, the rat was trained to withdraw its nose and insert it in the nose-poke hole on the left. A correct response to the left (AM) was rewarded with a food pellet; an incorrect response resulted in a time-out. If NBN or quiet was present during the hold-still period, the rat was trained to withdraw its nose and insert it into the right nose-poke hole. A correct response was rewarded with a food pellet and an incorrect response resulted in a time-out. When a normal hearing rat was fully trained, the rat would correctly identify the NBN trials by going to the right on more than 90% of the trials (Fig. 4b). In contrast, on AM and quiet trials, a well-trained rat would correctly identify these stimuli by poking its nose into the left nose-poke hole (Fig. 4b). When a well-trained rat was treated with a high dose of salicylate sufficient to induce tinnitus, it would continue to correctly identify the NBN or AM noise (Fig. 4c). However, if the rat was experiencing tinnitus on quiet trials, it would mistakenly respond to the nose-poke hole on the right, originally associated with the continuous NBN (see table in Fig. 4c). In this case, the continuous phantom sound of tinnitus was mistakenly categorized as a continuous NBN.

1.6 Noise-Induced Tinnitus Measured with 2AFC Paradigm

Additional experiments were conducted to compare the results from salicylate-induced tinnitus with noise-induced tinnitus. Individuals exposed to intense noise often develop temporary tinnitus immediately post-exposure, but in most cases the tinnitus gradually fades away if the exposure is not too severe (Loeb and Smith 1967; Atherley et al. 1968; Davis et al. 1950). To evaluate the onset and recovery of noise-induced tinnitus, rats were similarly trained with the aforementioned 2AFC paradigm (Fig. 4). The results showed that the percentages of correct responses on NBN, AM, and quiet trials were greater than 90% on pre-exposure days 1–5 (Fig. 5). Afterward, the rats were exposed to a 110-dB SPL narrow band noise centered at 16 kHz for 40 min. Approximately 15 min after the noise exposure, there was large reduction in percent correct responses only on quiet trials. On post-exposure days 1–3, the percent correct on quiet trials gradually recovered to more than 90% correct. During the entire testing period, percent correct performance on NBN and AM trials remained above 90%, indicating that the rat was under stimulus control. These results are consistent with human psychophysical studies in which tinnitus is present shortly after the exposure, but gradually fades away. The amount of time needed for tinnitus to fade away generally increases with the intensity and duration of the noise exposure (Loeb and Smith 1967; Atherley et al. 1968; Davis et al. 1950).

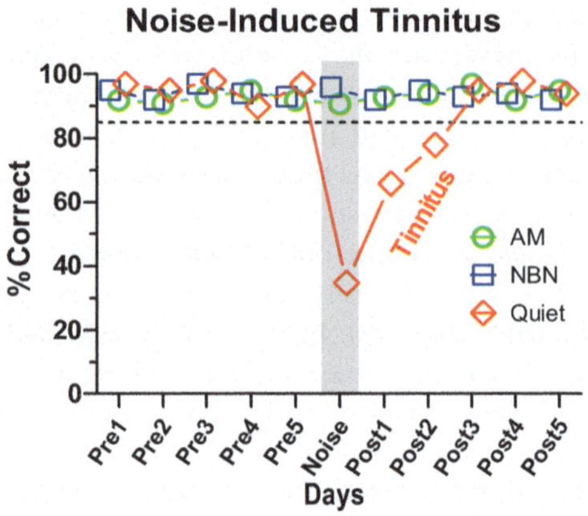

Fig. 5 Schematic showing the percent correct on amplitude-modulated (AM) trials, narrow band noise (NBN) trials, and quiet trials during testing. During pre-exposure days 1–5, percent correct >90% on all three stimulus conditions. On the day of the noise exposure and subsequent days, percent correct on NBN and AM trials >90% correct, indicating the rat's behavior is under stimulus control. However, on quiet trials, a large decrease in percent correct occurred on the day of the exposure and post-exposure days 1–2. On these days, the rat responded to the nose-poke hole associated with the NBN, behaviors suggesting that rat was perceiving the phantom sound of tinnitus

1.7 *Does Salicylate Induce Hyperacusis?*

High doses of salicylate have long been known to cause tinnitus, but it was unclear from human studies if salicylate could induce hyperacusis. Because hyperacusis is often associated with tinnitus, we speculated that salicylate might cause hyperacusis (Chen et al. 2015). Assessing loudness perception in nonverbal animals is difficult, but fortunately, human studies have found that the time it takes for a subject to respond to a sound (i.e., reaction time) decreases with stimulus intensity. Consequently, reaction time has been used to assess loudness in humans (Marshall and Brandt 1980; Seitz and Rakerd 1997; Schlittenlacher et al. 2014). Auditory neuroscientists have also used reaction time-intensity (RT-I) functions to estimate loudness growth in animal models with normal and impaired hearing (Moody 1970, 1973; Pfingst et al. 1975; Stebbins 1966). To determine if high doses of salicylate would induce hyperacusis, we measured RT-I functions in rats before and a few hours after administering various doses of salicylate (Radziwon et al. 2017). Figure 6a is a schematic of an RT-I function obtained pre- and post-salicylate. The baseline RT-I function displayed a gradual decrease in reaction time as noise burst intensity increased from 30 to 90 dB SPL. Following treatment with 100 mg/kg of salicylate, there was no change in the RT-I function (Fig. 6a) indicating that this low dose (Fig. 2) did not cause any change in loudness. However, when the salicylate dose was increased to 250 mg/kg, the RT-I function became steeper than normal. Reaction times from 60 to 90 dB SPL were much shorter than normal indicating that

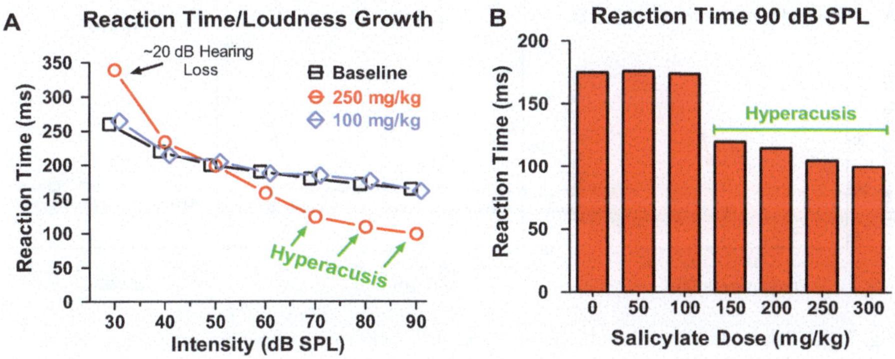

Fig. 6 (**a**) Schematic of reaction time-intensity (RT-I) function measured with noise bursts at baseline and several hours after treatment with 100 mg/kg of salicylate or 250 mg/kg of salicylate. At baseline, note steady decline in reaction time as intensity is increased from 3-0 to 90-dB SPL. RT-I function after treatment with 100 mg/kg of salicylate nearly identical to baseline. A few hours after treatment with 250 mg/kg, reaction times become longer than baseline at low intensities due to a hearing loss of approximately 20 that makes the stimulus less audible. However at intensities above 60-dB SPL, reaction times become shorter than normal, behaviors indicative of hyperacusis. (**b**) Schematic of reaction times measured at 90-dB SPL before (0 mg/gm) and several hours after treatment with 50–300 mg/kg of salicylate. Reaction times after the 50- and 100-mg/kg dose of salicylate similar to baseline, but reaction times become significantly shorter than normal for doses between 150 and 300 mg/kg, behaviors indicative of hyperacusis

they were now perceived as louder than normal, i.e., behavioral evidence of hyperacusis (Fig. 6a). However, at 30 dB SPL, reaction time was longer than normal. The increase was the results of a salicylate-induced hearing loss of approximately 20 dB which made the 30 dB SPL stimulus less audible and not as loud as it normally would be. RT-I functions were measured with different doses of salicylate to determine which doses would induce hyperacusis. The schematic in Fig. 6b shows the reaction times obtained at 90 SPL with different doses of salicylate. The reaction times with the 50- and 100-mg/kg dose were similar to baseline (0 mg/kg) indicating that they did not induce hyperacusis. In contrast, reaction times at 90 dB were much shorter than normal with doses from 150 to 300 mg/kg, indicating that salicylate doses ≥150 mg/kg consistently induced hyperacusis. From these results it appears that salicylate doses that induced hyperacusis were identical to those that caused tinnitus.

1.8 Noise-Induced Hyperacusis

We carried out additional experiments to determine if noise exposure could induce similar changes in RT-I function as high-dose salicylate. Previous animal studies that have used RT-I to assess loudness growth functions in animals with noise-induced hearing loss have observed evidence of loudness recruitment; in these cases, the

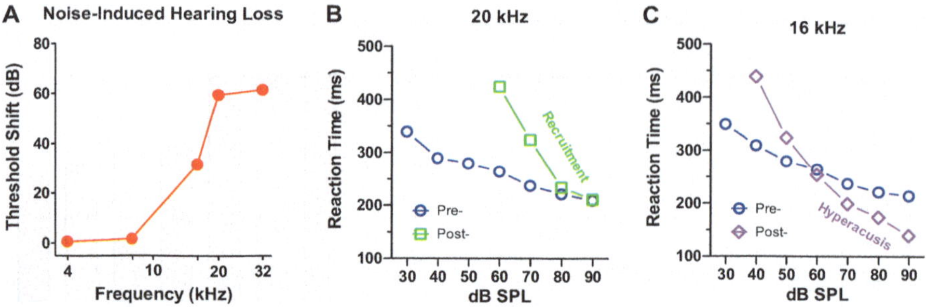

Fig. 7 (**a**) Schematic of noise-induced threshold shift following a prolonged high-frequency noise exposure. Thresholds normal at low frequencies, but threshold elevated approximately 30 dB at 16 kHz and roughly 60 dB at 20 and 24 kHz. (**b**) Schematic of RT-I functions measured at 20 kHz in the hearing loss region. RT-I function with recruitment-like function; reaction times at high intensities similar to baseline. (**c**) RT-I function measured at 16 kHz, at the edge of the hearing loss. Reaction times at high intensities shorter than baseline, behaviors indicative of hyperacusis

slope of the RT-I function was steeper than normal at low and moderate intensities, but the slopes returned to normal at high intensities so that reaction times never became shorter than normal (Moody et al. 1980; Stebbins 1966; Stebbins and Miller 1964). However, the noise exposures used previously were generally of short duration and reaction times were typically measured shortly after the exposure. To determine if long-duration noise exposure might give rise to hyperacusis, we exposed rats to an intense, high-frequency band of noise (16–20 kHz, 104 dB SPL, 12 weeks) and assessed hearing thresholds and RT-I functions several months after the noise exposure (Radziwon et al. 2019). As schematized in Fig. 7a, the noise exposure induced a large 60-dB threshold shift at 20–32 kHz, a moderate 30-dB threshold shift at 16 kHz, and no hearing loss at lower frequencies. As schematized in Fig. 6b, we found evidence of loudness recruitment at 20 kHz, consistent with previous animal studies (Moody et al. 1980; Stebbins 1966; Stebbins and Miller 1964). However, at 16 kHz, near the edge of the hearing loss, we found striking evidence of hyperacusis from 70 to 90 dB SPL where reaction times were shorter than normal. We also found evidence of hyperacusis at much lower frequencies where hearing thresholds were normal (Radziwon et al. 2019). These results demonstrated for the first time that it was possible to induce noise-induced hyperacusis by using an intense, long-duration noise and evaluating changes in loudness growth ~3 months post-exposure.

1.9 Noise-Induced Avoidance Hyperacusis

So far, we have described methods for assessing transient hyperacusis induced by salicylate and permanent hyperacusis induced by intense noise exposure. However, a recent review has suggested that humans experience other loudness intolerance

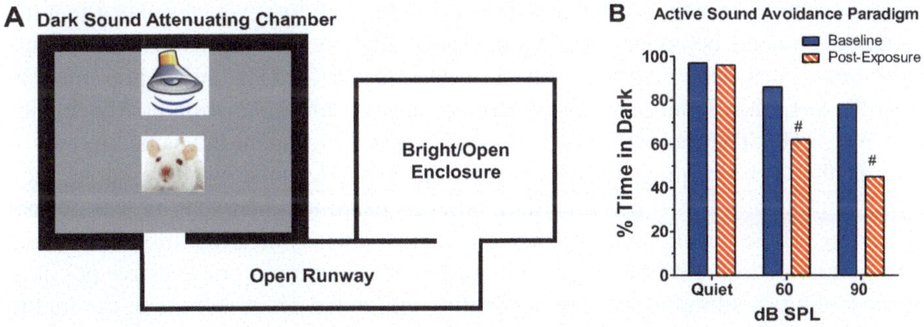

Fig. 8 Schematic of active sound avoidance paradigm (ASAP) used to test for annoying or aversive sounds. (**a**) Schematic of test apparatus consisting of a dark sound-attenuating chamber equipped with a loudspeaker. Dark chamber connected by an open runway to a bright open enclosure. Rats naturally prefer to stay in the dark sound-attenuating enclosure, but when noise is presented through the loudspeaker, rats move from the dark chamber to the bright open enclosure. (**b**) Schematic illustrating the percent time rats spend in the dark enclosure under quiet conditions and when a 2–8-kHz noise is presented at 60- or 90-dB SPL. Results presented under normal baseline conditions and after rats developed a significant noise-induced hearing loss above 12 kHz. Under baseline conditions, rats spent more than 90% of the time in the dark enclosure, but percent time in the dark declined to approximately 80% when the intensity of the noise was 90-dB SPL. In rats with high-frequency hearing loss, the percent time spent in the dark chamber under quiet conditions remained above 90%. However as the intensity of the noise increased, the rats with hearing loss spent significantly less time in the dark chamber (#) compared to baseline. These results suggest that rats with hearing loss above 12 kHz perceived the 2–8-kHz noise as more annoying or more aversive than when hearing was normal

problems such as annoyance, fear, and pain hyperacusis that likely activate parts of the limbic and pain pathways (Tyler et al. 2014; Pienkowski et al. 2014). In an attempt to address the non-auditory aspects of annoyance and/or fear hyperacusis in an animal model, we developed the active sound avoidance paradigm (ASAP) (Fig. 8a). The ASAP testing apparatus consists of a dark, sound-attenuating enclosure equipped with a loudspeaker. The dark box is connected to a clear open runway leading to a bright open enclosure (Manohar et al. 2017). When a rat is placed in the apparatus, it exhibits a natural preference to spend most of its time in the dark enclosure and to avoid the bright open enclosure, presumably an innate response to avoid predators. Rats typically spend more than 90% of the time in the dark enclosure. To determine if loud sounds would evoke annoyance or fear, a 2–8-kHz band of noise was presented through the loudspeaker at different intensities. If the noise provoked annoyance or fear, the rat could escape the noxious noise by moving into the bright open enclosure. In quiet, normal rats typically spend more than 90% of the time in the dark box, but when the 2–8-kHz noise was presented at a very moderate intensity of 60-dB SPL, rat spends less time (~80%) in the dark box (Fig. 8b, solid blue bar). When the noise intensity was increased to 90-dB SPL, time spent in the dark box dropped to ~70%. This sound avoidance behavior suggested that as the intensity of the noise increased, it became more annoying, noxious, or fear-provoking.

Because hyperacusis is often associated with hearing loss, we speculated that the sound avoidance behaviors might be exacerbated by hearing loss. To test this hypothesis, rats were exposed for 4 weeks to 16–20-kHz noise presented at 104-dB SPL that induced a 30–40-dB permanent threshold shift at frequencies above 12 kHz. The rats with noise-induced hearing loss spent the same amount of time in the dark on quiet trials, but when the 2–8-kHz noise was presented, they spent significantly less time in the dark box compared to when hearing was normal (Fig. 8b). These results suggest that the high-frequency hearing loss caused the low-frequency test stimulus to be perceived as more annoying or aversive possibly because the test stimulus was now activating emotional brain circuits in the limbic system. Thus, ASAP could provide a method for assessing the non-auditory emotional aspects of hyperacusis.

1.10 Salicylate-Induced Hyperexcitability

High doses of salicylate have long been known to cause temporary cochlear hearing loss. However, salicylate readily crosses the blood-brain barrier allowing the drug to potentially disrupt brain activity (Boettcher et al. 1990; Su et al. 2012). To more fully characterize the peripheral and central effects of salicylate, we measured the local field potentials and/or spike discharge from multiunit clusters in the cochlea and central nervous system before and after administering a high dose of salicylate. The schematic input/output functions in Fig. 9 provide a comprehensive summary of the main effects of high-dose salicylate on the cochlea and central auditory pathway (Chen et al. 2012, 2013a, 2014, 2015, 2017a; Jiang et al. 2017). In these schematics, the ordinate is expressed as percent normalized amplitude. Normalization was carried out by expressing all pre- and post-exposure values relative to the pre-exposure amplitude obtained at 90-dB SPL. Hence, all the pre-exposure input/output functions reach 100% at 90-dB SPL. The effects of salicylate on the amplitude of the cochlear compound action potential (CAP), which reflects the gross neural output of the auditory nerve (AN), are illustrated in Fig. 9a. In this example, the pretreatment CAP threshold was approximately 10-dB SPL. The normalized CAP amplitude gradually increased with intensity reaching 100% at 90-dB SPL. A few hours after high-dose salicylate treatment, the input/output function was shifted rightward resulting in threshold shift of approximately 20 dB. In addition, suprathreshold amplitudes were greatly reduced. At 90-dB SPL, the amplitude was reduced by 50%. The threshold shifts and amplitude reductions occurred over a broad range of frequencies unlike other ototoxic drugs that preferentially damage the high frequencies.

The salicylate-induced decrease in the neural output of cochlea is relayed from the auditory nerve to the cochlear nucleus. Fig. 9b illustrates the main effects of high-dose salicylate on neural activity in the cochlear nucleus (CN) (i.e., local field potentials or multiunit spike discharge rate). The CN input/output function, measured a few hours post-salicylate, was shifted to the right, resulting in an increase in

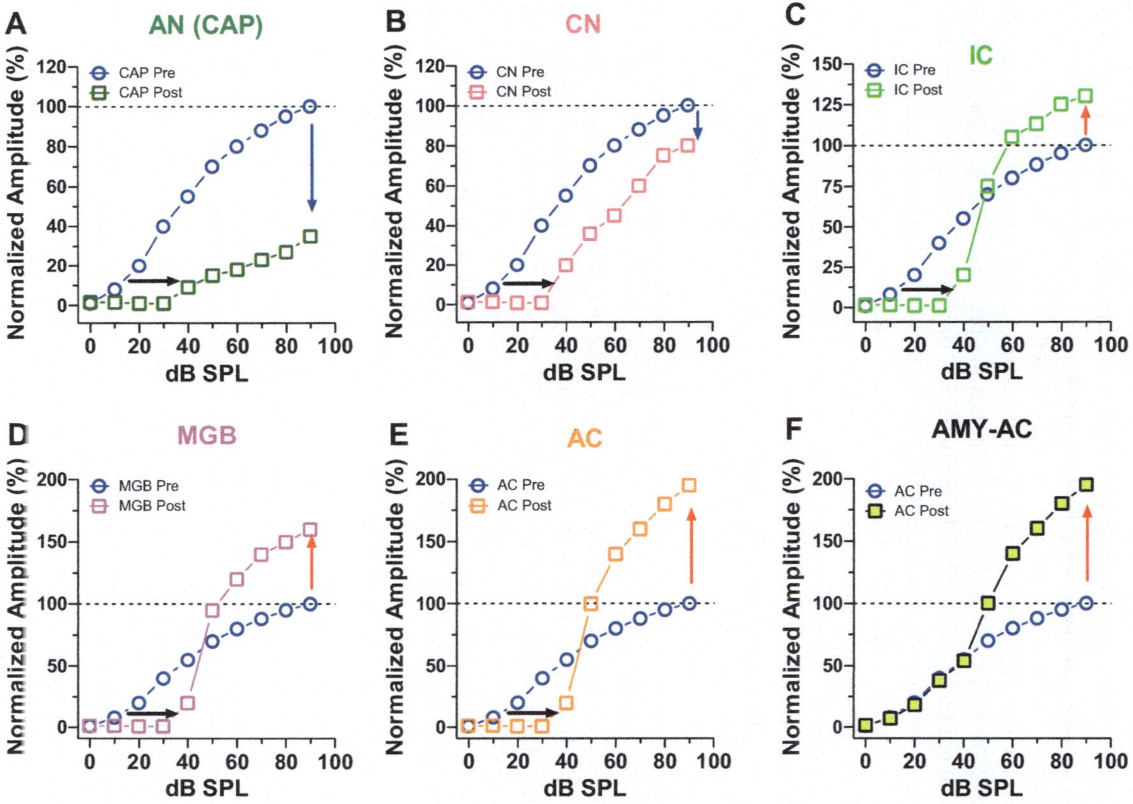

Fig. 9 Schematic illustrating the changes in input/output functions from the different levels of the auditory pathway following a high dose of sodium salicylate. Normalized amplitude expressed as % of maximum pre-exposure amplitude at 90-dB SPL. (**a**) Auditory nerve compound action potential (AN-CAP) input/

the CN threshold (i.e., threshold elevation) of approximately 20 dB. However, suprathreshold response amplitudes were reduced much less for the CN compared to the CAP for the cochlea. The CN and CAP threshold shifts were nearly identical suggesting that the CN threshold shift was inherited from the cochlea. However, the CN amplitudes were larger than the CAP, suggesting that the weaker signals inherited from the cochlea were amplified after reaching the CN, partially compensating for the reduced neural output of the cochlea.

The neural output from the CN is relayed both directly and indirectly to the inferior colliculus (IC), an important binaural processing center in the midbrain (Cant 1992; Webster 1992). Fig. 9c illustrates the principal changes that occur in IC input/output function (i.e., local field potentials or multiunit spike discharge rats) a few hours after salicylate treatment. The post-salicylate input/output function was again shifted to the right, resulting in an upward threshold shift of approximately 20 dB. These results suggest that the IC threshold shift was also inherited from the cochlea. However, IC responses rapidly increased with intensity so that post-salicylate responses are equal to or larger than normal at intensities above 60-dB SPL (Fig. 9c). The neural gain provided by the CN was amplified further by the IC resulting in suprathreshold IC responses that were larger than normal at high intensities.

The neural output of the IC is passed on to the medial geniculate body (MGB) where further neural processing takes place after which the output of the MGB is relayed to the auditory cortex (AC). The schematic input/output functions for the MGB (Fig. 9d) and AC (Fig. 9e) were both shifted rightward resulting in threshold shifts of approximately 20 dB indicating that these hearing losses were inherited from the cochlea. On the other hand, the suprathreshold amplitudes in the MGB (Fig. 9d) were greater than those in the IC (Fig. 9c), while those in the AC (Fig. 9e) were even larger than those in the MGB (Fig. 9d). Collectively, these results provide compelling evidence of multiple stages of central gain enhancement to boost the

---

Fig. 9 (continued) output function. Post-exposure input/output function shows a 20-dB threshold shift (black horizontal arrow) and large decrease in suprathreshold amplitude. (**b**) Post-exposure input/output function from the cochlear nucleus shows a 20-dB threshold shift (black horizontal arrow) and a modest amplitude reduction at suprathreshold intensities. (**c**) Post-exposure input/output function from the inferior colliculus (IC) displays a 20-dB threshold shift, but suprathreshold amplitudes are enhanced by approximately 20%. (**d**) Post-exposure input/output function from medial geniculate body (MGB) displays a 20-dB threshold shift, but suprathreshold amplitudes are enhanced by approximately 50%. (**e**) Post-exposure input/output function from auditory cortex (AC) displays a 20-dB threshold shift, but suprathreshold amplitudes are enhanced by approximately 100%. (**f**) Pre- and post-exposure input/output functions recorded from AC after sodium salicylate was infused into the amygdala (AMY). Post-exposure input/output function from the AC did not show a threshold shift after local drug injection into AMY. However, local application of salicylate into the AMY caused an increase in suprathreshold amplitude of approximately 100%

weak neural signal output originating in the cochlea (Qiu et al. 2000; Auerbach et al. 2014; Jiang et al. 2017; Sun et al. 2009; Salvi et al. 1990, 2000).

The auditory pathway does not process sounds in isolation, but instead sends and receives information from other so-called non-auditory areas such as the amygdala (AMY) (Doron and Ledoux 1999; Romanski et al. 1993; Marsh et al. 2002). The AMY can assign negative or positive emotional valence to sounds such as a baby crying or cooing (Brydges et al. 2013; Sander et al. 2003). Because salicylate readily crosses the blood-brain barrier, it could potentially alter sound-evoked activity in the AMY. To test this hypothesis, recordings (i.e., local field potentials or multiunit spike discharges) were made from the lateral AMY before and after systemic administration of salicylate. Salicylate enhanced sound-evoked responses in the lateral nucleus of the AMY and induced a threshold shift of approximately 20 dB, changes that are similar to those seen in the AC (Chen et al. 2014). To determine how the salicylate-induced changes in the lateral AMY affected other parts of the auditory pathway, we infused salicylate directly into the AMY while recording sound-evoked responses (local field potentials or multiunit spike discharges) from the AC pre- and post-salicylate (Chen et al. 2013b). As illustrated in Fig. 9f, suprathreshold amplitudes in the AC increased significantly after salicylate was applied locally to the AMY. However, unlike systemic treatment, local drug delivery failed to cause a threshold shift (i.e., pre- and post-salicylate response amplitudes at 10-dB SPL equal). These results confirm that the threshold shifts arising from systemic salicylate treatment originate in the cochlea. In contrast, the enhanced sound-evoked response seen in the AC presumably arises from salicylate-induced changes in the AMY which are relayed to the AC. A key take-home message from these results is that systemic salicylate likely had direct effect on the central nervous system, not just the cochlea, and that these contributions can be parceled out by local drug administration.

1.11 Salicylate-Induced Upshifts and Downshifts in Neural Tuning

Much of the tonotopic organization of the auditory system is determined by the mechanoelectrical properties of the cochlea. However, there is increasing evidence that tonotopy in higher levels of the central auditory pathway can be modified by cochlear damage due to neuroplastic changes occurring in high auditory centers (Robertson and Johnstone 1979; Kamke et al. 2003; Irvine and Rajan 1994). Neural tuning in higher auditory structures is not only determined by incoming excitatory signaling but also by complex inhibitory networks capable of retuning the system (Wang et al. 2000; Milbrandt et al. 2000; Tennigkeit et al. 1998; Park and Pollak 1993). High doses of salicylate could potentially disrupt inhibitory neurotransmission, thereby altering the tuning of neurons in the central auditory pathway (Butt et al. 2016; Patel and Zhang 2014; Su et al. 2009; Wang et al. 2008; Lu et al. 2009;

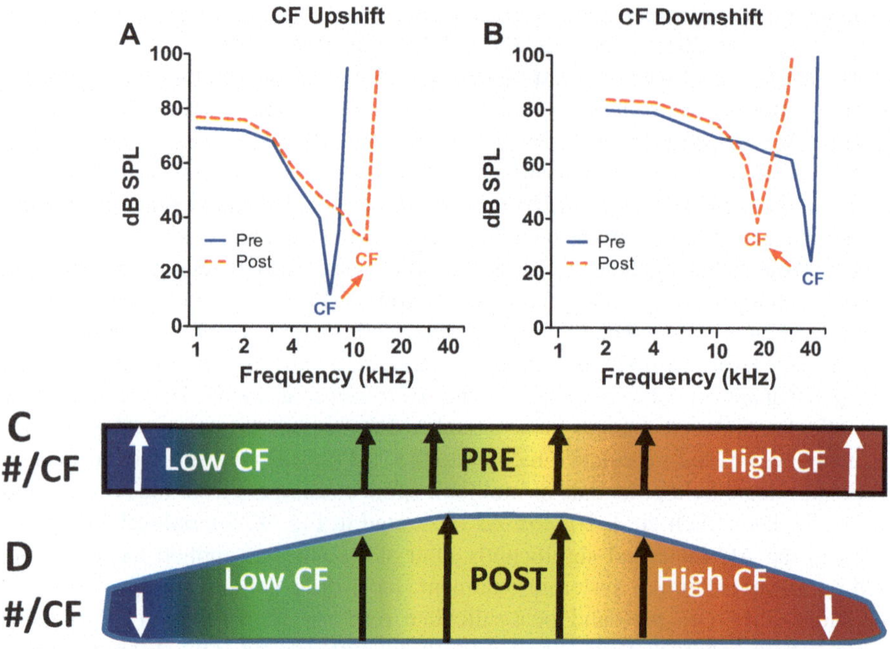

Fig. 10 Schematic illustrating the subpopulation of neurons in the IC, MGB, and AC that undergo a CF shift after high dose of salicylate. (**a**) Schematic illustrating the threshold-frequency tuning curve of a neuron with a low characteristic frequency (CF) pre-salicylate and that shows an upshift in its CF post-salicylate and an increase in threshold. (**b**) Schematic illustrating the threshold-frequency tuning curve of a neuron with a high CF pre-salicylate and that shows an upshift in its CF post-salicylate and an increase in threshold. (**c**) Hypothetical model showing roughly an equal # of neurons per CF (equal length up arrows) across the CF tonotopic map. (**d**) Hypothetical model showing an increase in the # of neurons per CF in the mid-frequency region (up arrow) and a decrease in the # of neurons per CF at the low- and high-CF regions (down arrows) following a high dose of salicylate. The increase in the number of mid-CF neurons on the CF tonotopic map results from a salicylate-induced CF upshift in a subpopulation of low-CF neurons (see panel **a**) and a CF downshift (see panel **b**) in a subpopulation of high-CF neurons

Zugaib et al. 2016). When we recorded from neurons in AC, MGB, and IC pre- and post-salicylate, we unexpectedly discovered dramatic changes in neural tuning curves that resulted in large upshifts and downshifts in the characteristic frequency (CF) of a subpopulation of neurons (Fig. 10a) (Jiang et al. 2017; Chen et al. 2013b; Stolzberg et al. 2011). In some low-CF neurons (CF < 12 kHz), salicylate caused neural thresholds around the original CF to increase and for threshold sensitivity to improve along the high-frequency edge of the tuning curve, thus resulting in an upshift in CF (Fig. 10a) toward 16 kHz. On the other hand, very high-CF neurons often showed increase in thresholds along the high-frequency edge of the original tuning curve combined with an improvement in threshold at lower frequencies that resulted in a downshift in CF toward 16 kHz (Fig. 10b). CF upshifts and downshifts were never observed among neurons with CFs near 16 kHz. Salicylate-induced CF upshifts and downshifts were most commonly observed in the AC; the magnitude of these CF upshifts and downshifts as well as the proportion of neurons that displayed such changes declined from AC to IC.

One consequence of these CF upshifts and CF downshifts was that it altered the distribution of CF along the tonotopic map as illustrated in Fig. 10c, d (note: ordinate arbitrarily scaled as number of neurons per CF (#/CF); abscissa scaled from low to high CF). In normal subject, the same numbers of #/CF are assigned to each CF location along the tonotopic map (Fig. 10c; equal length up-pointing arrows). However, the CF upshifts and downshifts that occur after salicylate treatment decreases (down arrows) the number of neurons in the low-CF and high-CF regions, whereas the number of neurons increases (up arrows) in the mid-frequency region (Fig. 10d). Because salicylate increases the number of neurons in the mid-frequency region (longer up arrows in mid-frequency region of Fig. 10d vs. Fig. 10c), the gain enhancement (total gain = # neurons per CF x increased response amplitude at suprathreshold levels as schematized in Fig. 10e) in the mid-frequency regions is greater than lower and higher frequencies (Jiang et al. 2017). The frequency-dependent enhancement of sound-evoked activity raises the possibility that salicylate-induced hyperacusis might be more prominent in the mid-frequency region.

1.12 Brain Imaging of Salicylate-Induced Tinnitus and Hyperacusis

Because salicylate crosses the blood-brain barrier, it could affect many different brain regions that would be nearly impossible to assess using conventional electrophysiological techniques. To surveil the entire brain, a resting state functional magnetic imaging (fMRI) study was carried out in rats to determine how spontaneous brain activity changed when a high dose of salicylate was administered (Chen et al. 2015). Spontaneous brain activity was assessed by measuring the fluctuations in the bold-oxygen-level-dependent (BOLD) fMRI responses before and after salicylate. Spontaneous brain activity was evaluated using the amplitude of low-frequency fluctuations (ALFF) in the BOLD signal. With the appropriate analysis, we identified the regions of the rat brain where the ALFF signal increased (red) or decreased (blue) significantly (Fig. 11). Significant increases in the spontaneous ALFF signal were seen in three regions in the classical auditory pathway, the AC, MGB, and IC, as well as the AMY which assigns emotional significance to auditory stimuli. Significant increases were also seen in the adjacent somatosensory cortex (SSC) and visual cortex (VC), which form multisensory connections with the AC (Wallace et al. 1993; Foxe et al. 2000; Kayser et al. 2007). Increases in ALFF also occurred in the superior colliculus (SC), which integrates auditory, visual, and somatosensory information (Skaliora et al. 2004; Meredith et al. 1992; Meredith and Stein 1986), and in the reticular formation (RF), an arousal center that responds

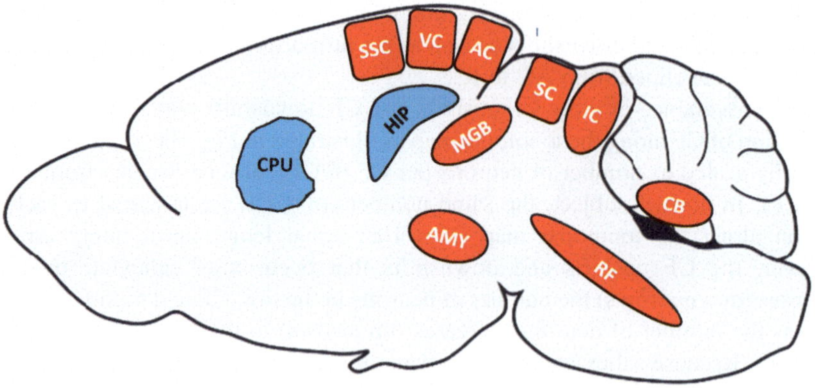

Fig. 11 Schematic showing the regions of the brain in which a high dose of sodium salicylate caused a significant increase (red) or decrease (blue) in the amplitude of low-frequency fluctuations (ALFF) of the blood-oxygen-level-dependent (BOLD) fMRI response. *CPU* caudate/putamen, *HIP* hippocampus, *SSC* somatosensory cortex, VC visual cortex, *AC* auditory cortex, *MGB* medial geniculate body, *SC* superior colliculus, *IC* inferior colliculus, *CB* cerebellum, *RF* reticular formation, *AMY* amygdala

robustly to acoustic stimulation and is an integral part of the acoustic startle reflex circuit (Lingenhohl and Friauf 1994; Paus 2000; Wu et al. 1988). Unexpectedly, ALFF activity increased in the parafloccular lobe and lobule 4 of the cerebellum, regions known respond to acoustic stimuli in humans and/or animals (Lockwood et al. 1999; Azizi et al. 1985; Azizi and Woodward 1990) and which have been implicated in tinnitus generation (Bauer et al. 2013). Decreased ALFF activity occurred in the hippocampus (HIP) and caudate/putamen (CPU), regions implicated in tinnitus (Lockwood et al. 1998; Chen et al. 2014, 2017b; Mirz et al. 2000; Cheung and Larson 2010; Ueyama et al. 2013).

ALFF imaging identified novel regions of salicylate-induced aberrant activity in the parafloccular lobe of the cerebellum and the caudal pontine reticular nucleus. To provide additional verification that salicylate was indeed affecting these brain regions, we conducted additional electrophysiological studies in which we measured the local field potential or multiunit spike discharges from these regions pre- and post-salicylate. We observed short-latency sound-evoked responses in the reticular nucleus and parafloccular lobe (Chen et al. 2017a); the long-latency component of the responses from the reticular nucleus and parafloccular lobe were significantly enhanced by high-dose salicylate, results confirming that salicylate significantly alters activity in these regions. Interestingly, corticosterone stress hormone levels were significantly increased by high-dose salicylate suggesting that stress may be a critical component in salicylate-induced tinnitus and hyperacusis (Chen et al. 2017a).

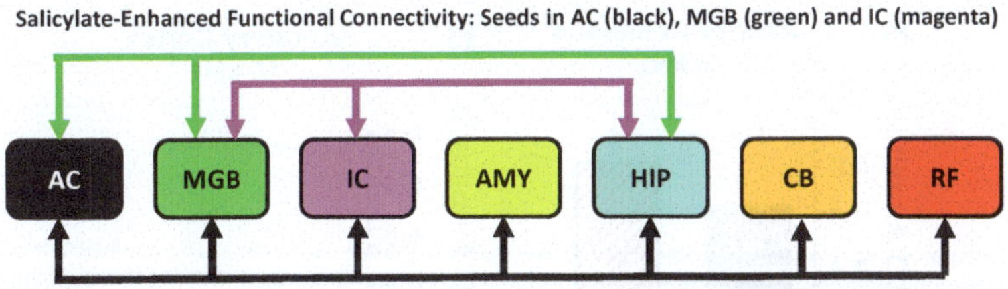

Fig. 12 Schematic illustrating resting state enhanced functional connectivity (FC) using the BOLD fMRI signal. With the seed region of interest (ROI) in the auditory cortex (AC, black line), increased FC occurred with the medial geniculate body (MGB), inferior colliculus (IC), amygdala (AMY), hippocampus (HIP), cerebellum (CB), and reticular formation (RF). With the seed ROI in the MGB (green line), increased FC occurred with the AC and HIP. With the seed ROI in the IC (magenta line), increased FC occurred with the MGB and HIP

1.13 Salicylate-Induced Increases in Functional Connectivity

Temporal fluctuations in the resting state BOLD response can be used to assess the functional connectivity (FC) between one region of interest (ROI) in the brain and other areas. FC is assessed by measuring the temporal correlation of individual time points in the BOLD signal from one ROI (e.g., AC) to other brain regions. FC can be used to identify those regions where there is a high degree of temporally correlated activity with a particular ROI (Friston 2011). FC was used to identify salicylate-induced changes in temporally correlated activity with three seed auditory ROIs, the AC, MGB, and IC, respectively (Chen et al. 2015). Using the AC as the seed ROI, high-dose salicylate increased FC between the AC and eight other brain regions as schematized by the black lines in Fig. 12. Increased FC was observed between the AC and two other auditory regions, the MGB and the IC. FC was also increased between the AMY, HIP, the paraflocculus, and lobule 4 of the CB and portions of the RF. With the seed ROI in the MGB, increased FC occurred between the MGB and the AC and the HIP. With the seed ROI in the IC, increased FC occurred between the IC and the HIP and the MGB. The increased FC connectivity between auditory regions and the AMY could add emotional valence to auditory percepts. The increased FC with the RF could enhance the arousal or attention component to a phantom sound, whereas activity in the HIP could evoke sound memories or assign a spatial location to a phantom sound. Increased FC between auditory regions and the CB could contribute to motor planning related to an auditory percept.

1.14 Sound-Evoked Hyperactivity Measured by fMRI

Task-based fMRI can be used to identify where changes in neural activity occur in the brain by comparing the magnitude of the BOLD signal at rest and during the presentation of an acoustic signal (Lockwood et al. 1999; Lau et al. 2015). To

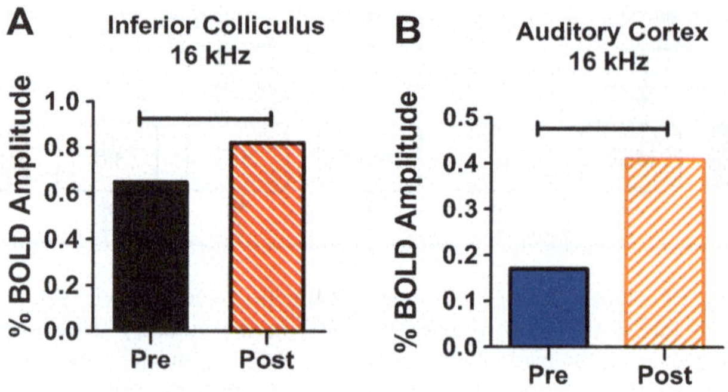

Fig. 13 Schematic illustrating the change in the 16-kHz sound-evoked BOLD fMRI signal pre- and post-salicylate. (**a**) % change in BOLD amplitude in the inferior colliculus was significantly greater after a high dose of salicylate than pre-salicylates. (**b**) % change in BOLD amplitude in the auditory cortex was significantly greater after a high dose of salicylate than pre-salicylates. Note that the magnitude of the increase was much greater in auditory cortex than inferior colliculus

determine if fMRI could be used to identify regions of salicylate-induced hyperactivity in the rat brain, we compared the BOLD signal obtained in the resting state versus data obtained during tone burst (20 s) stimulation. These measurements were obtained pre- and post-salicylate (Wong et al. 2020). Figure 13 illustrates the magnitude of the tone-evoked BOLD signal change (% difference between tone burst stimulation and resting state) pre- and post-salicylate. The 16-kHz tone caused a moderate increase in the pre-salicylate BOLD response in the IC (Fig. 13a). After the salicylate treatment, there was a small, but significant increase in the 16-kHz tone-evoked response compared to pretreatment. In the AC, the 16-kHz tone caused a small increase in the BOLD signal in the pretreatment period. After administering a high-dose of salicylate, the 16-kHz tone evoked a much larger BOLD response compared to pretreatment (Fig. 13b). The magnitude of the salicylate-induced changes in the tone-evoked BOLD response from the IC and AC mirrors those seen with electrophysiological measures (Fig. 9c, e). These sound-evoked imaging results reinforce many previous electrophysiological studies showing that the central auditory pathway amplifies the weak neural signals leaving a salicylate-damaged cochlea. The enhanced central gain seen in the AC could be a neural signature for hyperacusis. Future human studies that combine electrophysiology and psychophysical estimates of loudness hyperacusis could be conducted to test this model.

1.15 Synopsis

Clinical studies of rheumatoid arthritis led to the discovery that high doses of aspirin/salicylate could consistently induce tinnitus. Because of the strong cause and effect relationship, high-dose salicylate has become one of the most widely used tools to develop behavioral models of tinnitus and hyperacusis and to investigate the functional anatomical substrates for tinnitus and hyperacusis. From the behavioral and neurophysiological results presented here, it is possible to derive a conceptual understanding of the neurophysiological changes associated with salicylate-induced tinnitus and hyperacusis. From the resting state ALFF and FC data, it is clear that aberrant changes in spontaneous neural activity occur in a complex network that includes not only important regions in the central auditory pathway (AC, MGB, and IC) but also brain regions involved with multisensory integration (SSC, SC, and VC), arousal (RF), memory and spatial awareness (HIP), motor planning and control (CB), and emotion (AMY). The salicylate-induced changes in these regions not only account for auditory aspects of tinnitus but also the perception of its spatial location, emotional components, arousal, attention, and integration of auditory percepts with outer sensory systems. The hearing loss associated with salicylate, like other forms of sensorineural hearing loss, greatly reduced the neural output of the cochlea. If these weak neural signals were simply relayed up to the AC, the acoustic signals would likely be perceived as muffled. Fortunately, the weak signals leaving the cochlea are progressively amplified as they make their way up to the AC. This serial amplification process, often referred to as enhanced central gain, is the neural equivalent of a hearing aid. When the gain is adjusted properly, patients perceive suprathreshold sounds at the proper loudness, manifested clinically as loudness recruitment. However, if the gain is excessive, suprathreshold sounds are perceived much too loud, resulting in loudness intolerance disorders. The fear, annoyance, arousal, and emotional reaction to tinnitus and hyperacusis likely arise as information from the auditory pathway is relayed to other brain regions, most of which form reciprocal connections with the auditory pathway. While there is considerable evidence for enhanced central gain following cochlear damage, it remains to be seen whether it fully accounts for tinnitus or hyperacusis or whether it is an unrelated epiphenomenon (Sedley 2019).

References

Aazh H, Moore BCJ (2018) Prevalence and characteristics of patients with severe Hyperacusis among patients seen in a tinnitus and Hyperacusis clinic. J Am Acad Audiol 29(7):626–633. https://doi.org/10.3766/jaaa.17015

Atherley GR, Hempstock TI, Noble WG (1968) Study of tinnitus induced temporarily by noise. J Acoust Soc Am 44(6):1503–1506. https://doi.org/10.1121/1.1911288

Auerbach BD, Rodrigues PV, Salvi RJ (2014) Central gain control in tinnitus and hyperacusis. Front Neurol 5:206. https://doi.org/10.3389/fneur.2014.00206

Azizi SA, Woodward DJ (1990) Interactions of visual and auditory mossy fiber inputs in the paraflocculus of the rat: a gating action of multimodal inputs. Brain Res 533(2):255–262

Azizi SA, Burne RA, Woodward DJ (1985) The auditory corticopontocerebellar projection in the rat: inputs to the paraflocculus and midvermis. An anatomical and physiological study. Exp Brain Res 59(1):36–49. https://doi.org/10.1007/bf00237663

Baguley DM (2003) Hyperacusis. J R Soc Med 96(12):582–585. https://doi.org/10.1258/jrsm.96.12.582

Baguley D, McFerran D, Hall D (2013) Tinnitus. Lancet 382(9904):1600–1607. https://doi.org/10.1016/S0140-6736(13)60142-7
Bauer CA, Kurt W, Sybert LT, Brozoski TJ (2013) The cerebellum as a novel tinnitus generator. Hear Res 295:130–139. https://doi.org/10.1016/j.heares.2012.03.009
Bernstein JM, Weiss AD (1967) Further observations on salicylate ototoxicity. J Laryngol Otol 81 (8):915–925
Boettcher FA, Bancroft BR, Salvi RJ (1990) Concentration of salicylate in serum and perilymph of the chinchilla. Arch Otolaryngol Head Neck Surg 116(6):681–684
Brydges NM, Whalley HC, Jansen MA, Merrifield GD, Wood ER, Lawrie SM, Wynne SM, Day M, Fleetwood-Walker S, Steele D, Marshall I, Hall J, Holmes MC (2013) Imaging conditioned fear circuitry using awake rodent fMRI. PLoS One 8(1):e54197. https://doi.org/10.1371/journal.pone.0054197
Butt S, Ashraf F, Porter LA, Zhang H (2016) Sodium salicylate reduces the level of GABAB receptors in the rat's inferior colliculus. Neuroscience 316:41–52. https://doi.org/10.1016/j.neuroscience.2015.12.021
Cant NB (1992) Cochlear nuclei--cell types and connectivity. In: Popper A, Fay R (eds) The mammalian auditory pathway: neuroanatomy, Springer handbook of auditory research, vol 1. Springer, New York
Cazals Y (2000) Auditory sensori-neural alterations induced by salicylate. Prog Neurobiol 62 (6):583–631. https://doi.org/10.1016/s0301-0082(00)00027-7
Chen GD, Manohar S, Salvi R (2012) Amygdala hyperactivity and tonotopic shift after salicylate exposure. Brain Res 1485:63–76. https://doi.org/10.1016/j.brainres.2012.03.016
Chen GD, Sheppard S, Manohar S, Salvi R (2013a) Neural alterations in outside the classical auditory pathway after treatment with tinnitus inducers: noise and salicylate. In: Seventh international TRI tinnitus conference: tinnitus: a treatable disease, Valencia
Chen GD, Stolzberg D, Lobarinas E, Sun W, Ding D, Salvi R (2013b) Salicylate-induced cochlear impairments, cortical hyperactivity and re-tuning, and tinnitus. Hear Res 295(1–2):100–113. https://doi.org/10.1016/j.heares.2012.11.016
Chen GD, Radziwon KE, Kashanian N, Manohar S, Salvi R (2014) Salicylate-induced auditory perceptual disorders and plastic changes in nonclassical auditory centers in rats. Neural Plast 2014:658741. https://doi.org/10.1155/2014/658741
Chen YC, Li X, Liu L, Wang J, Lu CQ, Yang M, Jiao Y, Zang FC, Radziwon K, Chen GD, Sun W, Krishnan Muthaiah VP, Salvi R, Teng GJ (2015) Tinnitus and hyperacusis involve hyperactivity and enhanced connectivity in auditory-limbic-arousal-cerebellar network. elife 4:e06576. https://doi.org/10.7554/eLife.06576
Chen YC, Chen GD, Auerbach BD, Manohar S, Radziwon K, Salvi R (2017a) Tinnitus and hyperacusis: contributions of paraflocculus, reticular formation and stress. Hear Res 349:208–222. https://doi.org/10.1016/j.heares.2017.03.005
Chen YC, Xia W, Chen H, Feng Y, Xu JJ, Gu JP, Salvi R, Yin X (2017b) Tinnitus distress is linked to enhanced resting-state functional connectivity from the limbic system to the auditory cortex. Hum Brain Mapp 38(5):2384–2397. https://doi.org/10.1002/hbm.23525
Cheung SW, Larson PS (2010) Tinnitus modulation by deep brain stimulation in locus of caudate neurons (area LC). Neuroscience 169(4):1768–1778. https://doi.org/10.1016/j.neuroscience.2010.06.007
Davis H, Morgan CT, Hawkins JE, Galambos R, Smith FW (1950) Temporary deafness following exposures to loud tones and noise. Acta Otolaryngologica (Stockholm) 88:1–57
Dobie RA (1999) A review of randomized clinical trials in tinnitus. Laryngoscope 109 (8):1202–1211. https://doi.org/10.1097/00005537-199908000-00004
Doron NN, Ledoux JE (1999) Organization of projections to the lateral amygdala from auditory and visual areas of the thalamus in the rat. J Comp Neurol 412(3):383–409
Falk JL (1966) Schedule-induced polydipsia as a function of fixed interval length. J Exp Anal Behav 9(1):37–39. https://doi.org/10.1901/jeab.1966.9-37
Figueiredo RR, Langguth B, Mello de Oliveira P, Aparecida de Azevedo A (2008) Tinnitus treatment with memantine. Otolaryngol Head Neck Surg 138(4):492–496. https://doi.org/10.1016/j.otohns.2007.11.027

Foxe JJ, Morocz IA, Murray MM, Higgins BA, Javitt DC, Schroeder CE (2000) Multisensory auditory-somatosensory interactions in early cortical processing revealed by high-density electrical mapping. Brain Res Cogn Brain Res 10(1–2):77–83

Friston K (2011) Dynamic causal modeling and granger causality comments on: the identification of interacting networks in the brain using fMRI: model selection, causality and deconvolution. NeuroImage 58(2):303–305; author reply 310-301. https://doi.org/10.1016/j.neuroimage.2009.09.031

Gu JW, Halpin CF, Nam EC, Levine RA, Melcher JR (2010) Tinnitus, diminished sound-level tolerance, and elevated auditory activity in humans with clinically normal hearing sensitivity. J Neurophysiol 104(6):3361–3370. https://doi.org/10.1152/jn.00226.2010

Guitton MJ, Caston J, Ruel J, Johnson RM, Pujol R, Puel JL (2003) Salicylate induces tinnitus through activation of cochlear NMDA receptors. J Neurosci 23(9):3944–3952

Hayes SH, Radziwon KE, Stolzberg DJ, Salvi RJ (2014) Behavioral models of tinnitus and hyperacusis in animals. Front Neurol 5:179. https://doi.org/10.3389/fneur.2014.00179

Helfer TM, Jordan NN, Lee RB, Pietrusiak P, Cave K, Schairer K (2011) Noise-induced hearing injury and comorbidities among postdeployment U.S. Army soldiers: April 2003–June 2009. Am J Audiol 20(1):33–41. https://doi.org/10.1044/1059-0889(2011/10-0033)

Irvine DRF, Rajan R (1994) Plasticity of frequency organization in inferior colliculus of adult cats with restricted cochlear lesions. Assoc Res Otolaryngol 17:21

Jastreboff PJ, Brennan JF, Sasaki CT (1988) An animal model for tinnitus. Laryngoscope 98 (3):280–286. https://doi.org/10.1288/00005537-198803000-00008

Jiang C, Luo B, Manohar S, Chen GD, Salvi R (2017) Plastic changes along auditory pathway during salicylate-induced ototoxicity: hyperactivity and CF shifts. Hear Res 347:28–40. https://doi.org/10.1016/j.heares.2016.10.021

Johnson LB, Comeau M, Clarke KD (2001) Hyperacusis in Williams syndrome. J Otolaryngol 30 (2):90–92. https://doi.org/10.2310/7070.2001.20811

Kamke MR, Brown M, Irvine DR (2003) Plasticity in the tonotopic organization of the medial geniculate body in adult cats following restricted unilateral cochlear lesions. J Comp Neurol 459 (4):355–367. https://doi.org/10.1002/cne.10586

Kayser C, Petkov CI, Augath M, Logothetis NK (2007) Functional imaging reveals visual modulation of specific fields in auditory cortex. J Neurosci 27(8):1824–1835. https://doi.org/10.1523/JNEUROSCI.4737-06.2007

Khalfa S, Bruneau N, Roge B, Georgieff N, Veuillet E, Adrien JL, Barthelemy C, Collet L (2004) Increased perception of loudness in autism. Hear Res 198(1–2):87–92. https://doi.org/10.1016/j.heares.2004.07.006

Korsgaard MP, Hartz BP, Brown WD, Ahring PK, Strobaek D, Mirza NR (2005) Anxiolytic effects of Maxipost (BMS-204352) and retigabine via activation of neuronal Kv7 channels. J Pharmacol Exp Ther 314(1):282–292. https://doi.org/10.1124/jpet.105.083923

Lau C, Pienkowski M, Zhang JW, McPherson B, Wu EX (2015) Chronic exposure to broadband noise at moderate sound pressure levels spatially shifts tone-evoked responses in the rat auditory midbrain. NeuroImage 122:44–51. https://doi.org/10.1016/j.neuroimage.2015.07.065

Lingenhohl K, Friauf E (1994) Giant neurons in the rat reticular formation: a sensorimotor interface in the elementary acoustic startle circuit? J Neurosci 14(3 Pt 1):1176–1194

Lobarinas E, Sun W, Cushing R, Salvi R (2004) A novel behavioral paradigm for assessing tinnitus using schedule-induced polydipsia avoidance conditioning (SIP-AC). Hear Res 190 (1–2):109–114. https://doi.org/10.1016/S0378-5955(04)00019-X

Lobarinas E, Yang G, Sun W, Ding D, Mirza N, Dalby-Brown W, Hilczmayer E, Fitzgerald S, Zhang L, Salvi R (2006) Salicylate- and quinine-induced tinnitus and effects of memantine. Acta Otolaryngol Suppl 126(556):13–19. https://doi.org/10.1080/03655230600895408

Lobarinas E, Dalby-Brown W, Stolzberg D, Mirza NR, Allman BL, Salvi R (2011) Effects of the potassium ion channel modulators BMS-204352 Maxipost and its R-enantiomer on salicylate-induced tinnitus in rats. Physiol Behav 104(5):873–879. https://doi.org/10.1016/j.physbeh.2011.05.022

Lockwood AH, Salvi RJ, Coad ML, Towsley ML, Wack DS, Murphy BW (1998) The functional neuroanatomy of tinnitus: evidence for limbic system links and neural plasticity. Neurology 50 (1):114–120. https://doi.org/10.1212/wnl.50.1.114

Lockwood AH, Salvi RJ, Coad ML, Arnold SA, Wack DS, Murphy BW, Burkard RF (1999) The functional anatomy of the normal human auditory system: responses to 0.5 and 4.0 kHz tones at varied intensities. Cereb Cortex 9(1):65–76. https://doi.org/10.1093/cercor/9.1.65

Lockwood AH, Salvi RJ, Burkard RF (2002) Tinnitus. N Engl J Med 347(12):904–910. https://doi.org/10.1056/NEJMra013395

Loeb M, Smith RP (1967) Relation of induced tinnitus to physical characteristics of the inducing stimuli. J Acoust Soc Am 42(2):453–455. https://doi.org/10.1121/1.1910600

Lu YG, Tang ZQ, Ye ZY, Wang HT, Huang YN, Zhou KQ, Zhang M, Xu TL, Chen L (2009) Salicylate, an aspirin metabolite, specifically inhibits the current mediated by glycine receptors containing alpha1-subunits. Br J Pharmacol 157(8):1514–1522. https://doi.org/10.1111/j.1476-5381.2009.00321.x

Manohar S, Spoth J, Radziwon K, Auerbach BD, Salvi R (2017) Noise-induced hearing loss induces loudness intolerance in a rat active sound avoidance paradigm (ASAP). Hear Res 353:197–203. https://doi.org/10.1016/j.heares.2017.07.001

Marsh RA, Fuzessery ZM, Grose CD, Wenstrup JJ (2002) Projection to the inferior colliculus from the basal nucleus of the amygdala. J Neurosci 22(23):10449–10460

Marshall L, Brandt JF (1980) The relationship between loudness and reaction time in normal hearing listeners. Acta Otolaryngol 90(3–4):244–249

McFadden D, Plattsmier HS (1983) Aspirin can potentiate the temporary hearing loss induced by intense sounds. Hear Res 9(3):295–316

Meredith MA, Stein BE (1986) Visual, auditory, and somatosensory convergence on cells in superior colliculus results in multisensory integration. J Neurophysiol 56(3):640–662

Meredith MA, Wallace MT, Stein BE (1992) Visual, auditory and somatosensory convergence in output neurons of the cat superior colliculus: multisensory properties of the tecto-reticulo-spinal projection. Exp Brain Res 88(1):181–186

Milbrandt JC, Holder TM, Wilson MC, Salvi RJ, Caspary DM (2000) GAD levels and muscimol binding in rat inferior colliculus following acoustic trauma. Hear Res 147(1–2):251–260

Mirz F, Gjedde A, Ishizu K, Pedersen CB (2000) Cortical networks subserving the perception of tinnitus--a PET study. Acta Otolaryngol Suppl 120(543):241–243. https://doi.org/10.1080/000164800454503

Mongan E, Kelly P, Nies K, Porter WW, Paulus HE (1973) Tinnitus as an indication of therapeutic serum salicylate levels. JAMA 226(2):142–145

Moody DB (1970) Reaction time as an index of sensory function in animals. In: Stebbins WC (ed) Animal psychophysics: the design and conduct of sensory experiments. Appleton-Century-Crofts, New York, pp 277–301

Moody DB (1973) Behavioral studies of noise-induced hearing loss in primates: loudness recruitment. Adv Otorhinolaryngol 20:82–101

Moody DB, Winger G, Woods JH, Stebbins WC (1980) Effect of ethanol and of noise on reaction time in the monkey: variation with stimulus level. Psychopharmacology 69(1):45–51

Myers EN, Bernstein JM (1965) Salicylate ototoxicity; a clinical and experimental study. Arch Otolaryngol 82(5):483–493. https://doi.org/10.1001/archotol.1965.00760010485006

Oestreicher E, Arnold W, Ehrenberger K, Felix D (1998) Memantine suppresses the glutamatergic neurotransmission of mammalian inner hair cells. ORL J Otorhinolaryngol Relat Spec 60 (1):18–21

Park TJ, Pollak GD (1993) GABA shapes a topographic organization of response latency in the mustache bat's inferior colliculus. J Neurosci 13(12):5172–5187

Patel CR, Zhang H (2014) Local application of sodium salicylate enhances auditory responses in the Rat's dorsal cortex of the inferior Colliculus. Front Neurol 5:235. https://doi.org/10.3389/fneur.2014.00235

Paus T (2000) Functional anatomy of arousal and attention systems in the human brain. Prog Brain Res 126:65–77. https://doi.org/10.1016/S0079-6123(00)26007-X

Pfingst BE, Hienze R, Kimm J, Miller J (1975) Reaction-time procedure for measuring hearing. I. Suprathreshold functions. J Acoust Soc Am 57:421–430

Pienkowski M, Tyler RS, Roncancio ER, Jun HJ, Brozoski T, Dauman N, Coelho CB, Andersson G, Keiner AJ, Cacace AT, Martin N, Moore BC (2014) A review of hyperacusis and future directions: part II. Measurement, mechanisms, and treatment. Am J Audiol 23 (4):420–436. https://doi.org/10.1044/2014_AJA-13-0037

Qiu C, Salvi R, Ding D, Burkard R (2000) Inner hair cell loss leads to enhanced response amplitudes in auditory cortex of unanesthetized chinchillas: evidence for increased system gain. Hear Res 139(1–2):153–171

Radziwon K, Holfoth D, Lindner J, Kaier-Green Z, Bowler R, Urban M, Salvi R (2017) Salicylate-induced hyperacusis in rats: dose- and frequency-dependent effects. Hear Res 350:133–138. https://doi.org/10.1016/j.heares.2017.04.004

Radziwon K, Auerbach BD, Ding D, Liu X, Chen GD, Salvi R (2019) Noise-induced loudness recruitment and hyperacusis: insufficient central gain in auditory cortex and amygdala. Neuroscience 422:212–227. https://doi.org/10.1016/j.neuroscience.2019.09.010

Robertson D, Johnstone BM (1979) Aberrant tonotopic organization in the inner ear damaged by kanamycin. J Acoust Soc Am 66(2):466–469

Romanski LM, Clugnet MC, Bordi F, LeDoux JE (1993) Somatosensory and auditory convergence in the lateral nucleus of the amygdala. Behav Neurosci 107(3):444–450

Ruel J, Chabbert C, Nouvian R, Bendris R, Eybalin M, Leger CL, Bourien J, Mersel M, Puel JL (2008) Salicylate enables cochlear arachidonic-acid-sensitive NMDA receptor responses. J Neurosci 28(29):7313–7323. https://doi.org/10.1523/JNEUROSCI.5335-07.2008

Salvi RJ, Saunders SS, Gratton MA, Arehole S, Powers N (1990) Enhanced evoked response amplitudes in the inferior colliculus of the chinchilla following acoustic trauma. Hear Res 50 (1–2):245–257. https://doi.org/10.1016/0378-5955(90)90049-u

Salvi RJ, Wang J, Ding D (2000) Auditory plasticity and hyperactivity following cochlear damage. Hear Res 147(1–2):261–274. https://doi.org/10.1016/s0378-5955(00)00136-2

Sander K, Brechmann A, Scheich H (2003) Audition of laughing and crying leads to right amygdala activation in a low-noise fMRI setting. Brain Res Brain Res Protoc 11(2):81–91. https://doi.org/10.1016/s1385-299x(03)00018-7

Schlittenlacher J, Ellermeier W, Arseneau J (2014) Binaural loudness gain measured by simple reaction time. Atten Percept Psychophys 76(5):1465–1472. https://doi.org/10.3758/s13414-014-0651-1

Sedley W (2019) Tinnitus: does gain explain? Neuroscience 407:213–228. https://doi.org/10.1016/j.neuroscience.2019.01.027

Seitz PF, Rakerd B (1997) Auditory stimulus intensity and reaction time in listeners with longstanding sensorineural hearing loss. Ear Hear 18(6):502–512

Skaliora I, Doubell TP, Holmes NP, Nodal FR, King AJ (2004) Functional topography of converging visual and auditory inputs to neurons in the rat superior colliculus. J Neurophysiol 92 (5):2933–2946. https://doi.org/10.1152/jn.00450.2004

Stebbins WC (1966) Auditory reaction time and the derivation of equal loudness contours for the monkey. J Exp Anal Behav 9(2):135–142. https://doi.org/10.1901/jeab.1966.9-135

Stebbins WC, Miller JM (1964) Reaction time as a function of stimulus intensity for the monkey. J Exp Anal Behav 7:309–312

Stolzberg D, Chen GD, Allman BL, Salvi RJ (2011) Salicylate-induced peripheral auditory changes and tonotopic reorganization of auditory cortex. Neuroscience 180:157–164. https://doi.org/10.1016/j.neuroscience.2011.02.005

Stolzberg D, Hayes SH, Kashanian N, Radziwon K, Salvi RJ, Allman BL (2013) A novel behavioral assay for the assessment of acute tinnitus in rats optimized for simultaneous recording of oscillatory neural activity. J Neurosci Methods 219(2):224–232. https://doi.org/10.1016/j.jneumeth.2013.07.021

Su YY, Luo B, Wang HT, Chen L (2009) Differential effects of sodium salicylate on current-evoked firing of pyramidal neurons and fast-spiking interneurons in slices of rat auditory cortex. Hear Res 253(1–2):60–66. https://doi.org/10.1016/j.heares.2009.03.007

Su YY, Luo B, Jin Y, Wu SH, Lobarinas E, Salvi RJ, Chen L (2012) Altered neuronal intrinsic properties and reduced synaptic transmission of the rat's medial geniculate body in salicylate-induced tinnitus. PLoS One 7(10):e46969. https://doi.org/10.1371/journal.pone.0046969

Suhnan AP, Finch PM, Drummond PD (2017) Hyperacusis in chronic pain: neural interactions between the auditory and nociceptive systems. Int J Audiol 56(11):801–809. https://doi.org/10.1080/14992027.2017.1346303

Sun W, Lu J, Stolzberg D, Gray L, Deng A, Lobarinas E, Salvi RJ (2009) Salicylate increases the gain of the central auditory system. Neuroscience 159(1):325–334. https://doi.org/10.1016/j.neuroscience.2008.12.024

Tennigkeit F, Schwarz DW, Puil E (1998) GABA(B) receptor activation changes membrane and filter properties of auditory thalamic neurons. Hear Res 122(1–2):18–24

Tyler RS, Pienkowski M, Roncancio ER, Jun HJ, Brozoski T, Dauman N, Dauman N, Andersson G, Keiner AJ, Cacace AT, Martin N, Moore BC (2014) A review of hyperacusis and future directions: part I. definitions and manifestations. Am J Audiol 23(4):402–419. https://doi.org/10.1044/2014_AJA-14-0010

Ueyama T, Donishi T, Ukai S, Ikeda Y, Hotomi M, Yamanaka N, Shinosaki K, Terada M, Kaneoke Y (2013) Brain regions responsible for tinnitus distress and loudness: a resting-state FMRI study. PLoS One 8(6):e67778. https://doi.org/10.1371/journal.pone.0067778

Wallace MT, Meredith MA, Stein BE (1993) Converging influences from visual, auditory, and somatosensory cortices onto output neurons of the superior colliculus. J Neurophysiol 69 (6):1797–1809. https://doi.org/10.1152/jn.1993.69.6.1797

Wang J, Caspary D, Salvi RJ (2000) GABA-A antagonist causes dramatic expansion of tuning in primary auditory cortex. Neuroreport 11(5):1137–1140. https://doi.org/10.1097/00001756-200004070-00045

Wang HT, Luo B, Huang YN, Zhou KQ, Chen L (2008) Sodium salicylate suppresses serotonin-induced enhancement of GABAergic spontaneous inhibitory postsynaptic currents in rat inferior colliculus *in vitro*. Hear Res 236(1–2):42–51. https://doi.org/10.1016/j.heares.2007.11.015

Webster D (1992) An overview of mammalian auditory pathways with an emphasis on humans. In: Webster D, Popper A, Fay R (eds) The Mammalian auditory pathway: neuroanatomy. Springer, Berlin

Wier CC, Pasanen EG, McFadden D (1988) Partial dissociation of spontaneous otoacoustic emissions and distortion products during aspirin use in humans. J Acoust Soc Am 84 (1):230–237

Wong E, Radziwon K, Chen GD, Liu X, Manno FA, Manno SH, Auerbach B, Wu EX, Salvi R, Lau C (2020) Functional magnetic resonance imaging of enhanced central auditory gain and electrophysiological correlates in a behavioral model of hyperacusis. Hear Res 389:107908. https://doi.org/10.1016/j.heares.2020.107908

Wu MF, Suzuki SS, Siegel JM (1988) Anatomical distribution and response patterns of reticular neurons in relation to acoustic startle. Brain Res 457:399–406

Yankaskas K (2013) Prelude: noise-induced tinnitus and hearing loss in the military. Hear Res 295:3–8. https://doi.org/10.1016/j.heares.2012.04.016

Zheng Y, McNamara E, Stiles L, Darlington CL, Smith PF (2012) Evidence that Memantine reduces chronic tinnitus caused by acoustic trauma in rats. Front Neurol 3:127. https://doi.org/10.3389/fneur.2012.00127

Zugaib J, Ceballos CC, Leao RM (2016) High doses of salicylate reduces glycinergic inhibition in the dorsal cochlear nucleus of the rat. Hear Res 332:188–198. https://doi.org/10.1016/j.heares.2015.10.008

Neuroinflammation and Tinnitus

Abraham Shulman, Weihua Wang, Hao Luo, Shaowen Bao, Grant Searchfield, and Jinsheng Zhang

Contents

Abstract Neuroinflammation is the central nervous system's response to: injury, infection, and abnormal neural activity. Inflammatory processes are known to mediate many diseases, and recently evidence indicates that neuroinflammation underlies hearing disorders such as presbyacusis, middle-ear disease, ototoxicity, noise-induced hearing loss, and tinnitus. This chapter provides a review of the role of neuroinflammation in the etiology and treatment of tinnitus. Specifically, our research team has demonstrated that both tumor necrosis factor alpha (TNF-α) and calpain signaling pathways are involved in noise-induced tinnitus and that blocking them yielded therapeutic effects on tinnitus. Other efforts such as controlling acute inflammatory response via specialized pro-resolving mediators may help provide insight into preventing and treating tinnitus-related inflammatory processes.

A. Shulman (✉)
Department of Otolaryngology, State University New York-Downstate, Brooklyn, NY, USA

W. Wang and S. Bao
Department of Physiology, University of Arizona, Tucson, AZ, USA

H. Luo
Department of Otolaryngology, Wayne State University School of Medicine, Detroit, MI, USA

G. Searchfield
Section of Audiology, School of Population Health, University of Auckland, Auckland, New Zealand

J. Zhang
Department of Otolaryngology, Wayne State University School of Medicine, Detroit, MI, USA

Department of Communication Sciences and Disorders, Wayne State University College of Liberal Arts and Sciences, Detroit, MI, USA

Curr Topics Behav Neurosci (2021) 51: 161–174
https://doi.org/10.1007/7854_2021_238
Published Online: 20 July 2021

Keywords Inflammatopathy · Microglia specific pro-resolving mediator · Neuroinflammation · Tumor necrosis factor alpha (TNF-α) calpain

1 Introduction

Neuroinflammation is the central nervous system's response to: injury, infection, disease, and abnormal neural activity (Shabab et al. 2017). The concept of an "inflammatopathy" is introduced and defined as a particular clinical etiology, inflammation, which becomes clinically manifest as a disease or symptom of an organ system, of which the underlying initiating pathology is "inflammation." Tinnitus is the aberrant perception of sound unrelated to an external source of sound (Shulman and Goldstein 2005) and recently considered to be mediated by inflammatory processes (Szczepek et al. 2014; Wang et al. 2019). Inflammatory processes are involved in a wide variety of mental and physical health problems including addiction (Kohno et al. 2019), schizophrenia (Najjar and Pearlman 2015), depression (Troubat et al. 2021), dementia (Stefaniak and O'Brien 2016; Schain and Kreisl 2017), hearing loss (Tan et al. 2013), and tinnitus (Wang et al. 2019). Inflammation can affect cells within the central nervous system including neurons, macroglia, and microglia (Shabab et al. 2017). Markers of neuroinflammation include cytokines, chemokines, and reactive oxygen species. Microglial activation and the increased expression of cytokines are intended to protect the CNS but excessive or chronic can lead to pathological changes (Shabab et al. 2017).

In the auditory system inflammation has been associated with presbyacusis (Watson et al. 2017), middle-ear disease, ototoxicity, noise-induced hearing loss (Tan et al. 2013), and tinnitus (Wang et al. 2019). Presbyacusis is age related hearing loss. As we age we are subjective to mild inflammation, arising because of poorer control or downregulation of proinflammatory proteins (Watson et al. 2017). A buildup of reactive molecules and pathogen targeting cells damages sensory structures. Noise-induced hearing loss is also associated with inflammation. Cochlear pathology following noise depends on the intensity, frequency, and duration of noise, but can result in damage to a variety of sensory, supporting, and neural cells (Tan et al. 2013). Acute intense noise exposure (100 dB SPL, 8–16 kHz, 24 h) results in a rapid (at 6 h post exposure) expression of proinflammatory mediators and adhesion, followed by recruitment and infiltration of inflammatory cells into the cochlear from systemic circulation in mice (Tan et al. 2016). Chronic exposure to moderate noise (90 dB SPL, 8–16 kHz, 2 h per day for 6 weeks) also elicits an inflammatory response (Tan et al. 2016). Corticosteroids (anti-inflammatory medications) have been used to successfully treat sudden idiopathic sensory neural hearing loss (Trune and Canlon 2012). Given the link between hearing loss and inflammation and hearing loss and tinnitus, a role for inflammation in tinnitus is not surprising. Neuroinflammation mediates noise-induced synaptic imbalance and tinnitus in rodent models (Wang et al. 2019). Wang et al. (2019) studied the role of neuroinflammation in tinnitus by examining the pathophysiological changes in the rodent auditory cortices following noise-induced hearing loss (NIHL). The results indicated that NIHL is associated with elevated levels of proinflammatory cytokines and microglial activation in the primary auditory cortex (AI). The genetic knockout of tumor necrosis factor alpha (TNF-α) or pharmacologically blocking TNF-α expression prevented neuroinflammation and the behavioral phenotype associated with tinnitus in mice with NIHL was improved. Infusion of TNF-α into the AI resulted in behavioral signs of tinnitus in normal hearing wild-type and TNF-α knockout mice. In mice with NIHL and not tinnitus there was a pharmacological depletion of microglia. In animals with NIHL the frequency of miniature excitatory synaptic currents increased and that of miniature inhibitory synaptic currents decreased in AI pyramidal neurons. This excitatory-to-inhibitory synaptic imbalance was completely prevented by pharmacological blockade of TNF-α expression. These results implicate neuroinflammation as a therapeutic target for treating tinnitus and other hearing loss–related disorders (Wang et al. 2019). The following sections describe the work in this area in further details.

2 TNF-α Signaling Pathway in Tinnitus

Increasing evidence indicates that noise-induced hearing loss and conductive hearing loss can lead to inflammatory responses in the central auditory pathway (Fuentes-Santamaria et al. 2014, 2017; Baizer et al. 2015). For example, in a recent study, monaural exposure to a continuous 8-kHz tone at 112–114 dB SPL for 2 h (routinely used to introduce tinnitus in rodents) was found to rapidly increase the expression of proinflammatory cytokines such as TNF-α, IL-1β, and IL-18 (Fig. 1) (Wang et al. 2019). Along with the increased expression of those proinflammatory cytokines, microglia were also activated as exemplified by their morphological transition from ramified to deramified, amoeboid shapes (Wang et al. 2019; Ziebell and Morganti-Kossmann 2010; Donat et al. 2017; Dang et al. 2016; Hovens et al. 2015). In addition to noise trauma-induced increases in proinflammatory cytokines, salicylate-induced tinnitus is associated with increased expression of proinflammatory cytokines and activation of microglia in the central auditory pathway (Xia et al. 2020; Chen and Zheng 2017; Hwang et al. 2011; Hu et al. 2014). Neuroinflammatory responses profoundly influence neuronal functions. For example, microglia play an important role in neural development, maturation, plasticity, and aging (Wolf et al. 2017; Salter and Beggs 2014; Allen and Barres 2005). Proinflammatory cytokines also modulate neuronal functions such as synaptic transmission, plasticity, and membrane excitability (Di Filippo et al. 2008; Stellwagen and Malenka 2006; Steinmetz and Turrigiano 2010; Stellwagen et al. 2005; Bellinger et al. 1993). Many of these processes are implicated in tinnitus (Shore et al. 2016; Roberts et al. 2010), induced by noise trauma or salicylate administration. The notion that neuroinflammation is involved in tinnitus etiologies is supported by evidence that three separate manipulations that prevented

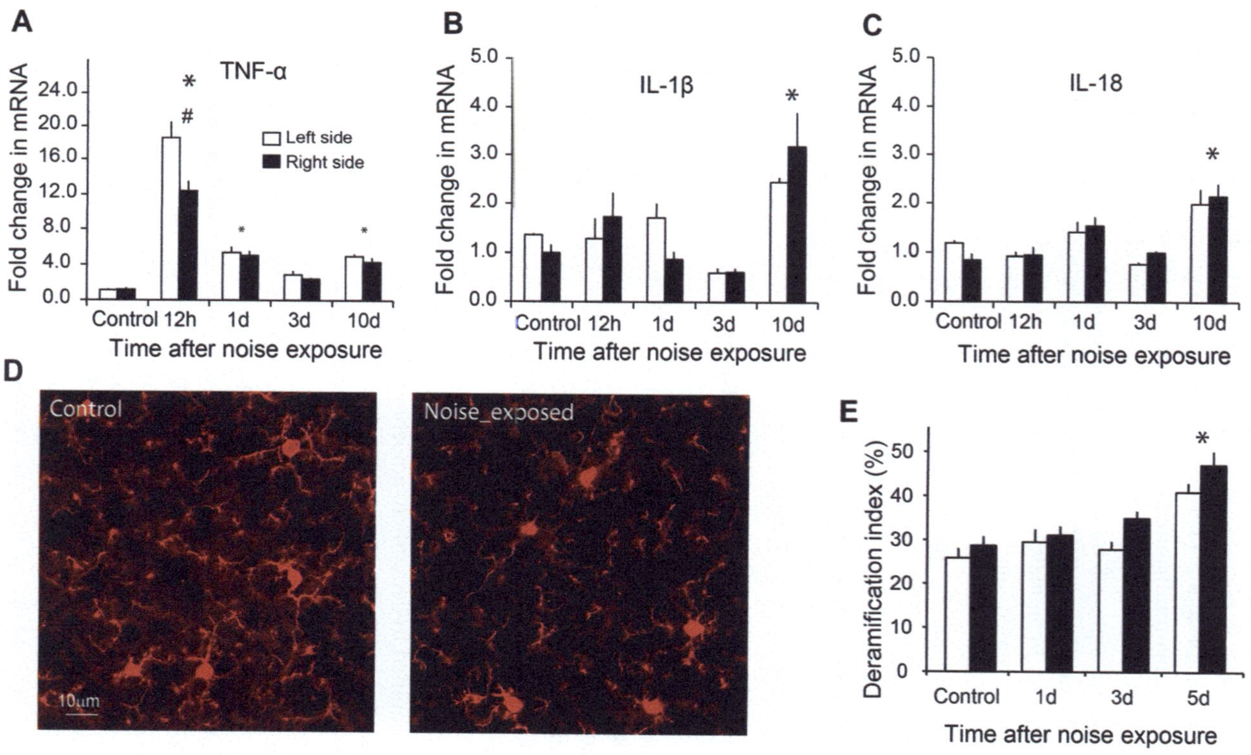

Fig. 1 Noise exposure results in elevated expression of proinflammatory cytokines and microglial deramification in the auditory cortex. (**a**) TNF-α mRNA level increased rapidly within 12 h of noise exposure, with a stronger ipsilateral than contralateral increase. The increase was also significant at 1 day and 10 days post noise exposure. (**b**, **c**) Expression of IL-1β and IL-18 was elevated 10 days after noise exposure. (**d**) Representative images of IBA1 antibody-stained microglia in AI of control and noise-exposed mice. (**e**) The soma-to-whole cell size ratio of microglia was used to measure microglial deramification as an index of microglial activation. There was a significant increase in the microglial deramification index 5 days after noise exposure. Error bars represent SEM. * depicts $p < 0.01$ compared to control; # indicates $p < 0.01$ comparing left and right hemispheres. Adapted from Wang et al. (2019)

noise-induced neuroinflammation in the auditory cortex – i.e., genetic knockout of TNF-α, blockade of TNF-α expression by 3,6′-dithiothalidomide, and removal of microglia by CSF1R blocker PLX3397 – all prevented noise-induced tinnitus assessed with a startle response-based gap detection test and an operant conditioned based lick suppression test (Wang et al. 2019). For example, administration of 3,6′-dithiothalidomide blocked noise-induced TNF-α expression, microglial deramification, excitation-inhibition synaptic imbalance, and tinnitus (Fig. 2) (Wang et al. 2019; Gonzalez-Gonzalez and Cazevieille 2020). Administration of TNF-α blockers also alleviated salicylate-induced tinnitus in mouse models (Gonzalez-Gonzalez and Cazevieille 2020; Hwang et al. 2017).

In addition to auditory pathologies that mediate tinnitus, several non-auditory pathologies and health conditions that are known to be associated with neuroinflammation can increase the risk of tinnitus (Simon et al. 2017; Lew et al. 2007; Calcia et al. 2016; Grippo and Scotti 2013; Mazurek et al. 2015; Isaacson et al. 2003; Walker et al. 2014; Wright and Gullickson 1996; Durai and Searchfield 2016; Langguth et al. 2011; Popeo et al. 2011; DeLeo and Yezierski 2001; Ellis and Bennett 2013; Kiguchi et al. 2012). All these diverse non-auditory risk factors could potentially increase proinflammatory cytokine concentration in the cerebrospinal fluid (Bianchi et al. 2007; Levine et al. 1999; Lerman et al. 2016; Juengst et al. 2014), providing a diffusible signal to influence the function of the central auditory system. We examined whether the infusion of recombinant TNF-α into the right lateral cerebral ventricle would increase the risk of tinnitus (Deng et al. 2020). We found that microglial activation and evidence of tinnitus were observed in the auditory cortex of mice that had received both TNF-α infusion and exposure to an 86-dB noise, but not in mice that had received either TNF-α infusion or noise exposure alone. These results suggest that disease-related increases in brain proinflammatory cytokine release could be a risk factor for tinnitus (Fig. 3).

Furthermore, strain differences in immune and neuroinflammatory responses have been widely reported in common mouse strains (Browne et al. 2012; Perez et al. 2013; Tacchini-Cottier et al. 2000). For example, the C57BL/6 and the FVB strains exhibit different immune response profiles, with C57BL/6 mice being Type 1 T helper cell-dominant (Th1-dominant), and FVB mice being Th2-dominant (Whitehead et al. 2003). Th1 cells promote the secretion of proinflammatory cytokines, and Th2 cells promote the secretion of anti-inflammatory cytokines (Arumugam et al. 2005). When challenged with experimentally induced stroke and reperfusion, the two strains showed different profiles of immune cell activation (Kim et al. 2014). In addition, C57BL/6 mice displayed greater neurological and motor deficits from the stroke than FVB mice did (Kim et al. 2014). Interestingly, strain differences were also reported between C57BL/6 and FVB mice in noise-induced neuroinflammation and tinnitus (Miyakawa et al. 2019; Zinsmaier et al. 2020). Exposure to loud noises resulted in elevated TNF-α expression in the auditory cortex and evidence of tinnitus in C57BL/6 but not FVB mice (Fig. 4). These results suggest that genetic variability in immune response profiles contributes to differences in tinnitus susceptibility.

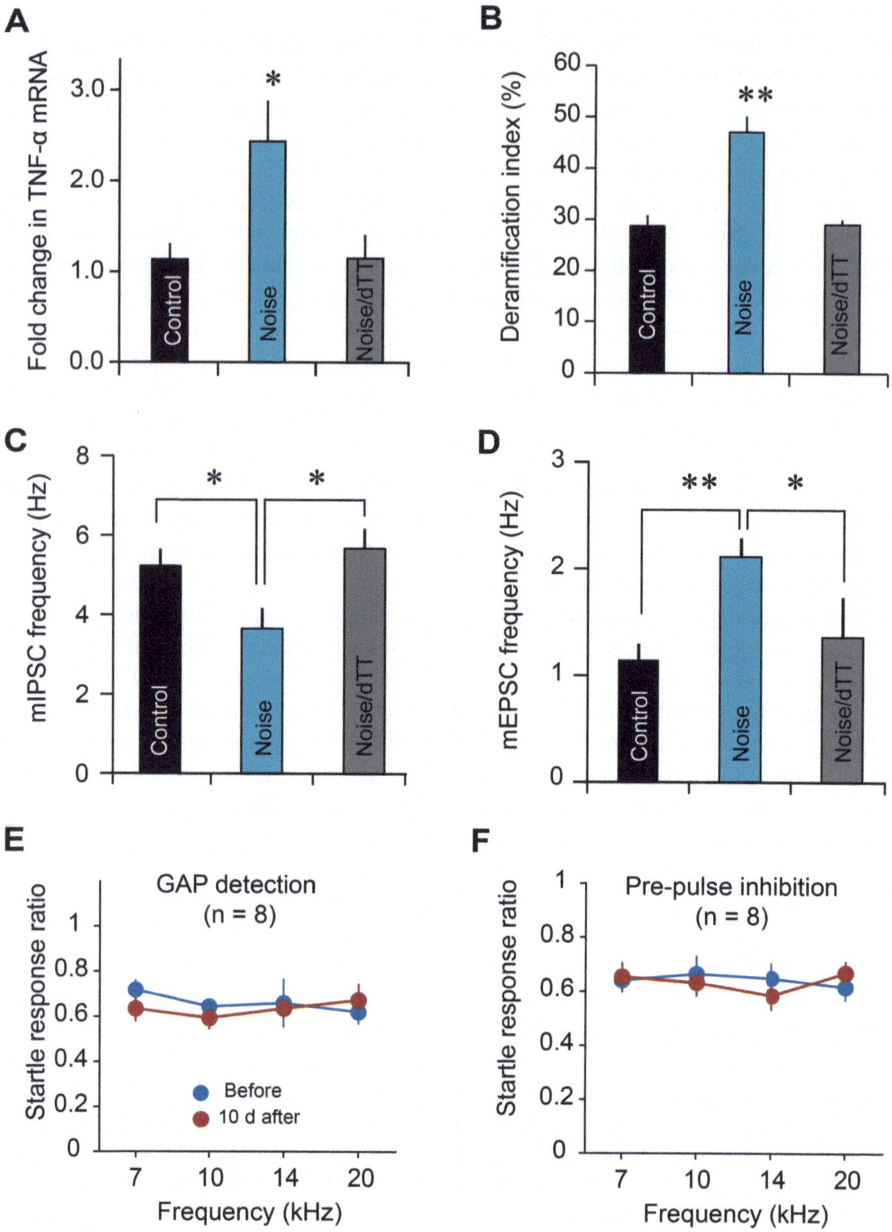

Fig. 2 Blocking TNF-α expression prevents cortical synaptic imbalance and behavioral evidence of tinnitus. Treatment with 3,6′-dithiothalidomide (dTT) prevented noise trauma-induced TNF-α expression (**a**), microglial deramification (**b**), inhibitory synaptic reduction (**c**), and excitatory synaptic enhancement (**d**) in the auditory cortex. Treatment with dTT prevented the development of behavioral evidence of tinnitus following noise trauma (**e**). Animals' ability to hear brief tones, which was measured with prepulse inhibition (PPI), was not altered by noise trauma/dTT treatment (**f**). *, ** and *** indicate $p < 0.05$, $p < 0.01$ and $p < 0.001$ respectively. Adapted from Wang et al. (2019)

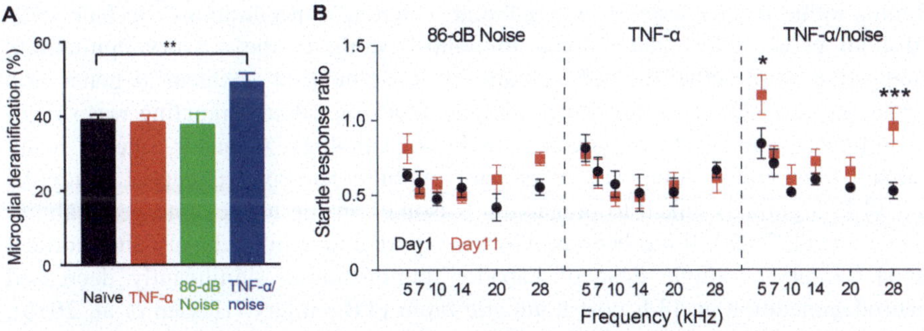

Fig. 3 TNF-α infusion and noise exposure synergistically induce microglial deramification and tinnitus. (**a**) Microglial deramification index was enhanced only in mice that had received both TNF-α infusion and exposure to a moderately loud noise at 86 dB SPL. (**b**) Animals' performances in gap detection before (Day 1) and after (Day 11) noise exposure (Noise), TNF-α infusion (TNF-α) and combined TNF-α infusion and noise exposure (TNF-α + Noise). Data are presented as mean ± SEM. * indicates $p < 0.05$, ** $p < 0.01$ and *** $p < 0.005$. Gap detection was impaired in mice that had received combined TNF-α infusion and noise exposure

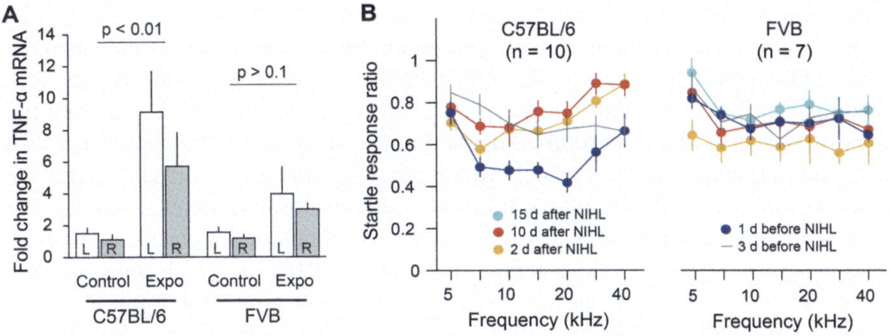

Fig. 4 Strain differences in noise trauma-induced TNF-α expression and tinnitus. (**a**) Elevated TNF-α expression was observed 10 days after noise trauma in C57BL/6 but not FVB mice. (**b**) Gap detection was impaired in C57BL/6 mice 2 and 10 days after noise trauma. The same noise trauma did not significantly alter gap detection performances in FVB mice 2, 10 and 15 days after the trauma. Adapted from Zinsmaier et al. (2020)

3 Calpain Signaling Pathway in Tinnitus

Calpain is an intracellular calcium activated proteases, which promote the breakdown of cellular proteins, kinases, phosphatases, and transcription factors (Wang 2000). Over activation of calpain is related to numerous neurodegenerative diseases, impairing brain plasticity, axonal transmission and contributing to necrotic and apoptotic cell death pathways (Bartus et al. 1995; Goll et al. 1992). As a result, calpain activation is involved in the process of numerous inflammation-associated disease (Perrin et al. 2003). Previous studies have demonstrated acoustic trauma-induced increases in intracellular calcium concentrations in hair cells (Bobbin et al. 2003) and calpain immuno-labeling in the sensory epithelium, suggesting a possible role of calpain in noise-induced cochlear degeneration (Wang et al. 1999). Thus downregulating or blocking calpain signaling was sought to mitigate trauma-induced cochlear or brain pathology. Gabadur, one of many calpain blockers, is composed of calpain inhibitor leupeptin (Acetyl-L-leucyl-L-leucyl-L–argininal) linked to pregabalin S-isomer analog to facilitate permeability (Dugue et al. 2018). It has been previously reported that, immediately after cortical impact injury, a single dose of gabadur administration significantly decreased neurodegeneration at 48 h post traumatic brain (TBI) injury (Hassen et al. 2018), which supports the notion that gabadur's protease inhibitor activity contributes to the neuroprotective effect. In addition, local infusion of leupeptin in the inner ear could significantly reduce the sensory cells loss after acoustic trauma (Wang et al. 1999). At the same time, many lines of evidence indicate that acoustic trauma and TBI are related to tinnitus (Dugue et al. 2018; Norena 2015). Thus, we hypothesize that the calpain mediates the etiology of tinnitus, and calpain inhibitors suppress tinnitus by attenuating calcium over activation and neuroinflammation after noise trauma.

To assess the gabadur's therapeutic effects on noise-induced tinnitus, 22 adult male Sprague Dawley rats (age = 110 days old, BW = 250–300 g) were purchased from Charles River Laboratories (Wilmington, MA). Rats were divided in three groups. Five rats were administrated with gabadur after noise exposure, 10 rats were administrated with PBS after noise exposure, and 7 rats received PBS treatment but no noise exposure (control group). All rats' tinnitus status was behaviorally assessed using our optimized conditioned licking suppression paradigm before and after noise exposure and continued for 5 days after drug administration (Pace et al. 2016). The details of the behavioral testing using the conditioned licking suppression paradigm were described in our previous report (Pace et al. 2016), in which animals were trained to lick waterspout during sound presence and tested for tinnitus during silence after tinnitus induction with noise trauma. The increased number of licks during silence indicates presence of tinnitus behavior. All procedures were approved by the Institutional Animal Care and Use Committee at Wayne State University and were in accordance with the regulations of the Federal Animal Welfare Act. In the gabadur treatment group, 3 weeks after noise exposure, the rats showed the chronic tinnitus behavior, as demonstrated by increased number of licks of waterspout during silence tests. During the continuous 5 days of gabadur administration, the licking rates during silent trials at 6–8, 10–12, or 14–16 kHz decreased compared to the pre-drug-injection period (Pre-inj), by the second day of gabadur injection. Their licking rates fell to 1 licks/per trial or below, which was considered negative tinnitus behavior. For silent trials at 22–24 or 30–32 kHz (Fig. 5), the overall licking rates only decreased to tinnitus negative levels by the fifth day of gabadur injection. Such results indicated tinnitus suppression with gabadur treatment, although with different effects at different frequency bands and different time windows.

In the PBS treatment and control groups, three rats showed chronic tinnitus behavior 3 weeks after noise exposure, with 5 days of PBS injections. The licking rates at all frequency bands did not consistently decrease relative to the pre-injection

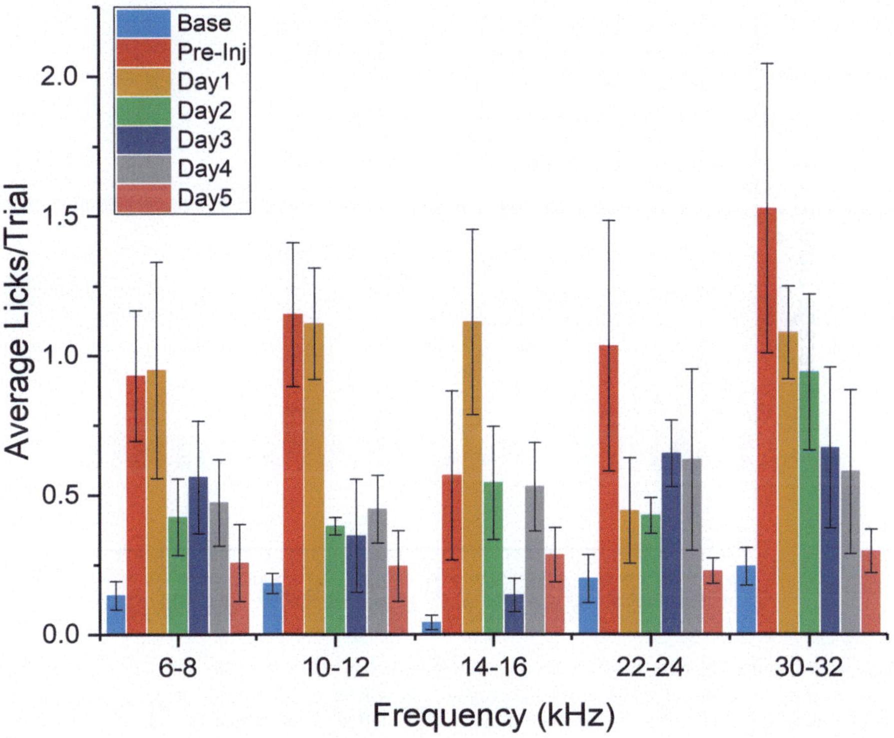

Fig. 5 Behavioral evidence of gabadur-suppressed noise-induced tinnitus. Behavioral testing results from 5 rats with chronic tinnitus after noise exposure. After gabadur administration, at 6–8, 10–12 and 14–16 kHz testing regions, licking rates significant decreased at second day after; at 22–24 and 30–32 kHz testing regions, licking rates decreased at day 5

(Pre-Inj) time point. This indicated that PBS injections alone did not have a therapeutic effect on the tinnitus behavior. Seven rats showed negative tinnitus behavior 3 weeks after noise exposure. After 5 days of PBS injection, the licking rates at all frequency bands remained the same level compared to the pre-injection period. In the control group with PBS injections, the average licks/sound trials at all frequency bands remained the same level at all testing time points, indicating no tinnitus behavior (Fig. 6).

The above results demonstrated that 5 days of gabadur administration induced tinnitus suppression at all testing frequency bands, suggesting that calpain is intimately involved in the etiology of tinnitus. First of all, calpains have been proposed to influence inflammatory processes via a variety of mechanisms (Ruetten and Thiemermann 1997; Cuzzocrea et al. 2000). Increasing evidence suggests that noise-induced hearing loss and conductive hearing loss can trigger inflammatory responses along the central auditory pathways (Fuentes-Santamaria et al. 2014, 2017; Baizer et al. 2015), and that altering trauma-related maladaptive plasticity could suppress noise-induced tinnitus with anti-inflammatory responses (Wang et al.

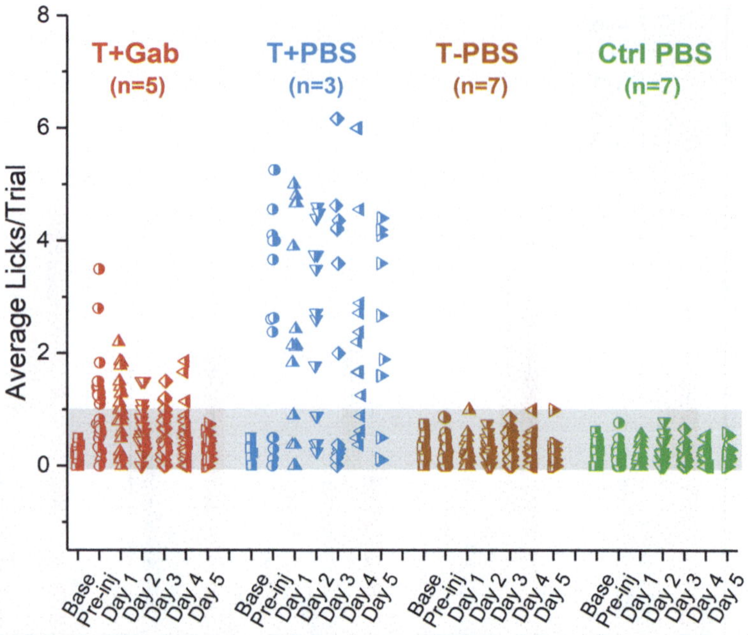

Fig. 6 Behavioral results from 4 groups. Behavioral data from the rats with chronic tinnitus (indicated by average licks/trial above the gray area in at least 1, silent trial category). The rats (T +Gab) were treated with gabadur, their licking rates fell to 1 licks/trial or below, which was considered tinnitus negative behavior. The tinnitus positive rats (T+PBS) treated with PBS remained same licking rates. The tinnitus negative rats treated with PBS (T-PBS) and control rats treated with PBS (Ctrl PBS) remained within the shaded area at all time points, indicating a lack of tinnitus behavior

2019). With the administration of a calpain inhibitor, the inflammatory responses related to the tinnitus generation could have been reset, and the noise-induced tinnitus suppression initiated. Considering the fact that noise trauma results in substantial increase in calpain immunostaining in the organ of Corti (Hamernik et al. 1984) and that calpain inhibitor (leupeptin) reduces the amount of hair cell loss after noise trauma (Wang et al. 1999), the currently induced tinnitus suppression may result from rescuing cells in the organ of Corti from noise-trauma-induced injury. The mitigation of the noise-trauma-induced peripheral deafferentation could have dampened the abnormally elevated central activity in the auditory brain, which has been considered to be the etiology of tinnitus. In future studies, there is a need to document the pathophysiological changes at both the cochlea and auditory brain structures. In addition, it is unclear how gabadur plays a role in both sensory and/or supporting cells. Based on the results, the suppression tinnitus at low and middle frequencies occurred at the second day after gabadur administration, whereas suppression of tinnitus at high frequency regions occurred at the fifth day after gabadur administration. These differential results suggest that the sensory or supporting cells along with their innovating auditory nerves at different frequency regions differentially benefit from the calpain inhibition with different time courses following gabadur administration.

In addition to the peripheral mechanisms that possibly underlie the gabadur-induced tinnitus suppression in our rat model, the results are reminiscent of the notion that tinnitus results from trauma-related maladaptive plasticity changes (Norena 2015; Llinas et al. 2005). It has been previously reported that TBI causes tinnitus by altering neural activity at the cortical and subcortical levels (Jury and Flynn 2001), which suggests that there is a correlation between TBI-related plasticity changes and tinnitus. Indeed, a single dose of gabadur has been previously proved to have therapeutic effects on TBI at 48 h post-TBI by reducing the calpain-2 level in the injured brain hemisphere (Dugue et al. 2018). Thus, there was a possibility that the gabadur's therapeutic effect on tinnitus was due to reset or attenuation of calpain-mediated central maladaptive plasticity.

Among numerous and complex underlying mechanisms of tinnitus, neuroinflammation is an important one that causes disruption of the homeostasis at molecular signaling, synaptic, cellular, systemic, and behavioral levels that mediate tinnitus. Blocking or regulating certain neuroinflammatory processes may have reset or attenuate the etiologies of tinnitus. Among numerous inflammatory processes, we have found that both TNF- α and calpain signaling pathways play important but different roles in the etiologies of tinnitus. Continued success in the effective downregulation of TNF- α and/or calpain signaling pathways may help identify reliable pharmacological solutions for treating tinnitus. In addition, efforts in controlling acute inflammatory response may have further potentials in elucidating the mechanism underlying tinnitus and in the development of effective treatments. The mechanisms that control the resolution of acute inflammation include specialized pro-resolving mediators (SPMs), lipoxins, resolvins, protectins, and maresins, which provide insight into preventing and treating inflammatory processes (Serhan and Levy 2018) that possibly involve tinnitus.

Acknowledgements This work was supported by the US Department of Defense (W81XWH-15-1-0028, W81XWH-15-1-0356, W81XWH-15-1-0357) and Martha Entenmann Tinnitus Research Center. The support of Richard M. Rosenfeld, M.D., and Matthew B. Hanson M.D. in tinnitus interest and efforts is appreciated.

References

Allen NJ, Barres BA (2005) Signaling between glia and neurons: focus on synaptic plasticity. Curr Opin Neurobiol 15(5):542–548

Arumugam TV, Granger DN, Mattson MP (2005) Stroke and T-cells. Neuromolecular Med 7 (3):229–242

Baizer JS et al (2015) Effects of acoustic trauma on the auditory system of the rat: the role of microglia. Neuroscience 303:299–311

Bartus RT et al (1995) Calpain as a novel target for treating acute neurodegenerative disorders. Neurol Res 17(4):249–258

Bellinger FP, Madamba S, Siggins GR (1993) Interleukin 1 beta inhibits synaptic strength and long-term potentiation in the rat CA1 hippocampus. Brain Res 628(1–2):227–234

Bianchi M et al (2007) Increased tumor necrosis factor-alpha and prostaglandin E2 concentrations in the cerebrospinal fluid of rats with inflammatory hyperalgesia: the effects of analgesic drugs. Anesth Analg 104(4):949–954

Bobbin RP, Parker M, Wall L (2003) Thapsigargin suppresses cochlear potentials and DPOAEs and is toxic to hair cells. Hear Res 184(1–2):51–60

Browne CA et al (2012) Differential lipopolysaccharide-induced immune alterations in the hippocampus of two mouse strains: effects of stress. Neuroscience 225:237–248

Calcia MA et al (2016) Stress and neuroinflammation: a systematic review of the effects of stress on microglia and the implications for mental illness. Psychopharmacology (Berl) 233 (9):1637–1650

Chen XH, Zheng LL (2017) Expression of pro-inflammatory cytokines in the auditory cortex of rats with salicylate-induced tinnitus. Mol Med Rep 16(4):5643–5648

Cuzzocrea S et al (2000) Calpain inhibitor I reduces the development of acute and chronic inflammation. Am J Pathol 157(6):2065–2079

Dang DK et al (2016) Apocynin prevents mitochondrial burdens, microglial activation, and pro-apoptosis induced by a toxic dose of methamphetamine in the striatum of mice via inhibition of p47phox activation by ERK. J Neuroinflammation 13:12

DeLeo JA, Yezierski RP (2001) The role of neuroinflammation and neuroimmune activation in persistent pain. Pain 90(1–2):1–6

Deng D, Wang W, Bao S (2020) Diffusible tumor necrosis factor-alpha (TNF-α) promotes noise-induced Parvalbumin-positive (PV+) neuron loss and auditory processing impairments. Front Neurosci 14:573047

Di Filippo M et al (2008) Neuroinflammation and synaptic plasticity: theoretical basis for a novel, immune-centred, therapeutic approach to neurological disorders. Trends Pharmacol Sci 29 (8):402–412

Donat CK et al (2017) Microglial activation in traumatic brain injury. Front Aging Neurosci 9:208

Dugue R et al (2018) Controlled cortical impact-induced neurodegeneration decreases after administration of the novel calpain-inhibitor Gabadur. Brain Res Bull 142:368–373

Durai M, Searchfield G (2016) Anxiety and depression, personality traits relevant to tinnitus: a scoping review. Int J Audiol 55(11):605–615

Ellis A, Bennett DL (2013) Neuroinflammation and the generation of neuropathic pain. Br J Anaesth 111(1):26–37

Fuentes-Santamaria V et al (2014) Glia-related mechanisms in the anteroventral cochlear nucleus of the adult rat in response to unilateral conductive hearing loss. Front Neurosci 8:319

Fuentes-Santamaria V et al (2017) The role of glia in the peripheral and central auditory system following noise overexposure: contribution of TNF-alpha and IL-1beta to the pathogenesis of hearing loss. Front Neuroanat 11:9

Goll DE et al (1992) Is calpain activity regulated by membranes and autolysis or by calcium and calpastatin? Bioessays 14(8):549–556

Gonzalez-Gonzalez S, Cazevieille C (2020) 3,6′-Dithiothalidomide reduces tinnitus phenotype in two different mouse preclinical models of tinnitus. J Community Prev Med 3(1)

Grippo AJ, Scotti MA (2013) Stress and neuroinflammation. Mod Trends Pharmacopsychiatry 28:20–32

Hamernik RP et al (1984) Anatomical correlates of impulse noise-induced mechanical damage in the cochlea. Hear Res 13(3):229–247

Hassen GW et al (2018) Effects of novel Calpain inhibitors in transgenic animal model of Parkinson's disease/dementia with Lewy bodies. Sci Rep 8(1):18083

Hovens IB et al (2015) Postoperative cognitive dysfunction and microglial activation in associated brain regions in old rats. Neurobiol Learn Mem 118:74–79

Hu SS et al (2014) Effects of salicylate on the inflammatory genes expression and synaptic ultrastructure in the cochlear nucleus of rats. Inflammation 37(2):365–373

Hwang JH et al (2011) Expression of tumor necrosis factor-alpha and interleukin-1beta genes in the cochlea and inferior colliculus in salicylate-induced tinnitus. J Neuroinflammation 8:30
Hwang JH et al (2017) Effects of tumor necrosis factor blocker on salicylate-induced tinnitus in mice. Int Tinnitus J 21(1):24–29
Isaacson JE et al (2003) Clinical associations between tinnitus and chronic pain. Otolaryngol Head Neck Surg 128(5):706–710
Juengst SB et al (2014) Exploratory associations with tumor necrosis factor-alpha, disinhibition and suicidal endorsement after traumatic brain injury. Brain Behav Immun 41:134–143
Jury MA, Flynn MC (2001) Auditory and vestibular sequelae to traumatic brain injury: a pilot study. N Z Med J 114(1134):286–288
Kiguchi N, Kobayashi Y, Kishioka S (2012) Chemokines and cytokines in neuroinflammation leading to neuropathic pain. Curr Opin Pharmacol 12(1):55–61
Kim HA et al (2014) Brain immune cell composition and functional outcome after cerebral ischemia: comparison of two mouse strains. Front Cell Neurosci 8:365
Kohno M et al (2019) Neuroinflammation in addiction: a review of neuroimaging studies and potential immunotherapies. Pharmacol Biochem Behav 179:34–42
Langguth B et al (2011) Tinnitus and depression. World J Biol Psychiatry 12(7):489–500
Lerman I et al (2016) Posttraumatic stress disorder influences the nociceptive and intrathecal cytokine response to a painful stimulus in combat veterans. Psychoneuroendocrinology 73:99–108
Levine J et al (1999) Cerebrospinal cytokine levels in patients with acute depression. Neuropsychobiology 40(4):171–176
Lew HL et al (2007) Auditory dysfunction in traumatic brain injury. J Rehabil Res Dev 44 (7):921–928
Llinas R et al (2005) Rhythmic and dysrhythmic thalamocortical dynamics: GABA systems and the edge effect. Trends Neurosci 28(6):325–333
Mazurek B, Szczepek AJ, Hebert S (2015) Stress and tinnitus. HNO 63(4):258–265
Miyakawa A et al (2019) Tinnitus correlates with downregulation of cortical glutamate decarboxylase 65 expression but not auditory cortical map reorganization. J Neurosci 39 (50):9989–10001
Najjar S, Pearlman DM (2015) Neuroinflammation and white matter pathology in schizophrenia: systematic review. Schizophr Res 161(1):102–112
Norena AJ (2015) Revisiting the cochlear and central mechanisms of tinnitus and therapeutic approaches. Audiol Neurootol 20(Suppl 1):53–59
Pace E et al (2016) A conditioned behavioral paradigm for assessing onset and lasting tinnitus in rats. PLoS One 11(11):e0166346
Perez CJ et al (2013) Several classical mouse inbred strains, including DBA/2, NOD/Lt, FVB/N, and SJL/J, carry a putative loss-of-function allele of Gpr84. J Hered 104(4):565–571
Perrin C et al (2003) Calpain and caspase-3 inhibitors reduce infarct size and post-ischemic apoptosis in rat heart without modifying contractile recovery. Cell Mol Biol (Noisy-le-Grand) 49 Online Pub:OL497–OL505
Popeo DM, Tobias KG, Kellner CH (2011) Tinnitus as a symptom of psychotic depression successfully treated with electroconvulsive therapy. J ECT 27(1):e7–e8
Roberts LE et al (2010) Ringing ears: the neuroscience of tinnitus. J Neurosci 30(45):14972–14979
Ruetten H, Thiemermann C (1997) Effect of calpain inhibitor I, an inhibitor of the proteolysis of I kappa B, on the circulatory failure and multiple organ dysfunction caused by endotoxin in the rat. Br J Pharmacol 121(4):695–704
Salter MW, Beggs S (2014) Sublime microglia: expanding roles for the guardians of the CNS. Cell 158(1):15–24
Schain M, Kreisl WC (2017) Neuroinflammation in neurodegenerative disorders – a review. Curr Neurol Neurosci Rep 17(3):25
Serhan CN, Levy BD (2018) Resolvins in inflammation: emergence of the pro-resolving superfamily of mediators. J Clin Invest 128(7):2657–2669

Shabab T et al (2017) Neuroinflammation pathways: a general review. Int J Neurosci 127 (7):624–633
Shore SE, Roberts LE, Langguth B (2016) Maladaptive plasticity in tinnitus – triggers, mechanisms and treatment. Nat Rev Neurol 12(3):150–160
Shulman A, Goldstein B (2005) Subjective idiopathic tinnitus. A review of clinical experience 1979–2005. Otorinolaringologia 55(1):23
Simon DW et al (2017) The far-reaching scope of neuroinflammation after traumatic brain injury. Nat Rev Neurol 13(3):171–191
Stefaniak J, O'Brien J (2016) Imaging of neuroinflammation in dementia: a review. J Neurol Neurosurg Psychiatry 87(1):21–28
Steinmetz CC, Turrigiano GG (2010) Tumor necrosis factor-alpha signaling maintains the ability of cortical synapses to express synaptic scaling. J Neurosci 30(44):14685–14690
Stellwagen D, Malenka RC (2006) Synaptic scaling mediated by glial TNF-alpha. Nature 440 (7087):1054–1059
Stellwagen D et al (2005) Differential regulation of AMPA receptor and GABA receptor trafficking by tumor necrosis factor-alpha. J Neurosci 25(12):3219–3228
Szczepek AJ et al (2014) Biological correlates of tinnitus-related distress: an exploratory study. Hear Res 318:23–30
Tacchini-Cottier F et al (2000) An immunomodulatory function for neutrophils during the induction of a CD4+ Th2 response in BALB/c mice infected with Leishmania major. J Immunol 165 (5):2628–2636
Tan WJ, Thorne PR, Vlajkovic SM (2013) Noise-induced cochlear inflammation. World J Otorhinolaryngol 3(3):89–99
Tan WJ, Thorne PR, Vlajkovic SM (2016) Characterisation of cochlear inflammation in mice following acute and chronic noise exposure. Histochem Cell Biol 146(2):219–230
Troubat R et al (2021) Neuroinflammation and depression: a review. Eur J Neurosci 53(1):151–171
Trune DR, Canlon B (2012) Corticosteroid therapy for hearing and balance disorders. Anat Rec Adv Integr Anat Evol Biol 295(11):1928–1943
Walker AK et al (2014) Neuroinflammation and comorbidity of pain and depression. Pharmacol Rev 66(1):80–101
Wang KK (2000) Calpain and caspase: can you tell the difference?, by kevin K.W. Wang Vol. 23, pp. 20–26. Trends Neurosci 23(2):59
Wang J et al (1999) Leupeptin protects sensory hair cells from acoustic trauma. Neuroreport 10 (4):811–816
Wang W et al (2019) Neuroinflammation mediates noise-induced synaptic imbalance and tinnitus in rodent models. PLoS Biol 17(6):e3000307
Watson N et al (2017) Chronic inflammation – inflammaging – in the ageing cochlea: a novel target for future presbycusis therapy. Ageing Res Rev 40:142–148
Whitehead GS et al (2003) Allergen-induced airway disease is mouse strain dependent. Am J Physiol Lung Cell Mol Physiol 285(1):L32–L42
Wolf SA, Boddeke HW, Kettenmann H (2017) Microglia in physiology and disease. Annu Rev Physiol 79:619–643
Wright EF, Gullickson DC (1996) Dental pulpalgia contributing to bilateral preauricular pain and tinnitus. J Orofac Pain 10(2):166–168
Xia C et al (2020) Neuroglial activation in the auditory cortex and medial geniculate body of salicylate-induced tinnitus rats. Am J Transl Res 12(10):6043–6059
Ziebell JM, Morganti-Kossmann MC (2010) Involvement of pro- and anti-inflammatory cytokines and chemokines in the pathophysiology of traumatic brain injury. Neurotherapeutics 7(1):22–30
Zinsmaier AK et al (2020) Resistance to noise-induced GAP detection impairment in FVB mice is correlated with reduced neuroinflammatory response and Parvalbumin neuron loss. Sci Rep 10:20445

Using Big Data to Develop a Clinical Decision Support System for Tinnitus Treatment

Winfried Schlee, Berthold Langguth, Rüdiger Pryss, Johannes Allgaier, Lena Mulansky, Carsten Vogel, Myra Spiliopoulou, Miro Schleicher, Vishnu Unnikrishnan, Clara Puga, Ourania Manta, Michalis Sarafidis, Ioannis Kouris, Eleftheria Vellidou, Dimitris Koutsouris, Konstantina Koloutsou, George Spanoudakis, Christopher Cederroth, and Dimitris Kikidis

Contents

W. Schlee (✉) and B. Langguth
Department of Psychiatry and Psychotherapy, University of Regensburg, Regensburg, Germany

R. Pryss, J. Allgaier, L. Mulansky, and C. Vogel
Institute of Clinical Epidemiology and Biometry, University of Würzburg, Würzburg, Germany

M. Spiliopoulou, M. Schleicher, V. Unnikrishnan, and C. Puga
Faculty of Computer Science, Otto von Guericke University Magdeburg, Magdeburg, Germany

O. Manta, M. Sarafidis, I. Kouris, E. Vellidou, and D. Koutsouris
Institute of Communications and Computer Systems (ICCS) – National Technical University of Athens, Athens, Greece

K. Koloutsou and G. Spanoudakis
Sphynx Technology Solutions AG, Zug, Switzerland

C. Cederroth
Laboratory of Experimental Audiology, Department of Physiology and Pharmacology, Karolinska Institute, Stockholm, Sweden

Hearing Sciences, Division of Clinical Neuroscience, School of Medicine, University of Nottingham, Nottingham, UK

National Institute for Health Research (NIHR) Nottingham Biomedical Research Centre, Nottingham University Hospitals NHS Trust, Nottingham, UK

D. Kikidis
First Department of Otorhinolaryngology, Head and Neck Surgery, National and Kapodistrian University of Athens, Hippokration General Hospital, Athens, Greece

Curr Topics Behav Neurosci (2021) 51: 175–190
https://doi.org/10.1007/7854_2021_229
Published Online: 11 April 2021

Abstract Tinnitus is a common symptom of a phantom sound perception with a considerable socioeconomic impact. Tinnitus pathophysiology is enigmatic and its significant heterogeneity reflects a wide spectrum of clinical manifestations, severity and annoyance among tinnitus sufferers. Although several interventions have been suggested, currently there is no universally accepted treatment. Moreover, there is no well-established correlation between tinnitus features or patients' characteristics and projection of treatment response. At the clinical level, this practically means that selection of treatment is not based on expected outcomes for the particular patient.

The complexity of tinnitus and lack of well-adapted prognostic factors for treatment selection highlight a potential role for a decision support system (DSS). A DSS is an informative system, based on big data that aims to facilitate decision-making based on: specific rules, retrospective data reflecting results, patient profiling and predictive models. Therefore, it can use algorithms evaluating numerous parameters and indicate the weight of their contribution to the final outcome. This means that DSS can provide additional information, exceeding the typical questions of superiority of one treatment versus another, commonly addressed in literature.

The development of a DSS for tinnitus treatment selection will make use of an underlying database consisting of medical, epidemiological, audiological, electrophysiological, genetic and tinnitus subtyping data. Algorithms will be developed with the use of machine learning and data mining techniques. Based on the profile features identified as prognostic these algorithms will be able to suggest whether additional examinations are needed for a robust result as well as which treatment or combination of treatments is optimal for every patient in a personalized level.

In this manuscript we carefully define the conceptual basis for a tinnitus treatment selection DSS. We describe the big data set and the knowledge base on which the DSS will be based and the algorithms that will be used for prognosis and treatment selection.

Keywords Big data · Clinical decision support system · Personalized treatment · Tinnitus

1 Introduction

Tinnitus is defined as the conscious awareness of a tonal or composite noise for which there is no identifiable corresponding external acoustic source (De Ridder et al. 2021). In many cases it is associated with emotional distress, cognitive dysfunction, or autonomic arousal and can lead to behavioural changes and functional disability (Langguth et al. 2013). Tinnitus has a high prevalence in western societies and is currently considered as a complex chronic disorder which can be caused by genetic or environmental factors or combinations of such factors (Cederroth et al. 2019). While tinnitus can be well tolerated by some individuals, it can cause levels of annoyance, significant impairment and severe impact on the quality of life in other affected individuals. Therefore, any clinical approach and decision-making developed to support tinnitus treatment should take into account the individual patient's reaction to tinnitus.

At the moment, a wide range of therapeutic interventions exist, some of which have been partly successful in some groups of patients. However, no optimal and universal tinnitus treatment has been reached yet. Existing tinnitus interventions include, but are not limited to, sound amplification via hearing aids, restoration of hearing by cochlear implants, different forms of sound therapy, various psychological interventions, magnetic and electrical stimulation of the brain or nerves, physiotherapy and a wide variety of different drugs, which are used off-label (Cima et al. 2019). Cognitive behavioural therapy (CBT) has also been repeatedly shown to reduce the psychological distress caused by the tinnitus.

Despite the existence of this range of interventions, it should be noted that none of them reliably reduces the loudness of the tinnitus. In recent years, new, innovative and improved treatments have been developed, showing promising results in subgroups of patients. There is a pattern across the tinnitus literature, in which a varying subgroup of responders is found in most studies. These responders show a clinical meaningful improvement as response to a particular treatment, which can also last until the follow-up measurement point, while the non-responders are not showing any clinical meaningful improvement (Elgoyhen et al. 2015). Therefore, it has been postulated that tinnitus is a heterogeneous condition. It is assumed that there exist different forms of tinnitus that differ in their pathophysiology and in their response to a given treatment. This is in accordance with the clinical experience of practitioners who treat chronic tinnitus patients: in many cases, patients are trying different treatment approaches, one after the other, with the hope to finally find the treatment that helps in this individual case (Cima et al. 2020; Sanchez et al. 2020; Simões et al. 2019; El-Shunnar et al. 2011). A recent analysis by Simoes and colleagues showed that chronic tinnitus patients at the University of Regensburg try on average three different treatments (mean 3.0, standard deviation 2.6, (Simões et al. 2019)). A recent analysis of survey data from more than 5,000 patients revealed that responders to different treatments differ from each other in clinical characteristics, underscoring the relevance of tinnitus subtyping for treatment selection (Simões et al. 2019).

The current trial-and-error approach causes large costs for the health system and the patients and increases the waiting list in the specialized tinnitus clinics and the time patients need to invest for clinical interventions. To address this, we are developing an approach based on data analytics and decision support technology. The core idea of our approach is the development of a decision support system which, based on the analysis of individual patient data available from large database of chronic tinnitus patients, will be able to predict the best treatment for the individual patient. Such a decision support system can be used to help the medical experts in the decision-making process to find the best treatment or best combination of treatments for individual patient (personalized treatment). However, a decision support system is only helpful for daily clinical routine if it is designed in a way that it can be easily integrated in the clinical workflow and helps the physician to make informed decisions. Likewise, big data is not necessarily big information unless the data set is sampled well to cover a representative patient group and the data is collected with high data quality. In the following sections, we are discussing the underlying idea and most important requirements for a decision support system in order to be of benefit for the treatment of tinnitus.

2 Decision Support System

In general, decision support systems are information systems that support the decision-making process in situations with complex, unstructured or rapidly changing decision problems. Such decision support systems can be fully computerized, fully human-powered or a mixture of both. In the field of tinnitus research, with the large heterogeneity of tinnitus subtypes, multiple possible aetiologies of tinnitus and complex interactions between them, we hypothesize that a decision support system could improve the clinical care of tinnitus patients by providing the physician with data-driven and knowledge-driven suggestions for the best treatment, or combination of treatments, for the individual patient. In the ideal case, the medical suggestion is based on two pillars: a large and high-quality database of similar tinnitus patients and a knowledge base with reliable scientific knowledge.

A clinical decision support system should be a user-friendly platform that informs the clinician which intervention or combination of interventions is the most suitable for a particular patient with respect to efficacy and tolerability. Suggestions are made based on patient's medical history, audiological findings, socioeconomic background and tinnitus features. In case of insufficient data of a given patient to make an informed decision, the clinical decision support system should propose appropriate additional diagnostic tests for the patient. The system should enable the specification of high-level data analysis processes, which will be associated with a dynamic decision-making process and the criteria for making alternative decisions based on the outcomes of these processes. They should be automatically updated and realized by the Decision Support System. The execution of the decision-making model incorporating them, along with different datasets is enabled by the processes' specifications. This capability will allow for the realization of continually and automatically updateable intervention-related decision-making models. In order to meet these goals, a decision support system for the clinical use should meet the requirements outlined in Table 1.

Table 1 Requirements for the clinical decision support system for tinnitus

No.	Decision support system requirements (DSSREQ)
DSSREQ 1	Individual patient data can be entered, including but not limited to the medical history, demographical data, clinical examination about the tinnitus, audiogram, regular medication, genetics and biomarkers, electrophysiological data, brain scans, current and previous treatments and the respective treatment outcome
DSSREQ 2	The decision support system will be able to suggest an optimal (combination of) treatment(s) for an individual patient, in dependence of the available data for this patient. For this process it is not relevant whether a therapeutic option is *in general* better than another one – the decision support system rather suggests the best treatment for the *given individual.* In case that the system requires additional information of a patient in order to propose a treatment, it informs which additional tests should be performed.
DSSREQ 3	A patient dashboard displays all the available data of the individual patient, which are taken into account when suggesting the optimal treatment (DSSREQ2). The only data of the patients that are used by the algorithm are those displayed in the dashboard. This is to ensure that the tinnitus expert in charge has a clear overview of the data that is used for the treatment suggestion.
DSSREQ 4	The individual data can be exported in a CSV format to make it available for further and additional analysis outside of the decision support system
DSSREQ 5	To ensure reliable use in the clinical context, the system should be operable and available daily without interruptions, fault-tolerant to enable continuous operation and fully recoverable in case of system failure
DSSREQ 6	The data handling of the decision support system should be fully compliant with the general data protection regulations (GDPR) by using pseudonymized data and encryption of personal data. The ongoing confidentiality, integrity, availability and resilience of processing systems and services should be ensured. Further requirements are the ability to restore the access to personal data in a timely manner in the event of a physical or technical incident, the support of a role-based access control, the anonymization of the data in case of data transmission or data export. The patients' right to be forgotten has to be implemented, the system should follow the data minimization principle, support the territorial identification of data controllers, monitor data breach and send automated notifications to the responsible data protection officer in case of data breach
DSSREQ 7	To enable data security and integrity, the system shall provide authenticated access for all users, all users shall treat the data with confidentiality and not share them without the patients' approval. All users shall be accountable for the correct usage of the system and their activity is recorded in the logs of the system. The system shall follow a modular approach to increase fault tolerance and recovery, and it shall be modifiable according to user feedback, and easily replaceable and portable. All data shall be backed up daily and system recovery should be possible after unplanned system downtime within two working days maximum
DSSREQ 8	In order to enable further tinnitus research with the increasing database, the system shall collect and store the consent of the individual patient, allowing the usage of the anonymized data for scientific purposes. This shall be realized by an online consent form, developed on opt-in principle, so without pre-ticked boxes, not bundled with other agreement: Participants are informed that they have the right to withdraw consent at any time but that this will not affect the lawfulness of processing based on consent before its withdrawal The anonymized data can be exported for scientific purposes in machine-readable

(continued)

Table 1 (continued)

No.	Decision support system requirements (DSSREQ)
	format (e.g. CSV) and filtered prior to data export based on the consent given by the patients
DSSREQ 9	In order to increase usability, the decision support system shall be configurable and localizable for each user, shall be easy to learn and easy to use, shall provide attractive and stimulating interfaces, adopt common standards and ensure compatibility and interoperability with other systems, and provide an appropriate error handling system. Furthermore, it shall maintain system logs that enable the administrators to check major functional and non-functional aspects and analyse problems, issues, failures, inconsistencies, time behaviour and resource utilization.
DSSREQ 10	The DSS may have too little data to make suggestions for a given patient. If this is the case, the DSS should indicate its uncertainty in what it suggests as 'optimal'

DSSREQ 1–4 are functional requirements. DSSREQ 5–9 are non-functional requirements

2.1 Requirements for the Underlying Database

To ensure effective usage, the decision support system must build upon an existing database, the processes and the profile of the respective clinical centre it is installed at, and at the same time exploit the body of knowledge contributed by other centres that are contributing data to the common data repository. Sharing data and models is mission-critical for sharing data and models in scientific research, and provenance must be similarly granted when the latest scientific advances from a Randomized Clinical Trial flow into personalized treatment choice inside a clinical centre.

In their recent work on the role of provenance for collaborative in silico scientific research, Jandre and colleagues (Jandre et al. 2020) introduce a taxonomy of collaboration for scientific research and highlight the role of sharing, whereby they distinguish between sharing data and sharing models. Among the systems they inspected 13 out of 18 do support sharing of "data and models", cf. Table 2 on 'Aspects of Collaboration in the surveyed Approaches'. Many of these systems feature a centralized database and thus an agreed-upon schema over the data.

We also anticipate a centralized database and derive a set of requirements for it. At the same time, we recognize that participating clinical centres might use different schemata for their clinical workflows, so that next to the access to the centralized database, further requirements concerning 'data and model' sharing must be specified and satisfied.

The challenge of model sharing is further exacerbated by two further facts. First, the patient populations are likely to differ among the centres. Moreover, each centre is likely to concentrate on different symptoms and outcomes of interest.

Hence, the challenges of sharing inside the decision support system led to several requirements towards the centres who want to incorporate the decision support system into their existing clinical workflows that are described in Table 2.

Table 2 Requirements for the database underlying the clinical decision support system

No.	Database requirements (DBREQ)
DBREQ 1	Legal agreement for data sharing among the participating clinical centres
DBREQ 2	Unique patient ID, allowing that all information appertaining to the same patient, including multiple treatments at different timepoints, as well as mHealth recordings if any, are linked to the same individual
DBREQ 3	Provenance for sharing of classifiers and regressors for single-output and multi-output learning (for multiple outcomes of interest)
DBREQ 4	Import/export utilities for data and for models
DBREQ 5	Machine learning algorithms for model transfer (a) from the database to the clinical centre and (b) from one centre to another while taking account of • discrepancies in the feature spaces • discrepancies in the data distributions
DBREQ 6	Next to requirements for sharing, the centralized database must satisfy further requirements, namely: Solution to following data heterogeneity issues • large spectrum of medical data • both cross-sectional and longitudinal data, including screening, baseline, visits during treatment, visit after treatment, follow-up visits • data on as many types of tinnitus as possible to cover the heterogeneity of tinnitus • data on different outcomes of interest, including, e.g. mental health indicators
DBREQ 7	High quality data, meaning data that is reliable and valid and has been recorded according to the current international standards of tinnitus research and with a low number of missing values.
DBREQ 8	Machine learning algorithms for outlier handling, whereby the term "outlier" needs to be formalized
DBREQ 9	Protocol for expansion of the database: As new patients are treated, some of them may be included into the database, while others may not, e.g. because of data quality issues or because of data protection constraints
DBREQ 10	Machine learning algorithms for adaptation of the models over the expanding database

Requirements DBREQ3, DBREQ5, DBREQ8 and DBREQ10, which concern algorithms

2.2 *Requirements for the Underlying Knowledge Base*

The decision support system derives the suggested optimal treatment strategy based on the individual data of the chronic tinnitus patient, already existing data from the database and already existing scientific knowledge. For the existing scientific knowledge, it is important that the knowledge is highly reliable and based on multiple evidence. Therefore, we propose certain requirements, which are outlined in Table 3.

Table 3 Requirements for algorithms underlying for the data models of the clinical decision support system

No.	Requirements for the knowledge base (KNOREQ)
KNOREQ 1	The knowledge base for the decision support system should be based on reliable scientific knowledge that is based on aggregated information of high methodological quality such as Cochrane reviews or clinical guidelines
KNOREQ 2	The clinician should be informed about which scientific knowledge has been used for the suggested decision. For each individualized decision, that is based on information from the knowledge base, the system should show the respective scientific reference – but not a long and exhaustive list of the entire knowledge base in the system
KNOREQ 3	The reference to the original scientific literature should be available and easily retrievable by the clinicians. This should allow quick cross-checks by the clinician and also encourage further reading and education

2.3 Requirements for the Algorithms

The algorithms underlying the decision support system will encompass models learned on the original data of the clinical centre and models learned on the data of the centralized database, and thus demands (a) algorithms that deliver these models and (b) algorithms that support the requirements towards the centralized database, cf. Sect. 2.2. As pointed out in Sect. 2.2, a major challenge to be solved at algorithmic level is the inherent disagreement between the data, models and outcomes in the centralized database and those in the clinical procedures of the centres involved. This challenge translates into the need for solutions to following requirements outlined in Table 4.

3 Decision Support System Within the UNITI Project

Research on decision support systems in the field of tinnitus is still in its infancy. With the EU-funded project 'Unification of treatments and interventions for tinnitus patients' (UNITI, (Schlee et al. 2021)) a first project on this topic was started in 2020. In this chapter, we describe the decision support within the UNITI project in more detail, which builds on the requirements outlined above.

In general, the utilization of decision support systems in clinical practice can reduce human cognitive deficiencies by integrating various sources of information, providing intelligent access to relevant knowledge and supporting the decision-making process. These systems are intended for assisting clinicians to overcome their knowledge limits and stand out for their ability to combine and factor multiple items of patient data in Dumitrescu Peculea and Ion Chitescu (2015). The promised benefits of clinical decision support systems rest in large part on their ability to use patient-specific data, learn by a large training dataset and, finally, provide personalized recommendations for care, by turning health observations into health

Table 4 Requirements for algorithms underlying for the data models of the clinical decision support system

No.	Requirements of the algorithms (ALGREQ)
ALGREQ 1	Management of missing values in an expert-approved way: Missing values emerge traditionally when questionnaire items are not filled by a patient; the use of different questionnaires leads, additionally, to systematic patterns of missingness, since each centre uses partially different questionnaires
ALGREQ 2	Identification of outliers, respectively quantification of what an outlier is, since in the context of a decision support system no patient is an outlier, but some patients may be so different from any other that no recommendations can be reliably done [refines DBREQ8]
ALGREQ 3	Quantification of the uncertainty of each model with respect to each treatment and to each outcome of interest, covering multiple outcomes of interest as multi-targets [complements DBREQ3]
ALGREQ 4	Exploitation of models learned in the centralized database and applied in each centre, since there are discrepancies between centralized database and centres with respect to feature space, data distribution and outcomes of interest [corresponds to DBREQ5]
ALGREQ 5	Learning on few data – models should achieve high prediction quality without demanding data that are too expensive to acquire; such algorithms also deliver a partial solution to the problem of systematic data missingness (cf. ALG1)
ALGREQ 6	Model adaptation, since the centralized database may grow with additional patient data from questionnaires and assessments, while the data distribution may change with or without effect on the targets [refines DBREQ10]
ALGREQ 7	The model induced for the formulations of suggestions must be explainable in the original feature space of questionnaires and assessments
ALGREQ 8	Models should be easy to query and should deliver answers without much waiting time in order to be of practical use for the clinical context
ALGREQ 9	Detection of low-quality data, e.g. typos and answers that are out of range, as well as detection of potential fraud and appropriate handling of the data in this case

knowledge (Kanatas et al. 2020). In other words, the clinical decision support system mimics the decision-making of an experienced and knowledgeable clinician with the difference that the information on which the decision is based is much larger than the number of patients that an experienced clinical expert can see in their lifetime.

The decision-making to manage and follow up the tinnitus patients is complex. There are a lot of tinnitus characteristics and different therapy care plans and guidelines to manage different individuals. This adds a high complexity and requires an active knowledge system in order to aid decision makers in the optimum option. The implementation of a decision support system will be designed to help experts in the tinnitus field to alleviate this problem, as it can provide timely information for the optimal treatment in an individualized level and out of the range of universally accepted treatments.

The EU project UNITI will develop a decision support system, which can process data from various heterogeneous sources, obtained statically or dynamically from systems monitoring lifestyles and behaviours of tinnitus affected chronic patients, bringing an advanced technologically and effective solution (Fig. 1). More

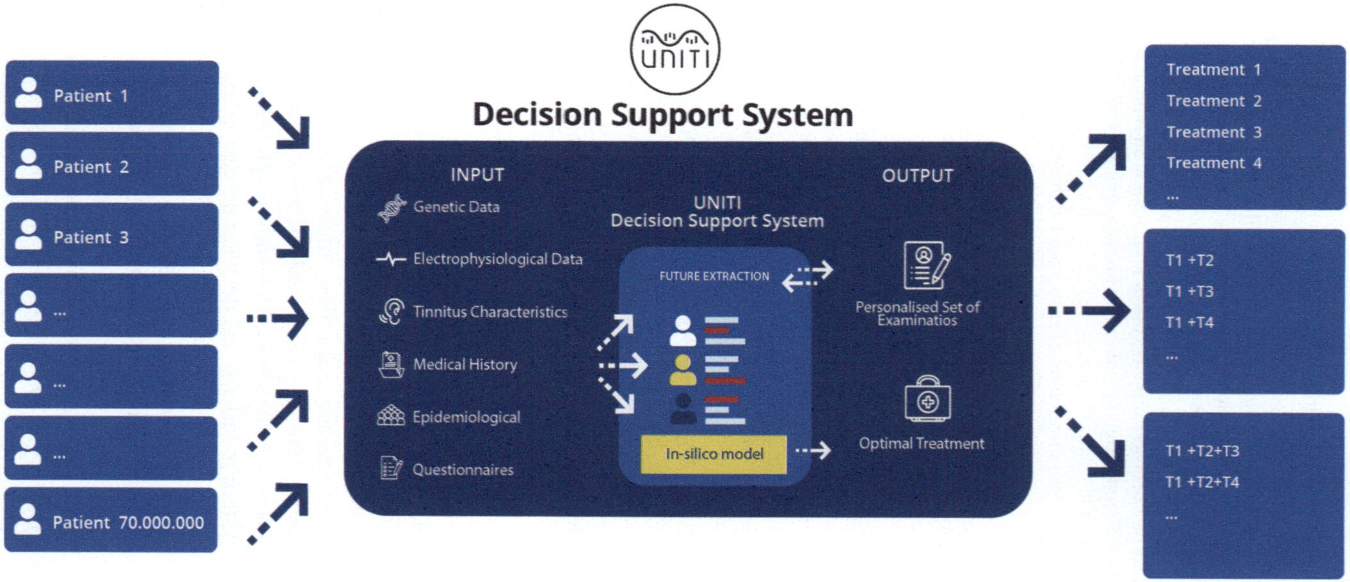

Fig. 1 The overall overview of UNITI's decision support system. Feedback about the efficacy of the proposed treatment will be stored in the database to allow updating of the models and improvement of the predictions over time

specifically, UNITI's decision support system will integrate epidemiological, clinical (e.g., tinnitus characteristics), and medical history data, along with more specialized data, if needed. Within this context, the decision support system will propose a set of additional specialized diagnostic tests, which can be electrophysiological measurements or genetic tests, if they are deemed essential according to the individual patient's profile. Last but not least, the decision support system will incorporate the patient's responses to the tinnitus clinical questionnaires, collected through dedicated mobile apps. These aforementioned data types will be gathered during the randomized clinical trial (RCT) which will be conducted during the UNITI project. All these collected data will contribute in creating a well-rounded patient profile and will be utilized in the analysis of the effects arising from the adoption of specific treatments to specific individuals. Therefore, UNITI aims to the development of such a decision support system which takes into consideration the wider profile of each individual end-user. The latter will enable the developed system to reason and identify treatments with poor results and adapt the relevant decision-making process according to an individual's profile, proposing the optimal treatment combination therapies for them.

The development of a fully functional decision support system at the end of the project will contribute to address some major issues preventing clinicians and researchers from obtaining a universal tinnitus treatment. These issues are briefed subsequently.

- *The inclusion of prior (clinical) knowledge on discovery learning*: decision support systems' relevance to theoretical and technical issues is a prominent factor in order to support the enhanced decision-making. The existing knowledge derived from the clinical guidelines and systematic reviews is also necessary to address the research question which is the optimal personalized treatment. UNITI's decision support system will be informed by the relevant literature on tinnitus management. Theory validation will be carefully addressed in the decision support system, enabling improvements in decision-making.

  - *Intelligent Data Mining and Analysis*: Improved data accessibility is often a major motivation for building a data-driven decision support system. Advocates of building data warehouses identify the possibility of more and better statistical analysis that can improve decision-making. The UNITI consortium has already stored more than 10,000 patient records in different databases. All these records will be anonymized and unified according to an enhanced database schema. Patients' data will be stored via user-friendly interfaces, facilitating quick, reliable, and guided data entry, providing error-pruning mechanisms. The decision support system aims at empowering new users to use data analysis effectively, starting with basic recommendations, through providing interactive assistance, and eventually giving contextual support also in the scope of a problem domain (http://www.salientworks.com/blog/2016/4/29/analysis-decision-support. Moreover, they can go further by enabling new analysis scenarios beyond the original configuration of an individual working with a dataset.

- *Personalized tinnitus treatment*: UNITI's decision support system will be based on patient's medical profile data (medical history, questionnaires responses, epidemiological data) and will generate, through a personalized user-friendly wizard, suggestions of the minimum required examinations necessary for the optimum treatment selection. The decision support system results will retrofit the predictive model to enhance the accuracy of the suggested treatments. Hence, the decision support system will contribute in shedding light on the heterogeneity of the tinnitus disease and in defining some certain tinnitus patients' groups (clusters). Moreover, the methodology of determining clear and quantitative therapeutic targets and outcome measures will be enhanced.
- *Patients explicit inclusion criteria*: All possible measures will be taken to ensure there is no discrimination or harm from the recruitment, exclusion or inclusion process. Specifically, UNITI project will adopt a non-discrimination policy in the recruitment, which means that all tinnitus patients will be included whether they are people with bothersome tinnitus or people with non-bothersome tinnitus. Having data from the full range of patients' tinnitus will lead to better decision support system training, which brings about more accurate personalized recommendations.

Conforming to GDPR, privacy and security requirements, such as privacy by design, user group access control, secure transmission channels, and anonymization prior to the unified data collection, will be adopted for transmitting and handling the data in the various layers of the UNITI platform.

4 Privacy and Ethical Concerns

The UNITI database will include different types of personal and clinical data the maintenance and use of which will need to be compliant with the General Data Protection Regulation (GDPR). Also, the use of decision support technology and the artificial intelligence (AI) for tinnitus treatment selection, based on the analysis of the underlying data can raise concerns in reference to the ethical AI directive. To address these, the design and development of the UNITI database will make use of different security and privacy control mechanisms, including pseudo-anonymization, outlier analysis, and continuous assurance.

Pseudo-anonymization will involve the substitution of data elements that can lead to the direct identification of individual patients, such as patient names and medical record numbers. Such items will be replaced with a sequence of successive pseudo-identifiers and an encrypted association system between these pseudo-identifiers that will make it impossible for any individual user of the system (other than the clinician in charge of the treatment of an individual patient) to establish a patient's real identifier through the information available to him/her. Also, data items that can indirectly identify individual patients (quasi-identifiers) will be substituted for with pseudo-identifiers and items that can lead to indirect identification (e.g., data of birth) will be partially removed as necessary to remove the possibility of direct identification (e.g., maintain only the year of a date of birth). A range of alternative techniques depending on the type of the particular data element will be used for this purpose, including data masking (e.g., removing data or part of data) and generalization (e.g., replacing years of age with age ranges). Pseudo-anonymization, through partial information removal and generalization will be done automatically in order to ensure that it is effective, given the entire data set at different instances of time. Before analysis of the data, all quasi-identifiers will be replaced by random numbers. This data anonymization procedure will be part of a continuous security and privacy assurance approach that will be used at all levels of the database, data analysis and decision support system. In this approach, a range of continuous system vulnerability analysis, penetration testing, and continuous runtime monitoring will be used to provide a continuous and multi-faceted assessment of the security and privacy posture of the UNITI system. Continuous monitoring of data access requests and instances of automated processing upon individual patient data that led to treatment decisions for the respective patients will be necessary in order to respond to GDPR requests.

5 Summary

The challenge of finding a treatment for chronic tinnitus is in large parts characterized by the challenge of dealing with the heterogeneity of tinnitus. There are multiple causes for tinnitus, multiple aetiologies exist, and the treatment of the chronic tinnitus patient needs to be tailored to the individual clinical case. Depending on the individual clinical case, a single treatment or a combination of treatments may be optimal.

Here, we suggest a new approach that takes advantage of big data and data analysis to develop a decision support system that helps the clinician in making the best treatment decision for the individual patient. Such a decision support system is expected to (1) identify factors that could predict treatment outcome and facilitate treatment selection, (2) identify patterns of response and rules (e.g., patients with hearing loss below a certain limit are better candidates for a certain treatment or females with recent onset for another, (3) identify a set of parameters (rather than one) that are important for treatment selection and projection of treatment outcomes, (4) weight the contribution of each one of the parameters (some could be important for certain profiles and some for other profiles), and (5) provide suggestions for a personalized treatment protocol in a meaningful, practical and implementable way.

It has to be said here that such a DSS comes with the limitation that it can only suggest therapeutic interventions, or combinations of interventions, where enough clinical data is already available. Therapeutic interventions that are new and innovative can only be suggested by the system as soon as there is a critical amount of data about this intervention available in the database. Such a system is therefore conservative by nature. This is an important limitation that should be kept in mind by the professionals that are using the system in their clinical routine.

Acknowledgements This work was partly funded by the ESIT (European School for Interdisciplinary Tinnitus Research, Schlee et al. 2018) project financed by European Union's Horizon 2020 research and innovation programme under the Marie Sklodowska-Curie grant agreement number 722046 and the UNITI (Unification of Treatments and Interventions for Tinnitus Patients, Schlee et al. 2021) project financed by the European Union's Horizon 2020 Research and Innovation Programme, Grant Agreement Number 848261.

Declaration of Interest C.R.C. is supported by the UK National Institute for Health Research (NIHR) Biomedical Research Centre but the views expressed herein are his own and do not represent those of NIHR nor the UK Department of Health and Social Care.

References

Cederroth CR, Gallus S, Hall DA et al (2019) Editorial: towards an understanding of tinnitus heterogeneity. Front Ag Neurosci 11:61–67

Cima RFF, Mazurek B, Haider H, Kikidis D, Lapira A, Norena A, Hoare DJ (2019) A multidisciplinary European guideline for tinnitus: diagnostics, assessment, and treatment. HNO 67:10–42

Cima RFF, Kikidis D, Mazurek B, Haider H, Cederroth CR, Norena A, Lapira A, Bibas A, Hoare DJ (2020) Tinnitus healthcare: a survey revealing extensive variation in opinion and practices across Europe. BMJ Open 10:e029346–e029312

De Ridder D, Schlee W, Vanneste S, Londero A, Weisz N, Kleinjung T, Shekhawat GS, Elgoyhen AB, Song JJ, Andersson G, Adhia D, de Azevedo AA, Baguley DM, Biesinger E, Binetti AC, Del Bo L, Cederroth CR, Cima R, Eggermont JJ, Figueiredo R, Fuller TE, Gallus S, Gilles A, Hall DA, Van de Heyning P, Hoare DJ, Khedr EM, Kikidis D, Kleinstaeuber M, Kreuzer PM, Lai JT, Lainez JM, Landgrebe M, Li LP, Lim HH, Liu TC, Lopez-Escamez JA, Mazurek B, Moller AR, Neff P, Pantev C, Park SN, Piccirillo JF, Poeppl TB, Rauschecker JP, Salvi R, Sanchez TG, Schecklmann M, Schiller A, Searchfield GD, Tyler R, Vielsmeier V, Vlaeyen JWS, Zhang J, Zheng Y, de Nora M, Langguth B (2021) Tinnitus and tinnitus disorder: theoretical and operational definitions (an international multidisciplinary proposal). Prog Brain Res 260:1–25. https://doi.org/10.1016/bs.pbr.2020.12.002. Epub 2021 Feb 1. PMID: 33637213

Dumitrescu Peculea A, Ion Chitescu R (2015) The role of decision support systems (DSS) in the optimization of public sector management. In: FINIZ 2015. Singidunum University, Belgrade, pp 137–141

Elgoyhen AB, Langguth B, De Ridder D, Vanneste S (2015) Tinnitus: perspectives from human neuroimaging. Nat Rev Neurosci 16:632–642

El-Shunnar SK, Hoare DJ, Smith S, Gander PE, Kang S, Fackrell K, Hall DA (2011) Primary care for tinnitus: practice and opinion among GPs in England. J Eval Clin Pract 17:684–692

Jandre E, Diirr B, Record VBAS (2020) Provenance in collaborative in silico scientific research: a survey. Dlacmorg 49:36–51

Kanatas P, Travlos IS, Gazoulis I, Tataridas A, Tsekoura A, Antonopoulos N (2020) Benefits and limitations of decision support systems (DSS) with a special emphasis on weeds. Agronomy 10:548–510

Langguth B, Kreuzer PM, Kleinjung T, De Ridder D (2013) Tinnitus: causes and clinical management. Lancet Neurol 12:920–930

Sanchez TG, Valim CC, Schlee W (2020) Long-lasting total remission of tinnitus: a systematic collection of cases. Progress Brain Res 260:269–282

Schlee W, Schoisswohl S, Staudinger S, Schiller A, Lehner A, Langguth B, Schecklmann M, Simoes J, Neff P, Marcrum SC, Spiliopoulou M, Niemann U, Schleicher M, Unnikrishnan V, Puga C, Mulansky L, Pryss R, Vogel C, Allgaier J, Giannopoulou E, Birki K, Liakou K, Cima R, Vlaeyen JWS, Verhaert N, Ranson S, Mazurek B, Brueggemann P, Boecking B, Amarjargal N, Specht S, Stege A, Hummel M, Rose M, Oppel K, Dettling-Papargyris J, Lopez-Escamez JA, Amanat S, Gallego-Martinez A, Escalera-Balsera A, Espinosa-Sanchez JM, Garcia-Valdecasas J, Mata-Ferron M, Martin-Lagos J, Martinez-Martinez M, Martinez-Martinez MJ, Müller-Locatelli N, Perez-Carpena P, Alcazar-Beltran J, Hidalgo-Lopez L, Vellidou E, Sarafidis M, Katrakazas P, Kostaridou V, Koutsouris D, Manta R, Paraskevopoulos E, Haritou M, Elgoyhen AB, Goedhart H, Koller M, Shekhawat GS, Crump H, Hannemann R, Holfelder M, Oberholzer T, Vontas A, Trochidis I, Moumtzi V, Cederroth CR, Koloutsou K, Spanoudakis G, Basdekis I, Gallus S, Lugo A, Stival C, Borroni E, Markatos N, Bibas A, Kikidis D (2021) Towards a unification of treatments and interventions for tinnitus patients: the EU research and innovation action UNITI. Prog Brain Res 260:441–451. https://doi.org/10.1016/bs.pbr.2020.12.005. Epub 2021 Feb 4. PMID: 33637231

Simões J, Neff P, Schoisswohl S, Bulla J, Schecklmann M, Harrison S, Vesala M, Langguth B, Schlee W (2019) Toward personalized tinnitus treatment: an exploratory study based on internet crowdsensing. Front Public Health 7:1600–1610

Part IV
Pharmacotherapy and Neuromodulation of Tinnitus

Pharmacotherapy of Tinnitus

Tobias Kleinjung and Berthold Langguth

Contents

Abstract Tinnitus is a common symptom for which there is in most cases no causal therapy. The search for an improvement of tinnitus through pharmacological interventions has a long tradition. The observation that tinnitus can be transiently suppressed by the use of lidocaine has shown that the symptom is susceptible to pharmacotherapy. So far, however, no medication has been found for either acute or chronic subjective tinnitus that reliably leads to a long-term reduction or even complete disappearance of the symptom for the majority of tinnitus sufferers. Nevertheless, in everyday clinical life, drugs are frequently used, usually off-label, to relieve tinnitus or tinnitus-associated symptoms (e.g. sleep disturbance, depression, anxiety disorder or hearing loss). This chapter shows the different approaches to acute and chronic subjective tinnitus by means of pharmacotherapeutic interventions. Furthermore, this review reports on the scientific studies carried out in this area in recent years and explains the difficulties in finding a suitable medication for most forms of tinnitus. In addition, it reports on the pharmacotherapeutic options for

T. Kleinjung (✉)
Department of Otorhinolaryngology, Head and Neck Surgery, University Hospital Zurich, University of Zurich, Zurich, Switzerland
e-mail: tobias.kleinjung@usz.ch

B. Langguth
Department of Psychiatry and Psychotherapy, Interdisciplinary Tinnitus Center, University of Regensburg, Regensburg, Germany
e-mail: berthold.langguth@medbo.de

Curr Topics Behav Neurosci (2021) 51: 193–212
https://doi.org/10.1007/7854_2020_169
Published Online: 7 August 2020

objective tinnitus and describes the development of tinnitus as a side effect of certain drugs. Finally, possible target structures are mentioned, which should possibly be addressed in pharmacological studies in the near future.

Keywords Acute tinnitus · Chronic tinnitus · Drug therapy · Local · Objective tinnitus · Pharmacology · Subjective tinnitus · Systemic

Abbreviations

ACE	Angiotensin-converting enzyme
AMPA	α-Amino-3-hydroxy-5-methyl-4-isoxazolepropionic acid
BT	Botulinum toxin
EMA	European Medicines Agency
FDA	US Food and Drug Administration
GABA	Gamma aminobutyric acid
$GABA_A$	Gamma aminobutyric acid type A
HMG-CoA	3-Hydroxy-3-methylglutaryl-coenzym-A
MEM	Middle ear myoclonus
NAC	n-Acetyl-l-cysteine
NMDA	n-Methyl-d-aspartate
NSAID	Non-steroidal anti-inflammatory drug
PM	Palatal myoclonus
SM	Stapedial myoclonus
SNRI	Serotonin-norepinephrine reuptake inhibitor
SSRI	Selective serotonin reuptake inhibitor
TCA	Tricyclic antidepressants
THI	Tinnitus Handicap Inventory
TTM	Tensor tympani muscle
TVP	Tensor veli palatini

1 Introduction

Currently no drug has been approved by the Food and Drug Administration (FDA) or the European Medicines Agency (EMA) for the treatment of tinnitus. At the same time, the majority of tinnitus patients would welcome a drug treatment for their tinnitus (Tyler 2012). Fifty-two percent of all patients would be very likely to try medication if it would offer tinnitus loudness and annoyance reduction of a half, rising to 62% if it would offer the chance of complete elimination of the percept. Obviously, there is a huge discrepancy between available evidence-based drug treatments and the patients' need. Thus, the question arises why there are no effective drug treatments for tinnitus available. In principle, there are two possibilities: Either tinnitus cannot be pharmacologically targeted or tinnitus is amenable to pharmacological treatment, but the right compound has not yet been identified.

Current pathophysiological models of tinnitus all converge in the assumption that tinnitus results from alterations of neuronal activity in the central nervous system. This implies that tinnitus can be pharmacologically assessed and treated. Further proof for this assumption is the transient dose-dependent reduction of tinnitus in up to 70% of patients after intravenous application of the voltage-gated sodium channel blocker lidocaine (Trellakis et al. 2007). Unfortunately, because of poor bioavailability after oral intake, lidocaine is only effective when applied intravenously. Moreover, its effect is short-lasting, and side effects are considerable, precluding the intravenous application of lidocaine as a long-term pharmacological treatment.

If tinnitus can be pharmacologically influenced, why is there no approved drug treatment? Different reasons may account for this situation. First, it is the lack of serendipity. Most pharmacological compounds for neuropsychiatric disorders were identified by chance. However, in the case of tinnitus, such a serendipitous discovery did not happen. A second reason is the still incomplete understanding of tinnitus pathophysiology. Third, there exists no validated assay for high-throughput screening of pharmacological compounds in order to assess their potential as a drug for tinnitus. Fourth, there probably exist different forms of tinnitus, which require different treatments. Finally, the measurement of tinnitus is not trivial, which makes assessment of potential treatment effects challenging.

In this chapter, we will give an overview about the current knowledge of pharmacological treatment of tinnitus. We will discuss acute and chronic tinnitus separately, as the mechanisms involved in the generation of tinnitus probably differ from the mechanisms that are relevant for the maintenance of chronic tinnitus. Presumably different treatment approaches may be needed for acute and chronic tinnitus, respectively.

Moreover, we will discuss the possible ways of administration. In addition to oral or intravenous administration, which leads to a systemic effect, also the topical administration to the cochlea or other structures around the ear has been investigated.

2 Rationale for Pharmacotherapy

In order to develop pharmacological compounds for a given disorder, the identification of appropriate drug targets is of utmost relevance (Morgan et al. 2018). Several approaches can be followed in order to identify potential drug targets. Very helpful are serendipitous discoveries of unexpected effects of existing pharmacological compounds for other indications. Based on the knowledge of the pharmacological mechanisms of this compound, one can identify potential drug targets.

As already mentioned in the introduction, the transient suppression of tinnitus after intravenous lidocaine injection is the proof of principle that tinnitus can be modulated by pharmacological treatment. As lidocaine cannot be used continuously because of its side effects and the need for intravenous application, researchers have tried to identify other compounds with similar mechanisms of action as lidocaine but with the possibility of an oral administration and a more favourable side effect profile. In order to identify whether the tinnitus suppressive effect of lidocaine depends on its action on the cochlea or the central nervous system, patients with tinnitus after vestibular schwannoma surgery have been investigated (Baguley et al. 2005). As lidocaine also suppresses tinnitus in this group, it has been concluded that the central nervous system is the relevant target for tinnitus suppression by lidocaine. Unfortunately, other drugs that mimic lidocaine effects on the central nervous system such as oral tocainide, flecainide, mexiletine and carbamazepine have not been proven to be effective against tinnitus (Dobie 1999; Kay 1981; Hoekstra et al. 2011).

Potential drug targets can also be derived from genetic research or from a detailed knowledge of the pathophysiological mechanisms. If a genetic polymorphism is related to tinnitus, the molecular structure for which the gene is coding represents a potential target structure for treatment. In the case of tinnitus, an increasing amount of genetic studies supports tinnitus heritability (Lopez-Escamez and Amanat 2020). These studies also identified possible candidate genes and associated molecular structures. However, further research is needed to validate these findings and to identify whether the relevant structures can be pharmacologically targeted.

Concerning the pathophysiology of tinnitus, several structures in the cochlea and the brain as well as certain neurotransmitter systems have been identified to be involved (Shore et al. 2016). In the central auditory pathways, glycinergic, GABAergic and glutamatergic neurotransmission are of relevance as well as potassium channels and hyperpolarization-activated cyclic nucleotide-gated channels.

For most of these targets, pharmacological compounds have been investigated. The result of these trials will be reported further below.

Recently, an unconventional approach has been made to identify potential drug target candidates (Elgoyhen et al. 2012, 2014). This approach has been based on the experience that tinnitus was reported as a side effect of many drugs. In order to identify molecular structures that are relevant to tinnitus generation, the information from a drug side-effect database and a drug-target database was integrated. A network of 1,313 drug-target pairs, based on 275 compounds that elicit tinnitus as side effect and their targets reported in databases, was constructed, and a quantitative score was applied to identify emergent targets that were more common than expected at random. In order to control for potential non-specific effects, these analyses were complemented with similar analyses for hearing impairment and hyperacusis. The analysis confirmed targets, known to be involved in tinnitus generation, include cyclooxygenase inhibitors and serotonin receptors and also identified novel emergent protein targets such as the angiotensin-converting enzyme (ACE).

3 Acute Tinnitus

Acute tinnitus is typically defined as a ringing in the ears that persists for less than 3 months. The pathophysiological mechanisms of tinnitus development in the acute phase and the mechanisms of maintenance and establishment of tinnitus perception in the chronic phase are not identical. Accordingly, the pharmacological therapy approaches in the respective phase also aim at different mechanisms of action. In the acute phase, the primary concern is to identify the causes, which are usually in the peripheral auditory system, and to treat them if possible. However, since acute tinnitus often disappears on its own, it must be noted that a certain period of 2–3 days may be waited with therapeutic considerations and that there is no medical emergency. With acute tinnitus, a distinction must be made between a form without measurable and perceivable hearing loss and a form with measurable hearing loss. The hearing loss itself can be sudden (sudden deafness, acute noise trauma) or long-term, gradual (presbycusis, chronic noise damage, genetic causes). In all cases, the cause is to be found in the peripheral auditory system. Depending on the cause, there may be reversible or irreversible damage to the hair cells, the spiral ganglion neurons or the auditory nerve, which are created according to the principles of excitotoxicity, apoptosis and oxidative stress (Malgrange et al. 2015; Shore and Wu 2019; Becatti et al. 2017; Gul et al. 2017). In acute tinnitus without measurable hearing loss, there are models that explain the damage in the peripheral auditory system. The "hidden or missed hearing loss" (Kara et al. 2020; Xiong et al. 2019; Lefeuvre et al. 2019) is explained as a result of a cochlear synaptopathy between the cochlea and the termination of the cochlear nerve or as damage in nerve fibres of the cochlear nerve (Liberman et al. 2016; Paul et al. 2017). Ultimately, the pharmacotherapy of acute tinnitus is thus aimed at improving the hearing loss and thus achieving a reduction in tinnitus. In addition, even with potentially reversible mechanisms, attempts can be made to prevent permanent damage to peripheral auditory structures in the sense of otoprotection and thus to have a positive effect on both tinnitus reduction and hearing retention. The pharmacotherapy of acute tinnitus symptoms, which are related to an affection of the middle ear (e.g. otitis media, otosclerosis), will not be discussed here.

For all of the pharmacological treatment options for acute tinnitus presented in the following, it must be declared that the literature references often differ from one another and are sometimes contradictory. Therefore, the drug therapy of acute tinnitus is not undisputed and does not meet the requirements of evidence-based medical practice (Hesse and Laubert 2010). As a consequence of this situation, the FDA or the EMA has not been able to approve any drugs for the treatment of acute tinnitus (Langguth et al. 2019). Nevertheless, various pharmacotherapeutic approaches should be mentioned at this point, some of which have found their way into medical literature and practice and are also mentioned in the guidelines of various medical-scientific professional societies as options with corresponding recommendation or rejection (Zenner et al. 2017; Lewis et al. 2020; Tunkel et al. 2014; Baguley et al. 2013; Langguth et al. 2013).

Systemic or local (intratympanic) steroid therapy for sudden-onset sensorineural hearing loss is considered the standard worldwide (Chandrasekhar et al. 2019), so tinnitus complaints arising in this context can be improved in the same way by treating the hearing loss (Zhao et al. 2015; Taha et al. 2019). According to a recent review paper (Ahmadzai et al. 2019), all types of steroid therapy are better than placebo in the treatment of acute hearing loss, with the combination of intratympanic and systemic steroids having the best results. However, there is no clear evidence for the treatment of acute tinnitus without acute accompanying hearing loss with steroids. While several small studies have shown a positive effect (Shim et al. 2017; An et al. 2014), another study found a negative effect (Lee et al. 2018), as did a review (Lavigne et al. 2016). In a systematic review (Wegner et al. 2018), the vasodilator betahistine, which is frequently used in practice, was not proven to be effective in acute tinnitus. For otoprotective reasons, drugs are used under certain circumstances to avoid or improve tinnitus and hearing loss directly after a noise trauma or already prophylactically in predictable situations (e.g. with ototoxic chemotherapy). Evidence for such an effect usually comes from animal experiments (Zhu et al. 2018; Kucharava et al. 2019; Bhatta et al. 2019; Prayuenyong et al. 2020; Le Prell 2019; Lynch et al. 2005; Tillinger et al. 2018). From such investigations, some applications have already been tested in clinical practice. Here, however, the results to date are very diverse, so that no therapeutic strategies have yet found their way into clinical practice. For n-acetyl-l-cysteine (NAC), positive effects were found in connection with noise trauma in connection with military missions (Rosenhall et al. 2019). Furthermore, a superiority in the preventive effect of intratympanic administration of NAC before cisplatin chemotherapy compared to intratympanic dexamethasone injections was found (Sarafraz et al. 2018). However, another study in head and neck tumour patients prior to cisplatin chemotherapy could not confirm the effect (Yoo et al. 2014).

The antioxidant trace element zinc has been used in various studies in noise-induced hearing loss. In one study in 38 ears, the tinnitus-associated suffering measured by the Tinnitus Handicap Inventory (THI) could be improved, but not the objective hearing test parameters (Yeh et al. 2019). However, a Cochrane review could not find a positive effect of zinc supplementation on tinnitus symptoms in adults (Person et al. 2016). Further, somewhat older, positive evidence for otoprotection in acoustic trauma exists for the intake of magnesium (Sendowski 2006; Attias et al. 1994).

Glutamate plays a decisive role as an excitatory neurotransmitter in traumatic inner ear damage (Puel et al. 2002) in terms of damage to the synapses and can thus be responsible for the development of acute tinnitus in this situation. The NMDA receptor therefore plays an important role in possible therapeutic considerations (Bing et al. 2015). The NMDA receptor antagonist esketamine hydrochloride (Keyzilen®, Auris Medical, Basel, Switzerland) has been tested in recent years for its effectiveness as an intratympanically applied drug up to a phase 3 study in acute trauma-induced tinnitus. After positive indications from the early phase studies (van de Heyning et al. 2014) and more positive indications regarding the safety of the new drug (Staecker et al. 2017), the phase 3 study could not provide positive proof of efficacy of the substance (https://clinicaltrials.gov/ct2/show/NCT01803646?term=AM-10).

In summary, it is shown that in cases of acute tinnitus with accompanying hearing loss a clear statement can be made regarding steroid therapy. In acute tinnitus without newly occurring hearing loss, the situation is much more difficult. There is a lack of randomized, placebo-controlled, multicentre studies that can generate clear evidence for or against a specific pharmacological therapy. Since many of those tinnitus subjects are severely affected by the acute ringing in the ear, a causal and effective therapy would be very desirable. Unfortunately, the pharmaceutical industry has not been very active in this area at the present time, so that in many cases the medical profession will have to make a difficult decision between therapeutic nihilism from a pharmacotherapeutic point of view and a decision for off-label treatment (e.g. with steroids) even if there is insufficient scientific evidence.

4 Chronic Tinnitus

A large variety of drugs with various mechanisms of action have been investigated for the treatment of chronic subjective tinnitus (>3 months).

Antidepressants are frequently proposed for the management of chronic tinnitus as many patients with tinnitus also suffer from depressive symptoms (Langguth et al. 2011). Tricyclic antidepressants (TCA) are among the most effective drugs for the treatment of chronic neuropathic pain syndromes, which resemble tinnitus in many aspects (De Ridder et al. 2011). Accordingly, studies with nortriptyline (Sullivan et al. 1993) and amitriptyline (Bayar et al. 2001) suggest beneficial effects. In contrast, trimipramine did not differ from placebo in its effects on tinnitus (Mihail et al. 1988).

The selective serotonin reuptake inhibitors (SSRI) paroxetine and sertraline have also been tested in tinnitus. In a randomized double-blind placebo-controlled study, sertraline was significantly more effective than placebo in reducing tinnitus loudness and tinnitus severity (Zoger et al. 2006). In a double-blind, placebo-controlled study, the paroxetine group showed little difference from placebo (Robinson et al. 2005). The serotonin reuptake enhancer tianeptine has shown beneficial effects on depressive symptoms and tinnitus handicap in an open study in patients with tinnitus and depression (Hwang et al. 2016). No studies have been performed with other antidepressants such as the serotonin-norepinephrine reuptake inhibitors (SNRI) duloxetine and venlafaxine, the norepinephrine-dopamine reuptake inhibitor bupropion, the dual-acting drug mirtazapine or the melatonin agonist agomelatine. In the interpretation of the effects of antidepressants on tinnitus, it has to be considered that the scales used for the measurement of tinnitus (e.g. Tinnitus Handicap Inventory) highly correlate with depression scales (Zeman et al. 2014). Thus, the reduction of tinnitus severity under antidepressant treatment could just reflect the antidepressant effect of the investigated drugs. Taken together, there is insufficient evidence to say that antidepressants improve tinnitus, as expressed in a recent Cochrane review (Baldo et al. 2012). From a clinical perspective, antidepressants should be considered in tinnitus patients, if they suffer from comorbid depression or anxiety.

Since tinnitus is related to an attenuation of inhibitory neurotransmission in auditory pathways (Yang et al. 2011), benzodiazepines, which are positive allosteric modulators of the $GABA_A$ (gamma aminobutyric acid type A) receptor, would be expected to alleviate tinnitus. Available clinical trials investigating benzodiazepines revealed mixed results. Whereas alprazolam had some beneficial effects on tinnitus loudness (Johnson et al. 1993; Jalali et al. 2009), diazepam had no effect (Kay 1981). Clonazepam, a long-acting benzodiazepine, has shown beneficial effects on both tinnitus loudness and annoyance as compared to placebo (Bahmad Jr. et al. 2006) or the traditional medicine *Ginkgo* (Han et al. 2012). The non-benzodiazepine hypnotics zopiclone, eszopiclone, zaleplon and zolpidem (Z-substances) have not yet been systematically investigated for the treatment of tinnitus.

Even if there are some hints for a potential benefit of benzodiazepines, caution is warranted in their prescription for the treatment of tinnitus because of their side effect profile, especially the risk of drug dependency (Bonnet 2014; Jufas and Wood 2015). The long-term treatment with benzodiazepines is not recommended. However, benzodiazepine can provide some short-term relief in severe cases, particularly in patients with co-morbid anxiety, depression and insomnia. In these cases, the use of benzodiazepines should be embedded in a multimodal treatment plan.

Anticonvulsants reduce neuronal excitability and are also used in the treatment of several non-epileptic conditions, including various psychiatric disorders and pain syndromes. Their pharmacological mechanisms of action include effects on voltage-gated sodium, calcium and potassium channels and on synaptic transmission – mainly mediated by $GABA_A$ receptors. Carbamazepine reduces neural firing through the stabilization of voltage-gated sodium channels resembling the mechanism of action of lidocaine. However, placebo-controlled studies using carbamazepine at a dosage of 600–1,000 mg daily failed (Hulshof and Vermeij 1985). A significant benefit from carbamazepine has only been reported for a rare group of patients who have intermittent tinnitus that sounds like a typewriter or ear clicking and which is caused by a neurovascular conflict (Mardini 1987).

The anticonvulsant gabapentin which acts on voltage-gated calcium channels and which is used for the treatment of seizures, neuropathic pain and migraine has not demonstrated convincing effects in tinnitus patients (Piccirillo et al. 2007; Dehkordi et al. 2011) despite promising effects in an animal study (Bauer and Brozoski 2001).

Lamotrigine, an inhibitor of voltage-sensitive sodium channels and a membrane stabilizer, has been investigated in a placebo-controlled cross-over study and failed to demonstrate a beneficial clinical effect on tinnitus (Simpson et al. 1999).

Modulators of voltage-gated potassium channels such as retigabine and other KCNQ2/3-specific channel activators have been shown to prevent the development of tinnitus in animals (Kalappa et al. 2015; Kumar et al. 2016). Anecdotal reports in patients suggest that retigabine may also reduce tinnitus severity in humans (Langguth et al. 2016). The investigational drug AUT00063, a modulator of voltage gated potassium channels (Kv3.1), has also been investigated in tinnitus but failed to demonstrate superiority versus placebo (https://www.clinicaltrialsregister.eu/ctr-search/trial/2014-002179-27/results).

A Cochrane meta-analysis reviewing clinical trials of anticonvulsants for tinnitus treatment has concluded that studies performed so far only show small clinical significance (Hoekstra et al. 2011). However, when tinnitus results from a neurovascular conflict, carbamazepine and also other anticonvulsants such as oxcarbazepine, gabapentin, pregabalin and lamotrigine can be tried clinically. This rather rare form of tinnitus differs in its pathophysiological mechanisms from other forms of tinnitus and is typically characterized by unilateral staccato tinnitus, described by patients with the adjectives "typewriter", "Morse code" or "machine gun" (Juan and Basura 2019; Reynard et al. 2019).

Various glutamate receptor antagonists have been investigated in tinnitus sufferers. In animal models increased glutamatergic neurotransmission has been demonstrated in the cochlea in acute tinnitus and in central auditory pathways in chronic tinnitus (Puel 1995; Brozoski et al. 2013). Caroverine, a non-specific calcium channel blocker and an antagonist of both non-NMDA and NMDA receptors, has been applied both systemically and topically, with inconclusive results (Domeisen et al. 1998; Denk et al. 1997; Ehrenberger 2005). Oral treatment with the putative non-selective NMDA receptor antagonist acamprosate, which is approved for the treatment of alcohol dependency, has been evaluated in two double-blind studies (Sharma et al. 2012; Azevedo and Figueiredo 2007), which both indicate beneficial effects. The non-selective NMDA antagonist memantine was no more effective than placebo in a prospective randomized double-blind crossover 90-day treatment study (Figueiredo et al. 2008). The memantine analogue neramexane, which blocks both NMDA and $\alpha9\alpha10$ nicotinic cholinergic receptors, showed promising effects in a phase 2 study (Suckfull et al. 2011), but could not be confirmed in the following phase 3 program. Recently, the AMPA antagonist selurampanel (BGG492) has been investigated in a placebo-controlled cross-over trial, where an acute effect on tinnitus loudness has been observed after a single dose and reduction of tinnitus handicap after 2 weeks of treatment (https://www.novctrd.com/CtrdWeb/displaypdf.nov?trialresultid=12123).

Both dopaminergic and antidopaminergic drugs have been investigated for the treatment of tinnitus. Two studies suggest a beneficial effect of the dopamine antagonist sulpiride (Lopez-Gonzalez et al. 2007a, b). Pramipexole, an agonist of D2/D3 receptors, has been shown to reduce both the THI score and tinnitus loudness significantly more than placebo (Sziklai et al. 2011), whereas the dopamine agonist piribedil was not superior to placebo (de Azevedo et al. 2009).

Baclofen, a $GABA_B$ agonist with muscle relaxant effects, was not more effective than placebo in a double-blind randomized placebo-controlled trial (Westerberg et al. 1996). In a recent open-label exploratory study, the effect of various muscle relaxants on tinnitus has been assessed, revealing beneficial effects for cyclobenzaprine at a dosage of 30 mg/day for 14 weeks, but not for orphenadrine (100 mg/day for 14 weeks), tizanidine (24 mg/day for 14 weeks), eperisone (50 mg/day for 12 weeks) and cyclobenzaprine at a dose of 10 mg/day for 12 weeks (Coelho et al. 2012).

Other drugs that have been tested with either limited or inconsistent efficacy include the oral antiarrhythmic drugs tocainide, flecainide and mexiletine; the HMG-CoA reductase inhibitor atorvastatin; the vasodilator cyclandelate; some herbal products like *Ginkgo biloba*, melatonin, oxytocin, naltrexone and ondansetron, the prostaglandin E1 analogue misoprostol; the L-type calcium blocker nimodipine; the phosphodiesterase inhibitors cilostazol and vardenafil; cannabinoids; MDMA; vitamin B12 and minerals including zinc (Langguth et al. 2019).

5 Objective Tinnitus

Objective tinnitus, in comparison with the much more frequent subjective tinnitus, represents a group of entities where the tinnitus is caused by sound sources in the body (so-called somatosounds) (Langguth et al. 2013). The affected person can hear these sounds directly or via bone conduction due to their proximity to the ear structures. In some cases the sound can also be heard by the examiner. The sound sources are involuntary muscle contractions (spasms) or vascular processes that alter the blood flow in the vessels near the ear. The pulse synchronicity makes it possible to differentiate between the vascular-induced objective tinnitus complaints caused by vessel processes and the more arrhythmic and salvo-like muscle-associated tinnitus complaints. Since the therapy of pulse-synchronous tinnitus forms is not a domain of pharmacotherapy but, depending on the findings, is surgical or interventional-radiological, it will not be discussed further here. The therapy of myocloni close to the ear, on the other hand, is usually primarily pharmacological, depending on the localization. There are two forms of objective tinnitus caused by myocloni, a middle ear myoclonus (MEM) and a palatal myoclonus (PM) (Salehi et al. 2019). The differentiation of the two forms from each other is easy in the case of palatal myoclonus with visible contractions of the palate on the affected side. In myocloni of the middle ear muscles, differentiation between the stapedius muscle (SM) and the tensor tympani muscle (TTM) is difficult. In the best case, visible contractions of the eardrum can also be recorded objectively over time using tympanometry. Only in very rare cases of perforations of the eardrum, a rhythmic movement of the stapes can be observed (Liu et al. 2011). Palatal myoclonus with clicking tinnitus is attributed to the tensor veli palatini (TVP) muscle, since the other palatal muscles have no connection to the Eustachian tube (Salehi et al. 2019). Because both the TTM and the TVP muscle have their origin in the wall area of the Eustachian tube and thus contribute to changes in the opening of the tube, a differentiation of the two muscles is not absolutely necessary for therapeutic reasons. The primary therapeutic approach lies in muscle relaxation. Here, topical therapy must be distinguished from systemic therapy. Various reviews come to the conclusion that despite the low evidence level of the studies in connection with several sources of bias, the topical application of botulinum toxin (BT) has to be regarded as therapy of first choice (Slengerik-Hansen and Ovesen 2016). The doses of BT can be administered as injections into the palatal muscles. These injections can be made under spray anaesthesia after palpation of the contraction or under electromyographic control in a position medial or lateral to the hamulus of the medial pterygoid plate (Sinclair et al. 2014; Dang and Carol Liu 2019; Krause et al. 2010). The doses appear to be highly variable in the present reviews. It is reported from 2.5 to 80 U BT (Sinclair et al. 2014; Slengerik-Hansen and Ovesen 2016; Kaffenberger et al. 2017). The clinical effect occurs a few days after injection and would last between 2 and 6 months (Kaffenberger et al. 2017). In the various reviews, a complete cessation of the objective click tinnitus is repeatedly reported. The possible side effects associated with a movement disorder of the palate (nasal regurgitation, hypernasality of the voice) are described as minimal and temporary (Slengerik-Hansen and Ovesen 2016; Sinclair et al. 2014). In MEM, if an injection of BT into the palate fails, local middle ear application may be considered. However, this is considerably more complicated, as it requires an opening of the tympanum in order to inject BT into the stapedial muscle under visual control or to place a gel sponge soaked in BT at appropriate points around the SM or TTM. Baclofen, carbamazepine, piracetam, orphenadrine citrate and benzodiazepines (clonazepam) were used as systemic methods of muscle relaxation in MEM or PM (Bhimrao et al. 2012; Kaffenberger et al. 2017). Most studies have reported a temporary, partial response to the therapy (Bhimrao et al. 2012). The systemic side effects of the medication must be taken into account, so that they are rather out of question as long-term therapy. In this case, repeated injections of BT should be preferred if they are successful. In any case of a constellation of objective tinnitus associated with MEM or PM, the effect of a 3-month drug therapy (BT or systemic) should be waited for before surgical measures such as sectioning of muscle tendons are offered (Dang and Carol Liu 2019).

6 Medications that Can Cause Tinnitus

In a chapter on pharmacotherapy of tinnitus, it should also be mentioned that there are drugs that can cause tinnitus. While noise-induced hearing loss, presbycusis or genetic hearing loss are responsible for a majority of tinnitus cases, there are about 130 drugs known to have an ototoxic effect (Seligmann et al. 1996). The consequences are varying degrees of hearing loss, possibly associated with tinnitus and/or vertigo. The main ototoxic drugs belong to the classes of antibiotics (especially aminoglycosides, macrolides, vancomycin), antimalarial drugs (quinine), anti-inflammatory drugs (salicylates, NSAIDs), loop diuretics (ethacrynic acid, furosemide) and antineoplastic drugs (platinum-containing chemotherapeutic agents) (Rybak 1993; Ding et al. 2002; Brien 1993; Santos et al. 2020; Radziwon et al. 2016). With these drugs, the mechanism of action of drug induced tinnitus is relatively clear, in the sense that damage to the peripheral auditory system leads to hearing loss and then tinnitus is induced consecutively via involvement of central auditory structures (De Ridder et al. 2015; Eggermont 2008). In addition to these medications, there are various drugs like antidepressants, gabapentin, antiarrhythmics and statins, which have tinnitus listed as side effect and the mechanisms for this effect is completely unknown. Interestingly enough, they are also partly used for the treatment of chronic tinnitus (see section on chronic tinnitus) (Cianfrone et al. 2011).

The main damage mechanisms of the above-mentioned drugs include an irreversible destruction of the outer hair cells at the base of the cochlea by apoptosis (e.g. platinum-containing chemotherapeutics, aminoglycosides) or a temporary impairment of the function of the outer hair cells but also of the spiral ganglion cells and the central auditory neurons (e.g. NSAIDs) (Tabuchi et al. 2011; Radziwon et al. 2016; Langer et al. 2013; Feng et al. 2011). The detailed mechanisms will not be discussed here. Since the occurrence of corresponding symptoms depends on various conditions, such as genetic susceptibility and pharmacokinetic parameters such as dose, absorption, metabolism, accumulation and elimination, a monitoring system is of crucial importance when using these, in part vital, drugs (Cianfrone et al. 2011; Baguley and Prayuenyong 2020; Clemens et al. 2019a, b; Landier 2016). For such a surveillance program, regular audiometric measurements are useful in addition to patient education and symptom monitoring. In addition to a standard pure tone audiometry, the high-frequency range up to 16 kHz should also be measured. In addition, otoacoustic emissions, brainstem electric response audiometry and speech hearing tests in quiet and in noise can provide valuable information (Baguley and Prayuenyong 2020). The importance of these measures in children, especially at an age when central auditory pathways are not yet fully developed, should be emphasized (Clemens et al. 2019b). As already reported in the section "acute tinnitus", there are approaches already tested in animal and human studies to reduce the ototoxicity of certain drugs by preventing the occurrence of hearing loss and tinnitus (Waissbluth 2020). These include measures such as the variation of infusion times of platinum-containing drugs (van As et al. 2018), as well as the administration of additional substances for potentially ototoxic drugs (van As et al. 2019; Berger et al. 2017; Waissbluth 2020; Sarafraz et al. 2018; Fransson et al. 2017; Sheth et al. 2017).

In addition to the ototoxic effect, some medication can cause tinnitus without causing hearing damage. A recent network analysis identified drug targets that are associated with an increased risk of inducing tinnitus as a side effect (Elgoyhen et al. 2014). These were drugs that target cyclooxygenase 1 and 2, the angiotensin-converting enzyme (ACE), the serotonin receptor 5HT1a and the sodium channel SCN5A.

7 Conclusion and Outlook

The options for evidence-based pharmacological treatment of tinnitus are currently very limited. A large number of clinical studies have been performed with various compounds and multiple rationales. Taken together the results are disappointing. The lack of evidence for any pharmacological compound is reflected by the fact that there is no FDA- or EMA-approved drug for tinnitus and no general recommendation for pharmacological treatment in tinnitus guidelines. In clinical practice, pharmacological treatment plays a role for the treatment of comorbidities of tinnitus such as anxiety, depression or insomnia and in rare cases where tinnitus is caused by a microvascular conflict or by myocloni of the palatal or middle ear muscles. Unfortunately, there is little hope that the situation will change in the near future. Recently performed phase 2 and 3 programs investigating various compounds (neramexane, the AMPA antagonist BCG 492, the potassium channel modulator AUT00063, the topical administration of esketamine hydrochloride) have been stopped (Cederroth et al. 2018). The difficulties related to the development of an effective drug therapy for tinnitus are manifold. This is due to the still incomplete understanding of the pathophysiology of tinnitus (Kleinjung and Langguth 2020), the lack of sufficient funding for tinnitus research (Cederroth et al. 2013), the lack of reliable animal models of tinnitus with a proven comparability to the humans (von der Behrens 2014), the clinical and pathophysiological heterogeneity of tinnitus (Cederroth et al. 2019) and the difficulties in quantifying treatment effects (Landgrebe et al. 2010).

Nevertheless, research activities have clearly increased in the last decade and have led to some promising new findings that warrant further research. These include still largely unexplored targets such as potassium channels, AMPA receptors and the ACE. In addition, the use of pharmacological treatment to enhance the effects of other treatments (e.g. sound therapy or cognitive behavioural therapy) might provide new opportunities. Finally, with the discovery of synaptopathy as a possible mechanism for hearing dysfunction and tinnitus generation (Liberman and Kujawa 2017; Bharadwaj et al. 2019), there are currently many efforts to develop pharmacological treatment options for this so-called hidden hearing loss. If this approach proves to be successful, it might also have beneficial effects for tinnitus.

Acknowledgements Conflict of interest: Nothing to declare.

References

Ahmadzai N, Kilty S, Cheng W, Esmaeilisaraji L, Wolfe D, Bonaparte JP et al (2019) A systematic review and network meta-analysis of existing pharmacologic therapies in patients with idiopathic sudden sensorineural hearing loss. PLoS One 14(9):e0221713. https://doi.org/10.1371/journal.pone.0221713

An YH, Yu KK, Kwak MY, Yoon SW, Shim HJ (2014) Prognostic factors for the outcomes of intratympanic dexamethasone in the treatment of acute subjective tinnitus. Otol Neurotol 35 (8):1330–1337. https://doi.org/10.1097/MAO.0000000000000526

Attias J, Weisz G, Almog S, Shahar A, Wiener M, Joachims Z et al (1994) Oral magnesium intake reduces permanent hearing loss induced by noise exposure. Am J Otolaryngol 15(1):26–32. https://doi.org/10.1016/0196-0709(94)90036-1

Azevedo AA, Figueiredo RR (2007) Treatment of tinnitus with acamprosate. Prog Brain Res 166:273–277. https://doi.org/10.1016/S0079-6123(07)66025-7

Baguley DM, Prayuenyong P (2020) Looking beyond the audiogram in ototoxicity associated with platinum-based chemotherapy. Cancer Chemother Pharmacol 85(2):245–250. https://doi.org/10.1007/s00280-019-04012-z

Baguley DM, Jones S, Wilkins I, Axon PR, Moffat DA (2005) The inhibitory effect of intravenous lidocaine infusion on tinnitus after translabyrinthine removal of vestibular schwannoma: a double-blind, placebo-controlled, crossover study. Otol Neurotol 26(2):169–176. https://doi.org/10.1097/00129492-200503000-00007

Baguley D, McFerran D, Hall D (2013) Tinnitus. Lancet 382(9904):1600–1607. https://doi.org/10.1016/S0140-6736(13)60142-7

Bahmad FM Jr, Venosa AR, Oliveira CA (2006) Benzodiazepines and GABAergics in treating severe disabling tinnitus of predominantly cochlear origin. Int Tinnitus J 12(2):140–144. https://www.ncbi.nlm.nih.gov/pubmed/17260879

Baldo P, Doree C, Molin P, McFerran D, Cecco S (2012) Antidepressants for patients with tinnitus. Cochrane Database Syst Rev 9:CD003853. https://doi.org/10.1002/14651858.CD003853.pub3

Bauer CA, Brozoski TJ (2001) Assessing tinnitus and prospective tinnitus therapeutics using a psychophysical animal model. J Assoc Res Otolaryngol 2(1):54–64. https://doi.org/10.1007/s101620010030

Bayar N, Boke B, Turan E, Belgin E (2001) Efficacy of amitriptyline in the treatment of subjective tinnitus. J Otolaryngol 30(5):300–303. https://doi.org/10.2310/7070.2001.19597

Becatti M, Marcucci R, Mannucci A, Gori AM, Giusti B, Sofi F et al (2017) Erythrocyte membrane fluidity alterations in sudden sensorineural hearing loss patients: the role of oxidative stress. Thromb Haemost 117(12):2334–2345. https://doi.org/10.1160/TH17-05-0356

Berger JI, Coomber B, Hill S, Alexander SPH, Owen W, Palmer AR et al (2017) Effects of the cannabinoid CB1 agonist ACEA on salicylate ototoxicity, hyperacusis and tinnitus in guinea pigs. Hear Res 356:51–62. https://doi.org/10.1016/j.heares.2017.10.012

Bharadwaj HM, Mai AR, Simpson JM, Choi I, Heinz MG, Shinn-Cunningham BG (2019) Non-invasive assays of cochlear synaptopathy – candidates and considerations. Neuroscience 407:53–66. https://doi.org/10.1016/j.neuroscience.2019.02.031

Bhatta P, Dhukhwa A, Sheehan K, Al Aameri RFH, Borse V, Ghosh S et al (2019) Capsaicin protects against cisplatin ototoxicity by changing the STAT3/STAT1 ratio and activating cannabinoid (CB2) receptors in the cochlea. Sci Rep 9(1):4131. https://doi.org/10.1038/s41598-019-40425-9

Bhimrao SK, Masterson L, Baguley D (2012) Systematic review of management strategies for middle ear myoclonus. Otolaryngol Head Neck Surg 146(5):698–706. https://doi.org/10.1177/0194599811434504

Bing D, Lee SC, Campanelli D, Xiong H, Matsumoto M, Panford-Walsh R et al (2015) Cochlear NMDA receptors as a therapeutic target of noise-induced tinnitus. Cell Physiol Biochem 35 (5):1905–1923. https://doi.org/10.1159/000374000

Bonnet U (2014) Decompensated chronic tinnitus and high-dose benzodiazepine dependence. Between Scylla and Charybdis. Nervenarzt 85(7):865–869. https://doi.org/10.1007/s00115-012-3670-y

Brien JA (1993) Ototoxicity associated with salicylates. A brief review. Drug Saf 9(2):143–148. https://doi.org/10.2165/00002018-199309020-00006

Brozoski TJ, Wisner KW, Odintsov B, Bauer CA (2013) Local NMDA receptor blockade attenuates chronic tinnitus and associated brain activity in an animal model. PLoS One 8(10):e77674. https://doi.org/10.1371/journal.pone.0077674

Cederroth CR, Canlon B, Langguth B (2013) Hearing loss and tinnitus--are funders and industry listening? Nat Biotechnol 31(11):972–974. https://doi.org/10.1038/nbt.2736

Cederroth CR, Dyhrfjeld-Johnsen J, Langguth B (2018) An update: emerging drugs for tinnitus. Expert Opin Emerg Drugs 23(4):251–260. https://doi.org/10.1080/14728214.2018.1555240

Cederroth CR, Gallus S, Hall DA, Kleinjung T, Langguth B, Maruotti A et al (2019) Editorial: towards an understanding of tinnitus heterogeneity. Front Aging Neurosci 11:53. https://doi.org/10.3389/fnagi.2019.00053

Chandrasekhar SS, Tsai Do BS, Schwartz SR, Bontempo LJ, Faucett EA, Finestone SA et al (2019) Clinical practice guideline: sudden hearing loss (update). Otolaryngol Head Neck Surg 161 (1_suppl):S1–S45. https://doi.org/10.1177/0194599819859885

Cianfrone G, Pentangelo D, Cianfrone F, Mazzei F, Turchetta R, Orlando MP et al (2011) Pharmacological drugs inducing ototoxicity, vestibular symptoms and tinnitus: a reasoned and updated guide. Eur Rev Med Pharmacol Sci 15(6):601–636. https://www.ncbi.nlm.nih.gov/pubmed/21796866

Clemens E, Meijer AJ, Broer L, Langer T, van der Kooi AL, Uitterlinden AG et al (2019a) Genetic determinants of ototoxicity during and after childhood cancer treatment: protocol for the PanCareLIFE study. JMIR Res Protoc 8(3):e11868. https://doi.org/10.2196/11868

Clemens E, van den Heuvel-Eibrink MM, Mulder RL, Kremer LCM, Hudson MM, Skinner R et al (2019b) Recommendations for ototoxicity surveillance for childhood, adolescent, and young adult cancer survivors: a report from the International Late Effects of Childhood Cancer Guideline Harmonization Group in collaboration with the PanCare Consortium. Lancet Oncol 20(1):e29–e41. https://doi.org/10.1016/S1470-2045(18)30858-1

Coelho C, Figueiredo R, Frank E, Burger J, Schecklmann M, Landgrebe M et al (2012) Reduction of tinnitus severity by the centrally acting muscle relaxant cyclobenzaprine: an open-label pilot study. Audiol Neurootol 17(3):179–188. https://doi.org/10.1159/000335657

Dang J, Carol Liu YC (2019) Treatment of objective tinnitus with transpalatal Botox((R)) injection in a pediatric patient with middle ear myoclonus: a case report. Int J Pediatr Otorhinolaryngol 116:22–24. https://doi.org/10.1016/j.ijporl.2018.09.024

de Azevedo AA, Langguth B, de Oliveira PM, Rodrigues Figueiredo R (2009) Tinnitus treatment with piribedil guided by electrocochleography and acoustic otoemissions. Otol Neurotol 30 (5):676–680. https://doi.org/10.1097/MAO.0b013e3181ab8fd5

De Ridder D, Elgoyhen AB, Romo R, Langguth B (2011) Phantom percepts: tinnitus and pain as persisting aversive memory networks. Proc Natl Acad Sci U S A 108(20):8075–8080. https://doi.org/10.1073/pnas.1018466108

De Ridder D, Vanneste S, Langguth B, Llinas R (2015) Thalamocortical dysrhythmia: a theoretical update in tinnitus. Front Neurol 6:124. https://doi.org/10.3389/fneur.2015.00124

Dehkordi MA, Abolbashari S, Taheri R, Einolghozati S (2011) Efficacy of gabapentin on subjective idiopathic tinnitus: a randomized, double-blind, placebo-controlled trial. Ear Nose Throat J 90 (4):150–158. https://doi.org/10.1177/014556131109000405

Denk DM, Heinzl H, Franz P, Ehrenberger K (1997) Caroverine in tinnitus treatment. A placebo-controlled blind study. Acta Otolaryngol 117(6):825–830. https://doi.org/10.3109/00016489709114208

Ding D, McFadden SL, Woo JM, Salvi RJ (2002) Ethacrynic acid rapidly and selectively abolishes blood flow in vessels supplying the lateral wall of the cochlea. Hear Res 173(1–2):1–9. https://doi.org/10.1016/s0378-5955(02)00585-3

Dobie RA (1999) A review of randomized clinical trials in tinnitus. Laryngoscope 109 (8):1202–1211. https://doi.org/10.1097/00005537-199908000-00004

Domeisen H, Hotz MA, Hausler R (1998) Caroverine in tinnitus treatment. Acta Otolaryngol 118 (4):606–608. https://doi.org/10.1080/00016489850154801

Eggermont JJ (2008) Role of auditory cortex in noise- and drug-induced tinnitus. Am J Audiol 17 (2):S162–S169. https://doi.org/10.1044/1059-0889(2008/07-0025)

Ehrenberger K (2005) Topical administration of Caroverine in somatic tinnitus treatment: proof-of-concept study. Int Tinnitus J 11(1):34–37. https://www.ncbi.nlm.nih.gov/pubmed/16419686

Elgoyhen AB, Langguth B, Vanneste S, De Ridder D (2012) Tinnitus: network pathophysiology-network pharmacology. Front Syst Neurosci 6:1. https://doi.org/10.3389/fnsys.2012.00001

Elgoyhen AB, Langguth B, Nowak W, Schecklmann M, De Ridder D, Vanneste S (2014) Identifying tinnitus-related genes based on a side-effect network analysis. CPT Pharmacometrics Syst Pharmacol 3:e97. https://doi.org/10.1038/psp.2013.75

Feng H, Yin SH, Tang AZ, Tan SH (2011) Salicylate initiates apoptosis in the spiral ganglion neuron of guinea pig cochlea by activating caspase-3. Neurochem Res 36(6):1108–1115. https://doi.org/10.1007/s11064-011-0455-9

Figueiredo RR, Langguth B, Mello de Oliveira P, Aparecida de Azevedo A (2008) Tinnitus treatment with memantine. Otolaryngol Head Neck Surg 138(4):492–496. https://doi.org/10.1016/j.otohns.2007.11.027

Fransson AE, Kisiel M, Pirttila K, Pettersson C, Videhult Pierre P, Laurell GFE (2017) Hydrogen inhalation protects against ototoxicity induced by intravenous cisplatin in the guinea pig. Front Cell Neurosci 11:280. https://doi.org/10.3389/fncel.2017.00280

Gul F, Muderris T, Yalciner G, Sevil E, Bercin S, Ergin M et al (2017) A comprehensive study of oxidative stress in sudden hearing loss. Eur Arch Otorhinolaryngol 274(3):1301–1308. https://doi.org/10.1007/s00405-016-4301-1

Han SS, Nam EC, Won JY, Lee KU, Chun W, Choi HK et al (2012) Clonazepam quiets tinnitus: a randomised crossover study with *Ginkgo biloba*. J Neurol Neurosurg Psychiatry 83(8):821–827. https://doi.org/10.1136/jnnp-2012-302273

Hesse G, Laubert A (2010) Pharmacotherapy of acute and chronic hearing loss. HNO 58 (10):990–998. https://doi.org/10.1007/s00106-010-2179-6

Hoekstra CE, Rynja SP, van Zanten GA, Rovers MM (2011) Anticonvulsants for tinnitus. Cochrane Database Syst Rev 7:CD007960. https://doi.org/10.1002/14651858.CD007960.pub2

Hulshof JH, Vermeij P (1985) The value of carbamazepine in the treatment of tinnitus. ORL J Otorhinolaryngol Relat Spec 47(5):262–266. https://doi.org/10.1159/000275781

Hwang SM, Lim SH, Oh DJ, Kim SK, Jung HH, Im GJ (2016) Effect of tianeptine on depressed tinnitus patients. J Audiol Otol 20(2):90–96. https://doi.org/10.7874/jao.2016.20.2.90

Jalali MM, Kousha A, Naghavi SE, Soleimani R, Banan R (2009) The effects of alprazolam on tinnitus: a cross-over randomized clinical trial. Med Sci Monit 15(11):PI55–PI60. https://www.ncbi.nlm.nih.gov/pubmed/19865063

Johnson RM, Brummett R, Schleuning A (1993) Use of alprazolam for relief of tinnitus. A double-blind study. Arch Otolaryngol Head Neck Surg 119(8):842–845. https://doi.org/10.1001/archotol.1993.01880200042006

Juan JS, Basura GJ (2019) Tinnitus management in lateral skull base lesions. J Neurol Surg B Skull Base 80(2):125–131. https://doi.org/10.1055/s-0038-1676308

Jufas NE, Wood R (2015) The use of benzodiazepines for tinnitus: systematic review. J Laryngol Otol 129(Suppl 3):S14–S22. https://doi.org/10.1017/S0022215115000808

Kaffenberger TM, Mandal R, Schaitkin BM, Hirsch BE (2017) Palatal botulinum toxin as a novel therapy for objective tinnitus in forced eyelid closure syndrome. Laryngoscope 127 (5):1199–1201. https://doi.org/10.1002/lary.26191

Kalappa BI, Soh H, Duignan KM, Furuya T, Edwards S, Tzingounis AV et al (2015) Potent KCNQ2/3-specific channel activator suppresses in vivo epileptic activity and prevents the development of tinnitus. J Neurosci 35(23):8829–8842. https://doi.org/10.1523/JNEUROSCI.5176-14.2015

Kara E, Aydin K, Akbulut AA, Karakol SN, Durmaz S, Yener HM et al (2020) Assessment of hidden hearing loss in normal hearing individuals with and without tinnitus. J Int Adv Otol 16:87. https://doi.org/10.5152/iao.2020.7062

Kay NJ (1981) Oral chemotherapy in tinnitus. Br J Audiol 15(2):123–124. https://doi.org/10.3109/03005368109081425

Kleinjung T, Langguth B (2020) Avenue for future tinnitus treatments. Otolaryngol Clin North Am. https://doi.org/10.1016/j.otc.2020.03.013

Krause E, Heinen F, Gurkov R (2010) Difference in outcome of botulinum toxin treatment of essential palatal tremor in children and adults. Am J Otolaryngol 31(2):91–95. https://doi.org/10.1016/j.amjoto.2008.11.007

Kucharava K, Sekulic-Jablanovic M, Horvath L, Bodmer D, Petkovic V (2019) Pasireotide protects mammalian cochlear hair cells from gentamicin ototoxicity by activating the PI3K-Akt pathway. Cell Death Dis 10(2):110. https://doi.org/10.1038/s41419-019-1386-7

Kumar M, Reed N, Liu R, Aizenman E, Wipf P, Tzounopoulos T (2016) Synthesis and evaluation of potent KCNQ2/3-specific channel activators. Mol Pharmacol 89(6):667–677. https://doi.org/10.1124/mol.115.103200

Landgrebe M, Zeman F, Koller M, Eberl Y, Mohr M, Reiter J et al (2010) The Tinnitus Research Initiative (TRI) database: a new approach for delineation of tinnitus subtypes and generation of predictors for treatment outcome. BMC Med Inform Decis Mak 10:42. https://doi.org/10.1186/1472-6947-10-42

Landier W (2016) Ototoxicity and cancer therapy. Cancer 122(11):1647–1658. https://doi.org/10.1002/cncr.29779

Langer T, am Zehnhoff-Dinnesen A, Radtke S, Meitert J, Zolk O (2013) Understanding platinum-induced ototoxicity. Trends Pharmacol Sci 34(8):458–469. https://doi.org/10.1016/j.tips.2013.05.006

Langguth B, Landgrebe M, Kleinjung T, Sand GP, Hajak G (2011) Tinnitus and depression. World J Biol Psychiatry 12(7):489–500. https://doi.org/10.3109/15622975.2011.575178

Langguth B, Kreuzer PM, Kleinjung T, De Ridder D (2013) Tinnitus: causes and clinical management. Lancet Neurol 12(9):920–930. https://doi.org/10.1016/S1474-4422(13)70160-1

Langguth B, Elgoyhen AB, Schlee W (2016) Potassium channels as promising new targets for pharmacologic treatment of tinnitus: can Internet-based 'crowd sensing' initiated by patients speed up the transition from bench to bedside? Expert Opin Ther Targets 20(3):251–254. https://doi.org/10.1517/14728222.2016.1125884

Langguth B, Elgoyhen AB, Cederroth CR (2019) Therapeutic approaches to the treatment of tinnitus. Annu Rev Pharmacol Toxicol 59:291–313. https://doi.org/10.1146/annurev-pharmtox-010818-021556

Lavigne P, Lavigne F, Saliba I (2016) Intratympanic corticosteroids injections: a systematic review of literature. Eur Arch Otorhinolaryngol 273(9):2271–2278. https://doi.org/10.1007/s00405-015-3689-3

Le Prell CG (2019) Otoprotectants: from research to clinical application. Semin Hear 40 (2):162–176. https://doi.org/10.1055/s-0039-1684045

Lee HJ, Kim MB, Yoo SY, Park SN, Nam EC, Moon IS et al (2018) Clinical effect of intratympanic dexamethasone injection in acute unilateral tinnitus: a prospective, placebo-controlled, multi-center study. Laryngoscope 128(1):184–188. https://doi.org/10.1002/lary.26541

Lefeuvre J, Chedeau J, Boulet M, Fain G, Papon JF, Nguyen Y et al (2019) Hidden hearing loss and tinnitus: utility of the high-definition audiograms in diagnosis. Clin Otolaryngol 44 (6):1170–1175. https://doi.org/10.1111/coa.13435

Lewis S, Chowdhury E, Stockdale D, Kennedy V, Guideline C (2020) Assessment and management of tinnitus: summary of NICE guidance. BMJ 368:m976. https://doi.org/10.1136/bmj.m976

Liberman MC, Kujawa SG (2017) Cochlear synaptopathy in acquired sensorineural hearing loss: manifestations and mechanisms. Hear Res 349:138–147. https://doi.org/10.1016/j.heares.2017.01.003

Liberman MC, Epstein MJ, Cleveland SS, Wang H, Maison SF (2016) Toward a differential diagnosis of hidden hearing loss in humans. PLoS One 11(9):e0162726. https://doi.org/10.1371/journal.pone.0162726

Liu HB, Fan JP, Lin SZ, Zhao SW, Lin Z (2011) Botox transient treatment of tinnitus due to stapedius myoclonus: case report. Clin Neurol Neurosurg 113(1):57–58. https://doi.org/10.1016/j.clineuro.2010.07.022

Lopez-Escamez JA, Amanat S (2020) Heritability and genetics contribution to tinnitus. Otolaryngol Clin North Am. https://doi.org/10.1016/j.otc.2020.03.003

Lopez-Gonzalez MA, Moliner-Peiro F, Alfaro-Garcia J, Esteban-Ortega F (2007a) Sulpiride plus hydroxyzine decrease tinnitus perception. Auris Nasus Larynx 34(1):23–27. https://doi.org/10.1016/j.anl.2006.09.021

Lopez-Gonzalez MA, Santiago AM, Esteban-Ortega F (2007b) Sulpiride and melatonin decrease tinnitus perception modulating the auditolimbic dopaminergic pathway. J Otolaryngol 36 (4):213–219. https://doi.org/10.2310/7070.2007.0018

Lynch ED, Gu R, Pierce C, Kil J (2005) Reduction of acute cisplatin ototoxicity and nephrotoxicity in rats by oral administration of allopurinol and ebselen. Hear Res 201(1–2):81–89. https://doi.org/10.1016/j.heares.2004.08.002

Malgrange B, Varela-Nieto I, de Medina P, Paillasse MR (2015) Targeting cholesterol homeostasis to fight hearing loss: a new perspective. Front Aging Neurosci 7:3. https://doi.org/10.3389/fnagi.2015.00003

Mardini MK (1987) Ear-clicking "tinnitus" responding to carbamazepine. N Engl J Med 317 (24):1542. https://doi.org/10.1056/nejm198712103172418

Mihail RC, Crowley JM, Walden BE, Fishburne J, Reinwall JE, Zajtchuk JT (1988) The tricyclic trimipramine in the treatment of subjective tinnitus. Ann Otol Rhinol Laryngol 97(2 Pt 1):120–123. https://doi.org/10.1177/000348948809700204

Morgan P, Brown DG, Lennard S, Anderton MJ, Barrett JC, Eriksson U et al (2018) Impact of a five-dimensional framework on R&D productivity at AstraZeneca. Nat Rev Drug Discov 17 (3):167–181. https://doi.org/10.1038/nrd.2017.244

Paul BT, Bruce IC, Roberts LE (2017) Evidence that hidden hearing loss underlies amplitude modulation encoding deficits in individuals with and without tinnitus. Hear Res 344:170–182. https://doi.org/10.1016/j.heares.2016.11.010

Person OC, Puga ME, da Silva EM, Torloni MR (2016) Zinc supplementation for tinnitus. Cochrane Database Syst Rev 11:CD009832. https://doi.org/10.1002/14651858.CD009832.pub2

Piccirillo JF, Finnell J, Vlahiotis A, Chole RA, Spitznagel E Jr (2007) Relief of idiopathic subjective tinnitus: is gabapentin effective? Arch Otolaryngol Head Neck Surg 133 (4):390–397. https://doi.org/10.1001/archotol.133.4.390

Prayuenyong P, Kasbekar AV, Baguley DM (2020) The efficacy of statins as otoprotective agents: a systematic review. Clin Otolaryngol 45(1):21–31. https://doi.org/10.1111/coa.13457

Puel JL (1995) Chemical synaptic transmission in the cochlea. Prog Neurobiol 47(6):449–476. https://doi.org/10.1016/0301-0082(95)00028-3

Puel JL, Ruel J, Guitton M, Wang J, Pujol R (2002) The inner hair cell synaptic complex: physiology, pharmacology and new therapeutic strategies. Audiol Neurootol 7(1):49–54. https://doi.org/10.1159/000046864

Radziwon KH, Hayes SH, Sheppard AM, Ding D, Salvi R (2016) Drug-induced tinnitus. In: Baguley D (ed) Tinnitus – clinical and research perspectives. Plural Publishing, San Diego

Reynard P, Ionescu E, Karkas A, Ltaeif-Boudrigua A, Thai-Van H (2019) Unilateral cochleovestibular nerve compression syndrome in a patient with bilateral IAC osteoma. Eur Ann Otorhinolaryngol Head Neck Dis 137(3):213–216. https://doi.org/10.1016/j.anorl.2019.12.004

Robinson SK, Viirre ES, Bailey KA, Gerke MA, Harris JP, Stein MB (2005) Randomized placebo-controlled trial of a selective serotonin reuptake inhibitor in the treatment of nondepressed tinnitus subjects. Psychosom Med 67(6):981–988. https://doi.org/10.1097/01.psy.0000188479.04891.74

Rosenhall U, Skoog B, Muhr P (2019) Treatment of military acoustic accidents with N-acetyl-L-cysteine (NAC). Int J Audiol 58(3):151–157. https://doi.org/10.1080/14992027.2018.1543961

Rybak LP (1993) Ototoxicity of loop diuretics. Otolaryngol Clin N Am 26(5):829–844. https://www.ncbi.nlm.nih.gov/pubmed/8233492

Salehi PP, Kasle D, Torabi SJ, Michaelides E, Hildrew DM (2019) The etiology, pathogeneses, and treatment of objective tinnitus: unique case series and literature review. Am J Otolaryngol 40 (4):594–597. https://doi.org/10.1016/j.amjoto.2019.03.017

Santos N, Ferreira RS, Santos ACD (2020) Overview of cisplatin-induced neurotoxicity and ototoxicity, and the protective agents. Food Chem Toxicol 136:111079. https://doi.org/10.1016/j.fct.2019.111079

Sarafraz Z, Ahmadi A, Daneshi A (2018) Transtympanic injections of N-acetylcysteine and dexamethasone for prevention of cisplatin-induced ototoxicity: double blind randomized clinical trial. Int Tinnitus J 22(1):40–45. https://doi.org/10.5935/0946-5448.20180007

Seligmann H, Podoshin L, Ben-David J, Fradis M, Goldsher M (1996) Drug-induced tinnitus and other hearing disorders. Drug Saf 14(3):198–212. https://doi.org/10.2165/00002018-199614030-00006

Sendowski I (2006) Magnesium therapy in acoustic trauma. Magnes Res 19(4):244–254. https://www.ncbi.nlm.nih.gov/pubmed/17402292

Sharma DK, Kaur S, Singh J, Kaur I (2012) Role of acamprosate in sensorineural tinnitus. Indian J Pharmacol 44(1):93–96. https://doi.org/10.4103/0253-7613.91876

Sheth S, Mukherjea D, Rybak LP, Ramkumar V (2017) Mechanisms of cisplatin-induced ototoxicity and otoprotection. Front Cell Neurosci 11:338. https://doi.org/10.3389/fncel.2017.00338

Shim HJ, Lee ES, An YH, Kim DH (2017) Comparison of long-term outcome of intratympanic dexamethasone therapy between acute noise-induced tinnitus and acute idiopathic tinnitus. J Int Adv Otol 13(1):53–60. https://doi.org/10.5152/iao.2016.2632

Shore SE, Wu C (2019) Mechanisms of noise-induced tinnitus: insights from cellular studies. Neuron 103(1):8–20. https://doi.org/10.1016/j.neuron.2019.05.008

Shore SE, Roberts LE, Langguth B (2016) Maladaptive plasticity in tinnitus--triggers, mechanisms and treatment. Nat Rev Neurol 12(3):150–160. https://doi.org/10.1038/nrneurol.2016.12

Simpson JJ, Gilbert AM, Weiner GM, Davies WE (1999) The assessment of lamotrigine, an antiepileptic drug, in the treatment of tinnitus. Am J Otol 20(5):627–631. https://www.ncbi.nlm.nih.gov/pubmed/10503585

Sinclair CF, Gurey LE, Blitzer A (2014) Palatal myoclonus: algorithm for management with botulinum toxin based on clinical disease characteristics. Laryngoscope 124(5):1164–1169. https://doi.org/10.1002/lary.23485

Slengerik-Hansen J, Ovesen T (2016) Botulinum toxin treatment of objective tinnitus because of essential palatal tremor: a systematic review. Otol Neurotol 37(7):820–828. https://doi.org/10.1097/MAO.0000000000001090

Staecker H, Morelock M, Kramer T, Chrbolka P, Ahn JH, Meyer T (2017) Safety of repeated-dose intratympanic injections with AM-101 in acute inner ear tinnitus. Otolaryngol Head Neck Surg 157(3):478–487. https://doi.org/10.1177/0194599817711378

Suckfull M, Althaus M, Ellers-Lenz B, Gebauer A, Gortelmeyer R, Jastreboff PJ et al (2011) A randomized, double-blind, placebo-controlled clinical trial to evaluate the efficacy and safety of neramexane in patients with moderate to severe subjective tinnitus. BMC Ear Nose Throat Disord 11:1. https://doi.org/10.1186/1472-6815-11-1

Sullivan M, Katon W, Russo J, Dobie R, Sakai C (1993) A randomized trial of nortriptyline for severe chronic tinnitus. Effects on depression, disability, and tinnitus symptoms. Arch Intern Med 153(19):2251–2259. https://www.ncbi.nlm.nih.gov/pubmed/8215728

Sziklai I, Szilvassy J, Szilvassy Z (2011) Tinnitus control by dopamine agonist pramipexole in presbycusis patients: a randomized, placebo-controlled, double-blind study. Laryngoscope 121(4):888–893. https://doi.org/10.1002/lary.21461

Tabuchi K, Nishimura B, Nakamagoe M, Hayashi K, Nakayama M, Hara A (2011) Ototoxicity: mechanisms of cochlear impairment and its prevention. Curr Med Chem 18(31):4866–4871. https://doi.org/10.2174/092986711797535254

Taha A, Shlamkovitch N, Abu-Eta R, Yeheskeli E, Muallem-Kalmovich L, Gavriel H et al (2019) High dose of intratympanic steroids for sudden sensorineural hearing loss salvage. Otol Neurotol 40(9):1134–1138. https://doi.org/10.1097/MAO.0000000000002386

Tillinger JA, Gupta C, Ila K, Ahmed J, Mittal J, Van De Water TR et al (2018) l-N-acetylcysteine protects outer hair cells against TNFalpha initiated ototoxicity in vitro. Acta Otolaryngol 138(8):676–684. https://doi.org/10.1080/00016489.2018.1440086

Trellakis S, Lautermann J, Lehnerdt G (2007) Lidocaine: neurobiological targets and effects on the auditory system. Prog Brain Res 166:303–322. https://doi.org/10.1016/S0079-6123(07)66028-2

Tunkel DE, Bauer CA, Sun GH, Rosenfeld RM, Chandrasekhar SS, Cunningham ER Jr et al (2014) Clinical practice guideline: tinnitus. Otolaryngol Head Neck Surg 151(2 Suppl):S1–S40. https://doi.org/10.1177/0194599814545325

Tyler RS (2012) Patient preferences and willingness to pay for tinnitus treatments. J Am Acad Audiol 23(2):115–125. https://doi.org/10.3766/jaaa.23.2.6

van As JW, van den Berg H, van Dalen EC (2018) Different infusion durations for preventing platinum-induced hearing loss in children with cancer. Cochrane Database Syst Rev 7: CD010885. https://doi.org/10.1002/14651858.CD010885.pub4

van As JW, van den Berg H, van Dalen EC (2019) Medical interventions for the prevention of platinum-induced hearing loss in children with cancer. Cochrane Database Syst Rev 5: CD009219. https://doi.org/10.1002/14651858.CD009219.pub5

van de Heyning P, Muehlmeier G, Cox T, Lisowska G, Maier H, Morawski K et al (2014) Efficacy and safety of AM-101 in the treatment of acute inner ear tinnitus--a double-blind, randomized, placebo-controlled phase II study. Otol Neurotol 35(4):589–597. https://doi.org/10.1097/MAO.0000000000000268

von der Behrens W (2014) Animal models of subjective tinnitus. Neural Plast 2014:741452. https://doi.org/10.1155/2014/741452

Waissbluth S (2020) Clinical trials evaluating transtympanic otoprotectants for cisplatin-induced ototoxicity: what do we know so far? Eur Arch Otorhinolaryngol. https://doi.org/10.1007/s00405-020-06003-w

Wegner I, Hall DA, Smit AL, McFerran D, Stegeman I (2018) Betahistine for tinnitus. Cochrane Database Syst Rev 12:CD013093. https://doi.org/10.1002/14651858.CD013093.pub2

Westerberg BD, Roberson JB Jr, Stach BA (1996) A double-blind placebo-controlled trial of baclofen in the treatment of tinnitus. Am J Otol 17(6):896–903. https://www.ncbi.nlm.nih.gov/pubmed/8915419

Xiong B, Liu Z, Liu Q, Peng Y, Wu H, Lin Y et al (2019) Missed hearing loss in tinnitus patients with normal audiograms. Hear Res 384:107826. https://doi.org/10.1016/j.heares.2019.107826

Yang S, Weiner BD, Zhang LS, Cho SJ, Bao S (2011) Homeostatic plasticity drives tinnitus perception in an animal model. Proc Natl Acad Sci U S A 108(36):14974–14979. https://doi.org/10.1073/pnas.1107998108

Yeh CW, Tseng LH, Yang CH, Hwang CF (2019) Effects of oral zinc supplementation on patients with noise-induced hearing loss associated tinnitus: a clinical trial. Biomed J 42(1):46–52. https://doi.org/10.1016/j.bj.2018.10.009

Yoo J, Hamilton SJ, Angel D, Fung K, Franklin J, Parnes LS et al (2014) Cisplatin otoprotection using transtympanic L-N-acetylcysteine: a pilot randomized study in head and neck cancer patients. Laryngoscope 124(3):E87–E94. https://doi.org/10.1002/lary.24360

Zeman F, Koller M, Langguth B, Landgrebe M, Tinnitus Research Initiative database study group (2014) Which tinnitus-related aspects are relevant for quality of life and depression: results from a large international multicentre sample. Health Qual Life Outcomes 12:7. https://doi.org/10.1186/1477-7525-12-7

Zenner HP, Delb W, Kroner-Herwig B, Jager B, Peroz I, Hesse G et al (2017) A multidisciplinary systematic review of the treatment for chronic idiopathic tinnitus. Eur Arch Otorhinolaryngol 274(5):2079–2091. https://doi.org/10.1007/s00405-016-4401-y

Zhao H, Dong H, Cheng Y, Ma Y, Lin P, Zhong S et al (2015) Multi-center study on the treatment of sudden deafness accompanied with tinnitus. Zhonghua Er Bi Yan Hou Tou Jing Wai Ke Za Zhi 50(6):453–457. https://www.ncbi.nlm.nih.gov/pubmed/26695794

Zhu C, Gausterer JC, Schopper H, Nieratschker M, Saidov N, Ahmadi N et al (2018) Evaluation of sustained-release steroid hydrogels in a guinea pig model for noise-induced hearing loss. Audiol Neurootol 23(2):73–81. https://doi.org/10.1159/000487662

Zoger S, Svedlund J, Holgers KM (2006) The effects of sertraline on severe tinnitus suffering--a randomized, double-blind, placebo-controlled study. J Clin Psychopharmacol 26(1):32–39. https://doi.org/10.1097/01.jcp.0000195111.86650.19

Sense and Sensibility: A Review of the Behavioral Neuroscience of Tinnitus Sound Therapy and a New Typology

Grant D. Searchfield

Contents

Abstract Tinnitus Sound Therapy is not a single strategy. It consists of many different sound types, targeting many different mechanisms. Therapies that use sound to cover, reduce attention to, or facilitate habituation of tinnitus are among the most common tinnitus treatment paradigms. Recent history has seen a

G. D. Searchfield (✉)
Section of Audiology, Auckland, New Zealand
e-mail: g.searchfield@auckland.ac.nz

Curr Topics Behav Neurosci (2021) 51: 213–248
https://doi.org/10.1007/7854_2020_183
Published Online: 6 February 2021

proliferation of sound therapies, but they have each been criticized for having limited empirical support. In this review, Sound Therapy's modern history will be described, and a typology will be introduced and discussed in light of current behavioral neuroscience research. It will be argued that contributing factors to the limited evidence for the efficacy of Sound Therapy are its diversity, plural modes of action, and absence of a clear typology. Despite gaps in understanding the efficacy of sound's effects on tinnitus, there is compelling evidence for its multiple, but related, neurophysiological mechanisms. Evidence suggests that sound may reduce tinnitus through its presence, context, reaction, and potentially adaptation. This review provides insights into the neurocognitive basis of these tinnitus Sound Therapy modes. It concludes that a unifying classification is needed to secure and advance arguments in favor of Sound Therapy.

Keywords Review · Sound therapy · Tinnitus · Typology

1 Introduction to the Sense and Sensibility of Sound Therapy

Sound Therapy is the use of sound to manage or treat tinnitus. Sound has been a common tinnitus therapy tool since the late 1970s. Yet Sound Therapy for tinnitus has been metaphorized as a sacred cow (Mckenna and Irwin 2008) and as the auditory equivalent of the cobra effect (a cure worse than the problem (Attarha et al. 2018)). Systematic reviews are highly critical of evidence used to support tinnitus Sound Therapy (Hoare et al. 2010; Hobson et al. 2012; Phillips and McFerran 2010; Sereda et al. 2018). Is the use of Sound Therapy blind faith, or has the evidence for tinnitus been misinterpreted? Why is there ambiguity as to the benefits and basis of Sound Therapy? In this review, I will attempt to answer these questions and will propose a typology to test the behavioral neuroscience of Sound Therapy.

Rather than using an animal (serpent or sacred) as a metaphor for Sound Therapy, I will use a literary analogy. Jane Austen (1775–1817) was the author of half a dozen novels describing life among English country aristocracy at the end of the eighteenth century. In Austen's *Sense and Sensibility* (Austen 1811), the two chief characters, sisters, represent qualities of "sense" and "sensibility." In the novel, "sense" means good judgment, wisdom, and prudence, while "sensibility" means sensitivity, sympathy, or emotionality. Tinnitus can also be considered as being both sense, in this case the sense of hearing, and sensibility, the emotional and cognitive aspects of tinnitus.

Tinnitus treatments are also seen through the sense and sensibility lenses of different professional traditions: audiology (sense) and psychology (sensibility). Audiologists see the value in using sound to provide tinnitus relief (Vernon and

Schleuning 1978) and partial masking (Tyler and Bentler 1987) and as an aid to habituation (Jastreboff and Jastreboff 2000). Psychologists, on the other hand, are critical of the use of sound as the basis for therapy and instead focus on the reactions and behaviors that clients adopt in response to tinnitus (Kroener-Herwig et al. 2000). A theme of Jane Austen's novel was the need for the sisters' "sense" and "sensibility" to cooperate. A common classification for Sound Therapy would remove some confusion and aid interdisciplinary approaches. Typologies are ways of organizing and categorizing testable theories (Doty and Glick 1994). As an example, a typology of attention comprised of alerting, orienting, and executive function provides a structure within and against which behavioral neuroscience evidence can be sought (Raz and Buhle 2006). In proposing a typology for tinnitus Sound Therapy, the intent is to create a conceptual grouping of Sound Therapy mechanisms based on known behavioral neuroscientific events that can then be tested.

2 A Modern History of Sound Therapy

The purpose of this section is to outline the development of the most common Sound Therapy for the period 1945–present.

2.1 Hearing Aids

Hearing aids were the first practical wearable solution to provide Sound Therapy (Saltzman and Ersner 1949; Bentzen 1958). They have been proposed to provide immediate and long-term relief by improving communication, redirecting attention from tinnitus to real sound, and reducing auditory gain (Schaette and Kempter 2006; Searchfield 2006; Shekhawat et al. 2013). In addition to being a treatment solution in their own right, they became important tools in other sound therapies including masking (Vernon and Schleuning 1978) and Tinnitus Retraining Therapy (Jastreboff and Jastreboff 2000). As hearing aid technology has improved, so have outcomes (Trotter and Donaldson 2008), and protocols for hearing aid fitting specifically for tinnitus have been developed (Searchfield 2006). A systematic review reported only low-level evidence for clinically significant improvement in tinnitus with hearing aids, equivalent to that found with noise generators and combination aids (hearing aid with noise generator) (Sereda et al. 2018). A scoping review concluded that the weight of evidence (17 research studies for, 1 against) supported their use (Shekhawat et al. 2013). For patients with persistent bothersome tinnitus, which accompanies a hearing loss, hearing aid use is recommended for clinical use according to the published American Academy of Otolaryngology-Head and Neck Surgery guidelines (Tunkel et al. 2014).

2.2 Masking

Clinically practical tinnitus masking began in the mid-1970s with Vernon (Vernon and Schleuning 1978) pioneering the use of hearing aid style noise-based "maskers." The goal was to provide relief by covering, effectively replacing, tinnitus with a more pleasant sound. By the mid-1980s, partial masking had become the primary application (Terry and Jones 1986; Coles and Hallam 1987; Tyler and Bentler 1987). Partial masking, as the name suggests, partially covers the tinnitus without totally replacing it. This change improved acceptance of the masking sound for some users (Terry and Jones 1986). Variations on masking have been trialed including the use of tinnitus-matched narrowband noise and harmonics (Mahboubi et al. 2012), noise adjusted for reduced hearing sensitivity ("threshold adjusted noise" Searchfield et al. 2002), and perceived location of tinnitus (Searchfield et al. 2016). Different devices are used for masking including hearing aids (Shekhawat et al. 2013), ear-level "maskers" (Vernon and Schleuning 1978), combinations of masker and hearing aids (Henry et al. 2015), bedside desktop sound generators (Handscomb 2006), and apps (Sereda et al. 2019). A systematic review of Sound Therapy indicated there was low-level evidence to support masking as a tinnitus therapy (Hobson et al. 2012). Recent clinical guidelines suggest that Sound Therapy (including masking) is a safe therapy option, with the proviso that patients are provided with realistic expectations of therapeutic benefit (Tunkel et al. 2014).

2.3 Tinnitus Retraining Therapy (TRT)

TRT was born from Jastreboff's neurophysiological model (Jastreboff 1990) and Hazel's experience with masking (Hazell et al. 1985). In their initial publications, the therapy wasn't called Tinnitus Retraining Therapy (Jastreboff and Hazell 1993; Jastreboff et al. 1996), but by the early 2000s, TRT had become one of the most well-known therapies incorporating sound use. The therapy used a combination of directive counselling (instruction) and partial masking sound (hearing aids or sound generators (maskers)) in which the masking sound and tinnitus mixed, so the tinnitus was audible but was less obvious (Jastreboff 1999). The sound was intended to be used for 6–8 h per day in order that it might yield, along with the instruction that addressed fear and understanding, habituation to the tinnitus. The method was very popular among audiologists but was strongly criticized by psychologists. Psychologists were critical of the directive counselling approach used (Wilson et al. 1998; Goebel 1997) and the value of the Sound Therapy component, suggesting it may act as a "technical placebo" (Kroener-Herwig et al. 2000). The importance of using sound at the mixing point with tinnitus has also been questioned (Tyler et al. 2012). The authors of TRT have updated recommendations as new information has become available (Jastreboff 2015), and it has also been modified by others (see below). A systematic review (Phillips and McFerran 2010) identified one study meeting the reviewer's criteria; this compared one form of masking with TRT. It showed that masking had a greater benefit after 3 months of therapy, but after 12–18 months of TRT, the results were superior (Henry et al. 2006). A more recent trial found similar results for TRT using sound generators and counselling with sound enrichment (but not using sound generators) (Scherer and Formby 2019). TRT appears most effective when applied by the developers (Jastreboff 2015).

2.4 Modified Tinnitus Retraining Therapy

Based on the criticism of TRT, some clinicians developed modified forms (Tyler and Bergan 2001; Mazurek et al. 2006; Aazh et al. 2008; Park et al. 2013). These modified methods replaced TRT's directive counselling with cognitive psychology approaches (Tyler and Bergan 2001), muscle relaxation and physiotherapy (Seydel et al. 2010), group therapy (Park et al. 2013), and simplified counselling, omitting sensory neuroscience instruction and reducing length of counselling from 90 to 30 min (Aazh et al. 2008). At what point a modification to a therapy becomes sufficient to be considered a new therapy is unclear, and this may have contributed to some ambiguity in reviews of TRT and Sound Therapy more generally.

2.5 Tinnitus Activities Treatment (TAT)

Building on their modifications to TRT (Tyler and Bergan 2001), Tyler et al. (2006) developed a new treatment approach they named Tinnitus Activities Treatment (TAT). TAT incorporated partial masking at "the lowest level that provides relief" along with client-centered counselling using pictures to explain key concepts. No clinical trials or systematic reviews were found in a literature search of TAT. An early version of TAT was used to test the sensitivity of the Tinnitus Primary Function Questionnaire; the therapy was associated with a significant improvement in scores on the questionnaire (Tyler et al. 2014a).

2.6 Music Therapy

Music therapy is a collective name for activities undertaken by a patient with a music therapist to improve a clients' quality of life. Music therapy is primarily a psychological approach to tinnitus, but as sound features strongly, it is included here. Music therapists often mix theories and techniques, creating their own versions (Hillecke et al. 2005); however, a manualized form of music therapy featured in the tinnitus literature is the Heidelberg model (Argstatter et al. 2012). This consists of nine 50-min sessions over 5 days, using counselling, resonance training (vocal exercises), "neuroauditive cortex training" (music listening training), and tinnitus reconditioning (music relaxation training) (Argstatter et al. 2012). A review of multiple treatments for tinnitus concluded that therapy conducted by a music therapist was a moderately validated approach (Zenner et al. 2017). The research primarily evidenced in the review compared music therapy to a single session of counselling, in which both groups showed improvement in the Tinnitus Questionnaire but a greater number improved with the music therapy (Argstatter et al. 2015).

2.7 Neuromonics®

Neuromonics® therapy is a habituation-based passive music Sound Therapy. The therapy uses counselling and music modified for audibility and loudness (Davis 2006). It was originally conceptualized as an auditory equivalent of systematic desensitization similar to the treatment of phobias (Davis 2006). In a two-stage process, the tinnitus is covered and then gradually introduced by reducing masking. Stage 1 is noise with modified music, while stage 2 is modified music alone. Trials report success in reducing the Tinnitus Reaction Questionnaire in a majority of participants (Tavora Vieira et al. 2011; Davis et al. 2007). A review of the Neuromonics® trials was critical of research methodology and, to that date, reliance on developer-led research (Henry and Istvan 2010). A comparison in outcomes between Neuromonics® and masking using noise-based sound generators were similar, both proving equally beneficial (Newman and Sandridge 2012).

2.8 Notched: Noise, Music, and Amplification

Several therapies apply notched filters to sound at the frequency of tonal tinnitus pitch match. Windowed Sound Therapy applied a notch in broadband noise (BBN) centered at tinnitus pitch match and with a width of twice the equivalent rectangular bandwidth (2x estimated critical band) (Lugli et al. 2009). The purpose was masking, but the authors believed sound near tinnitus pitch was deleterious to outcome, so filtered this region. A reduction in tinnitus loudness was reported for the notched noise compared to unfiltered broadband noise and a waterfall sound (Lugli et al. 2009). An exploratory study compared 1-octave notched noise around the tinnitus pitch match with 1-octave wide band of noise centered around the pitch match and a placebo of low-frequency noise; all three sounds showed benefits (Schad et al. 2018).

Notched music uses music that is spectrally flattened and then notched at tinnitus pitch, and in some studies, energy is added to the sidebands in an effort to deepen the notch (Stein et al. 2016). The sharp spectral edges bordering the tinnitus match are proposed to inhibit activity generating the tinnitus (Okamoto et al. 2010). Music is used to gather positive attention (Stein et al. 2016). Several studies provide low-level support for notched music's benefits on tinnitus (Okamoto et al. 2010); however, a recent double-blinded controlled trial only found change in a secondary measure of tinnitus loudness rating; there was no significant change in the primary outcome of the Tinnitus Questionnaire (Stein et al. 2016). A review of tinnitus therapies indicated that there was an insufficient evidence base for recommendation of notched music (Zenner et al. 2017).

Notched amplification takes a similar approach to the use of notched music, but in this case, the notch is applied to amplified sound. There is currently limited evidence to support notched amplification having benefit above conventional amplification (Haab et al. 2019).

2.9 *Fractal Sounds*

Fractal sounds are complex, digitally rendered, unpredictable, but self-similar patterns of sound. As applied to tinnitus, they resemble musical chimes. One hearing aid manufacturer introduced fractal Sound Therapy ("Zen") onboard some of their hearing aids approximately a decade ago (Sweetow and Sabes 2010). The Zen therapy uses five fractal patterns that differ in combination of pitch tempo and intensity. Most trials have been open label but have shown benefit (Simonetti et al. 2018). The relative contributions of amplification and fractal sounds to the total final benefit are unclear (Johansen et al. 2014). As with most other treatments, there are large individual differences in response (Tyler et al. 2017b).

2.10 *Acoustic CR® Neuromodulation: Desyncra™*

Acoustic CR® (Coordinated Reset) Neuromodulation was developed from an electrical stimulation paradigm to treat Parkinson's disease (Adamchic et al. 2014a). The tinnitus treatment consists of temporally patterned tones of frequencies that span a tonal pitch match. The treatment goal is desynchronizing aberrant spontaneous activity. Uncontrolled trials in a research (Hauptmann et al. 2015) and in clinic private practice population (Williams et al. 2015) showed reductions in tinnitus questionnaire scores. A controlled trial showed no benefit of the Desyncra treatment over a control; however, this has not been published in full in a peer-reviewed journal, due to methodological concerns as to the pitch-matching method used (Adamchic et al. 2014a). A systematic review of CR Neuromodulation concluded that acoustic CR Neuromodulation may have positive effects on tinnitus symptoms, but available evidence is insufficient yet to support clinical implementation (Williams et al. 2015).

2.11 The Levo® System: Otoharmonics®

The Levo System differs from most other Sound Therapy approaches in focusing on sound use during sleep and uses a synthesized tinnitus copy. It is based on the hypothesis that tinnitus emerges to replace an input deficit and that stimulation with a tinnitus replica should interrupt or reverse this (Pedemonte et al. 2010). A trial showed clinically meaningful change in the TFI after 3 months in groups using the tinnitus-matched stimulus and participant-selected "soothing" noise, and the matched sound showed greater benefit on a loudness rating scale (Theodoroff et al. 2017).

2.12 Timed Bimodal and Multimodal Stimulation

As the name suggests, therapies using multiple modes are not Sound Therapy alone but couple sound with other modulation including sounds paired with vagal nerve (MicroTransponder, Serenity® (Tyler et al. 2017a)), tonguetip™ trigeminal stimulation (Neuromod (Hamilton et al. 2016)), and auditory and somatosensory stimulation (Shore et al. 2016). These are relatively new concepts, and evidence for clinical efficacy is just beginning to emerge. The Neuromod and Serenity® systems are available clinically in some countries. These concepts are described in detail in another chapter in this volume.

2.13 Active Auditory Perceptual Training

Aside from music therapy, the therapies mentioned to this point have been "passive." Auditory perceptual training is an active therapy. Auditory training requires listeners to undertake specific listening tasks and respond to instructions. Frequency discrimination tasks have been the primary training mode (Flor et al. 2004), with attention (Searchfield et al. 2007) and categorization training being alternative methods (Jepsen et al. 2010). A review of auditory perceptual training identified that nine out of ten studies found improvement in outcome measures after auditory training, but all studies provided low or moderate levels of evidence (Hoare et al. 2010). Perceptual training games based on attention and conditioning (Wise et al. 2016) have shown promising results as have multisensory training tasks (Spiegel et al. 2015) and virtual reality (Londero et al. 2010). Clinical and self-help application of auditory training will probably be available in apps in the near future.

3 Treatment Frameworks and Protocol Strategy

The different sound therapies that have been described may suit different patients. Some approaches require more investment of client and clinician time and/or are more expensive. To manage these variables, treatment frameworks have been established that select therapy plans based on hierarchies or goal setting. Progressive Tinnitus Management (PTM) is a method based on five hierarchical levels of clinical management developed through several iterations through Veterans Affairs Hospitals in the USA (Henry et al. 2005, 2008). The five levels are (1) triage, (2) audiological evaluation, (3) group education, (4) tinnitus evaluation, and (5) individualized therapy (Henry et al. 2008). If a simple intervention is unsuccessful, the therapy is elevated to a more complex solution. The PTM method has been adapted to suit different institutions by adopting individual sessions instead of group education and modified counselling content (Tuepker et al. 2018; Beck et al. 2019). In this framework, broadband noise and environmental, music, and speech sounds can be used for soothing, as background sound, or to engage interest (Henry et al. 2008).

The author's Sound Therapy and Aural Rehabilitation (START) framework (Searchfield et al. 2019) does not use a hierarchical approach; instead, it employs needs assessment and goal setting to personalize a treatment plan (Searchfield 2017). The framework is based on a thematic analysis of Sound Therapy and patient goals (Searchfield et al. 2019) and a review of customization of Sound Therapy (Searchfield et al. 2017). That analysis identified four modes of therapy; three primary effects related to the presence, context, and reaction to sound; and a fourth secondary mode of adaptation. In the following sections, the current understanding of the behavioral neuroscience of tinnitus will be related to the typology that emerged from this research.

4 Typology: Evidence from Behavioral Neuroscience

An inclusive, but unspecific, definition of Sound Therapy is "the use of sound to manage or treat tinnitus." The plurality of methods and proposed mechanisms for Sound Therapy creates a dilemma in definition: Do you try and encapsulate all methods with a succinct general definition or provide such detail that a definition is unpragmatic? The use of a single definition leads to ambiguity; sound therapies using different sounds, with different modes of delivery and different proposed mechanisms, are being reviewed together under a single gross classification (Hobson et al. 2012). Tinnitus Sound Therapy is complex and cannot be easily defined in a single sentence. In order for the field to develop and address criticism, we need clarity on what Sound Therapy is and how it works. Here I propose a typology for Sound Therapy and provide retrospective evidence to support it. A typology is a means to organize and classify multiple theories that can be tested and proven or

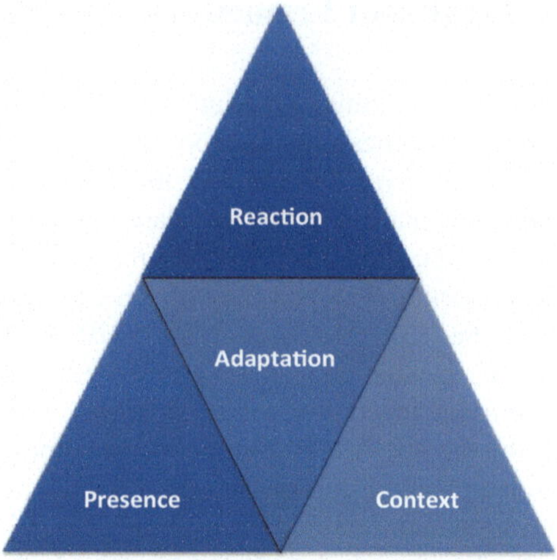

Fig. 1 Proposed tinnitus Sound Therapy typology. Sound is proposed to act therapeutically by one or more of these modes. The primary modes are sound's presence, its context, and its effect on reaction to tinnitus perception. The secondary, central mode, of adaptation is a potential outcome if tinnitus is diminished sufficiently across the three primary modes

falsified (Doty and Glick 1994). A typology should be comprehensive, but simple, free as possible of language that infers superiority or negative connotations that would limit its use. The typology described and explored here through the neuroscience literature is based on four modes of Sound Therapy (Fig. 1).

4.1 Presence of Sound Effect

The absence of sound facilitates tinnitus perception (Heller and Bergman 1953), while its presence can reduce tinnitus perception (Vernon and Schleuning 1978). The presence of sound effect is the reduction or removal of tinnitus perception by passive exposure to a sound. Its effects are hypothesized to primarily be due to bottom-up effects on sensory processes (Fig. 2). This mode includes masking (Vernon and Schleuning 1978), more effective gating (Han et al. 2019b), residual inhibition (Roberts 2007), desynchronization (Eggermont and Tass 2015), peripheral re-afferentation reversing abnormal gain (Norena 2011), and lateral inhibition (Okamoto et al. 2010).

4.1.1 Masking

The introduction of sound generally has positive effects on reducing ease of tinnitus detection (Feldmann 1971). One mechanism for change in tinnitus perception in the presence of sound is masking. Tinnitus masking is the process of partially or totally covering tinnitus, replacing its perception with that of another sound. Sound-on-sound masking may occur through a "line-busy" effect in which the sound activates neurons, preventing firing to another sound, or by a suppression effect in

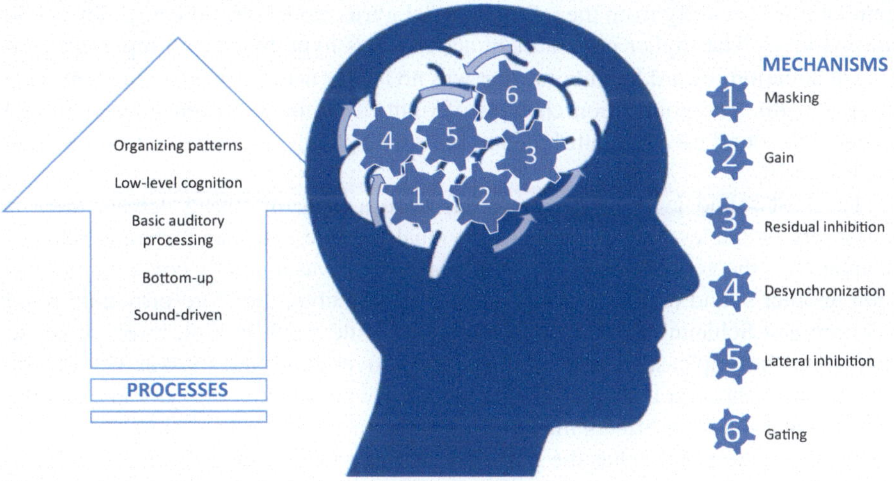

Fig. 2 Schematic of potential processes and mechanisms within the presence of sound effect

which the travelling wave of the cochlea to a probe tone is replaced by the vibration to the masker leading to probe tone suppression (Eggermont 2012; Delgutte 1990). In psychoacoustics, these peripheral mechanisms are referred to as *energetic masking* because the energy in the masker interferes with the target in a frequency- and intensity-specific manner (Brungart et al. 2006; Ihlefeld and Shinn-Cunningham 2008). The presence of sound may also disrupt the streaming of tinnitus-related activity (Durai et al. 2019), interfering prior to central processes of schema-based analysis that would aid figure-ground separation. Tinnitus masking at the level of the cochlear mechanics and early processing cannot explain why a masker frequency that does not overlap with tinnitus pitch can be effective (Burns 1984; Feldmann 1971; Mitchell 1983). Tinnitus heard as a broadband sound may also be masked by a single tone, and a masker contralateral to the tinnitus ear can be effective (Tyler and Conrad-Armes 1984; Feldmann 1971). Psychoacoustics classifies these higher-order "central" or "neural" masking effects as *informational masking* (Arbogast et al. 2002). Some direct central interference effects may help better explain tinnitus masking (see also Context of Sound 4.2). Central effects observed, presumed to be downstream effects of masking, include fMRI changes in five brain regions (right insula, right inferior parietal lobule (IPL), bilateral thalami, and left middle temporal gyrus) (Han et al. 2019a). Narrowband masking has been shown to modify a frontoparietal-cingulate network (Han et al. 2019b), suggesting sensory gating (Rauschecker et al. 2015). Sensory gating is the neural process of identifying goal-irrelevant information based on previous coding and then filtering this from further processing. Gating can be likened to a gatekeeping or noise-cancelling process. Both bottom-up and top-down processing are required for gating, but it is under limited cognitive control (Jones et al. 2016). The central gatekeeper appears impaired with tinnitus (Campbell et al. 2018). Han et al. (2019b) observed in rs-fMRI that after 12 weeks of three 20-min sessions per day of narrowband noise (NBN) sound generator use, activity from the inferior frontal gyrus reduced from being elevated to normal level. The higher pretreatment activity was hypothesized to represent controlled attention toward the tinnitus that was no longer necessary posttreatment (Han et al. 2019b). The connection changes of right thalamus and right inferior frontal gyrus with sound use were attributed to improved gating and noise cancelling (Han et al. 2019b).

The best sound level for masking has been a topic of strong debate; there is variability on the level of sound that individuals choose to use and whether relief or adaptation is the goal. The early focus of masker use was to totally cover the tinnitus with another sound resulting in tinnitus being inaudible, therefore providing relief (Vernon and Schleuning 1978). However, this could result in high levels of sound that might be unpleasant, and in some cases, over time, the masker would lose effectiveness through adaptation, or the tinnitus would increase (Tyler and Bentler 1987). Total masking effects may not be maintained due to the action of a central contrast gain control that has the goal of representing subtle changes in low contrast situations (Rabinowitz et al. 2011; Robinson and McAlpine 2009). Gain increases as stimulus contrast decreases in narrow frequency bands (Rabinowitz et al. 2011); this implies activity representing tinnitus might be increased with high levels of competing activity.

In clinical practice, partial masking was found to be a more comfortable solution than total masking (Terry and Jones 1986). The level of partial masking recommended has varied from the lowest level that provides relief (Tyler et al. 2006) to a comfortable level where the tinnitus and masker mix without total masking (Jastreboff 1999). Tinnitus perception in the presence of partial masker can be considered akin to a figure-ground process (Teki et al. 2011). As the level of masker is raised, the figure (tinnitus) is more difficult to distinguish from the background (masker). The ease of extracting the tinnitus signal from ongoing background neural activity is reduced as the level of sound is increased. Partial masking of sound may also affect magnitude estimation due to a reduction in contrast of tinnitus against other activity (Searchfield et al. 2012). Raising the overall stimulation of the auditory system through sound should raise internal perception criteria (the adaptation level) relative to tinnitus, thus decreasing the perceived magnitude of tinnitus over time (Searchfield et al. 2012).

Masking, at different intensities, can affect tinnitus in different ways across time. Masking can have an almost immediate benefit on perceived tinnitus loudness, before plateauing, while tinnitus handicap and distress may continue to reduce with time (Fig. 3). Masking has been reported to have a stronger initial outcome than TRT, but TRT's effects are greater with more time (Henry et al. 2006). In another study, patients whose tinnitus was masked at the fitting of hearing aids had greater long-term reductions in negative reactions to tinnitus than those with partial or no masking at aid fitting (McNeill et al. 2012). Another study found that the level of sound does not affect treatment outcomes (Tyler et al. 2012). Several therapies for managing tinnitus using a transition from high to lower levels of masking via two (Davis et al. 2007) or three (Lopez-Gonzalez and Lopez-Fernandez 2004) stages. These therapies attempt to exploit the posited immediate relief from higher masking

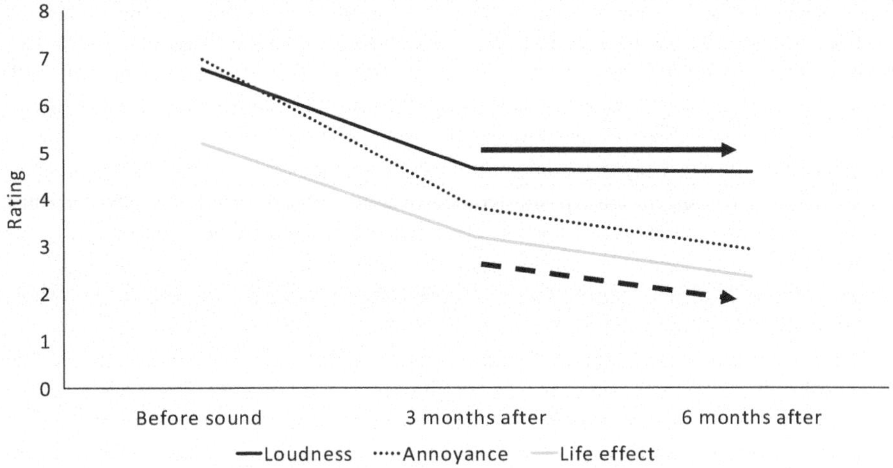

Fig. 3 Schematic showing the time course of Sound Therapy with TRT. Tinnitus magnitude (loudness rating scale) is decreased after 3 months and stabilizes at 6 months following commencing use of sound. Annoyance and life effect ratings continue to improve. This suggests two-time frames for different effects, short-term magnitude reduction (solid arrow), and longer-term psychosocial benefit (dashed arrow) (data from (Carraba et al. 2008))

and use lower and sustaining sound stimulation to achieve lasting tinnitus reduction. It is difficult to separate sound benefit from counselling in these studies, but masking may have an effect through two phases: relief (reducing tinnitus magnitude) and adaptation (reducing handicap and impact of tinnitus) (Fig. 3, Carraba et al. 2008). Prolonged, and maintained, periods of reduced magnitude through the presence of sound may contribute to tinnitus adaptation and its relief. The ideal starting sound level for masking may be goal-driven (Searchfield 2017) and based on individual adaptation levels and listening preference (Durai and Searchfield 2017). Clinicians should be aware that a neural contrast gain control (Rabinowitz et al. 2011) may act to extract tinnitus from noise requiring further adjustment in level or type of sound to avoid such an effect.

4.1.2 Residual Inhibition

Experimental masking studies have demonstrated post-masking suppression of tinnitus, which has become known as auditory residual inhibition (ARI). ARI manifests as diminished or abolished tinnitus sensation for a short period of time after the cessation of a masker (Roberts 2007). Complete ARI is the absence of tinnitus following the offset of masking sound, and partial ARI is the reduction but not complete absence of tinnitus (Vernon and Meikle 1988). For example, 1 min of broadband noise at 10 dB above the minimum masking level (MML) results in ARI in 80–90% of patients (Vernon and Meikle 2003). ARI usually lasts for less than a minute (Roberts 2007). Because ARI usually requires intense stimulation and has short-lasting effects, it has limited benefit as a treatment. However, its marked effect on tinnitus, and the ability to elicit the effect and return to baseline, is an attractive model for exploring the physiological basis of Sound Therapy. The reduction with ARI is also independent of the counselling that normally confounds clinical Sound Therapy studies. The time course of the ARI effect is consistent with desynchronization of hypersynchronous neural networks in humans (Roberts 2007) and with suppressed spontaneous firing in the inferior colliculus in mice (Galazyuk et al. 2017). ARI is most effective when the effector sound spans the frequency region matched to tinnitus pitch (Roberts et al. 2008). Roberts et al. (2015) investigated ARI effects using evoked 40 Hz auditory steady-state response (ASSR) and transient N1 component of the auditory evoked potentials. Among persons with ARI-induced tinnitus reduction, the amplitude of ASSR was found to increase while the amplitude of N1 decreased. It was concluded that the increase in ASSR was due to improved phase locking to the 40 Hz stimulus accompanying suppression of hypersynchronous neurons in the tonotopic primary auditory cortex, with reduced N1 amplitude coinciding with ARI, but not predicting to the strength of suppression (Roberts et al. 2015). N1 is associated with the activity of the secondary auditory cortex and with parieto-frontal filtering of irrelevant stimuli (Zhang et al. 2011). ARI may suppress spontaneous hyperactivity to below a detection threshold (Galazyuk et al. 2019). Frequency-specific suppression probably occurs in the tonotopically organized primary auditory cortex, with a potential, but less specific suppression in the secondary auditory cortex related to attention mechanisms (Roberts et al. 2015). This localized auditory network can be extended to include areas not specifically auditory. Study results of intracranial recordings from an awake tinnitus patient during ARI support a temporal lobe "tinnitus driving network," a parahippocampal and inferior parietal cortex "tinnitus memory network," and a widespread "tinnitus perception network" with roles in predictive processing (Sedley et al. 2015). A recent study by our group (King et al. 2021) identified ARI in 17 of 30 participants; these participants had significant increases in the power spectral density of alpha and gamma bands of the EEG. This finding is consistent with the theory that an increase in alpha activity is a signature of ARI and that this may represent inhibitory control of synchronized spontaneous activity, potentially decreasing activity of a tinnitus precursor or altering precision of perception (Sedley et al. 2015). In a proof-of-concept study using a spiking neural network model of EEG, our group found that higher connections were created between the temporal and centroparietal regions of the spiking neural network models after ARI stimulation (Sanders et al. 2020). It is possible that ARI interrupts activity in memory and perception networks secondary to the desynchronization of the peripheral-driven tinnitus activity.

4.1.3 Entrainment and Desynchronization

It has been proposed that amplitude modulated (AM) sounds enhance ARI (Neff et al. 2017, 2019; Reavis et al. 2012; Tyler et al. 2014b) and that may this occur through entrainment (Neff et al. 2017). Entrainment occurs when neural patterns align with temporal changes in sound and has been identified as a modifiable substrate of attention (Calderone et al. 2014). Modulating tones within background noise used for tinnitus therapy have been observed to change activity in the right middle frontal gyrus and right superior gyrus, consistent with changes to primary auditory and attention processing (Liu et al. 2018). However, two recent studies (Sanders et al. 2020; Neff et al. 2019) did not find significant differences in tinnitus suppression between tinnitus pitch-matched AM and constant stimuli.

Tonal patterns have also been proposed as potential modifiers of tinnitus. Some tone-based sound therapies set out to desynchronize neural assemblies in a tonotopically focused area around tinnitus pitch to weaken neural connections (Eggermont and Tass 2015). In a different study, 12 weeks of this acoustic coordinated reset (CR) neuromodulation therapy was found to increase EEG alpha band activity in temporal and frontal cortices and reduce delta activity strength in primary and secondary auditory cortices (Tass et al. 2012). Alongside a general gamma reduction, theta was reduced at the anterior cingulate and frontal regions, with a reduction in the superior temporal gyrus (Tass et al. 2012). In follow-up studies, the EEG following therapy resembled that of control participants (Silchenko et al. 2013; Adamchic et al. 2014b). Prior to CR therapy and compared to the healthy controls, the good responders showed a significantly increased connectivity between the left primary auditory cortex and the posterior cingulate cortex in the gamma and delta bands and in the alpha band a significantly decreased effective connectivity between the right primary auditory cortex and the dorsolateral prefrontal cortex. After 12 weeks of CR therapy, most of the pathological interactions were absent. The results were attributed to the therapy restoring a posterior cingulate cortex "salience-based cognitive auditory comparator" (Silchenko et al. 2013). The therapy has been further explored, acutely, by comparing EEG and visual analogue scales before, during, and after 16 minutes of CR in and around tinnitus pitch, a "noisy CR" without stimulation at the tinnitus pitch, and a low-frequency range stimulation outside of the tinnitus frequency region; there were differences also in repetition rates (Adamchic et al. 2017). The low-frequency stimulation range had little effect on ratings, and the EEG was reduced in the delta band. CR and noisy CR reduced a loudness scale, while the noisy CR also reduced tinnitus on an annoyance scale. CR had a stronger effect in reduction of delta and gamma and increase in alpha power in the auditory cortex (Adamchic et al. 2017). The current measurement resolution limits the ability to draw together the specific putative mechanism of effect for this therapy with physiological and behavioral outcomes.

4.1.4 Lateral Inhibition

An alternative to applying energy at the tinnitus pitch is to exclude sound frequencies in the tinnitus pitch region, using a notch filter to simulate neighboring frequencies. The sharp spectral edges created by notching sound have been hypothesized to result in lateral inhibition of overactive neurons at the tinnitus pitch (Teismann et al. 2011). Physiological studies have been undertaken on the most commonly cited lateral inhibition method known as “tailor-made notch music” (TMNM) (Okamoto et al. 2010). The use of TMNM for 12 months showed reduced magnetoencephalography N1m responses associated with decreased tinnitus loudness (Okamoto et al. 2010); a similar result was observed after just 5 days, considered to be consistent with fewer active neurons or reduced synchrony (Teismann et al. 2011). The use of TMNM for 3 h per day for 3 days was sufficient to reduce subjective tinnitus loudness and N1m amplitude in the temporal, parietal, and frontal regions. These results were consistent with TMNM effects on a broad sensory-cognitive network, reflecting the presence of sound along with music’s benefit on emotion and reaction to tinnitus (Stein et al. 2015a). The ability of notched music effect to create lateral inhibition has been questioned due to music’s spectrum (Noreña 2012). Although the music in TMNM is compensated to provide a flat spectrum to reduce a low-frequency bias, the therapy is less effective for pitch-matched tinnitus than high-pitch tinnitus (>8 kHz) due to the limited high-frequency energy in music (Teismann et al. 2011). Greater depth of the notch relative to side bands of noise improves the effectiveness, thus supporting lateral inhibition as a mechanism of effect (Stein et al. 2015b), whereas the absence of effects of altering notch width (1 octave, ½ and ¼ octave widths) appears incongruent with lateral inhibition (Wunderlich et al. 2015).

The spectral limitations of notching music also apply to the concept of notched amplification (Haab et al. 2019) but should not apply to notched broadband noise. The equal benefit of notched noise at tinnitus pitch, narrowband noise at tinnitus pitch, and sound remote from tinnitus pitch casts some doubt on the merit of tailoring frequency spectrum for tinnitus inhibition (Schad et al. 2018). The equivalency of results suggests that either there are multiple Sound Therapy mechanisms that play an equal role in reducing tinnitus or there is a common mechanism that is independent of tinnitus pitch.

4.1.5 Peripheral Re-afferentation and Change in Gain

Changes in gain have been modelled as a contributor to tinnitus generation (Schaette and Kempter 2009; Norena 2011). Increased gain within the auditory pathways is seen as necessary to maintain homeostasis following hearing loss and deafferentation. The link between hearing thresholds and spectrum of tinnitus suggests that if the hearing loss could be compensated for, the driver creating the spectrum would be eliminated (Schaette et al. 2010). Reducing hearing loss as the solution to tinnitus is not a new concept (Fowler 1948), but modern hearing aids or cochlear implants are tools that now enable this. Raising cochlear neural activity in a frequency-specific manner through hearing aids (Shekhawat et al. 2013) and sound tailored to hearing loss (threshold-adjusted noise (Searchfield et al. 2002)) have been proposed to counteract gain by reducing the drive for increased spontaneous activity and hence a reduction in tinnitus (Schaette et al. 2010). The presence of sound has been demonstrated to reverse gain associated with auditory deprivation (Munro 2008), possibly reducing tinnitus in the process (Schaette et al. 2012). Changes in gain appear to explain increases in sound sensitivity (Auerbach et al. 2019), but its role specifically in generating tinnitus is less clear (Sedley 2019). The gain model may require concomitant mechanisms including a source of neural noise to be amplified (Zeng 2013), interaction with prediction (Sedley 2019), or memory mechanisms (De Ridder et al. 2013; De Ridder et al. 2014) for sound to be perceived. The gain model is primarily a passive bottom-up response to deafferentation but may also be under a degree of top-down control (Robinson and McAlpine 2009).

Recent developments in understanding of neuroplasticity mechanisms have resulted in innovative methods to build on sounds' innate effects. Efforts have been made to potentiate neuroplasticity to sound using vagal nerve (Tyler et al. 2017a) or trigeminal nerve (Hamilton et al. 2016) stimulation. Although innovative in their use of neuromodulators, the effector of change in these multimodal methods is sound. If the Sound Therapy is mistargeted, the therapies should not be effective.

In summary, the presence of sound results in many neurophysiological changes, some of which may account for the behavioral benefits seen as response to basic sounds. The effect encompasses a number of treatment modes that affect tinnitus by the passive presence of sound affecting sensory aspects of tinnitus perception.

4.2 *Context of Sound Effect*

The context of sound effect is a sensory-cognitive mode reflecting how tinnitus can be altered by the information contained within sound. The context of sound effect is hypothesized to reflect top-down knowledge-based processes. We are active explorers of our soundscape and attempt to match sounds to visual objects in the environment. Categorization and recognition of auditory objects are as critical to survival as identifying objects of visual perception (Bregman 1990; Brefczynski-Lewis and Lewis 2017). Tinnitus lacks visual context and in the absence of meaning engenders attention (Feldmann 1992), and so it is a challenge to our learned reality of normal soundscapes (Feldmann 1992). We need to classify sounds as being real or unreal or of self-nonself (Brefczynski-Lewis and Lewis 2017). The neural activity forming tinnitus conflicts with memory and expectations of true sounds (Searchfield et al. 2007). To manage the complexity of our auditory soundscapes, auditory objects need to be selectively attended and then to be filtered and followed over time. Attention mechanisms enhance the processing of relevant auditory inputs and suppress those that are irrelevant (Eramudugolla et al. 2005). If the source and its meaning are identified and there is no reason to attend to it, sensory resources can be applied to other sensory tasks. If activity fails to be matched to existing templates or predictions, further attention resources may be allocated to extract features (Searchfield et al. 2007). In this view, the unusual aspects of tinnitus engender attention and tinnitus are under top-down executive control (Heeren et al. 2014). The meaning and relevance of sound and the context in which it is heard will have strong effects on whether it is listened to or recedes into the background (Escera et al. 2000); the perception of tinnitus is likely to follow a similar process (Fig. 4).

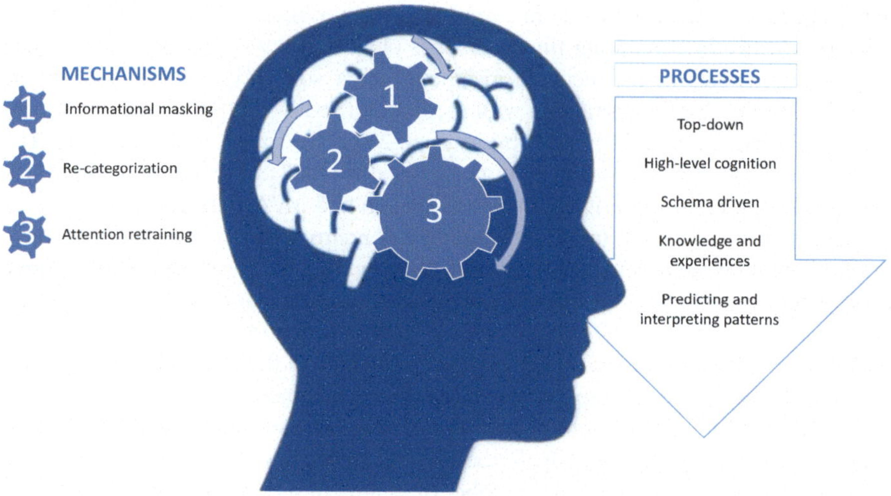

Fig. 4 Schematic of proposed mechanisms within the context of sound effect

The therapeutic benefit of the context mode may occur in response to the information contained in the sound competing with, recategorizing, or defocusing attention on tinnitus. Schema-driven grouping relies on prior knowledge of familiar patterns in acoustic data (Treisman and Gelade 1980). It is possible that, with tinnitus, spontaneous activity in the primary afferents is not ignored, as usually would be the case, but instead becomes a default prediction of sound (Sedley et al. 2016). Cochlear output is incongruent with expectations causing a shift in the balance of excitation and inhibition in coding at the primary and secondary auditory cortices (Roberts et al. 2015). Treatment could target processes used for prediction of sound, or disrupt regularities within the neural signal, rather than suppressing global neural firing (Searchfield et al. 2007; Sedley 2019). Our ability to select tinnitus above more relevant information in surrounding sounds should be heavily influenced by attention (Giard et al. 2000), and interesting sounds may tax our limited attention capacity interfering with figure-ground segregation (Treisman 1964), reducing resources available to listen to tinnitus.

4.2.1 Informational Masking

Tinnitus can be masked by information-bearing sounds that demand central processing (Searchfield et al. 2016). Such informational "central" masking of sound can be achieved by complex patterns of sound including speech, temporally, and frequency-varying tones (Kidd et al. 2002; Oh and Lutfi 1999). Hearing aids may reduce neural gain or provide frequency-specific masking through the presence of sound effect, but their main effect may be through amplification of speech and its demand on cognitive processing leaving limited resources to listen to tinnitus. For humans, speech is a signal of great importance, strongly associated with memories and complex in its tonal and temporal nature as well as meaning. Speech should be the focus of attention. Hearing aid users have been found to be more likely to successfully reduce their tinnitus if they had higher functional connectivity between auditory and default mode networks, consistent with a predisposition to effectively process and interpret speech (Han et al. 2020).

Although less ecologically important than speech sounds, the auditory cortex also responds strongly to other modulated sounds. Sounds that vary in temporal characteristics along with frequency can capture attention and result in informational masking (Kidd et al. 2008). Durai et al. (2018) undertook a short-term adaptation study exploring the feasibility of simulated surf-like sounds for tinnitus therapy and compared predictable amplitude modulated sound with ones unpredictable in amplitude and timing. The results indicated unpredictable sounds provided greater reduction in loudness and annoyance (Durai et al. 2018). Rhythmic sounds are more likely to be habituated to, while sounds with unpredictable rhythms may interfere with prediction and engage attention. Modulated noise may suppress tinnitus through dual mechanisms of prediction disruption and engaging attention. A Sound Therapy using modulating tones with background noise has been associated with rs-fMRI changes in the right middle frontal and right superior gyri, consistent with changes to primary auditory activity and attention processing (Liu et al. 2018). Recordings from the auditory thalamus in rats have demonstrated that modulated noise attenuates probe sounds by two mechanisms: (1) reducing the total number of evoked spikes (masking) and (2) context-specific changes in response timing (scrambling) (Martin et al. 2004).

The informational masking concept can be extended to include spatial cues. At a cognitive level, the auditory system places a premium on whether sound is caused by a living agent and is perceived to be near or far from the listener (Brefczynski-Lewis and Lewis 2017). Near objects are considered as more relevant and potentially threatening (Brefczynski-Lewis and Lewis 2017). Tinnitus is perceived as a sound in or in the near field around the head (Searchfield et al. 2015). Like a real sound heard in the near field around the head, tinnitus demands attention. Sounds that are perceived as coming from the same spatial location are better maskers than those heard as being separated in space (Arbogast et al. 2002; Ihlefeld and Shinn-Cunningham 2008). In some individuals, tinnitus informational maskers may interfere with both "where and what" sound processing pathways with spatial information requiring greater cognitive processing demand, and so there are fewer cognitive resources to extract tinnitus from the complex sounds (Searchfield 2014). The therapy sounds may have provided context for the perception of sound, enabling participants to disengage attention from their tinnitus. A behavioral and electrophysiological modelling study found the most effective masker combined energy at tinnitus pitch with spatial information (Durai et al. 2020). Large changes in strength of connectivity were observed, centered around parietal-occipital regions. The parietal cortex, implicated in Durai et al. (2020) results as a locus for masking effects, is involved in spatial attention and association (Brefczynski-Lewis and Lewis 2017). There were also changes at frontal, central, and temporal sites (Durai et al. 2020). These observations support the theory that the perception of tinnitus relies on a distributed neural network (De Ridder et al. 2014) that includes the processing of spatial cues.

4.2.2 Attention and Categorization Retraining

A change from focusing on tinnitus to focusing on real sounds and a recategorization of tinnitus from an unreal (phantom) sound to an auditory object that can be ignored may be facilitated through auditory training. Auditory training involves discrimination, categorization, and attention to simple (tones) or complex (auditory object) sounds. Discrimination training at tinnitus pitch, learning to distinguish between or separate sound, has received most attention (Hoare et al. 2010). The benefits for tinnitus of training using discrimination tasks have been limited, and change is best predicted by the amount of regular training undertaken and psychological variables (Flor et al. 2004), suggesting a role for focused attention (Hoare et al. 2010). Categorization training is an alternative to discrimination training. Categorization is the process of grouping; training can be done on the basis of sounds' perceptual, functional, or emotional aspects (Bergman et al. 2009). If tinnitus categorization is able to "embody the sound," that is, match sound with actions (Brefczynski-Lewis and Lewis 2017), tinnitus may be classified as irrelevant and be ignored. Such categorization should be aided by the addition of multisensory cues. Persons with tinnitus are more sensitive to cross-modal interference; they have increased difficulty in ignoring irrelevant stimuli (Araneda et al. 2015). Exposure to complex tinnitus avatars alongside a visual representation (e.g., virtual reality (Londero et al. 2010)) or tactile presentation (Spiegel et al. 2015) may diminish or desensitize the experience of a nearfield, but unidentified sound. Auditory-visual-tactile tasks of either integration or attention diversion were similarly effective in improving Tinnitus Functional Index scores and rating scales after 20 daily sessions of 20–30 min (Spiegel et al. 2015). The multisensory training also improved auditory and visual attention and eye tracking (Spiegel et al. 2015). A further study repeated the integration training task with and without fluoxetine (Searchfield et al. 2020a). It was hypothesized that fluoxetine would improve auditory plasticity in the manner reported for vision (Vetencourt et al. 2008). The fluoxetine did not improve outcomes (consistent with similar human vision studies (Lagas et al. 2014; Huttunen et al. 2018)), but the training task improved ratings on problem and annoyance scales and was associated with increased rs-fMRI connectivity between the auditory, visual, and somatosensory brain areas and decreased connectivity within attention and memory networks, consistent with hypothesized modes of action (Searchfield et al. 2020a).

If tinnitus was totally under top-down executive control (Heeren et al. 2014), training might enable tinnitus simply to be "put out of mind" (Attarha et al. 2018). By focusing on stimuli other than their tinnitus, patients may learn skills that can assist in their ignoring tinnitus (Searchfield et al. 2007). Evidence from perceptual training studies suggest that attention to tinnitus is malleable. Perceptual training may enable a degree of control over attention to tinnitus. Training with the goal of greater attention to external sounds appeared to increase the ability for sounds to interfere with tinnitus (Searchfield et al. 2007). After several weeks of training at home using an auditory object identification and localization task, minimum masking levels were reduced, indicative of the task increasing the ease to which external sounds interfered with tinnitus (Searchfield et al. 2007). Gamification of training may be another important consideration for tinnitus therapy in order to maintain motivation and compliance with tasks (Wise et al. 2016). After 20 days of 30-min training on a game requiring the user to find target sounds (unlike tinnitus) resulting in points reward (while ignoring distractor sounds resembling tinnitus), Tinnitus Functional Index scores improved as did performances on audio and visual attention tasks. The N1 auditory evoked potential latency was also reduced for sounds remote from tinnitus pitch (Wise et al. 2016). However, care needs to be undertaken in interpreting neural changes with training which might be assumed to be a correlate of tinnitus but instead may be a more general measure of attention (Sedley 2019).

The spatial similarity between a masker and tinnitus may aid in the process of recategorization of tinnitus as a controllable sound, leading to less attention and processing. The processing of nonliving sounds activates visual episodic networks such as parietal-occipital, parahippocampal gyri, and posterior cingulate cortices (Engel et al. 2009). As an alternative action to informational masking, blending of a true sound with tinnitus may aid in the reclassification of tinnitus from being a stimulus of interest to a background noise. If tinnitus is embedded within a noise and the noise-tinnitus blend is perceived as a single entity, it may be categorized as unimportant, which may facilitate its habituation (Jastreboff and Hazell 1993). The Levo treatment attempts this during sleep through a sound matched to tinnitus, the concept being that adaptation to the tinnitus-like sound will also capture the tinnitus it is matched to (Pedemonte et al. 2010).

4.3 *Reaction to Sound Effect*

4.3.1 Gating

The reaction to sound effect is the reduction or removal of negative reactions to tinnitus by exposure to sound. Like the context of sound effect, reaction to sound is governed by cognitive processes. Emotion affects perception (Siegel and Stefanucci 2011), for example, the putative "gatekeeping" system for sensory information is under frontostriatal control (Rauschecker et al. 2015; see also the presence of sound effect). Very few people find tinnitus pleasant (Stouffer and Tyler 1990). Negative emotional association of the tinnitus perception might lead to more disabling tinnitus through processes that drive attention toward the tinnitus signal, increasing distress and preventing adaptive responses. If the negative emotions could be decoupled

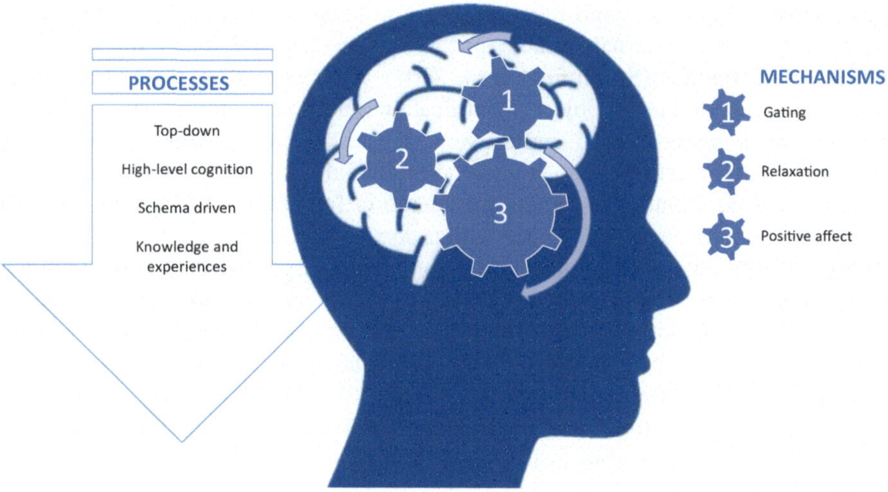

Fig. 5 Schematic of proposed mechanisms within the reaction to sound effect. Gating is a mechanism present for both presence and reaction to sound effects

from tinnitus, attention may be reduced, and the ability for tinnitus-related neural noise may be improved through gating (Rauschecker et al. 2015) (Fig. 5).

4.3.2 Affect

A negative mood may result in tinnitus being louder. Durai et al. (2017a, b) exposed persons with tinnitus to brief emotional sounds and images. The lowest tinnitus loudness ratings were recorded after exposure to positive valence stimuli, highest after the negative sounds. Effects were modality-specific; visual images did not result in changes in tinnitus, suggesting auditory attention facilitated the negative and positive responses (Durai et al. 2017a). A positive reaction to sound may counteract negative affect created by tinnitus, improving mood. Individual psychology factors, personality, memory, prediction, cognition, and emotion play strong roles in this driver of affect to tinnitus (Searchfield et al. 2012). Personal preferences to sound associated with memories or negative events also contribute to reaction to sound.

The neural effects of the Heidelberg Neuro-Music Therapy (HNMT) have been compared to controls (Krick et al. 2015, 2017). Gray matter increases were found in the precuneus, medial superior frontal areas, and auditory cortex. Using fMRI, increased default mode network activity particularly in the posterior cingulate cortex was associated with the improvement's tinnitus-related distress related to the HNMT therapy. This therapy offers the possibility to evaluate the neural changes associated with the improvements in tinnitus distress (depression and anxiety) (Krick et al. 2015, 2017).

4.3.3 Relaxation

Therapies that include music (e.g., Neuromonics, notched music) or fractal tones may promote relaxation (Simonetti et al. 2018). Although relaxation is a stated aim of these therapies, research investigating their benefits do not appear to have included physiological or behavioral indicators of emotion or stress to test this hypothesis.

Brefczynski-Lewis and Lewis (2017) speculated that sounds, which did not require physical action, aided relaxation by reducing inward thoughts. Sounds matching this description include surf-like sound (Searchfield 2019). While widely adopted for tinnitus control by hearing aid manufacturers, there is, however, little evidence for their efficiency (Sereda et al. 2017). Minor changes in temporal characteristics of these sounds can have dramatic effects on acceptance. Individual preferences for sound rhythms mirror known psychological benefits (Searchfield 2019). Damped sounds (increase rapidly and then decrease gradually with time) are preferred as short-term aids to tinnitus by more people than the opposite ramped sounds (slow rise, rapid decrease). Slow oscillations are preferred to rapid changes (Searchfield 2019). The asymmetry in response is believed to be due to ramped sounds as appearing to approach the listener (rather than recede away – damped); ramped sounds are more arousing or unpleasant as a consequence (Bach et al. 2011). The ramp archetype has been used to capture attention in music (Huron 1992); consideration of such psychoacoustic features would aid clinical implementation of sounds.

BBN is not a sound that is considered overtly pleasant or soothing, yet is more effective in reducing short-term tinnitus-related stress (Aydin and Searchfield 2019) and long-term tinnitus (Durai and Searchfield 2017) than sounds considered more pleasant. Direct measures of stress biomarkers suggest BBN is superior in reducing stress than self-selected nature sounds (Aydin and Searchfield 2019). Streamed nature sounds and BBN were found to be equally effective over 6 months (Barozzi et al. 2016), while BBN was superior to nature sounds after 3 months use of each in a crossover trial (Durai and Searchfield 2017). The immediate relief provided by BBN masking appears to surpass the more pleasant nature sound effects. If tinnitus magnitude is reduced or it is removed from perception and the reaction to the introduced masker is at least neutral, reduced stress should be a downstream result. Categorizing tinnitus as an innocuous or pleasant sound may be accelerated using imagery (Searchfield et al. 2020b) or may be aided by pairing the tinnitus with a pleasant sound.

4.3.4 Clinical Implications of Reaction to Sound

Sounds that appear promising as a treatment may not turn out to be as effective as initially anticipated, not because they fail to affect tinnitus perception but rather because they are poorly tolerated (Durai et al. 2018; Durai and Searchfield 2017).

For example, tones may be considered a useful sound to change tinnitus activity but may be poorly accepted (Terry and Jones 1986), and notched sound might be effective in inhibiting tinnitus but may be unpleasant to listen too (Manabe et al. 2019). Annoying or negative sounds may drive attention toward tinnitus, defeating efforts in habituation or adapting to tinnitus (Durai and Searchfield 2017). A piece of music may elicit positive or negative reactions depending on personal tastes and associated memories (Hann et al. 2008). Outcomes of Sound Therapy might be strengthened if patients were able to choose from a range of treatment sounds.

4.4 Adaptation

Adaptation is a two-way process allowing both an increase or a decrease in response (Helson 1964) that includes psychological adaptations (Wilson et al. 1998) and changes in sensory response that might be through mechanisms of habituation (Jastreboff et al. 1996; Jastreboff and Hazell 1993). The ideal outcome of tinnitus treatment would be the complete elimination of tinnitus perception, with no side effects. Complete elimination of perception may be unrealistic, given that many people, who do not complain of tinnitus, experience tinnitus-like sounds (Heller and Bergman 1953). Most current treatments aim for a significant decrease in suffering and/or some reduction in tinnitus perception.

Tinnitus can be plausibly reduced by either the presence, context, or reaction to sound. If the desired goal is adaptation, positive effects from all three sound effect modes may be required. Tinnitus needs to be "fought on many fronts" (Schlee et al. 2009); a reduction in auditory cortex hyperactivity, as demonstrated through acoustic residual inhibition, may not eliminate tinnitus unless there is long-term control of input from the global workspace that drives tinnitus distress (Schlee et al. 2009). A consequence of effective Sound Therapy may be preventing tinnitus-related activity being integrated into the global workspace (across sensory, frontal, and partial brain regions). Activity above a threshold will enter this network and result in conscious perception (Schlee et al. 2009). Activity may be altered to lay below the threshold, or potentially the threshold criteria may be raised. This process is consistent with the concepts of habituation to reaction and perception (Jastreboff et al. 1996; Jastreboff and Hazell 1993) signal detection theory (Welch and Dawes 2008) and the adaptation-level theory of tinnitus (Searchfield et al. 2012). Habituation is a reduction in response to expected or prolonged exposure to stimuli. Failure to habituate to tinnitus is a theme common to psychological (Hallam et al. 1984) and sound-based therapies (Jastreboff et al. 1996; Jastreboff and Hazell 1993). Tinnitus complainers show less habituation of N1 and P1 amplitudes of ERPs to tone pips (Hallam et al. 1984). Several MRI studies suggest that habituation of tinnitus is associated with interactions of attention, emotion, and auditory networks (Husain 2016), consistent with the typology suggested in this chapter. Signal detection theory has been used to explain the predisposition for tinnitus detection for certain personality types (Welch and Dawes 2008) and might explain transition of tinnitus with Sound Therapy from detected to undetected state. For persons without tinnitus, auditory activity may not exceed the threshold for tinnitus perception, while for persons with tinnitus, it would.

The adaptation-level theory of tinnitus (ALT, (Searchfield et al. 2012) extends the concept of a threshold of detection to a criterion for magnitude estimation. It also provides a psychoacoustic explanation on how tinnitus adaptation may work given the multiple factors contributing to its perception. The ALT model is influenced by personality and psychosocial factors as well as the perception of tinnitus (Durai et al. 2015). The adaptation level is an internal reference point (criteria) for sensory magnitude estimations and approximates the magnitude of tinnitus (Schmidt et al. 2014). The model predicts that positive adaptation to tinnitus can occur if the environment is helpful to the individual and the perceived magnitude and reaction to tinnitus is reduced (Searchfield et al. 2012). With respect to the typology's three primary modes of effect, external sound can (1) suppress tinnitus-related activity and reduce perceived magnitude of tinnitus, (2) interfere with processes that extract tinnitus from other neuronal activity and reduce attention to tinnitus, and (3) elicit positive psychological response to reduce tinnitus distress. If the perceived magnitude of tinnitus can be maintained at a low level, a learned change in criteria applied to signal detection may occur. Adaptation may account for some people becoming unaware or unconscious of tinnitus. The time course of adaptation to chronic tinnitus is unclear, although many clinical trials of tinnitus Sound Therapy use repeated measures over 6–12 months (Jastreboff 2015).

5 Sound Therapy with Precision

Few sound therapies are personalized with any great deal of precision (Searchfield et al. 2017). The terms "customized" and "tailored" have been used to describe changes within a single dimension of sound (usually frequency) rather than viewing tinnitus as a complex combination of effects (Searchfield 2014). Many fields in health are moving from "one-size-fits-all" models of treatment to tailored solutions (Schleidgen et al. 2013; Tutton 2012). A multifactorial treatment that is not wedded to a single treatment paradigm is a pragmatic solution to the diversity in treatment goals that patients report. In the absence of a single effective solution and with physiological predictors of treatment at an early stage, weight needs to be given to the goals of individuals in the context of well-informed clinical decision-making. Knowing modes of action and communicating them in a transparent manner may facilitate collective knowledge and collaboration. Work is underway to test a Precision Sound Therapy™ that examines individual differences and treatment goals to aid selection from different Sound Therapy modules. Eventually behavioral measures (Durai et al. 2017b; Kleinstäuber et al. 2018) and physiological measures such as fMRI and EEG (Han et al. 2019a; Durai et al. 2020) may be biomarkers for therapy success and enable a priori therapy selection.

6 Is Sound Therapy Harmful?

It has been speculated that the use of passive Sound Therapy may be harmful (Attarha et al. 2018). Harm did not emerge as a significant theme in reviewing the research. It is possible that harm has been insufficiently investigated. But there are good reasons as to why BBN Sound Therapy in humans is unlikely to be harmful. The controlled environment in animal studies is not replicable in normal human activity. Persons using BBN Sound Therapy are not solely exposed to this sound, they receive patterned sounds as speech and environment sounds, and the time spent exposed to treatment sound is limited (Durai and Searchfield 2017; Durai et al. 2018). If harm is possible, I believe BBN is less likely to be the candidate than other manipulations of sound. In our modern world, BBN is common, computer drive fans, air conditioning, and traffic hum, yet we seldom experience our world through notched filters or frequently listen to tonal stimuli. In the majority of tinnitus cases, the underlying pathology is more likely to disrupt brain function than therapies attempting to reverse the effects of pathology. Hearing loss, dead regions, and deafferentation due to their very persistence are likely to be the primary drivers of harmful auditory plasticity and tinnitus.

7 Conclusion

At the beginning of this chapter, the questions I asked were: Is the use of Sound Therapy simply blind faith, or has the evidence for tinnitus Sound Therapy been misinterpreted? And why is there ambiguity about the benefits and basis of Sound Therapy? The answers are complex. Sound Therapy is generally inadequately defined; this has engendered ambiguity and propensity for misunderstanding. I do not believe that Sound Therapy should hold the mantel of being a sacred cow (Mckenna and Irwin 2008), nor do I think that there is great risk of a cobra effect (Attarha et al. 2018). The majority of evidence, albeit sometimes low level, indicates sound is a useful contributor to tinnitus therapy. However, there are gaps in our knowledge, particularly related to mechanisms of effect that need to be filled. Acoustic residual inhibition demonstrates that sound can, albeit temporarily, reduce activity below a threshold that delimits perception. When tinnitus detection is reduced, relief ensues; if this can be maintained, this suppressed activity may fall below a criterion for conscious processing. Alternatively, sound may act to raise the criteria for detection, and even though the tinnitus driving activity is unchanged, tinnitus may not enter consciousness. Most likely complete adaptation requires sound to act on both sense (auditory activity) and sensibility (cognition and conscious perception). At present, all sound therapies appear to have similar effectiveness. Hillecke et al. (2005) describe this effect with regard to music therapies as "the dodo bird verdict" (in Alice in Wonderland, the dodo bird says "everyone has won and all must have prizes"). It is possible that many, theoretically different, sound therapies share action on a common final pathway. Reduction of primary auditory cortex activity with a reduction in connectivity to perception and consciousness networks was commonly reported in the EEG and fMRI studies reviewed; that is, Sound Therapy acts on both sense and sensibility.

This review has provided evidence and theoretical basis for different sound therapies within a new typology. The typology is intended to bridge the knowledge gap between sensory and cognitive neuroscience, theory, and clinical practice, through the use of a common unbiased language. It remains to be seen if it is effective in this goal.

Acknowledgments I am greatly appreciative of the detailed editing on a draft of this chapter from colleagues at the LisTin Lab at the University of Auckland, Dr. Zohreh Doborjeh and Dunja Vajsakovic, and of the support of clinicians in our Hearing and Tinnitus Clinic, Dr. Gavin Coad, Tania Linford, Dr. Alice Smith. The authors' research cited in this chapter was supported through generous grants from the Auckland Medical Research Foundation's JM Cathie Trust for tinnitus research, the American Tinnitus Association, the Tinnitus Research Initiative, the Deafness Research Foundation of NZ, and the Eisdell Moore Centre for Research in Hearing and Balance.

References

Aazh H, Moore BC, Glasberg BR (2008) Simplified form of tinnitus retraining therapy in adults: a retrospective study. BMC Ear Nose Throat Dis 8:7

Adamchic I, Hauptmann C, Barnikol UB, Pawelczyk N, Popovych O, Barnikol TT, Silchenko A, Volkmann J, Deuschl G, Meissner WG (2014a) Coordinated reset neuromodulation for Parkinson's disease: proof-of-concept study. Mov Disord 29:1679–1684

Adamchic I, Toth T, Hauptmann C, Tass PA (2014b) Reversing pathologically increased EEG power by acoustic coordinated reset neuromodulation. Hum Brain Mapp 35:2099–2118

Adamchic I, Toth T, Hauptmann C, Walger M, Langguth B, Klingmann I, Tass PA (2017) Acute effects and after-effects of acoustic coordinated reset neuromodulation in patients with chronic subjective tinnitus. Neuroimage Clin 15:541–558

Araneda R, De Volder AG, Deggouj N, Renier L (2015) Altered inhibitory control and increased sensitivity to cross-modal interference in tinnitus during auditory and visual tasks. PLoS One 10:e0120387

Arbogast TL, Mason CR, Kidd G (2002) The effect of spatial separation on informational and energetic masking of speech. J Acoust Soc Am 112:2086–2098

Argstatter H, Grapp M, Hutter E, Plinkert P, Bolay HV (2012) Long-term effects of the "Heidelberg model of music therapy" in patients with chronic tinnitus. Int J Clin Exp Med 5:273

Argstatter H, Grapp M, Hutter E, Plinkert PK, Bolay H-V (2015) The effectiveness of neuro-music therapy according to the Heidelberg model compared to a single session of educational counseling as treatment for tinnitus: a controlled trial. J Psychosom Res 78:285–292

Attarha M, Bigelow J, Merzenich MM (2018) Unintended consequences of white noise therapy for tinnitus – otolaryngology's cobra effect: a review. JAMA Otolaryngol Head Neck Surg 144:938–943

Auerbach BD, Radziwon K, Salvi R (2019) Testing the central gain model: loudness growth correlates with central auditory gain enhancement in a rodent model of hyperacusis. Neuroscience 407:93–107

Austen J (1811) Sense and sensibility, reprinted 1992. Wordsworth Classics, Hertfordshire

Aydin N, Searchfield GD (2019) Changes in tinnitus and physiological biomarkers of stress in response to short-term broadband noise and sounds of nature. Complement Ther Med 46:62–68

Bach DR, Buxtorf K, Strik WK, Neuhoff JG, Seifritz E (2011) Evidence for impaired sound intensity processing in schizophrenia. Schizophr Bull 37:426–431

Barozzi S, Del Bo L, Crocetti A, Dyrlund O, Passoni S, Zolin A, Panicucci E, Mancuso A, Kaur M, Searchfield G (2016) A comparison of nature and technical sounds for tinnitus therapy. Acta Acust United Acustica 102:540–546

Beck JE, Zaugg TL, Egge JL, Lima EN, Thielman EJ (2019) Progressive tinnitus management at two veterans affairs medical centers: clinical implementation with modified protocols. Am J Audiol 28:162–173

Bentzen O (1958) Audiology and hearing-aid therapy. Acta Otolaryngol 49:24–32

Bergman P, Sköld A, Västfjäll D, Fransson N (2009) Perceptual and emotional categorization of sound. J Acoust Soc Am 126:3156–3167

Brefczynski-Lewis JA, Lewis JW (2017) Auditory object perception: a neurobiological model and prospective review. Neuropsychologia 105:223–242

Bregman AS (1990) Auditory scene analysis. The perceptual organization of sound. MIT Press, Cambridge

Brungart DS, Chang PS, Simpson BD, Wang D (2006) Isolating the energetic component of speech-on-speech masking with ideal time-frequency segregation. J Acoust Soc Am 120:4007–4018

Burns EM (1984) A comparison of variability among measurements of subjective tinnitus and objective stimuli. Audiology 23(4):426–440

Calderone DJ, Lakatos P, Butler PD, Castellanos FX (2014) Entrainment of neural oscillations as a modifiable substrate of attention. Trends Cogn Sci 18:300–309

Campbell J, Bean C, Labrec A (2018) Normal hearing young adults with mild tinnitus: reduced inhibition as measured through sensory gating. Audiol Res:8

Carraba L, Costantini M, Del Bo L, Drylund O, Forti S, Jastreboff M, Searchfield G, Coad G (2008) Combination open ear instrument for tinnitus treatment. American Academy of Audiology, Charlotte

Coles R, Hallam R (1987) Tinnitus and its management. Br Med Bull 43:983–998

Davis P (2006) Music and the acoustic desensitization protocol for tinnitus. In: Tyler RS (ed) Tinnitus treatment. Thieme, New York

Davis PB, Paki B, Hanley PJ (2007) Neuromonics tinnitus treatment: third clinical trial. Ear Hear 28:242–259

De Ridder D, Vanneste S, Weisz N, Londero A, Schlee W, Elgoyhen AB, Langguth B (2014) An integrative model of auditory phantom perception: tinnitus as a unified percept of interacting separable subnetworks. Neurosci Biobehav Rev 44:16–32

Delgutte B (1990) Physiological mechanisms of psychophysical masking: observations from auditory-nerve fibers. J Acoust Soc Am 87:791–809

Doty DH, Glick WH (1994) Typologies as a unique form of theory building: toward improved understanding and modeling. Acad Manag Rev 19:230–251

Durai M, Searchfield GD (2017) A mixed-methods trial of broad band noise and nature sounds for tinnitus therapy: group and individual responses modeled under the adaptation level theory of tinnitus. Front Aging Neurosci 9:44

Durai M, Kobayashi K, Searchfield G (2015) A preliminary examination of the roles of contextual stimuli and personality traits under the adaptation level theory model of tinnitus. Acta Acust United Acustica 101:543–551

Durai M, O'keeffe MG, Searchfield GD (2017a) Examining the short term effects of emotion under an adaptation level theory model of tinnitus perception. Hear Res 345:23–29

Durai M, O'keeffe MG, Searchfield GD (2017b) The personality profile of tinnitus sufferers and a nontinnitus control group. J Am Acad Audiol 28:271–282

Durai M, Kobayashi K, Searchfield GD (2018) A feasibility study of predictable and unpredictable surf-like sounds for tinnitus therapy using personal music players. Int J Audiol 57:707–713

Durai M, Sanders M, Kobayashi K, Searchfield GD (2019) Auditory streaming and prediction in tinnitus sufferers. Ear Hear 40:345–357

Durai M, Sanders P, Doborjeh Z, Doborjeh M, Wendt A, Kasabov N, Searchfield GD (2020) Prediction of tinnitus masking benefit within a case series using a spiking neural network model. Prog Brain Res. https://doi.org/10.1016/bs.pbr.2020.08.003

Eggermont JJ (2012) The neuroscience of tinnitus. Oxford University Press, Oxford

Eggermont JJ, Tass PA (2015) Maladaptive neural synchrony in tinnitus: origin and restoration. Front Neurol 6:29

Engel LR, Frum C, Puce A, Walker NA, Lewis JW (2009) Different categories of living and non-living sound-sources activate distinct cortical networks. NeuroImage 47:1778–1791

Eramudugolla R, Irvine DR, Mcanally KI, Martin RL, Mattingley JB (2005) Directed attention eliminates 'change deafness' in complex auditory scenes. Curr Biol 15:1108–1113

Escera C, Alho K, Schroger E, Winkler I (2000) Involuntary attention and distractibility as evaluated with event-related brain potentials. Audiol Neurootol 5:151–166

Feldmann H (1971) Homolateral and contralateral masking of tinnitus by noise-bands and by pure tones. Audiology 10(3):138–144

Feldmann H (1992) Tinnitus – reality or phantom. In: Aran JM, Dauman R (eds) Tinnitus 91: proceedings of the fourth international tinnitus seminar. Kugler Publications, Amsterdam

Flor H, Hoffman D, Struve M, Diesch E (2004) Auditory discrimination training for the treatment of tinnitus. Appl Psychophysiol Biofeedback 29:113–120

Fowler EP (1948) Nonvibratory tinnitus: factors underlying subaudible and audible irritations. Arch Otolaryngol 47:29–36

Galazyuk AV, Voytenko SV, Longenecker RJ (2017) Long-lasting forward suppression of spontaneous firing in auditory neurons: implication to the residual inhibition of tinnitus. J Assoc Res Otolaryngol 18:343–353

Galazyuk A, Longenecker R, Voytenko S, Kristaponyte I, Nelson G (2019) Residual inhibition: from the putative mechanisms to potential tinnitus treatment. Hear Res 375:1–13

Giard MH, Fort A, Mouchetant-Rostaing Y, Pernier J (2000) Neurophysiological mechanisms of auditory selective attention in humans. Front Biosci 5:1

Goebel G (1997) Retraining therapy in tinnitus. Paradigm change or old wine in new bottles? HNO 45:664–667

Haab L, Lehser C, Corona-Strauss FI, Bernarding C, Seidler H, Hannemann R, Strauss DJ (2019) Implementation and long-term evaluation of a hearing aid supported tinnitus Treatment using notched environmental sounds. IEEE J Trans Eng Health Med 7:1–9

Hallam RS, Rachman S, Hinchcliffe R (1984) Psychological aspects of tinnitus. In: Rachman S (ed) Contributions to medical psychology, 1st edn. Pergamon Press, New York

Hamilton C, D'arcy S, Pearlmutter BA, Crispino G, Lalor EC, Conlon BJ (2016) An investigation of feasibility and safety of bi-modal stimulation for the Treatment of tinnitus: an open-label pilot study. Neuromodulation Technol Neural Interf 19:832–837

Han L, Na Z, Chunli L, Yuchen C, Pengfei Z, Hao W, Xu C, Peng Z, Zheng W, Zhenghan Y (2019a) Baseline functional connectivity features of neural network nodes can predict improvement after sound therapy through adjusted narrow band noise in tinnitus patients. Front Neurosci:13

Han L, Yawen L, Hao W, Chunli L, Pengfei Z, Zhengyu Z, Zhaodi W, Zhenghan Y, Shusheng G, Zhenchang W (2019b) Effects of sound therapy on resting-state functional brain networks in patients with tinnitus: a graph-theoretical-based study. J Magn Reson Imaging 50:1731–1741

Han JJ, Ridder DD, Vanneste S, Chen Y-C, Koo J-W, Song J-J (2020) Pre-treatment ongoing cortical oscillatory activity predicts improvement of tinnitus after partial peripheral reafferentation with hearing aids. Front Neurosci 14:410

Handscomb L (2006) Use of bedside sound generators by patients with tinnitus-related sleeping difficulty: which sounds are preferred and why? Acta Otolaryngol 126:59–63

Hann D, Searchfield GD, Sanders M, Wise K (2008) Strategies for the selection of music in the short-term management of mild tinnitus. Aust NZ J Audiol 30:129

Hauptmann C, Ströbel A, Williams M, Patel N, Wurzer H, von Stackelberg T, Brinkmann U, Langguth B, Tass PA (2015) Acoustic coordinated reset neuromodulation in a real life patient population with chronic tonal tinnitus. Bio Med Res Int 2015:569052. https://doi.org/10.1155/2015/569052

Hazell JW, Wood SM, Cooper HR, Stephens SD, Corcoran AL, Coles RR, Baskill JL, Sheldrake JB (1985) A clinical study of tinnitus maskers. Br J Audiol 19:65–146

Heeren A, Maurage P, Perrot H, De Volder A, Renier L, Araneda R, Lacroix E, Decat M, Deggouj N, Philippot P (2014) Tinnitus specifically alters the top-down executive control sub-component of attention: evidence from the attention network task. Behav Brain Res 269:147–154

Heller MF, Bergman M (1953) VII tinnitus aurium in normally hearing persons. Ann Otol Rhinol Laryngol 62:73–83

Helson H (1964) Adaptation-level theory: an experimental and systematic approach to behavior. Harper and Row, New York

Henry JA, Istvan J (2010) An independent review of neuromonics tinnitus treatment controlled clinical trials. Aust NZ J Audiol 32:41

Henry JA, Schechter MA, Loovis CL, Zaugg TL, Kaelin C, Montero M (2005) Clinical management of tinnitus using a "progressive intervention" approach. J Rehabil Res Dev 42:95

Henry J, Schechter M, Zaugg T, Griest S, Jastreboff P, Vernon J, Kaelin C, Meikle M, Lyons K, Stewart B (2006) Clinical trial to compare tinnitus masking and tinnitus retraining therapy. Acta Otolaryngol 126:64–69

Henry JA, Zaugg TL, Myers PJ, Schechter MA (2008) Using therapeutic sound with progressive audiologic tinnitus management. Trends Amplif 12:188–209

Henry JA, Frederick M, Sell S, Griest S, Abrams H (2015) Validation of a novel combination hearing aid and tinnitus therapy device. Ear Hear 36:42–52

Hillecke T, Nickel A, Bolay HV (2005) Scientific perspectives on music therapy. Ann N Y Acad Sci 1060:271–282

Hoare DJ, Stacey PC, Hall DA (2010) The efficacy of auditory perceptual training for tinnitus: a systematic review. Ann Behav Med 40:313–324

Hobson J, Chisholm E, EL Refaie A (2012) Sound therapy (masking) in the management of tinnitus in adults. Cochr Database Syst Rev 11:Cd006371

Huron D (1992) The ramp archetype and the maintenance of passive auditory attention. Music Percept 10:83–91

Husain FT (2016) Neural networks of tinnitus in humans: elucidating severity and habituation. Hear Res 334:37–48

Huttunen HJ, Palva JM, Lindberg L, Palva S, Saarela V, Karvonen E, Latvala M-L, Liinamaa J, Booms S, Castrén E (2018) Fluoxetine does not enhance the effect of perceptual learning on visual function in adults with amblyopia. Sci Rep 8:1–12

Ihlefeld A, Shinn-Cunningham B (2008) Spatial release from energetic and informational masking in a selective speech identification task. J Acoust Soc Am 123:4369–4379

Jastreboff PJ (1990) Phantom auditory perception (tinnitus): mechanisms of generation and perception. Neurosci Res 8:221–254

Jastreboff PJ (1999) Optimal sound use in TRT – theory and practice. In: Hazell J (ed) 6th international tinnitus seminar, 1999 Cambridge. The Tinnitus and Hyperacusis Centre, London, pp 491–494

Jastreboff P (2015) 25 years of tinnitus retraining therapy. HNO 63:307–311

Jastreboff PJ, Hazell JW (1993) A neurophysiological approach to tinnitus: clinical implications. Br J Audiol 27:7–17

Jastreboff PJ, Jastreboff MM (2000) Tinnitus retraining therapy (TRT) as a method for treatment of tinnitus and hyperacusis patients. J Am Acad Audiol 11:162–177

Jastreboff PJ, Gray WC, Gold SL (1996) Neurophysiological approach to tinnitus patients. Am J Otol 17:236–240

Jepsen K, Sanders M, Searchfield G, Kobayashi K (2010) Perceptual training of tinnitus. In: Proceedings of 'tinnitus discovery': Asia-Pacific tinnitus symposium 11–12th September 2009. The New Zealand Medical Journal, Auckland

Johansen JD, Skellgaard PH, Caporali S (2014) Effect of counseling, amplification and fractal tones in tinnitus management. J Commun Dis Deaf Stud Hear Aids

Jones L, Hills PJ, Dick KM, Jones SP, Bright P (2016) Cognitive mechanisms associated with auditory sensory gating. Brain Cogn 102:33–45

Kidd G, Arbogast TL, Mason CR, Walsh M (2002) Informational masking in listeners with sensorineural hearing loss. J Assoc Res Otolaryngol JARO 3:107–119

Kidd G, Mason CR, Richards VM, Gallun FJ, Durlach NI (2008) Informational masking. Auditory perception of sound sources. Springer, Berlin

King RO, Shekhawat GS, King C, Chan E, Kobayashi K, Searchfield GD (2021) The effect of auditory residual inhibition on tinnitus and the electroencephalogram. Ear Hear 42(1):130–141

Kleinstäuber M, Weise C, Andersson G, Probst T (2018) Personality traits predict and moderate the outcome of internet-based cognitive behavioural therapy for chronic tinnitus. Int J Audiol 57:538–544

Krick CM, Grapp M, Daneshvar-Talebi J, Reith W, Plinkert PK, Bolay HV (2015) Cortical reorganization in recent-onset tinnitus patients by the Heidelberg model of music therapy. Front Neurosci 9:49

Krick CM, Argstatter H, Grapp M, Plinkert PK, Reith W (2017) Heidelberg neuro-music therapy enhances task-negative activity in tinnitus patients. Front Neurosci 11:384

Kroener-Herwig B, Biesinger E, Gerhards F, Goebel G, Greimel KV, Hiller W (2000) Retraining therapy for chronic tinnitus: a critical analysis of its status. Scand Audiol 29:67–78

Lagas AK, Stinear CM, Byblow WD, Russel BR, Kydd RR, Thompson B (2014) The effects of fluoxetine and lorazepam on visual perceptual learning in healthy adults. Invest Ophthalmol Vis Sci 55:783–783

Liu C, Lv H, Jiang T, Xie J, He L, Wang G, Liu J, Wang Z, Gong S (2018) The Cochleural alternating acoustic beam therapy (CAABT): a pre-clinical trial. Am J Otolaryngol 39:401–409

Londero A, Viaud-Delmon I, Baskind A, Delerue O, Bertet S, Bonfils P, Warusfel O (2010) Auditory and visual 3D virtual reality therapy for chronic subjective tinnitus: theoretical framework. Virtual Reality 14:143–151

Lopez-Gonzalez M, Lopez-Fernandez R (2004) Sequential sound therapy in tinnitus. Int Tinnitus J 10:150–155

Lugli M, Romani R, Ponzi S, Bacciu S, Parmigiani S (2009) The windowed sound therapy: a new empirical approach for an effective personalized treatment of tinnitus. Int Tinnitus J 15

Mahboubi H, Ziai K, Djalilian HR (2012) Customized web-based sound therapy for tinnitus. Int Tinnitus J 17

Manabe Y, Sato K, Fukuda S, Miyashita T (2019) Unpleasantness of amplified environmental sound used in tinnitus sound therapy: a preliminary study of clinical assessment. J Int Adv Otol 15:94

Martin EM, West MF, Bedenbaugh PH (2004) Masking and scrambling in the auditory thalamus of awake rats by Gaussian and modulated noises. Proc Natl Acad Sci 101:14961–14965

Mazurek B, Fischer F, Haupt H, Georgiewa P, Reisshauer A, Klapp BF (2006) A modified version of tinnitus retraining therapy: observing long-term outcome and predictors. Audiol Neurotol 11:276–286

Mckenna L, Irwin R (2008) Sound therapy for tinnitus–sacred cow or idol worship?: An investigation of the evidence. Audiol Med 6:16–24

Mcneill C, Távora-Vieira D, Alnafjan F, Searchfield GD, Welch D (2012) Tinnitus pitch, masking, and the effectiveness of hearing aids for tinnitus therapy. Int J Audiol 51:914–919

Mitchell C (1983) The masking of tinnitus with pure tones. Audiology 22(1):73–87

Munro KJ (2008) Reorganization of the adult auditory system: perceptual and physiological evidence from monaural fitting of hearing aids. Trends Amplif 12:254–271

Neff P, Michels J, Meyer M, Schecklmann M, Langguth B, Schlee W (2017) 10 Hz amplitude modulated sounds induce short-term tinnitus suppression. Front Aging Neurosci 9:130

Neff P, Zielonka L, Meyer M, Langguth B, Schecklmann M, Schlee W (2019) Comparison of amplitude modulated sounds and pure tones at the tinnitus frequency: residual tinnitus suppression and stimulus evaluation. Trends Hear 23:2331216519833841

Newman CW, Sandridge SA (2012) A comparison of benefit and economic value between two sound therapy tinnitus management options. J Am Acad Audiol 23:126–138

Norena AJ (2011) An integrative model of tinnitus based on a central gain controlling neural sensitivity. Neurosci Biobehav Rev 35:1089–1109

Noreña AJ (2012) Stimulating the auditory system to treat tinnitus: from alleviating the symptoms to addressing the causes. Tinnitus. Springer

Oh EL, Lutfi RA (1999) Informational masking by everyday sounds. J Acoust Soc Am 106:3521–3528

Okamoto H, Stracke H, Stoll W, Pantev C (2010) Listening to tailor-made notched music reduces tinnitus loudness and tinnitus-related auditory cortex activity. Proc Natl Acad Sci 107:1207–1210

Park S-N, Bae S-C, Kim D-K, Park Y-S, Yeo S-W, Park S-Y (2013) Small-group counseling in a modified tinnitus retraining therapy for chronic tinnitus. Clin Exp Otorhinolaryngol 6:214

Pedemonte M, Drexler D, Rodio S, Geisinger D, Bianco A, Pol-Fernandes D, Bernhardt V (2010) Tinnitus treatment with sound stimulation during sleep. Int Tinnitus J 16

Phillips JS, Mcferran D (2010) Tinnitus retraining therapy (TRT) for tinnitus. Cochr Database Syst Rev 3:CD007330

Rabinowitz NC, Willmore BD, Schnupp JW, King AJ (2011) Contrast gain control in auditory cortex. Neuron 70:1178–1191

Rauschecker JP, May ES, Maudoux A, Ploner M (2015) Frontostriatal gating of tinnitus and chronic pain. Trends Cogn Sci 19:567–578

Raz A, Buhle J (2006) Typologies of attentional networks. Nat Rev Neurosci 7:367–379

Reavis KM, Rothholtz VS, Tang Q, Carroll JA, Djalilian H, Zeng F-G (2012) Temporary suppression of tinnitus by modulated sounds. J Assoc Res Otolaryngol 13:561–571

Roberts LE (2007) Residual inhibition. Prog Brain Res 166:487–495

Roberts LE, Moffat G, Baumann M, Ward LM, Bosnyak DJ (2008) Residual inhibition functions overlap tinnitus spectra and the region of auditory threshold shift. J Assoc Res Otolaryngol 9:417–435

Roberts LE, Bosnyak DJ, Bruce IC, Gander PE, Paul BT (2015) Evidence for differential modulation of primary and nonprimary auditory cortex by forward masking in tinnitus. Hear Res 327:9–27

Robinson BL, Mcalpine D (2009) Gain control mechanisms in the auditory pathway. Curr Opin Neurobiol 19:402–407

Saltzman M, Ersner MS (1949) Tinnitus aurium in otosclerosis. Arch Otolaryngol 50:440–442

Sanders P, Doborjeh Z, Doborjeh M, Kasabov NK, Searchfield GD (2020) Exploring the neural basis of acoustic residual inhibition with constant and amplitude modulated white noise using a brain-inspired spiking neural network (in press)

Schad ML, Mcmillan GP, Thielman EJ, Groon K, Morse-Fortier C, Martin JL, Henry JA (2018) Comparison of acoustic therapies for tinnitus suppression: A preliminary trial. Int J Audiol 57:143–149

Schaette R, Kempter R (2006) Development of tinnitus-related neuronal hyperactivity through homeostatic plasticity after hearing loss: a computational model. Eur J Neurosci 23:3124–3138

Schaette R, Kempter R (2009) Predicting tinnitus pitch from patients' audiograms with a computational model for the development of neuronal hyperactivity. J Neurophysiol 101:3042–3052

Schaette R, König O, Hornig D, Gross M, Kempter R (2010) Acoustic stimulation treatments against tinnitus could be most effective when tinnitus pitch is within the stimulated frequency range. Hear Res 269:95–101

Schaette R, Turtle C, Munro KJ (2012) Reversible induction of phantom auditory sensations through simulated unilateral hearing loss. PLoS One 7:e35238

Scherer RW, Formby C (2019) Effect of tinnitus retraining therapy vs standard of care on tinnitus-related quality of life: a randomized clinical trial. JAMA Otolaryngol Head Neck Surg 145:597–608

Schlee W, Mueller N, Hartmann T, Keil J, Lorenz I, Weisz N (2009) Mapping cortical hubs in tinnitus. BMC Biol 7:80

Schleidgen S, Klingler C, Bertram T, Rogowski WH, Marckmann G (2013) What is personalized medicine: sharpening a vague term based on a systematic literature review. BMC Med Ethics 14:55

Schmidt CJ, Kerns RD, Griest S, Theodoroff SM, Pietrzak RH, Henry JA (2014) Toward development of a tinnitus magnitude index. Ear Hear 35:476–484

Searchfield GD (2006) Hearing aids and tinnitus. In: Tyler R (ed) Tinnitus treatment. Thieme, New York

Searchfield GD (2014) Tinnitus what and where: an ecological framework. Front Neurol 5:271

Searchfield GD (2017) A client oriented scale of improvement in tinnitus (COSIT) for therapy goal planning and assessing outcomes. (Submitted)

Searchfield GD (2019) Good vibrations: a proof-of-concept study of the preferred temporal characteristics in surf-like sounds for tinnitus therapy les bonnes vibrations: Une étude de démonstration de faisabilité des caractéristiques temporelles des sons. Canadian J Speech Lang Pathol Audiol

Searchfield GD, Warr AA, Kuklinski JV, Purdy SC (2002) Digital instruments for tinnitus: mixing point identification and threshold-adjusted noise. In: Patuzzi R (ed) Proceedings of the seventh international tinnitus seminar. The University of Western Australia, Perth

Searchfield GD, Morrison-Low J, Wise K (2007) Object identification and attention training for treating tinnitus. Prog Brain Res 166:441–460

Searchfield G, Kobayashi K, Sanders M (2012) An adaptation level theory of tinnitus audibility. Front Syst Neurosci 6:46

Searchfield G, Kobayashi K, Proudfoot K, Tevoitdale H, Irving S (2015) The development and test–retest reliability of a method for matching perceived location of tinnitus. J Neurosci Methods 256:1–8

Searchfield GD, Kobayashi K, Hodgson S-A, Hodgson C, Tevoitdale H, Irving S (2016) Spatial masking: development and testing of a new tinnitus assistive technology. Assist Technol 28:115–125

Searchfield GD, Durai M, Linford T (2017) A state-of-the-art review: personalization of tinnitus sound therapy. Front Psychol 8:1599

Searchfield GD, Linford T, Durai M (2019) Sound therapy and aural rehabilitation for tinnitus: a person centred therapy framework based on an ecological model of tinnitus. Disabil Rehabil 41:1966–1973

Searchfield G, Spiegel D, Poppe T, Durai M, Jensen M, Kobayashi K, Park J, Russell B, Shekhawat G, Sundram F (2020a) A proof-of-concept study comparing tinnitus and neural connectivity changes following multisensory perceptual training with and without a low-dose of fluoxetine. Int J Neurosci:1–12

Searchfield GD, Boone M, Bensam J, Durai M, Hodgson S-A, Linford T, Vogel D (2020b) A proof-of-concept study of the benefits of a single-session of tinnitus instruction and counselling with homework on tinnitus. Int J Audiol 59:374–382

Sedley W (2019) Tinnitus: does gain explain? Neuroscience 407:213–228

Sedley W, Gander PE, Kumar S, Oya H, Kovach CK, Nourski KV, Kawasaki H, Howard MA, Griffiths TD (2015) Intracranial mapping of a cortical tinnitus system using residual inhibition. Curr Biol 25:1208–1214

Sedley W, Friston KJ, Gander PE, Kumar S, Griffiths TD (2016) An integrative tinnitus model based on sensory precision. Trends Neurosci 39:799–812

Sereda M, Davies J, Hall DA (2017) Pre-market version of a commercially available hearing instrument with a tinnitus sound generator: feasibility of evaluation in a clinical trial. Int J Audiol 56:286–294
Sereda M, Xia J, EL Refaie A, Hall DA, Hoare DJ (2018) Sound therapy (using amplification devices and/or sound generators) for tinnitus. Cochrane Database Syst Rev
Sereda M, Smith S, Newton K, Stockdale D (2019) Mobile apps for Management of Tinnitus: users' survey, quality assessment, and content analysis. JMIR Mhealth Uhealth 7:e10353
Seydel C, Haupt H, Szczepek AJ, Klapp BF, Mazurek B (2010) Long-term improvement in tinnitus after modified tinnitus retraining therapy enhanced by a variety of psychological approaches. Audiol Neurotol 15:69–80
Shekhawat GS, Searchfield GD, Stinear CM (2013) Role of hearing aids in tinnitus intervention: a scoping review. J Am Acad Audiol 24:747–762
Shore SE, Roberts LE, Langguth B (2016) Maladaptive plasticity in tinnitus – triggers, mechanisms and treatment. Nat Rev Neurol 12:150
Siegel EH, Stefanucci JK (2011) A little bit louder now: negative affect increases perceived loudness. Emotion 11:1006
Silchenko AN, Adamchic I, Hauptmann C, Tass PA (2013) Impact of acoustic coordinated reset neuromodulation on effective connectivity in a neural network of phantom sound. NeuroImage 77:133–147
Simonetti P, Vasconcelos LG, Oiticica J (2018) Effect of fractal tones on the improvement of tinnitus handicap inventory functional scores among chronic tinnitus patients: an open-label pilot study. Int Arch Otorhinolaryngol 22:387–394
Spiegel DP, Linford T, Thompson B, Petoe MA, Kobayashi K, Stinear CM, Searchfield GD (2015) Multisensory attention training for treatment of tinnitus. Sci Rep 5:10802
Stein A, Engell A, Junghoefer M, Wunderlich R, Lau P, Wollbrink A, Rudack C, Pantev C (2015a) Inhibition-induced plasticity in tinnitus patients after repetitive exposure to tailor-made notched music. Clin Neurophysiol 126:1007–1015
Stein A, Engell A, Lau P, Wunderlich R, Junghoefer M, Wollbrink A, Bruchmann M, Rudack C, Pantev C (2015b) Enhancing inhibition-induced plasticity in tinnitus–spectral energy contrasts in tailor-made notched music matter. PLoS One 10:e0126494
Stein A, Wunderlich R, Lau P, Engell A, Wollbrink A, Shaykevich A, Kuhn J-T, Holling H, Rudack C, Pantev C (2016) Clinical trial on tonal tinnitus with tailor-made notched music training. BMC Neurol 16:38
Stouffer J, Tyler RS (1990) Characterization of tinnitus by tinnitus patients. J Speech Hear Disord 55:439–453
Sweetow RW, Sabes JH (2010) Effects of acoustical stimuli delivered through hearing aids on tinnitus. J Am Acad Audiol 21:461–473
Tass PA, Adamchic I, Freund HJ, von Stackelberg T, Hauptmann C (2012) Counteracting tinnitus by acoustic coordinated reset neuromodulation. Restor Neurol Neurosci 30:137–159
Tavora Vieira D, Eikelboom R, Ivey G, Miller S (2011) A multi-centre study on the long-term benefits of tinnitus management using Neuromonics tinnitus treatment. Int Tinnitus J 16
Teismann H, Okamoto H, Pantev C (2011) Short and intense tailor-made notched music training against tinnitus: the tinnitus frequency matters. PLoS One 6:e24685
Teki S, Chait M, Kumar S, von Kriegstein K, Griffiths TD (2011) Brain bases for auditory stimulus-driven figure–ground segregation. J Neurosci 31:164–171
Terry A, Jones D (1986) Preference for potential tinnitus maskers: results from annoyance ratings. Br J Audiol 20:277–297
Theodoroff SM, Mcmillan GP, Zaugg TL, Cheslock M, Roberts C, Henry JA (2017) Randomized controlled trial of a novel device for tinnitus sound therapy during sleep. Am J Audiol 26:543–554
Treisman AM (1964) Selective attention in man. Br Med Bull 20:12–16
Treisman AM, Gelade G (1980) A feature-integration theory of attention. Cogn Psychol 12:97–136

Trotter M, Donaldson I (2008) Hearing aids and tinnitus therapy: a 25-year experience. J Laryngol Otol 122:1052–1056
Tuepker A, Elnitsky C, Newell S, Zaugg T, Henry JA (2018) A qualitative study of implementation and adaptations to progressive tinnitus management (PTM) delivery. PLoS One 13
Tunkel DE, Bauer CA, Sun GH, Rosenfeld RM, Chandrasekhar SS, Cunningham ER Jr, Archer SM, Blakley BW, Carter JM, Granieri EC (2014) Clinical practice guideline: tinnitus. Otolaryngol Head Neck Surg 151:S1–S40
Tutton R (2012) Personalizing medicine: futures present and past. Soc Sci Med 75:1721–1728
Tyler RS, Bentler RA (1987) Tinnitus maskers and hearing aids for tinnitus. Seminars in hearing. Thieme Medical Publishers, pp 49-60
Tyler RS, Bergan CJ (2001) Tinnitus retraining therapy: a modified approach. Hear J 54:36–42
Tyler RS, Conrad-Armes D (1984) Masking of tinnitus compared to masking of pure tones. J Speech Hear Res 27:106–111
Tyler R, Gehringer A, Noble W, Dunn C, Witt S, Bardia A (2006) Chapter 9: Tinnitus activities treatment. In: Tinnitus treatment: clinical protocols. Thieme, New York, pp 116–132
Tyler RS, Noble W, Coelho CB, Ji H (2012) Tinnitus retraining therapy: mixing point and total masking are equally effective. Ear Hear 33:588–594
Tyler R, Ji H, Perreau A, Witt S, Noble W, Coelho C (2014a) Development and validation of the tinnitus primary function questionnaire. Am J Audiol 23:260–272
Tyler R, Stocking C, Secor C, Slattery WH III (2014b) Amplitude modulated S-tones can be superior to noise for tinnitus reduction. Am J Audiol 23:303–308
Tyler R, Cacace A, Stocking C, Tarver B, Engineer N, Martin J, Deshpande A, Stecker N, Pereira M, Kilgard M (2017a) Vagus nerve stimulation paired with tones for the treatment of tinnitus: a prospective randomized double-blind controlled pilot study in humans. Sci Rep 7:1–11
Tyler RS, Deshpande AK, Lau CC, Kuk F (2017b) The effectiveness of the progression of Widex Zen tinnitus therapy: A pilot study. Am J Audiol 26:283–292
Vernon JA, Meikle MB (1988) Measurement of tinnitus: an update. In: Kitahara M (ed) Tinnitus: pathophysiology and management. Igaku-Shoin, Tokyo
Vernon JA, Meikle MB (2003) Tinnitus: clinical measurement. Otolaryngol Clin N Am 36:293–305, vi
Vernon J, Schleuning A (1978) Tinnitus: a new management. Laryngoscope 88:413–419
Vetencourt JFM, Sale A, Viegi A, Baroncelli L, De Pasquale R, O'leary OF, Castrén E, Maffei L (2008) The antidepressant fluoxetine restores plasticity in the adult visual cortex. Science 320:385–388
Welch D, Dawes PJ (2008) Personality and perception of tinnitus. Ear Hear 29:684–692
Williams M, Hauptmann C, Patel N (2015) Acoustic CR neuromodulation therapy for subjective tonal tinnitus: a review of clinical outcomes in an independent audiology practice setting. Front Neurol 6:54
Wilson PH, Henry JL, Andersson G, Hallam RS, Lindberg P (1998) A critical analysis of directive counselling as a component of tinnitus retraining therapy. Br J Audiol 32:273–286
Wise K, Kobayashi K, Magnusson J, Welch D, Searchfield GD (2016) Randomized controlled trial of a perceptual training game for tinnitus therapy. Games Health J 5:141–149
Wunderlich R, Lau P, Stein A, Engell A, Wollbrink A, Rudack C, Pantev C (2015) Impact of spectral notch width on neurophysiological plasticity and clinical effectiveness of the tailor-made notched music training. PLoS One:10
Zeng FG (2013) An active loudness model suggesting tinnitus as increased central noise and hyperacusis as increased nonlinear gain. Hear Res 295:172–179
Zenner H-P, Delb W, Kröner-Herwig B, Jäger B, Peroz I, Hesse G, Mazurek B, Goebel G, Gerloff C, Trollmann R (2017) A multidisciplinary systematic review of the treatment for chronic idiopathic tinnitus. Eur Arch Otorhinolaryngol 274:2079–2091
Zhang F, Deshpande A, Benson C, Smith M, Eliassen J, Fu Q-J (2011) The adaptive pattern of the auditory N1 peak revealed by standardized low-resolution brain electromagnetic tomography. Brain Res 1400:42–52

Tinnitus and Brain Stimulation

Dirk De Ridder, Divya Adhia, and Berthold Langguth

Contents

Abstract The pathophysiological mechanisms that underlie the generation and maintenance of tinnitus are being unraveled progressively. Based on this knowledge, a large variety of different neuromodulatory interventions have been developed and are still being designed, adapting to the progressive mechanistic insights in the pathophysiology of tinnitus. rTMS targeting the temporal, temporoparietal, and the frontal cortex has been the mainstay of non-invasive neuromodulation. Yet, the evidence is still unclear, and therefore systematic meta-analyses are needed for drawing conclusions on the effectiveness of rTMS in chronic tinnitus. Different forms of transcranial electrical stimulation (tDCS, tACS, tRNS), applied over the frontal and temporal cortex, have been investigated in tinnitus patients, also without robust evidence for universal efficacy. Cortex and deep brain stimulation with implanted electrodes have shown benefit, yet there is insufficient data to support their routine clinical use. Recently, bimodal stimulation approaches have revealed promising results and it appears that targeting different sensory modalities in temporally combined manners may be more promising than single target approaches.

While most neuromodulatory approaches seem promising, further research is required to help translating the scientific outcomes into routine clinical practice.

D. De Ridder (✉) and D. Adhia
Neurosurgery, Department of Surgical Sciences, University of Otago, Dunedin, New Zealand
e-mail: dirk.deridder@otago.ac.nz

B. Langguth
Department of Psychiatry and Psychotherapy, University of Regensburg, Regensburg, Germany

Curr Topics Behav Neurosci (2021) 51: 249–294
https://doi.org/10.1007/7854_2021_219
Published Online: 8 April 2021

Keywords Implant · Neuromodulation · Neurostimulation · rTMS · tDCS · tES · Tinnitus

1 Introduction

Tinnitus can be defined as the conscious awareness of a non-complex sound for which there is no identifiable corresponding external sound source (Jastreboff 1990). Tinnitus occurs in 5–15% of the population (Axelsson and Ringdahl 1989; Heller 2003). Whereas most patients (80%) can habituate to this sound, quality of life is severely disrupted in about 20–25% of the patients who cannot cope with the tinnitus (Axelsson and Ringdahl 1989; Vanneste et al. 2014). In these patients, tinnitus is frequently associated with anxiety, depression, cognitive impairment, and sleep disturbances (Bhatt et al. 2017; Vanneste et al. 2016; Y. Wang et al. 2018b), and tinnitus becomes a mental disorder. Tinnitus disorder can therefore be diagnosed in cases of tinnitus with tinnitus-associated suffering.

Although many treatments have been proposed, both pharmacological and non-pharmacological, some of which are commercially available, evidence for a successful therapy that benefits everybody with tinnitus is lacking (Dobie 1999; Langguth et al. 2013). The lack of efficacious treatments for tinnitus likely originates from the heterogeneity of tinnitus and an incomplete understanding of the pathophysiology of the different forms of tinnitus (Elgoyhen et al. 2015).

Since 2011 pharmaceutical interest in developing neuropharmacological products for neurological or psychiatric indications has dramatically declined. Big Pharma has invested 50% less for brain related disorders (Yokley et al. 2017), because it is too expensive and too risky. Developing medication for brain disorders has 50% less chance of making it to the market (6.2% vs 13.3%) (Gribkoff and Kaczmarek 2017), takes 30% longer to develop (19.3 vs 14.7 months) (Gribkoff and Kaczmarek 2017), and costs 30% more than heart medication (Gribkoff and Kaczmarek 2017). Eighty percent of medications fail phase III trials (Kesselheim et al. 2015), predominantly because they do not outperform placebo (46%) (Kesselheim et al. 2015). Thus, notwithstanding repurposing of existing medication (Pushpakom et al. 2019), the likelihood of a pharmaceutical solution for tinnitus and tinnitus disorder is therefore limited. As such, other treatment options need to be investigated, one of which is brain stimulation. The history and evolution of the understanding of the pathophysiology of tinnitus goes hand in hand with the development of brain stimulation approaches.

2 History and Evolution of the Pathophysiology of Tinnitus

Tinnitus is an enigmatic symptom. Even the earliest historic reference, dating back to the 19th CBC, in the Ebers papyrus, in which tinnitus was considered the consequence of a bewitched ear, is controversial (Dietrich 2004) (Fig. 1). Tinnitus as a

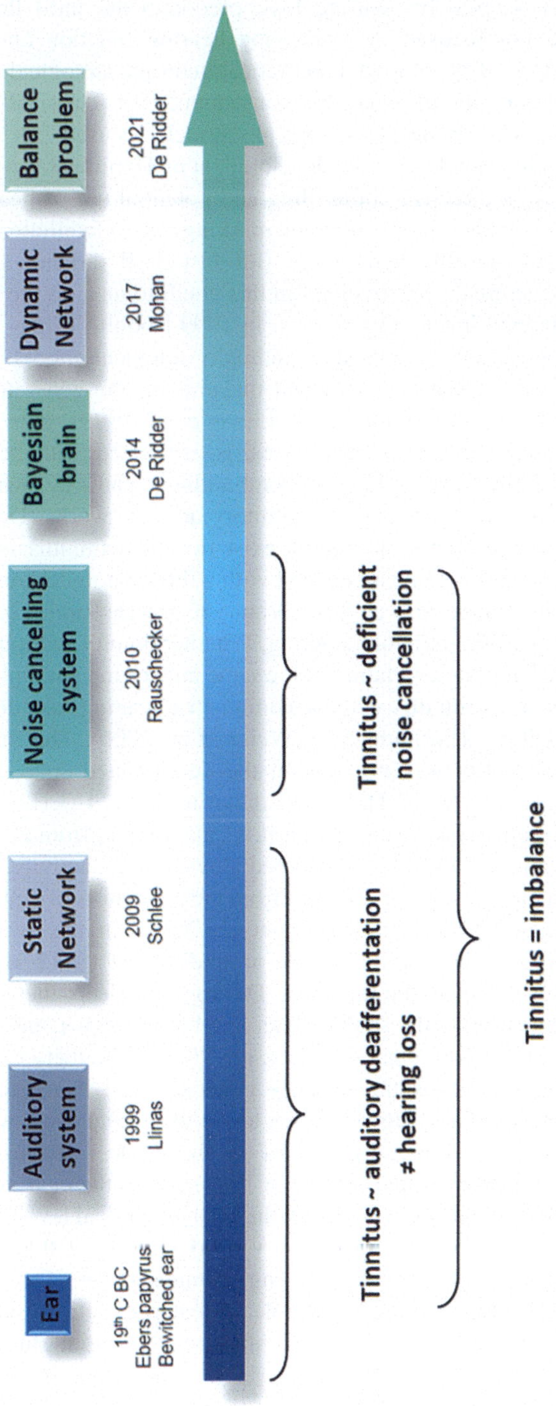

Fig. 1 History of the pathophysiological models for tinnitus

problem of the ear, caused by hearing loss, predominated until the 1990s, and consequently treatment focused on improving hearing function and masking of sound. In 1990 Jastreboff developed a theoretical neurophysiological model implying that the phantom sound was the consequence of classical conditioning (Jastreboff 1990). The neurophysiological model focused on the distress associated with tinnitus, and lead to the development of Tinnitus Retraining Therapy (TRT). TRT comprises directive counselling and potential use of hearing aids and sound therapy, not with the goal of complete masking, but to facilitate habituation of tinnitus reaction and possibly perception. But Jastreboff did not specify which neuroanatomical structures were involved in this conditioning, so the brain components involved remained inside a black box. In 1999 Rodolfo Llinas published the first paper that specifically proposed a pathophysiological model implying the interaction between the auditory cortex and thalamus as the generator of tinnitus, called thalamocortical dysrhythmia (R. R. Llinas et al. 1999). It is based on the clinical evidence that tinnitus, in most cases, is associated with hearing loss or abnormalities in the inner ear or the peripheral auditory pathway, which results in neuroplastic changes in auditory and non-auditory brain networks (De Ridder et al. 2014d). However, some forms of tinnitus exist in which tinnitus is unrelated to hearing loss, and this seems to be associated with a different underlying pathophysiological mechanism (Vanneste et al. 2019; Vanneste and De Ridder 2015), possibly requiring different neuromodulation targeting. Whether tinnitus without hearing loss is fundamentally different or a variation of the same underlying pathophysiology still remains to be proven, as tinnitus without audiometric hearing loss does not imply that there is no auditory deafferentation (Weisz et al. 2006). The thalamocortical dysrhythmia model of Llinas states that in the deafferented state, the dominant resting-state alpha rhythm (8–12 Hz) slows down to theta (4–7 Hz) (R. R. Llinas et al. 1999) band frequencies. Even though it is called theta, from an electrophysiological point of view it should be considered slowed alpha activity, as the neural generators of true theta activity are different from the generators of alpha and slowed alpha activity (Tsanov 2015; Whittington et al. 2018). In the awake state, the thalamocortical loops within the brain idle around 10 Hz (8–12 Hz), called the alpha frequency, generated by the thalamus. During light sleep, theta band activity is predominant, generated by the septal nuclei and hippocampus, and in deep sleep delta activity (1–3 Hz) predominates (G Buzsaki 2006). The theta that emerges in thalamocortical dysrhythmia is thalamically generated and therefore more likely represents alpha activity that is slowed down to the theta frequency range. Conceptually, this can be explained as a decrease of the firing rate when less information needs to be processed (Borst and Theunissen 1999) due to the deafferentation, and that firing and oscillation rate are coupled at the thalamocortical level (Crunelli et al. 2018; R. Llinas et al. 2005). As a consequence, $GABA_A$ mediated lateral inhibition weakens (R. Llinas et al. 2005), inducing gamma (>30 Hz) band activity (R. R. Llinas et al. 1999) surrounding the deafferented theta area, also known as the edge effect (R. Llinas et al. 2005). Confirming the initial studies by Llinas (R. Llinas et al. 2005; R. R. Llinas et al. 1999) a decrease in alpha power is associated with an increase in gamma power (Lorenz et al. 2009), and gamma is coupled to theta activity (Weisz et al. 2007). Other studies have since demonstrated the presence of both low (delta or theta) and high frequency activity in the auditory cortex of tinnitus patients (Adamchic et al. 2012, 2014; Adjamian et al. 2012; Ramirez et al. 2009), and the theta-gamma cross-frequency coupling has also been identified on electrode recordings of implanted patients (De Ridder et al. 2011b), confirming the MEG data.

Theta activity is associated with negative symptoms such as hearing loss (and hypoesthesia in the somatosensory system), and gamma activity reflects the positive symptoms, tinnitus (pain in the somatosensory system) in diseases characterized by thalamocortical dysrhythmia (R. Llinas et al. 2005; R. R. Llinas and Steriade 2006). Thus negative symptoms (e.g., hearing loss, hypoesthesia) are linked to decreased information processing and therefore slowed alpha activity, as if the deafferented thalamocortical columns are "as asleep"(R. Llinas et al. 2005). It has been proposed that this theta could then act as a long range carrier wave (Freeman 2003, 2005; Freeman and Rogers 2002) on which the tinnitus information can be nested by means of high frequency oscillatory activity (De Ridder et al. 2015b; De Ridder et al. 2014d). If this model is universal, then that would imply that all tinnitus patients could be treated by auditory cortex stimulation, whether non-invasively or invasively. The first attempts to test this theory were performed with transcranial magnetic stimulation and brain implants (De Ridder et al. 2004; Eichhammer et al. 2003a; Plewnia et al. 2003).

However, early research of TMS and implants demonstrated that not every person suffering from tinnitus could be helped with auditory cortex stimulation. Different explanations were postulated for this including the duration of the tinnitus (De Ridder et al. 2005; Kleinjung et al. 2007), the gender of the tinnitus patients (De Ridder et al. 2007d), the amount of hearing loss (Kleinjung et al. 2007), the tinnitus pitch (De Ridder et al. 2007d), the perceived laterality, left or right side, of the tinnitus (De Ridder et al. 2007d), laterality of the tinnitus generator in the brain (De Ridder 2010; Langguth et al. 2006), and also TMS related aspects such as the delivered dose (Plewnia et al. 2007). The failure in obtaining predictable and clinically satisfying results prompted the proposal of a novel pathophysiological model, in which the tinnitus percept would not be phrenologically limited to the auditory cortex but to a tinnitus-associated network (Schlee et al. 2009a, b). Even though that network dynamically changes, adapting to externally and internally generated triggers, on average it remains fairly stable, so that it could be averaged over 5–20 min recording intervals, with the current MEG and EEG technology (De Ridder et al. 2011a; Schlee et al. 2009a, b). Translating theoretical concepts from network science (Albert et al. 2000) to brain stimulation, it was hypothesized that directing stimulation at more than one target could disrupt the tinnitus network more effectively than one single target. As the frontal cortex was already a somewhat successful target for transcranial direct current stimulation (tDCS) (De Ridder and Vanneste 2012; Frank et al. 2011; Song et al. 2012; To et al. 2017; Vanneste and De Ridder 2011; Vanneste et al. 2011) and TMS (De Ridder et al. 2012a, c; Vanneste and De Ridder 2012a), the first multitarget stimulations selected the frontal cortex and temporoparietal junction (auditory cortex) (Kreuzer et al. 2011), but later attempts also used auditory cortex and high cervical C2 area (De Ridder and Vanneste 2015), using the same reasoning. Because the success rate was not high, these network science based models were modified for tinnitus to employ both random and targeted modulation on the tinnitus network (Mohan et al. 2017). The restricted benefits of neurostimulation based on the distributed network model led to investigation of an alternative model, in which tinnitus was the result of a deficient noise cancelling mechanism (Leaver et al. 2011; Rauschecker et al. 2010), analogous to pain (Fields 2004; Kong et al. 2010). However, according to this model the subgenual anterior cingulate cortex was critically involved in tinnitus generation, a difficult neuroanatomical target for current brain stimulation technologies. Furthermore, other groups failed to replicate the imaging data on which the model was based. A further model was proposed that tinnitus could actually be the result of a Bayesian prediction error (De Ridder et al. 2014a, c). According to this model tinnitus can be understood as the result of a mismatch between the predicted sound percept and the actual neural input from the ear, which is reduced because of hearing loss. It was hoped that the underlying anatomical and oscillatory correlates could be targeted with brain stimulation. It also theorized that neuroplastic changes involved in tinnitus were multiphasic, suggesting a theoretical difference between the neural generators of tinnitus in tinnitus with and without deafferentation. As such, the model attempted to combine and integrate both the deafferentation based theories and the noise cancelling theories (De Ridder et al. 2014a, c, d).

Using a sliding window analysis method combined with graph theoretical analyses permitted the development of a dynamical tinnitus network, in which the tinnitus sound is the consequence of the auditory cortex constantly looking for missing information, in keeping with the Bayesian tinnitus concept. Distress is associated with the loss of temporal flexibility (Mohan et al. 2018a, b). This model has not yet led to novel therapeutic approaches.

In a further attempt to integrate both the deafferentation based and noise cancelling models it has recently been proposed that tinnitus could be the result of an imbalance between bottom-up and top-down influences (Vanneste et al. 2019), which would make very clear predictions on what targets to apply which kind of brain stimulation (Fig. 1).

3 A Theory of Symptom Generation in the Brain

The brain can be seen as a complex adaptive system (Freeman et al. 2001; Sporns et al. 2004), similar to the world wide web, the climate, the economy, or an ant society (Holland 2014). In order to qualify as a complex adaptive system, a system has to fulfill two criteria (Amaral et al. 2004): (1) its structure follows a small world topology and (2) the system has to embed noise (Amaral et al. 2004). The reason for these criteria relates to the adaptiveness of the system. Network systems can be topologically structured in three ways (Bullmore and Sporns 2012). At one extreme, the network can have a lattice or regular topology, which means that every stimulus will always result in exactly the same processing, which is both predictive and efficient but not adaptive whatsoever (Catania 2009). At the other extreme, a system can be completely random, which is inefficient and disadvantageous because every stimulus will always have a completely random response. An intermediate structure between regular and random has a small world topology, which permits flexibility and adaptiveness to changing environments through variability. As such, such a system can learn (Bassett et al. 2006; Karuza et al. 2016). The brain has a small world structure and thus fulfills the first criterion (Achard et al. 2006; Bassett et al. 2006; Bullmore and Sporns 2009; Eguiluz et al. 2005). The brain is also noisy, fulfilling the second criterion, but the noise is structured, generally following a power law distribution (G. Buzsaki and Mizuseki 2014), i.e. a pink (1/f) or brown ($1/f^2$) noise structure. This structured noise has an advantage that it has memory and can carry information, in contrast to white noise which is an unstructured completely random noise (Keshner 1982). Thus, such a system can learn, is flexible, while still maintaining stability. Since small worlds are adaptive, implanting electrodes in an adaptive system such as the brain makes intuitive sense as a means to modify its structure and thus its function. In a regular network or completely random system, the same concept would make little to no sense: a regular system would respond identical with or without stimulation and a random system would respond differently to every stimulus. One of the most important fundamental characteristics of every complex adaptive system is emergence: the whole is more than the sum of its components and cannot be predicted from its constituent parts. The whole has new properties that depend on the very specific connectivity between the parts. A collection of car parts is not a car. Only when all parts are put together (i.e., connected) in a very specific way does a functional car emerge. Yet, a simple car is complex but non-adaptive. It doesn't reconfigure itself based on a changing environment. Bees do, ants do, our brain does and so does the internet. Emergentism in the philosophy of mind supports the belief that consciousness is an emergent property of brain function, and by extension, that every thought, feeling, action is the consequence of specific connectivity patterns resulting from adaptive neuroplasticity, and every symptom or disease is the result of maladaptive plasticity (Fornito and Bullmore 2014). Thus, using TMS or transcranial electrical stimulation (tES), or implanting electrodes on the cortex should change or use the brain's connectivity in order to create a change in symptoms (De Ridder et al. 2017).

Neuroplasticity can operationally be defined as the brain's capacity to modify its structure and function in order to adjust to a changing environment. However, these adaptive brain changes can be both adaptive and maladaptive, i.e. can lead to learning how to adjust to a changing environment but can also lead to symptoms. Any changes in the external or internal environment lead to neuroplasticity, i.e. both deprivation of input or increased environmental stimulation. These adaptive changes can be modulated by adrenal and gonadal hormones, neurotransmitters, growth factors, certain drugs, and aging (Fuchs and Flugge 2014). This results in adaptive changes at multiple scales: molecular and neurobiochemical changes, synaptic adjustments, neurogenesis, connectivity, and network changes (Fuchs and Flugge 2014). From a clinical perspective, neuroplasticity can be visualized by structural and functional brain imaging as changes in structure, activity, and connectivity. Changes in connectivity can be differentiated in structural, functional, and effective connectivity (Bassett et al. 2006; Lewis et al. 2009). Structural connectivity refers to anatomical changes in the brain (Hagmann et al. 2008), functional connectivity refers to co-activation of different brain areas (Friston 2011), and effective connectivity identifies from where to where the information flows in the brain, and can thus be seen as a form of directional functional connectivity (Friston 2011). This reorganization facilitates stability in constantly changing functional and effective connectivity networks, which results in changing emergent properties, like altered percepts, thoughts, emotions, actions, symptoms, etc.

3.1 Neuromodulation as Targeted Neuroplasticity

Neuromodulation and neurostimulation or brain stimulation are being used interchangeably. Yet, neuromodulation is becoming the preferred term because it doesn't carry the connotation of activation, which seems to be inherently implied with neurostimulation. Indeed, if the electrical or magnetic stimuli reach a brain area in which GABA receptors predominate, the clinical effect can be inhibitory. And neuromodulation means influencing brain activity (and connectivity), without suggesting the influence is excitatory.

Neuromodulation can be operationally defined as the induction of neuroplastic changes via targeted application of electrical, magnetic, sound (including ultrasound), optical or pharmacological stimuli. This is a broader definition than the one used by the International Neuromodulation Society: "Neuromodulation is technology that acts directly upon nerves. It is the alteration – or modulation – of nerve activity by delivering electrical or pharmaceutical agents directly to a target area" (http://www.neuromodulation.com/about-neuromodulation).

Neuromodulation can be performed on any part of the nervous system, from the peripheral nerve field, to specific peripheral or autonomic nerves, to the dorsal root ganglion (DRG), the spinal cord, the brainstem, or the brain. In brain stimulation, a distinction can be made between cortex stimulation and deep brain stimulation, but even here the terminology is not always uniform (De Ridder et al. 2017). For example, wire electrodes have been implanted inside the anterior cingulate gyrus and this procedure was called deep brain stimulation (DBS) of the anterior cingulate (Boccard et al. 2014), whereas paddle electrodes have been implanted onto the same target (De Ridder et al. 2016a, b; Leong et al. 2020) and this qualifies as cortex stimulation. The same can be said for deep brain stimulation of the subgenual anterior cingulate cortex (Brodmann area 25) for major depressive disorder. In essence, it is intracortical stimulation (with wire electrodes) of the subgenual anterior cingulate cortex, in which the electrodes are inserted inside the cortex rather than onto the cortex (Mayberg et al. 2005). The same terminological confusion exists for tinnitus. Whereas in most studies the electrodes are implanted extradurally or intradurally overlying the primary or secondary auditory cortex, respectively

(De Ridder et al. 2004, 2006a, 2007a, b, 2008, 2010, 2011b; Friedland et al. 2007; Litre et al. 2009), some patients have been treated with wire electrodes implanted inside the auditory cortex (Seidman et al. 2008). We will consider any form of cortical stimulation, whether intracortical or onto the cortex, as cortex stimulation, and deep brain stimulation as specifically targeting deep nuclei, rather than cortical structures.

3.2 Mechanism of Action of Cortex Stimulation

A better understanding of how cortex stimulation exerts its beneficial effect is essential; this requires a better understanding of the pathophysiological mechanisms involved in tinnitus generation. Symptoms are emergent properties resulting from maladaptive network activity, and not phrenological activity of one area in the brain (Barabasi et al. 2011; Fornito and Bullmore 2014; Fornito et al. 2015). Indeed, when patients in vegetative state, who have no conscious awareness, are presented with a sound, their auditory cortex is functionally and metabolically activated, yet there is no conscious sound perception (Boly et al. 2005). This suggests that activity in the auditory cortex per se is insufficient for conscious awareness, as has been discussed and demonstrated in the visual cortex (Crick and Koch 1995; Melloni et al. 2007). Only if the auditory cortex is functionally connected to a consciousness supporting network do auditory stimuli become accessible for conscious perception, analogous to what has been shown for painful stimuli (Demertzi et al. 2012; Laureys et al. 2000, 2002). These consciousness supporting networks consist of the self-representational default mode network and the attentional frontal parietal network (Akeju et al. 2014). Therefore it is plausible that the auditory cortex has to be functionally connected to the default mode network and frontoparietal attention network to permit conscious awareness of the presented stimulus.

It has been proposed that the presence of functional connections might be an essential requirement for transmitting the cortically applied stimulus into a wider network associated with the emergent property, i.e. tinnitus, that requires treatment (De Ridder et al. 2016a; Fox et al. 2014). When comparing success and failures to auditory cortex stimulation via implanted electrodes, the functional connectivity between the auditory cortex and the parahippocampus was critical for obtaining a beneficial result (De Ridder and Vanneste 2014). The importance of functional connectivity is similar to what was suggested for anterior cingulate implants (De Ridder et al. 2016a) and is in keeping with what has been shown for non-invasive brain stimulation in multiple neurological and psychiatric disorders (Fox et al. 2014). Indeed, it was shown that a lack of functional connectivity identified sites where stimulation was ineffective, and the sign of the correlation related to whether excitatory or inhibitory non-invasive stimulation was found to be clinically effective. These results suggested that resting-state functional connectivity may be useful for both optimizing treatment and identifying new stimulation targets (Fox et al. 2014).

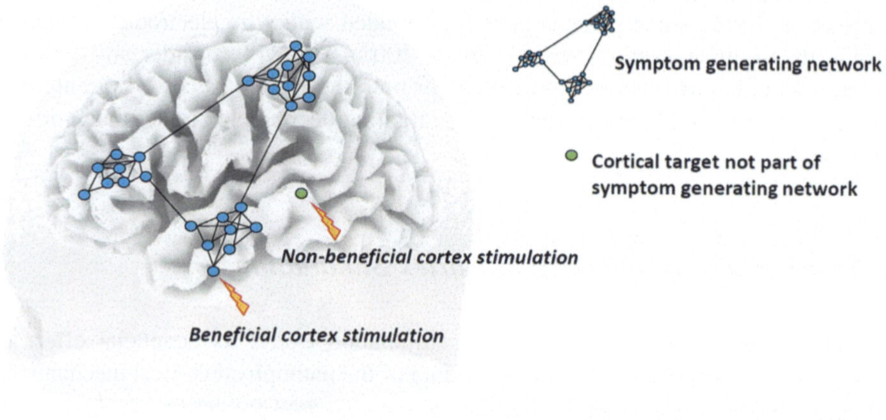

Fig. 2 The hypothesized mechanism of action of cortex stimulation (De Ridder et al. 2017). The electrode or non-invasive stimulation device has to be positioned at a cortical target where the symptom generating network reaches the brain surface. The stimulation is thought to change the functional connectivity of the network, thereby changing its topology and its related emergent property, i.e. the symptom

In summary, if the cortical target is part of a symptom generating network, the stimulation might be beneficial, whereas stimulation of a cortical target that is not functionally connected to the symptom generating network might not be beneficial (De Ridder et al. 2017) (Fig. 2).

3.2.1 Brain Stimulation Techniques in the Treatment of Tinnitus

Brain stimulation can be performed in two ways, non-invasively and invasively, in other words with or without surgery involved. Non-invasive brain stimulation encompasses transcranial magnetic stimulation (Barker 1999; Barker et al. 1985) and transcranial electrical stimulation (Moreno-Duarte et al. 2014; Paulus 2011; Vanneste et al. 2013a).

3.2.2 TMS for Treating Tinnitus

In 1985 it was demonstrated that it is possible to depolarize neurons in the brain using external magnetic stimulation (Barker et al. 1985). Unlike electrical stimuli, magnetic stimuli are not attenuated by the bone of the skull. The skull has 8–15 times the resistivity of soft tissues (Barker et al. 1985). Indeed, for electrical stimuli it has been calculated that 75% of the applied current is blocked by the skull (Voroslakos et al. 2018). Magnetic stimulation of the cortex is particularly effective because of the ability of the magnetic field to pass through high-resistance structures. TMS produces a magnetic field of the same size as that of an MRI scanner, i.e. 1–3 Tesla, but that only lasts for about a millisecond (Walsh and Rushworth 1999). Because the magnetic field changes very rapidly (from zero to a very high value, then back to zero again in 1 ms), based on Faraday's law, it induces electrical currents in the area of the brain beneath the coil. Effectively, the magnetic field "carries" the electrical stimulus across the barrier of the skull and scalp into the brain (Ridding and Rothwell 2007). The induced current pulse lasts for about 200 μs and is similar in amplitude to that produced by a conventional stimulator applied directly to the surface of the brain (Ridding and Rothwell 2007). It is thought to activate the axons of neurons in the cortex and subcortical white matter, rather than the cell bodies of cortical neurons (which have a much higher threshold) (Ridding and Rothwell 2007) a longer pulse width (>1,000 μs) is required (Ranck 1975).

There are two important limitations of TMS: (1) the magnetic field falls off rapidly with distance from the coil surface, limiting direct stimulation to the superficial parts of the cerebral cortex immediately under the skull, and (2) the site of stimulation is not as focal as direct electrical stimulation of the brain with inserted electrodes (Ridding and Rothwell 2007), even though a 2 mm precision can be reached when neuronavigation is used (Schonfeldt-Lecuona et al. 2005), improving further to 1.3 mm when robotic neuronavigated positioning is used (Goetz et al. 2019). This reaches the intrinsic resolution of the structural and functional imaging itself: At 3 T, MRI machines can resolve details of the brain as small as 1 mm. That resolution can be as fine as 0.5 mm in a 7-T machine (Nowogrodzki 2018).

Magnetic coils can have different shapes: round, figure of eight, double cone coil, and H-coils. Round coils are relatively powerful but less focal. Figure-eight-shaped coils are more focal with a maximal current at the intersection of the two round components (Rossini et al. 2015). Double cone coils and H-coils penetrate deeper and can, for example, reach the anterior and posterior cingulate cortex (Carmi et al. 2019; Hayward et al. 2007). Due to the strong decline of the magnetic field with increasing distance from the coil, direct stimulation effects are limited to superficial cortical areas. However, stimulation effects can propagate to functionally connect remote areas: low frequency TMS can increase functional connectivity, whereas high frequency TMS can decrease functional connectivity (Fox et al. 2012).

Based on TMS studies of the motor cortex it has been shown through electromyographic recordings of the activated muscles that TMS has a double effect. A single TMS stimulus evokes a burst of activity that can last for 5–10 ms after the pulse (Day et al. 1987), which is followed by a period lasting 100–200 ms during which activity is suppressed (Ridding and Rothwell 2007). The effect of the TMS pulse is brain state dependent, as well as dependent on the position and orientation of the stimulation coil and the exact site of stimulation (Ridding and Rothwell 2007). For example, if a TMS stimulus is given during sleep, anesthesia or coma, the stimulus will only exert a local effect and will not spread through the brain, in contrast to an awake state (Massimini et al. 2005). Furthermore, TMS efficacy seems to be dependent on the stimulated person's genetic polymorphism. Certain genes such as BDNF and 5-HT(1A) influence the sensitivity to non-invasive stimulation, both TMS and tDCS (Cheeran et al. 2008; Malaguti et al. 2011). In view of the interindividual anatomical variability of the brain, it has been suggested that the efficacy can be improved by using neuronavigated TMS based on the individual's brain structure obtained by structural or functional imaging (Fleming et al. 2012). The interindividual variability between the location of the sylvian sulcus and superior temporal sulcus that borders the auditory cortex is 1.5–2 cm (Steinmetz et al. 1990), suggesting that indeed a navigated approach makes sense. Concomitant intake of medication such as benzodiazepines (Deppe et al. 2020; Hunter et al. 2019), psychostimulants (Hunter et al. 2019), or neuroleptics (Hebel et al. 2020) can influence the effect of rTMS, but other medication such as anti-migraine medication does not seem to influence the treatment effect by TMS (Almaraz et al. 2010). For example, whereas concomitant benzodiazepines decrease the efficacy of TMS as a treatment for depression (Deppe et al. 2020; Hunter et al. 2019), concomitant psychostimulants increase the efficacy (Hunter et al. 2019).

Device related parameters determine the effect and side effects of TMS. The type of coil used also influences the reliability of the TMS (Fleming et al. 2012). Stimulation parameters such as stimulation frequency and amplitude influence the effect of the TMS as well (Speer et al. 2000).

Whereas single magnetic pulses only exert an immediate effect, the application of multiple pulses repetitively, called repetitive transcranial magnetic stimulation (rTMS), can have long-lasting effects that outlast the stimulation period.

In summary, many patient-dependent and device-related factors determine the outcome of transcranial magnetic stimulation (Table 1)

Table 1 Factors influencing TMS and tES effects

	TMS influencing factors	tES influencing factors
Person/patient dependent	Brain structure	Brain structure
	Brain state/function	Brain state/function
	History of activity in the stimulated Area	History of activity in the stimulated area
	Brain area	Brain area
	Genetic polymorphism	Genetic polymorphism
	Medication	Medication
	Skull-cortex distance	Soft tissue and bone structure
Device/protocol dependent	Coil type	Electrodes size
	Coil orientation	Electrodes positions
	Frequency	Frequency in tACS/tRNS
	Intensity	Intensity
	Stimulation pattern (burst/tonic)	Electrode polarity
	Duration	Duration
	Inter-train interval (intermittent/continuous)	
	Number of sessions	Number of sessions
	Interval between sessions	Interval between sessions
	Pulse form	

Table 2 Different effects of high (≥5 Hz) and low (1 Hz) frequency transcranial magnetic stimulation

	High frequency TMS	Low frequency TMS
Frequency	>5 Hz	1 Hz
Excitability	Increased	Decreased
PET metabolism	Increased metabolism	Decreased metabolism
EEG	Upper alpha and beta synchronization	Lower alpha and beta synchronization
Molecular	GABA and glutamate unchanged	GABA and glutamate increased
Plasticity	LTP-like	LTD-like

It is assumed that high frequency stimulation (HFS) and low frequency stimulation (LFS) have opposite effects (Table 2), as demonstrated by functional imaging, with HFS exerting an increased metabolism and LFS a decreased metabolism (Kimbrell et al. 1999; Speer et al. 2000). The effects on excitability and plasticity are opposite as well: whereas HFS seems to exert an increase in excitability (Pascual-Leone et al. 1994) and an LTP-like effect (Wang et al. 1996, 1999), LFS seems to generate a decrease in excitability (R. Chen et al. 1997) and an LTD-like effect (Wang et al. 1996, 1999). Furthermore, the effect on oscillatory activity is different (Brignani et al. 2008; Fuggetta et al. 2008; Paus et al. 2001), as is the effect on neurotransmitter release (Keck et al. 2001, 2002; Yue et al. 2009) (Table 2).

The widespread effect of TMS is beyond the area of stimulation (Kimbrell et al. 2002). This is mediated by transmission of the stimulus via structural connectivity (Momi et al. 2020) thereby influencing functional and effective connectivity (Shen et al. 2015). It has indeed been shown that low and high frequency stimulation exert a different effect on functional connectivity. Furthermore, TMS changes directional functional connectivity, in other words effective connectivity (Grefkes et al. 2010). By altering the functional and effective connectivity TMS can change the emergent property of the stimulated network and thereby exert its clinical effect.

The Regimen, Parameter, and Efficacy of TMS

rTMS has been used in tinnitus research in two different ways: single and repeated sessions.

Single Sessions

Single sessions of rTMS have been used for three reasons: (1) Pathophysiology and anatomy – to evaluate whether cortical areas that are identified on functional imaging are pathophysiologically involved in the generation of tinnitus, or just spurious associations. The assumption is that if rTMS can induce transient reductions in tinnitus perception by targeting the brain areas that are associated with tinnitus, these areas are causally involved in the tinnitus generation. As such the temporal cortex (De Ridder et al. 2005; Eichhammer et al. 2003a; Plewnia et al. 2003), the frontal cortex (De Ridder et al. 2012; Vanneste and De Ridder 2012a), the parietal cortex (Vanneste et al. 2012), and the anterior cingulate cortex (Vanneste and De Ridder 2012b) have been shown to be implicated in the generation of tinnitus. (2) To verify if TMS could be a prognostic test to select proper candidates for more invasive permanent solutions such as cortical brain implants (De Ridder et al. 2004, 2006a, 2011c, 2016a). (3) To delineate optimal stimulation parameters (De Ridder et al. 2005; Schoisswohl et al. 2019; Kreuzer et al. 2017) that can be employed for multiple rTMS sessions as a possible treatment.

Repeated Sessions

Repeated sessions of low frequency rTMS of the temporal (auditory) cortex have been proposed as a novel treatment approach for tinnitus based on the assumption that tinnitus is related to increased neuronal activity in the auditory cortex (Eichhammer et al. 2003b). A large number of studies have looked at low frequency TMS for treatment of tinnitus, with inconclusive results (Schoisswohl et al. 2019), likely due to the high variability and the fact that women respond better than men to rTMS, as shown in a meta-analysis (Lefebvre-Demers et al. 2020). Whereas most meta-analyses do show an improvement for tinnitus associated suffering (J. J. Chen et al. 2020; Lefebvre-Demers et al. 2020; Liang et al. 2020; Soleimani et al. 2016), other meta-analyses do not (Dong et al. 2020). Comprehensive analyses of the literature therefore identify a possible therapeutic efficacy in terms of reduction of tinnitus suffering, but the effect at clinical level is usually partial and temporary (Lefaucheur et al. 2020), with a THI improvement of about 7–8 points (Liang et al. 2020), which is the minimum for clinical efficacy (Zeman et al. 2011), lasting up to 6 months (Liang et al. 2020). Before the publication of the recent meta-analyses already a level C recommendation (possible efficacy) using evidence-based guidelines was proposed (Lefaucheur et al. 2014), stating that low frequency (1 Hz) rTMS of the left temporoparietal cortex in tinnitus is possibly efficacious for tinnitus (Lefaucheur et al. 2014). A more recent follow-up study confirmed these results (Londero et al. 2018)

Apart from the variability in study design, and the fact that women respond better than men, also different stimulation parameters are being used. A recent systematic analysis of the relationship between stimulation parameters and treatment outcome revealed a higher success rate for lower stimulation intensities (Schoisswohl et al. 2019), confirming the evidence based guidelines (Lefaucheur et al. 2014).

More recently, in an attempt to reduce the individual variability, a personalized approach has been investigated, taking into account the heterogeneity of tinnitus generators in the brain, by performing a stimulation protocol tailored to the individual patient. In a pilot study this concept was explored, by delivering rTMS at various frequencies over the left and right dorsolateral prefrontal (DLPFC) or temporoparietal (TPC) cortex in a preliminary test session to select the type of protocol subsequently applied for several days (Kreuzer et al. 2017). The personalized protocol yielded a larger benefit than the standard protocol (20 Hz-rTMS over left DLPFC followed by 1 Hz-rTMS over the left TPC). This suggests that a "tailored" rTMS protocol may be clinically more beneficial.

In summary, rTMS likely provides benefit in the treatment of tinnitus. However, if rTMS is to become routine clinical practice, it is essential to look for mechanisms that may boost or potentiate the therapeutic effect. One such way is to enhance the efficacy of rTMS pharmacologically, e.g. by adding psychostimulants or ketamine. It has been shown that psychostimulants improve the benefit of rTMS for depression (Hunter et al. 2019). Another approach could be to test ketamine enhanced rTMS, which in a preliminary trial in the treatment of depression, a frequent co-morbidity in tinnitus, shows it is feasible and exerts long-term (2 years) beneficial effect (Best et al. 2019).

Thus, rTMS may benefit especially those patients who severely suffer from the tinnitus, expressing anxiety or depression.

Reports of side effects were rare. In a meta-analysis of rTMS for tinnitus, side effects consisted of headache, discomfort at the area of stimulation, muscle twitching, neck contractions and worsening of tinnitus (Dong et al. 2020). No seizures were reported. This could be related to the fact that TMS delivered within published guidelines in individuals without risk factors results in fewer than 1 seizure per 60,000 sessions (Lerner et al. 2019).

tES for Treating Tinnitus

Three different versions exist of tES, depending on how the current is delivered to the brain (Fig. 3): transcranial direct current stimulation (tDCS), transcranial alternating current stimulation (tACS), and transcranial random noise stimulation (tRNS), which is a special version of tACS (Paulus 2011; Vanneste et al. 2013a).

3.2.3 Transcranial Direct Current Stimulation (tDCS)

Conventional tDCS

Medical use of direct electric current to the scalp has a long history. In 43–48 A.D., Scribonius Largus (the physician of Roman Emperor Claudius) reported the treatment of pain by placing a live torpedo fish – delivering a strong direct current – over the scalp (Fodstad and Hariz 2007). In the eleventh century, Ibn-Sidah suggested the placement of a live electrical catfish on the frontal bone for the treatment of patients suffering from epilepsy (Kelloway 1946). In the eighteenth century, with the introduction of the electrical battery by Galvani (Galvani 1791, 1797), it was recognized that electrical stimuli of varying duration can evoke different physiological effects (Zago et al. 2008). In honor of Galvani, direct current stimulation is often called Galvanic stimulation. One of the first clinical applications of galvanic currents was in the nineteenth century when Aldini (Aldini 1804) (Galvani's nephew) and other

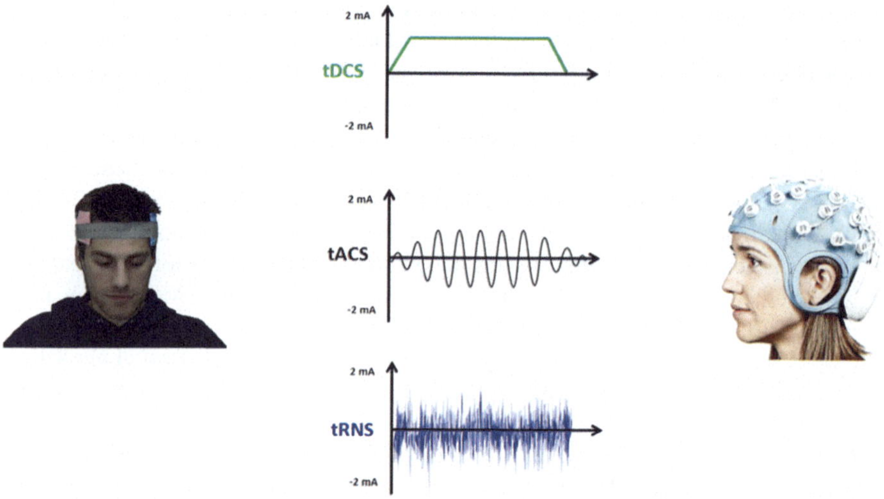

Fig. 3 Different forms of tES: tDCS delivers direct current, tACS and tRNS deliver alternating current. tES can be applied with two electrodes or more. When more electrodes are used it is called high-definition transcranial electrical stimulation (HD-tES)

researchers (Arndt 1869) used transcranial electrical simulation to treat melancholia and depression. In the 1960s and 1970s, this method had a brief comeback (Lippold and Redfearn 1964; Redfearn et al. 1964), with a more sustained revival at the turn of the millennium (Guleyupoglu et al. 2013).

Conventional tDCS uses one anode electrode and one cathode electrode on the scalp to modulate a particular brain area by inducing a controlled electrical current which flows from the anode to the cathode. Due to the high electrical resistance of the skull (Barker et al. 1985), only 25 (to 50%) of the transcranially applied direct current reaches the brain, the rest being shunted through the extracranial soft tissues, as demonstrated by calculations on realistic head models, validated both in animal (Rush and Driscoll 1968; Voroslakos et al. 2018) and human (Dymond et al. 1975; Voroslakos et al. 2018) experiments.

tDCS modulates the cellular membrane potential facilitating or inhibiting spontaneous neuronal activity (Moreno-Duarte et al. 2014; M. A. Nitsche et al. 2008). Anodal stimulation will produce inward current flow, resulting in depolarization of pyramidal cortical neurons and apical dendrite hyperpolarization, while cathodal stimulation will typically produce outward current flow resulting in somatic hyperpolarization of pyramidal cortical neurons and apical dendrite depolarization (Radman et al. 2009; Zaghi et al. 2010a). The depolarization under the anode will result in an increase of firing and excitability under the anode, whereas the firing rate and excitability are decreased under the cathode (Bindman et al. 1964; M. A. Nitsche and Paulus 2001).

However, tDCS often results in a delayed clinical effect (Fujiyama et al. 2014; Stramaccia et al. 2015), which cannot be explained by the immediate effect of tDCS on pyramidal or interneuron cell firing. Therefore, two other mechanisms have been proposed to be involved in tDCS: glial and stem cell modulation. One type of glial cells, astrocytes are possibly modulated by tDCS (Monai and Hirase 2018; Ruohonen and Karhu 2012). Astrocytes control the formation, maturation, function (and elimination) of synapses through various secreted and contact-mediated signals (Clarke and Barres 2013) and can thereby regulate neural circuit development and function (Clarke and Barres 2013). Furthermore, another type of glial cell, microglia, who prune synapses, might also be involved (Mishima et al. 2019). It has indeed been shown that tDCS activates microglia both under anode and cathode (Rueger et al. 2012). Thus, glial cells might be modulated by tDCS resulting in synapse formation and/or elimination, which takes time and can therefore better explain the delayed effects of tDCS.

Furthermore, microglial and astrocytic activation by tDCS may have a neuroimmunomodulatory effect (Goerigk et al. 2021) by altering the expression of immune-mediating genes (Rabenstein et al. 2019).

But apart from modulating neurons, both pyramidal and interneurons and glial cells, both astrocytes and microglia, tDCS could exert its delayed effects via stem cell activation. Indeed, tDCS seems to recruit proliferating neural stem cell under the cathode (Rueger et al. 2012) thereby opening the possibility of regenerative capacities for tDCS and an even more delayed clinical effect of tDCS. The neurogenesis and improved rehabilitation effects of tDCs have been shown in animal models (Zhang et al. 2020).

The effects of tDCS depend on a lot of factors, both patient-related and device-related factors (Table 1). Some factors cannot be controlled, such as the resistance of several cephalic structures including the skin, skull, blood vessels, and brain tissue (Brunoni et al. 2012; Medeiros et al. 2012; Moreno-Duarte et al. 2014; Wagner et al. 2007). Device related factors include (1) polarity of the electrodes, (2) size of the electrodes, (3) the position of the electrodes, (4) the intensity of stimulation or the amount of current delivered (in mA), and (5) the duration of the stimulation (varies between 20–40 min in most studies) (Brunoni et al. 2012; Moreno-Duarte et al. 2014; Nitsche et al. 2015; Wagner et al. 2007). By varying these tDCS parameters, stimulation protocols can be customized to a certain extent to achieve the desired direction, strength, focality, and duration of effects on cortical activity and excitability (Brunoni et al. 2012; Nitsche et al. 2015).

In general, no serious adverse events are seen by tDCS application, as evaluated in more than 10,000 subjects investigated in the contemporary tDCS literature (1998–2014) (A. R. Brunoni et al. 2011; Fregni et al. 2015). The safety of tDCS depends on the strength of current, the size of the electrodes, the contact media and the duration of the stimulation (Iyer et al. 2005; Poreisz et al. 2007).

There have been no reports of a serious adverse effect or irreversible injury across over 33,200 sessions and 1,000 subjects with repeated sessions protocols in human trials ($\leq$40 min, $\leq$4 mA, $\leq$7.2 C). tDCS has not produced any severe side effects (Bikson et al. 2016). The threshold for adverse events has been investigated in a safety study in rats, where the current density needed to induce brain damage in rats was found to be at least 100-times higher than the current density used in tDCS trials

(Liebetanz et al. 2009). The most frequent side effects include a tingling sensation during stimulation, predominantly under the anode (A. R. Brunoni et al. 2011; Poreisz et al. 2007), an itching sensation (A. R. Brunoni et al. 2011; Fertonani et al. 2015; Poreisz et al. 2007) right under the electrodes, headache (A. R. Brunoni et al. 2011; Poreisz et al. 2007), moderate fatigue (Poreisz et al. 2007) and burning sensation (A. R. Brunoni et al. 2011; Fertonani et al. 2015; Poreisz et al. 2007).

tDCS has been shown to modulate not only the areas underlying anodal and cathodal stimulation (Antal et al. 2004; Brunoni et al. 2012; Dieckhofer et al. 2006; Matsunaga et al. 2004; Zaehle et al. 2011), but also functional and effective connectivity (Alon et al. 2011; Chib et al. 2013; Keeser et al. 2011a, b; Meinzer et al. 2012, 2013; Pena-Gomez et al. 2012; Polania et al. 2011, 2012b; Stagg et al. 2013; Vanneste and De Ridder 2011; Weber et al. 2014), thereby possibly changing the emergent property of the symptom-generating network (Luft et al. 2014). These functional and effective connectivity changes permit modulation of areas beyond the effects of tDCS under the stimulation electrodes (Lang et al. 2005).

The effect of tDCS has been investigated on the physiology of the motor cortex (Brunoni et al. 2012), the visual cortex (Antal et al. 2004), the somatosensory cortex (Dieckhofer et al. 2006; Matsunaga et al. 2004), and the auditory cortex (Zaehle et al. 2011).

Single sessions of tDCS, tACS, and tRNS have been used for elucidating the involvement of specific brain networks in tinnitus pathophysiology and repeated sessions have been investigated as a potential therapeutic approach for tinnitus patients. In comparison with the large number of studies performed by rTMS, tES has been investigated less frequently and most of these studies focused on the effects of single sessions of tDCS. The targets for stimulation have been either the auditory cortex, the temporoparietal cortex or the dorsolateral prefrontal cortex.

Initial studies demonstrated transient tinnitus reduction for anodal, but not cathodal tDCS over the temporoparietal cortex, but this effect could not be consistently replicated by further studies (Yuan et al. 2018). Thus, the most promising approaches in single session studies over the auditory cortex were left anodal tDCS and bilateral tRNS. Both approaches were tested as a potential treatment in controlled studies, in which repeated sessions were applied, but unfortunately the results were disappointing, as there was no superiority of tDCS over sham stimulation (Lefaucheur et al. 2017) or a control treatment (Kreuzer et al. 2019) respectively.

In addition to the studies focusing on temporal and temporoparietal areas, several studies have targeted the DLPFC, mostly by using a bifrontal electrode montage. A single session of bifrontal anode right/cathode left tDCS reduced tinnitus intensity or distress in about one third of the patients, whereas anode left/cathode right tDCS had no effect (Vanneste et al. 2010). In a further study, the same group investigated whether the efficacy of bifrontal tDCS can be increased, if the electrode polarity is informed by gamma connectivity in EEG measurements (De Ridder and Vanneste 2012), but this was not the case. The bifrontal tDCS protocol with anode right and cathode left was also investigated as therapeutic approach with repeated sessions in uncontrolled pilot studies, which revealed preliminary promising effects (Frank et al. 2012; Shekhawat and Vanneste 2018).

A meta-analysis demonstrates that tDCS can improve tinnitus distress, but not loudness perception in tinnitus patients (Wang et al. 2018a), confirming an earlier meta-analytic study (Song et al. 2012). Yet, another meta-analysis found no benefit (Lefebvre-Demers et al. 2020). In stark contrast, a recent network meta-analysis revealed that from all non-invasive neuromodulatory approaches, cathodal tDCS over the left DLPFC combined with tRNS over the bilateral auditory cortex was associated with the greatest improvement in tinnitus severity and quality of life compared with the controls (J. J. Chen et al. 2020)

tDCS has also been combined with different forms of auditory stimulation (Lee et al. 2017; Shekhawat et al. 2014; Teismann et al. 2014). However tDCS could enhance neither the therapeutic effects of hearing aids on tinnitus complaints (Shekhawat et al. 2014) nor the effects of notched music training, a specific form of individualized auditory stimulation (Teismann et al. 2014).

In summary, tDCS effects on tinnitus are promising but also variable, analogous to what is seen in rTMS. Due to the many influencing factors this is not surprising, and further development of the more promising tDCS approaches is mandatory.

High-Definition tDCS

HD-tDCS has been recently introduced to improve the spatial accuracy, by using arrays of smaller "high-definition" electrodes, instead of the two large pad electrodes of conventional tDCS (Datta et al. 2009; Dmochowski et al. 2011; Guleyupoglu et al. 2013; Heimrath et al. 2015; Shekhawat et al. 2015; Villamar et al. 2013) (Fig. 3, right image). For high-definition tDCS, studies using 4 × 1 ring configuration with intensities up to 2 mA for up to 20 min have demonstrated its tolerability in both healthy (Borckardt et al. 2012; Kuo et al. 2013; Nikolin et al. 2015) and patient populations (Donnell et al. 2015; Villamar et al. 2013). No significant differences were found between 2 and 3 mA, suggesting the safety limits can be extended to 3 mA (Reckow et al. 2018). High-definition tDCS has a higher spatial resolution which allows more focal targeting (Borckardt et al. 2012). Furthermore, HD-tDCS permits simultaneous multifocal stimulation, permitting to develop network stimulation (Ruffini et al. 2014).

HD-tDCS has been used in tinnitus, with higher benefit associated with longer stimulation duration and higher stimulation intensities (Shekhawat et al. 2015), but no clinical difference has been noted between the benefit of conventional and HD-tDCS (Jacquemin et al. 2018).

3.2.4 Transcranial Alternating Current Stimulation (tACS)

The main mechanisms by which tACS influences brain physiology have been attributed to frequency-specific entrainment, i.e. phase alignment of endogenous brain oscillations to externally applied oscillating electric currents (Thut et al. 2011; Witkowski et al. 2015; Zaehle et al. 2010), and modulation of spike-timing dependent plasticity (Polania et al. 2012a; Vossen et al. 2015). It has been shown that alpha-tACS enhances the individual alpha frequency (Zaehle et al. 2010), but the functional effect depends on background activity (Kanai et al. 2008). tACS boosts motor excitability at 140 Hz (Moliadze et al. 2010) and decreases excitability at 15 Hz (Zaghi et al. 2010b). It has been demonstrated that tACS can influence perception (Kanai et al. 2008), memory (Marshall et al. 2006), motor function (Brignani et al. 2013), and higher-order cognition (Santarnecchi et al. 2013).

tACS of the auditory cortex does not seem to improve tinnitus (Vanneste et al. 2013a, b), neither in single nor in multiple sessions (Claes et al. 2014).

3.2.5 Transcranial Random Noise Stimulation (tRNS)

tRNS is a modification of tACS (see Fig. 3). The tRNS device generates alternating current, which follows a white noise structure, i.e. all frequencies (0.1–640 Hz) have equal power, with a Gaussian amplitude structure. Low frequency tRNS is defined as 0.1–100 Hz random noise stimulation, whereas high frequency tRNS is limited to frequencies between 100 and 640 Hz. tRNS has a higher perception threshold than tDCS (1,200 μA vs 400 μA) (Ambrus et al. 2010).

Importantly, both for tACS and tRNS low and high amplitudes seem to have an opposite effect. Both tACS and HF-tRNS at 0.4 mA are inhibitory but switch to excitatory modulation at 1 mA (Moliadze et al. 2012).

High frequency tRNS seems to increase excitability (Terney et al. 2008), and its mechanisms of action are still unknown. It could theoretically increase excitability by a stochastic resonance effect mediated through repeated subthreshold stimulations (Terney et al. 2008) that prevent homeostasis of the system (Fertonani et al. 2011). Another possible working mechanism is through desynchronization of (pathological) rhythms (Paulus 2011), but none of the abovementioned mechanisms of action have been proven. tRNS modulates perception (Romanska et al. 2015), memory (Mulquiney et al. 2011), learning (Herpich et al. 2015; S. Tyler et al. 2015), and other cognitive functions (Cappelletti et al. 2013) possibly by NMDA-receptor independent but sodium-channel blocker and benzodiazepines sensitive plasticity (Chaieb et al. 2015).

Low and high frequency tRNS have been clinically used for tinnitus, both with beneficial results on loudness perception and distress (Joos et al. 2015; Vanneste et al. 2013a). In a head to head comparison of tDCS, tACS, and tRNS, it was shown that tRNS was the only efficacious transcranial electrical stimulation for tinnitus suppression (Vanneste et al. 2013a). Interestingly, both low and high frequency tRNS were beneficial but the combined low + high frequency tRNS was inefficacious for tinnitus suppression (Joos et al. 2015). In a further pilot study tACS and tRNS, respectively, were applied bilaterally over the temporal cortices. This study revealed transient suppression of tinnitus for tRNS, but not for tACS. Furthermore, repeated sessions were more beneficial than single sessions (Claes et al. 2014).

Yet, when performing daily sessions of HF-tRNS, in contrast to the more traditional 2–3 weekly sessions, no benefit was obtained (Kreuzer et al. 2019), suggesting that too much stimulation may be counterproductive. Adding bifrontal DLPFC tDCS to auditory cortex tRNS was superior to only auditory cortex tRNS (To et al. 2017). And this multisite tRNS approach has indeed shown promise by confirmatory studies (Mohsen et al. 2018, 2019).

3.2.6 Vagus Nerve Stimulation (VNS) and Transcutaneous Vagal Nerve Stimulation (tVNS)

The combination of auditory stimulation with vagal stimulation via implanted electrodes has demonstrated highly impressive results in an animal model of tinnitus (Engineer et al. 2011). Based on the rationale that vagal stimulation renders the simultaneously presented sounds more salient, the combined treatment almost completely reversed neurophysiological and behavioral signs of tinnitus, which was not the case with auditory stimulation alone. In subsequent human pilot studies the efficacy of the invasive VNS + auditory stimulation treatment was confirmed (De Ridder et al. 2014b, 2015a; R. Tyler et al. 2017) albeit the effects were clearly less pronounced than in animals. Furthermore, a study in which VNS stimulation was performed via implanted electrodes to treat epilepsy found equally good tinnitus attenuating results (Wichova et al. 2018). But since these patients did not receive the paired auditory stimulation, the relevance of the paired auditory stimulation is yet unclear.

In these abovementioned studies, vagus nerve stimulation has been performed by the implantation of a neurostimulation device connected to an electrode located along the cervical branch of the vagus nerve. However, recently a non-invasive approach for stimulating the vagus nerve has been developed. This so-called transcutaneous vagus nerve stimulation (tVNS) takes advantage of the fact that the vagus nerve has an afferent branch that is located medially to the tragus at the entry of the acoustic meatus and that innervates the external ear canal. For reliable stimulation of the auricular branch of the vagal nerve specific devices have been developed that provide electrical stimulation. Analogous to invasive VNS, stimulation of the vagus with tVNS is typically performed on the left side to minimize potential effects on cardiac rhythm (Kreuzer et al. 2012). In a pilot study the feasibility and safety of 6 months of tVNS were investigated in patients with tinnitus. The stimulation was well tolerated, but did not lead to a relevant improvement of tinnitus complaints (Kreuzer et al. 2014). However, there was a trend towards a reduction of depressive symptoms in treated patients (Kreuzer et al. 2014). Further support for a tVNS induced stress attenuation comes from another study (Ylikoski et al. 2017). In this study the effect of tVNS on heart rate activity was investigated in 173 tinnitus patients and it was found that tVNS can attenuate the sympathetic activation of tinnitus patients

In two small pilot studies tVNS was combined with auditory stimulation, and these studies showed promising effects on tinnitus (Lehtimaki et al. 2012; Shim et al. 2015).

These findings are in line with the animal data (Engineer et al. 2011), where the pairing of vagus nerve stimulation with tones excluding the tinnitus frequency was critical for the reduction of tinnitus, whereas vagal nerve stimulation alone had no effect on tinnitus related behavior.

But tVNS has also been performed without auditory pairing, also demonstrating some benefit (Suk et al. 2018), again questioning the relevance of the paired sound presentation.

3.2.7 Bimodal (Auditory and Sensory) Stimulation

Central auditory circuits are influenced by the somatosensory system, a relationship that may underlie tinnitus generation (Basura et al. 2015; Dehmel et al. 2008; Dehmel et al. 2012; Shore 2005; Shore et al. 2008; Stefanescu et al. 2015).

Bi- or multimodal stimulation is presumably more effective for the induction of neuroplastic effects than unimodal stimulation, as synchronicity of events is an important criterion for the induction of neuroplastic effects. This was first expressed by Donald Hebb many decades ago: "Neurons that fire together, wire together" (Hebb 1949). Unimodal stimulation can induce activity-dependent neuroplastic changes such as long-term potentiation or long-term depression, whereas bimodal stimulation provides the additional opportunity to induce alterations of neuronal activity by the mechanisms of spike-timing dependent plasticity (Basura et al. 2015). However, the experimental investigation of bimodal stimulation is more challenging, because of the much larger parameter space.

In recent years different approaches of bi- or multimodal stimulation have been proposed for the treatment of tinnitus. Apart from the combination of vagal stimulation, tDCS and rTMS with auditory stimulation, the combination of auditory stimulation and somatosensory stimulation has been investigated. The somatosensory stimulation was either performed via the trigeminal nerve or via C2 afferents. The combined auditory and somatosensory modulation is motivated by an increasing understanding of the relevance of the somatosensory system for tinnitus pathophysiology (Shore et al. 2016). The clinical phenomenon, that many patients can modulate their tinnitus by face or neck movements, can be explained by the interaction between somatosensory and auditory afferents on the level of the dorsal cochlear nucleus (Levine 1999). This knowledge in turn motivated two different approaches of combined somatosensory and auditory stimulation.

One approach aims at the modulation of activity in the central auditory pathway by exerting an inhibitory effect on the level of the dorsal cochlear nucleus. The somatosensory and auditory stimuli were presented at specific intervals, derived from basic neurophysiological studies in animals describing stimulus timing dependent plasticity in the dorsal cochlear nucleus (Marks et al. 2018). These findings were translated into a clinical pilot trial with 20 patients, in which the combination of sounds with transcutaneous electrical stimulation of the C2 nerve at the level of the neck was applied over 28 days. The bimodal treatment reduced tinnitus loudness and intrusiveness, whereas the control condition (auditory stimulation alone) did not deliver benefit (Marks et al. 2018). Yet, the effect was short-lasting, especially for loudness perception, which did not outlast the 3-week bimodal stimulation period, and the distress improvement only lasted for 3 weeks.

In another approach, sounds are simultaneously applied in combination with electrical stimulation of the trigeminal nerve at the level of the tongue (Conlon et al. 2019; D'Arcy et al. 2017). This approach is based on the idea that tinnitus is caused by auditory deafferentation and that bimodal stimulation might compensate the auditory deafferentation by providing stimuli over the somatosensory system. The combined application of sounds and electrical stimulation of the tongue, for a period of 3 months, was investigated in two large trials (with more than 500 patients) (Conlon et al. 2020). Similar to other studies the bimodal stimulation resulted in a clinical benefit for tinnitus distress, but data for loudness improvement were not provided. In contrast to the C2-auditory bimodal stimulation study, long-lasting results (1 year) were obtained (Conlon et al. 2020).

3.2.8 Invasive Brain Stimulation for Tinnitus

Auditory Cortex Stimulation

The procedure is based on a pathophysiological model that the auditory cortex is involved in a pathologically functioning neuronal network that generates tinnitus and that interference with this network activity by auditory cortex stimulation can alleviate tinnitus and follows a four-step rationale (De Ridder et al. 2012b):

1. Tinnitus is related to increased activity in the auditory and frontal cortex.
2. The anatomical location of the tinnitus generator can be determined by fMRI (De Ridder et al. 2011b).
3. The abnormal neuronal activity can be modulated by neuronavigated TMS resulting in transient tinnitus reduction (De Ridder et al. 2004).
4. If TMS can transiently suppress the tinnitus, electrical stimulation through an electrode implanted on the same area can provide permanent tinnitus suppression (De Ridder et al. 2004, 2006a, 2007a, 2011c).

Multiple small and one larger series of patients with auditory cortex electrodes have been published. A series of 43 patients with severe treatment resistant tinnitus (grade 3 and 4 tinnitus according to the tinnitus questionnaire (Goebel and Hiller 1994)) were implanted with a cortical electrode overlying the secondary auditory cortex (De Ridder et al. 2011c). Surgical eligibility was based on 2 positive TMS sessions. Although all patients reacted to TMS, 1 out of 3 patients did not respond to the cortical stimulation with tonic stimuli after implantation. Among the responders to cortical stimulation there was an average decrease in the perceived tinnitus loudness of 51.3%. There was a significant but weak positive correlation ($r = 0.34$, $p < 0.05$) between the amount of the suppression effect from the test TMS and cortical stimulation after implantation, even though TMS could not predict who would and who would not respond to the implant (De Ridder et al. 2011c).

This may be due to the fact that the mechanism of action of TMS and implanted electrodes is different.

When switching tonic stimulation to burst stimulation (De Ridder et al. 2010) half of the non-responding patients demonstrated change, thereby improving the total response rate from 1 in 3 to 2 out of 3 patients. Burst stimulation was specifically superior to tonic stimulation for suppressing noise-like tinnitus (De Ridder et al. 2011c), analogous to what has been described for TMS (De Ridder et al. 2007c). In contrast to TMS, where the suppression effect decreases with longer tinnitus duration, no correlation was found between the effect of electrical cortical stimulation and tinnitus duration for the same study population, again suggesting that electrical cortical stimulation acts on tinnitus by a different mechanism than TMS.

Treatment effects also depended on tinnitus type. Pure tone tinnitus could be improved more than narrow band noise or the combination of pure tone and narrow band noise, and unilateral tinnitus better than bilateral tinnitus. This surgical neuromodulatory approach of the auditory cortex has been repeated by other centers. A French case study obtained persisting 65% tinnitus reduction in a women using an fMRI based extradural auditory cortex implant (Litre et al. 2009, 2010). Another study of eight patients using a similar technique but different hardware found no permanent tinnitus suppression (Friedland et al. 2007). Temporary effects on tinnitus perception were observed in six out of the eight patients. However, tinnitus distress decreased slowly over time, even without suppression of tinnitus intensity. An electrode with only two contacts was used which limits the way the electrodes can be programmed.

In four patients an intradural electrode on the primary auditory cortex was inserted in the Sylvian fissure, stimulating gray matter of the primary auditory cortex (De Ridder et al. 2004, 2006a). In two patients the purpose was to obtain stabilization of tinnitus suppression, because the stimulus parameters had to be reprogrammed every 2–3 days. In both patients the intradural positioning resulted in a stabilized suppression of their tinnitus.

Another approach has been proposed, inserting a wire electrode in the white matter beneath layer 6 of the primary auditory cortex. This has been performed successfully, using magnetic source imaging for target localization, resulting in tinnitus suppression (Seidman et al. 2008). Interestingly, in patients with tinnitus, intracortical stimulation does not generate a sound percept associated with the delivered current. This is in contrast to patients with epilepsy, in whom intracortical electrical stimulation within Heschl's gyrus does generate sound percepts, the loudness of which correlates with the delivered amplitude (Donovan et al. 2015).

From a mechanistic point of view it was shown that the success of the auditory cortex implant critically depended on activity in the parahippocampal area, which is related to auditory memory (De Ridder et al. 2006b). Responders to the implant were characterized by high beta3 and gamma band activity in the parahippocampal area, even though the electrodes were overlying the auditory cortex. Only those patients who had functional connections between the area of the implant, i.e. the auditory cortex and the parahippocampal area, benefited from the auditory cortex implant (De Ridder and Vanneste 2014).

Multisite stimulation may benefit tinnitus perception, analogous to what is noted in non-invasive neuromodulation. In a partial responder to auditory cortex implantation, complete resolution of the pure tone component of his tinnitus was obtained, without any beneficial effect on the noise-like component of the tinnitus, even after switching to burst stimulation (De Ridder and Vanneste 2015). After an initial successful treatment of his noise-like component with transcutaneus electrical nerve stimulation, a wire electrode was inserted subcutaneously and connected to his internal pulse generator. With the dual stimulation his pure tone tinnitus remains abolished after 5 years of stimulation and his noise-like tinnitus is improved by 50%, from 8/10 to 4/10. This case report suggested that multitarget stimulation might be better than single target implantation (De Ridder and Vanneste 2015).

In some case reports implants were also performed on the dorsolateral prefrontal cortex (De Ridder et al. 2012b), anterior cingulate cortex (De Ridder et al. 2016a) and parahippocampal area (De Ridder et al. 2012b) following the same 4-step approach described above. In the two anterior cingulate implants one patient responded whereas another patient did not benefit from the electrode insertion. The responder also had increased functional connectivity to a tinnitus distress network in contrast to the non-responder (De Ridder et al. 2016a). This suggests that analogous to non-invasive stimulation, brain stimulation via implanted electrodes requires functional connectivity to carry the delivered stimulus throughout the symptom generating network (Fox et al. 2014).

Deep Brain Stimulation

Deep brain stimulation (DBS) has been performed as well, in an attempt to treat tinnitus. This was based on a case report of a woman who became tinnitus-free after a stroke in the locus coeruleus (LC) area of the caudate nucleus while undergoing DBS for Parkinson's disease (Larson and Cheung 2013). Initially, tinnitus was evaluated in patients in whom DBS was performed to alleviate movement disorders. In a first study, tinnitus loudness reductions were found in 4/7 patients, of which most clearly by ventral intermediate nucleus (VIM) stimulation for tremor [256]. In another observational study in six patients with comorbid tinnitus, the concomitant effect on tinnitus perception was evaluated: In five participants where the DBS lead tip traversed area LC, tinnitus loudness in both ears was suppressed to a nadir of level 2 or lower on a 0–10 rating scale. In one subject where the DBS lead was outside area LC, tinnitus was not modulated (Cheung and Larson 2010; Larson and Cheung 2011).

A large multicenter study evaluated the clinical impact of DBS on tinnitus in patients undergoing DBS for movement disorders: the THI tinnitus questionnaire improved only after subthalamic nucleus stimulation [254], suggesting this target may be selected to treat tinnitus related distress. After encouraging results from these observational studies, a phase I study was performed targeting the caudate nucleus as goal to treat severe intractable tinnitus. Tinnitus distress measures improved for three of five patients and one patient had a profound loudness suppression (7.8 points improvement on NRS). This suggests that the caudate nucleus may be a target worthwhile of further exploration, using different stimulation designs and different electrode configurations. Even though the target space may be narrowed down, the stimulation parameter space for optimal improvement is still large.

4 Conclusion

In the last decades, neuroscientific research has contributed to an increasingly better understanding of the pathophysiological mechanisms that underlie the generation and maintenance of tinnitus. Based on this knowledge, a large variety of different neuromodulatory interventions have been developed.

Most studies for rTMS have targeted the temporal, temporoparietal, and the frontal cortex. Recent meta-analyses have shown that rTMS may be beneficial for tinnitus, improving the suffering, but not the loudness perception. The recent rTMS European guideline (Lefaucheur et al. 2020) recommended that repeated sessions of low frequency-rTMS of the temporoparietal cortex of the left hemisphere or contralateral to the affected ear have a *possible effect* in tinnitus. Many questions remain concerning the use of this technique in everyday practice, such as what could be the optimal treatment target(s) protocol and what could be the role of individual susceptibility to auditory cortex stimulation.

Different forms of transcranial electrical stimulation (tDCS, tACS, tRNS), applied over the frontal and temporal cortex, have been investigated in tinnitus patients. Recent meta-analysis suggests that also tES may be beneficial in chronic tinnitus, and that especially the combination of bifrontal tDCS and auditory cortex tRNS may attenuate tinnitus. Cortex and deep brain stimulation with implanted electrodes have shown benefit but there is insufficient data to support their routine clinical use.

Two decades of research in non-invasive neuromodulatory interventions in tinnitus have yet to result in regular clinical routine use. The most recent meta-analyses do suggest that a transition from experimental to clinical applications of non-invasive stimulation may be in view. Furthermore, research has revealed important insights in the pathophysiology of tinnitus, in particular in the relevance of non-auditory brain areas as well in the heterogeneous nature of tinnitus. Recently, bimodal stimulation approaches have also revealed promising results and it appears that targeting different sensory modalities in temporally combined manners may also be a promising avenue.

References

Achard S, Salvador R, Whitcher B, Suckling J, Bullmore E (2006) A resilient, low-frequency, small-world human brain functional network with highly connected association cortical hubs. J Neurosci 26(1):63–72. https://doi.org/10.1523/JNEUROSCI.3874-05.2006

Adamchic I, Hauptmann C, Tass PA (2012) Changes of oscillatory activity in pitch processing network and related tinnitus relief induced by acoustic CR neuromodulation. Front Syst Neurosci 6:18. https://doi.org/10.3389/fnsys.2012.00018

Adamchic I, Toth T, Hauptmann C, Tass PA (2014) Reversing pathologically increased EEG power by acoustic coordinated reset neuromodulation. Hum Brain Mapp 35(5):2099–2118. https://doi.org/10.1002/hbm.22314

Adjamian P, Sereda M, Zobay O, Hall DA, Palmer AR (2012) Neuromagnetic indicators of tinnitus and tinnitus masking in patients with and without hearing loss. J Assoc Res Otolaryngol 13(5): 715–731. https://doi.org/10.1007/s10162-012-0340-5

Akeju O, Loggia ML, Catana C, Pavone KJ, Vazquez R, Rhee J et al (2014) Disruption of thalamic functional connectivity is a neural correlate of dexmedetomidine-induced unconsciousness. Elife 3:e04499. https://doi.org/10.7554/eLife.04499

Albert R, Jeong H, Barabasi AL (2000) Error and attack tolerance of complex networks. Nature 406(6794):378–382. https://doi.org/10.1038/35019019

Aldini G (1804) Essai theoretique et experimental sur le galvanisme. Paris

Almaraz AC, Dilli E, Dodick DW (2010) The effect of prophylactic medications on TMS for migraine aura. Headache 50(10):1630–1633. https://doi.org/10.1111/j.1526-4610.2010.01787.x

Alon G, Roys SR, Gullapalli RP, Greenspan JD (2011) Non-invasive electrical stimulation of the brain (ESB) modifies the resting-state network connectivity of the primary motor cortex: a proof of concept fMRI study. Brain Res 1403:37–44. https://doi.org/10.1016/j.brainres.2011.06.013

Amaral LA, Diaz-Guilera A, Moreira AA, Goldberger AL, Lipsitz LA (2004) Emergence of complex dynamics in a simple model of signaling networks. Proc Natl Acad Sci U S A 101(44): 15551–15555. https://doi.org/10.1073/pnas.0404843101

Ambrus GG, Paulus W, Antal A (2010) Cutaneous perception thresholds of electrical stimulation methods: comparison of tDCS and tRNS. Clin Neurophysiol 121(11):1908–1914. https://doi.org/10.1016/j.clinph.2010.04.020

Antal A, Kincses TZ, Nitsche MA, Bartfai O, Paulus W (2004) Excitability changes induced in the human primary visual cortex by transcranial direct current stimulation: direct electrophysiological evidence. Invest Ophthalmol Vis Sci 45(2):702–707. Retrieved from http://www.ncbi.nlm.nih.gov/pubmed/14744917

Arndt R (1869) Die electricitat in der psychiatrie. Arch Psychiat Nervenkrankh 2:259–261

Axelsson A, Ringdahl A (1989) Tinnitus – a study of its prevalence and characteristics. Br J Audiol 23(1):53–62. Retrieved from http://www.ncbi.nlm.nih.gov/entrez/query.fcgi?cmd=Retrieve&db=PubMed&dopt=Citation&list_uids=2784987

Barabasi AL, Gulbahce N, Loscalzo J (2011) Network medicine: a network-based approach to human disease. Nat Rev Genet 12(1):56–68. https://doi.org/10.1038/nrg2918

Barker AT (1999) The history and basic principles of magnetic nerve stimulation. Electroencephalogr Clin Neurophysiol Suppl 51:3–21. Retrieved from http://www.ncbi.nlm.nih.gov/entrez/query.fcgi?cmd=Retrieve&db=PubMed&dopt=Citation&list_uids=10590931

Barker AT, Jalinous R, Freeston IL (1985) Non-invasive magnetic stimulation of human motor cortex. Lancet 1(8437):1106–1107

Bassett DS, Meyer-Lindenberg A, Achard S, Duke T, Bullmore E (2006) Adaptive reconfiguration of fractal small-world human brain functional networks. Proc Natl Acad Sci U S A 103(51): 19518–19523. https://doi.org/10.1073/pnas.0606005103

Basura GJ, Koehler SD, Shore SE (2015) Bimodal stimulus timing-dependent plasticity in primary auditory cortex is altered after noise exposure with and without tinnitus. J Neurophysiol 114(6): 3064–3075. https://doi.org/10.1152/jn.00319.2015

Best SRD, Pavel DG, Haustrup N (2019) Combination therapy with transcranial magnetic stimulation and ketamine for treatment-resistant depression: A long-term retrospective review of clinical use. Heliyon 5(8):e02187. https://doi.org/10.1016/j.heliyon.2019.e02187

Bhatt JM, Bhattacharyya N, Lin HW (2017) Relationships between tinnitus and the prevalence of anxiety and depression. Laryngoscope 127(2):466–469. https://doi.org/10.1002/lary.26107

Bikson M, Grossman P, Thomas C, Zannou AL, Jiang J, Adnan T et al (2016) Safety of transcranial direct current stimulation: evidence based update 2016. Brain Stimul 9(5):641–661. https://doi.org/10.1016/j.brs.2016.06.004

Bindman LJ, Lippold OC, Redfearn JW (1964) The action of brief polarizing currents on the cerebral cortex of the rat (1) during current flow and (2) in the production of long-lasting after-effects. J Physiol 172:369–382. Retrieved from http://www.ncbi.nlm.nih.gov/pubmed/14199369

Boccard SG, Fitzgerald JJ, Pereira EA, Moir L, Van Hartevelt TJ, Kringelbach ML et al (2014) Targeting the affective component of chronic pain: a case series of deep brain stimulation of the anterior cingulate cortex. Neurosurgery. https://doi.org/10.1227/NEU.0000000000000321

Boly M, Faymonville ME, Peigneux P, Lambermont B, Damas F, Luxen A et al (2005) Cerebral processing of auditory and noxious stimuli in severely brain injured patients: differences between VS and MCS. Neuropsychol Rehabil 15(3–4):283–289. Retrieved from http://www.ncbi.nlm.nih.gov/entrez/query.fcgi?cmd=Retrieve&db=PubMed&dopt=Citation&list_uids=16350972

Borckardt JJ, Bikson M, Frohman H, Reeves ST, Datta A, Bansal V et al (2012) A pilot study of the tolerability and effects of high-definition transcranial direct current stimulation (HD-tDCS) on pain perception. J Pain 13(2):112–120. https://doi.org/10.1016/j.jpain.2011.07.001

Borst A, Theunissen FE (1999) Information theory and neural coding. Nat Neurosci 2(11):947–957. https://doi.org/10.1038/14731

Brignani D, Manganotti P, Rossini PM, Miniussi C (2008) Modulation of cortical oscillatory activity during transcranial magnetic stimulation. Hum Brain Mapp 29(5):603–612. https://doi.org/10.1002/hbm.20423

Brignani D, Ruzzoli M, Mauri P, Miniussi C (2013) Is transcranial alternating current stimulation effective in modulating brain oscillations? PLoS One 8(2):e56589. https://doi.org/10.1371/journal.pone.0056589

Brunoni AR, Amadera J, Berbel B, Volz MS, Rizzerio BG, Fregni F (2011) A systematic review on reporting and assessment of adverse effects associated with transcranial direct current stimulation. Int J Neuropsychopharmacol 14(8):1133–1145. https://doi.org/10.1017/S1461145710001690

Brunoni AR, Nitsche MA, Bolognini N, Bikson M, Wagner T, Merabet L, Edwards DJ, Valero-Cabre A, Rotenberg A, Pascual-Leone A, Ferricci R, Priori A, Boggio PS, Fregni F (2012) Clinical research with transcranial direct current stimulation (tDCS): challenges and future directions. Brain Stimul 5:175–195

Bullmore E, Sporns O (2009) Complex brain networks: graph theoretical analysis of structural and functional systems. Nat Rev Neurosci 10(3):186–198. https://doi.org/10.1038/nrn2575

Bullmore E, Sporns O (2012) The economy of brain network organization. Nat Rev Neurosci 13(5): 336–349. https://doi.org/10.1038/nrn3214

Buzsaki G (2006) Rhythms of the brain. Oxford University Press, Oxford

Buzsaki G, Mizuseki K (2014) The log-dynamic brain: how skewed distributions affect network operations. Nat Rev Neurosci 15(4):264–278. https://doi.org/10.1038/nrn3687

Cappelletti M, Gessaroli E, Hithersay R, Mitolo M, Didino D, Kanai R et al (2013) Transfer of cognitive training across magnitude dimensions achieved with concurrent brain stimulation of the parietal lobe. J Neurosci 33(37):14899–14907. https://doi.org/10.1523/JNEUROSCI.1692-13.2013

Carmi L, Tendler A, Bystritsky A, Hollander E, Blumberger DM, Daskalakis J et al (2019) Efficacy and safety of deep transcranial magnetic stimulation for obsessive-compulsive disorder: a prospective multicenter randomized double-blind placebo-controlled trial. Am J Psychiatry., appiajp201918101180. https://doi.org/10.1176/appi.ajp.2019.18101180

Catania KC (2009) Tentacled snakes turn C-starts to their advantage and predict future prey behavior. Proc Natl Acad Sci U S A 106(27):11183–11187. https://doi.org/10.1073/pnas.0905183106

Chaieb L, Antal A, Paulus W (2015) Transcranial random noise stimulation-induced plasticity is NMDA-receptor independent but sodium-channel blocker and benzodiazepines sensitive. Front Neurosci 9:125. https://doi.org/10.3389/fnins.2015.00125

Cheeran B, Talelli P, Mori F, Koch G, Suppa A, Edwards M et al (2008) A common polymorphism in the brain-derived neurotrophic factor gene (BDNF) modulates human cortical plasticity and the response to rTMS. J Physiol 586(Pt 23):5717–5725. https://doi.org/10.1113/jphysiol.2008.159905

Chen R, Classen J, Gerloff C, Celnik P, Wassermann EM, Hallett M, Cohen LG (1997) Depression of motor cortex excitability by low-frequency transcranial magnetic stimulation. Neurology 48(5):1398–1403. Retrieved from http://www.ncbi.nlm.nih.gov/entrez/query.fcgi?cmd=Retrieve&db=PubMed&dopt=Citation&list_uids=9153480

Chen JJ, Zeng BS, Wu CN, Stubbs B, Carvalho AF, Brunoni AR et al (2020) Association of central noninvasive brain stimulation interventions with efficacy and safety in tinnitus management: a meta-analysis. JAMA Otolaryngol Head Neck Surg. https://doi.org/10.1001/jamaoto.2020.1497

Cheung SW, Larson PS (2010) Tinnitus modulation by deep brain stimulation in locus of caudate neurons (area LC). Neuroscience 169(4):1768–1778. https://doi.org/10.1016/j.neuroscience.2010.06.007

Chib VS, Yun K, Takahashi H, Shimojo S (2013) Noninvasive remote activation of the ventral midbrain by transcranial direct current stimulation of prefrontal cortex. Transl Psychiatry 3:e268. https://doi.org/10.1038/tp.2013.44

Claes L, Stamberger H, Van de Heyning P, De Ridder D, Vanneste S (2014) Auditory cortex tACS and tRNS for tinnitus: single versus multiple sessions. Neural Plast 2014:436713. https://doi.org/10.1155/2014/436713

Clarke LE, Barres BA (2013) Emerging roles of astrocytes in neural circuit development. Nat Rev Neurosci 14(5):311–321. https://doi.org/10.1038/nrn3484

Conlon B, Hamilton C, Hughes S, Meade E, Hall DA, Vanneste S et al (2019) Noninvasive bimodal neuromodulation for the treatment of tinnitus: protocol for a second large-scale double-blind randomized clinical trial to optimize stimulation parameters. JMIR Res Protoc 8(9):e13176. https://doi.org/10.2196/13176

Conlon B, Langguth B, Hamilton C, Hughes S, Meade E, Connor CO et al (2020) Bimodal neuromodulation combining sound and tongue stimulation reduces tinnitus symptoms in a large randomized clinical study. Sci Transl Med 12(564). https://doi.org/10.1126/scitranslmed.abb2830

Crick F, Koch C (1995) Are we aware of neural activity in primary visual cortex? Nature 375(6527):121–123. https://doi.org/10.1038/375121a0

Crunelli V, Lorincz ML, Connelly WM, David F, Hughes SW, Lambert RC et al (2018) Dual function of thalamic low-vigilance state oscillations: rhythm-regulation and plasticity. Nat Rev Neurosci. https://doi.org/10.1038/nrn.2017.151

D'Arcy S, Hamilton C, Hughes S, Hall DA, Vanneste S, Langguth B, Conlon B (2017) Bi-modal stimulation in the treatment of tinnitus: a study protocol for an exploratory trial to optimise stimulation parameters and patient subtyping. BMJ Open 7(10):e018465. https://doi.org/10.1136/bmjopen-2017-018465

Datta A, Bansal V, Diaz J, Patel J, Reato D, Bikson M (2009) Gyri-precise head model of transcranial direct current stimulation: improved spatial focality using a ring electrode versus conventional rectangular pad. Brain Stimul 2(4):201–207., 207.e201. https://doi.org/10.1016/j.brs.2009.03.005

Day BL, Rothwell JC, Thompson PD, Dick JP, Cowan JM, Berardelli A, Marsden CD (1987) Motor cortex stimulation in intact man. 2. Multiple descending volleys. Brain 110(Pt 5):1191–1209. Retrieved from http://www.ncbi.nlm.nih.gov/pubmed/3676698

De Ridder D (2010) Should rTMS for tinnitus be performed left-sided, ipsilaterally or contralaterally, and is it a treatment or merely investigational? Eur J Neurol 17(7):891–892. https://doi.org/10.1111/j.1468-1331.2010.02967.x

De Ridder D, Vanneste S (2012) EEG driven tDCS versus bifrontal tDCS for tinnitus. Front Psych 3:84. https://doi.org/10.3389/fpsyt.2012.00084
De Ridder D, Vanneste S (2014) Targeting the parahippocampal area by auditory cortex stimulation in tinnitus. Brain Stimul. https://doi.org/10.1016/j.brs.2014.04.004
De Ridder D, Vanneste S (2015) Multitarget surgical neuromodulation: combined C2 and auditory cortex implantation for tinnitus. Neurosci Lett 591:202–206. https://doi.org/10.1016/j.neulet.2015.02.034
De Ridder D, De Mulder G, Walsh V, Muggleton N, Sunaert S, Moller A (2004) Magnetic and electrical stimulation of the auditory cortex for intractable tinnitus. Case report. J Neurosurg 100(3): 560–564. Retrieved from http://www.ncbi.nlm.nih.gov/entrez/query.fcgi?cmd=Retrieve&db=PubMed&dopt=Citation&list_uids=15035296
De Ridder D, Verstraeten E, Van der Kelen K, De Mulder G, Sunaert S, Verlooy J et al (2005) Transcranial magnetic stimulation for tinnitus: influence of tinnitus duration on stimulation parameter choice and maximal tinnitus suppression. Otol Neurotol 26(4):616–619. Retrieved from http://www.ncbi.nlm.nih.gov/entrez/query.fcgi?cmd=Retrieve&db=PubMed&dopt=Citation&list_uids=16015156
De Ridder D, De Mulder G, Verstraeten E, Van der Kelen K, Sunaert S, Smits M et al (2006a) Primary and secondary auditory cortex stimulation for intractable tinnitus. ORL J Otorhinolaryngol Relat Spec 68(1):48–54. Retrieved from http://www.ncbi.nlm.nih.gov/entrez/query.fcgi?cmd=Retrieve&db=pubmed&dopt=Citation&list_uids=16514263
De Ridder D, Fransen H, Francois O, Sunaert S, Kovacs S, Van De Heyning P (2006b) Amygdalohippocampal involvement in tinnitus and auditory memory. Acta Otolaryngol (556):50–53. Retrieved from http://www.ncbi.nlm.nih.gov/entrez/query.fcgi?cmd=Retrieve&db=PubMed&dopt=Citation&list_uids=17114143
De Ridder D, De Mulder G, Menovsky T, Sunaert S, Kovacs S (2007a) Electrical stimulation of auditory and somatosensory cortices for treatment of tinnitus and pain. Prog Brain Res 166:377–388. https://doi.org/10.1016/S0079-6123(07)66036-1
De Ridder D, De Mulder G, Verstraeten E, Seidman M, Elisevich K, Sunaert S et al (2007b) Auditory cortex stimulation for tinnitus. Acta Neurochir Suppl 97(Pt 2):451–462. Retrieved from http://www.ncbi.nlm.nih.gov/entrez/query.fcgi?cmd=Retrieve&db=PubMed&dopt=Citation&list_uids=17691335
De Ridder D, van der Loo E, Van der Kelen K, Menovsky T, van de Heyning P, Moller A (2007c) Do tonic and burst TMS modulate the lemniscal and extralemniscal system differentially? Int J Med Sci 4(5):242–246. Retrieved from http://www.ncbi.nlm.nih.gov/entrez/query.fcgi?cmd=Retrieve&db=PubMed&dopt=Citation&list_uids=17952200
De Ridder D, van der Loo E, Van der Kelen K, Menovsky T, van de Heyning P, Moller A (2007d) Theta, alpha and beta burst transcranial magnetic stimulation: brain modulation in tinnitus. Int J Med Sci 4(5):237–241. Retrieved from http://www.ncbi.nlm.nih.gov/entrez/query.fcgi?cmd=Retrieve&db=PubMed&dopt=Citation&list_uids=17952199
De Ridder D, Menovsky T, van de Heyning P (2008) Auditory cortex stimulation for tinnitus suppression. Otol Neurotol 29(4):574–575.; author reply 575. https://doi.org/10.1097/MAO.0b013e3181719773
De Ridder D, Vanneste S, van der Loo E, Plazier M, Menovsky T, van de Heyning P (2010) Burst stimulation of the auditory cortex: a new form of neurostimulation for noise-like tinnitus suppression. J Neurosurg 112(6):1289–1294. https://doi.org/10.3171/2009.10.JNS09298
De Ridder D, Elgoyhen AB, Romo R, Langguth B (2011a) Phantom percepts: tinnitus and pain as persisting aversive memory networks. Proc Natl Acad Sci U S A 108(20):8075–8080. https://doi.org/10.1073/pnas.1018466108
De Ridder D, van der Loo E, Vanneste S, Gais S, Plazier M, Kovacs S et al (2011b) Theta-gamma dysrhythmia and auditory phantom perception. J Neurosurg 114(4):912–921. https://doi.org/10.3171/2010.11.JNS10335

De Ridder D, Vanneste S, Kovacs S, Sunaert S, Menovsky T, van de Heyning P, Moller A (2011c) Transcranial magnetic stimulation and extradural electrodes implanted on secondary auditory cortex for tinnitus suppression. J Neurosurg 114(4):903–911. https://doi.org/10.3171/2010.11.JNS10197

De Ridder D, Song JJ, Vanneste S (2012a) Frontal cortex TMS for tinnitus. Brain Stimul 6(3):355–362. https://doi.org/10.1016/j.brs.2012.07.002

De Ridder D, Vanneste S, Menovsky T, Langguth B (2012b) Surgical brain modulation for tinnitus: the past, present and future. J Neurosurg Sci 56(4):323–340

De Ridder D, Vanneste S, Plazier M, Menovsky T, van de Heyning P, Kovacs S, Sunaert S (2012c) Dorsolateral prefrontal cortex transcranial magnetic stimulation and electrode implant for intractable tinnitus. World Neurosurg 77(5-6):778–784. https://doi.org/10.1016/j.wneu.2011.09.009

De Ridder D, Joos K, Vanneste S (2014a) The enigma of the tinnitus-free dream state in a Bayesian world. Neural Plast 2014:612147. https://doi.org/10.1155/2014/612147

De Ridder D, Vanneste S, Engineer ND, Kilgard MP (2014b) Safety and efficacy of vagus nerve stimulation paired with tones for the treatment of tinnitus: a case series. Neuromodulation 17(2):170–179. https://doi.org/10.1111/ner.12127

De Ridder D, Vanneste S, Freeman W (2014c) The Bayesian brain: phantom percepts resolve sensory uncertainty. Neurosci Biobehav Rev 44C:4–15. https://doi.org/10.1016/j.neubiorev.2012.04.001

De Ridder D, Vanneste S, Weisz N, Londero A, Schlee W, Elgoyhen AB, Langguth B (2014d) An integrative model of auditory phantom perception: tinnitus as a unified percept of interacting separable subnetworks. Neurosci Biobehav Rev 44:16–32. https://doi.org/10.1016/j.neubiorev.2013.03.021

De Ridder D, Kilgard M, Engineer N, Vanneste S (2015a) Placebo-controlled vagus nerve stimulation paired with tones in a patient with refractory tinnitus: a case report. Otol Neurotol 36(4):575–580. https://doi.org/10.1097/MAO.0000000000000704

De Ridder D, Vanneste S, Langguth B, Llinas R (2015b) Thalamocortical dysrhythmia: a theoretical update in tinnitus. Front Neurol 6:124. https://doi.org/10.3389/fneur.2015.00124

De Ridder D, Joos K, Vanneste S (2016a) Anterior cingulate implants for tinnitus: report of 2 cases. J Neurosurg 124(4):893–901. https://doi.org/10.3171/2015.3.JNS142880

De Ridder D, Manning P, Glue P, Cape G, Langguth B, Vanneste S (2016b) Anterior cingulate implant for alcohol dependence. Neurosurgery. https://doi.org/10.1227/NEU.0000000000001248

De Ridder D, Perera S, Vanneste S (2017) State of the art: novel applications for cortical stimulation. Neuromodulation 20(3):206–214. https://doi.org/10.1111/ner.12593

Dehmel S, Cui YL, Shore SE (2008) Cross-modal interactions of auditory and somatic inputs in the brainstem and midbrain and their imbalance in tinnitus and deafness. Am J Audiol 17(2):S193–S209. https://doi.org/10.1044/1059-0889(2008/07-0045)

Dehmel S, Pradhan S, Koehler S, Bledsoe S, Shore S (2012) Noise overexposure alters long-term somatosensory-auditory processing in the dorsal cochlear nucleus – possible basis for tinnitus-related hyperactivity? J Neurosci 32(5):1660–1671. https://doi.org/10.1523/JNEUROSCI.4608-11.2012

Demertzi A, Soddu A, Laureys S (2012) Consciousness supporting networks. Curr Opin Neurobiol. https://doi.org/10.1016/j.conb.2012.12.003

Deppe M, Abdelnaim M, Hebel T, Kreuzer P, Poeppl T, Langguth B, Schecklmann M (2020) Concomitant lorazepam use and antidepressive efficacy of repetitive transcranial magnetic stimulation in a naturalistic setting. Eur Arch Psychiatry Clin Neurosci.1–7

Dieckhofer A, Waberski TD, Nitsche M, Paulus W, Buchner H, Gobbele R (2006) Transcranial direct current stimulation applied over the somatosensory cortex – differential effect on low and high frequency SEPs. Clin Neurophysiol 117(10):2221–2227. https://doi.org/10.1016/j.clinph.2006.07.136

Dietrich S (2004) Earliest historic reference of 'tinnitus' is controversial. J Laryngol Otol 118(7): 487–488. https://doi.org/10.1258/0022215041615182

Dmochowski JP, Datta A, Bikson M, Su Y, Parra LC (2011) Optimized multi-electrode stimulation increases focality and intensity at target. J Neural Eng 8(4):046011. https://doi.org/10.1088/1741-2560/8/4/046011

Dobie RA (1999) A review of randomized clinical trials in tinnitus. Laryngoscope 109(8): 1202–1211. Retrieved from http://www.ncbi.nlm.nih.gov/entrez/query.fcgi?cmd=Retrieve&db=PubMed&dopt=Citation&list_uids=10443820

Dong C, Chen C, Wang T, Gao C, Wang Y, Guan X, Dong X (2020) Low-frequency repetitive transcranial magnetic stimulation for the treatment of chronic tinnitus: a systematic review and meta-analysis of randomized controlled trials. Biomed Res Int 2020:3141278. https://doi.org/10.1155/2020/3141278

Donnell A, Nascimento TD, Lawrence M, Gupta V, Zieba T, Truong DQ et al (2015) High-definition and non-invasive brain modulation of pain and motor dysfunction in chronic TMD. Brain Stimul. https://doi.org/10.1016/j.brs.2015.06.008

Donovan C, Sweet J, Eccher M, Megerian C, Semaan M, Murray G, Miller J (2015) Deep brain stimulation of Heschl gyrus: implantation technique, intraoperative localization, and effects of stimulation. Neurosurgery 77(6):940–947. https://doi.org/10.1227/NEU.0000000000000969

Dymond AM, Coger RW, Serafetinides EA (1975) Intracerebral current levels in man during electrosleep therapy. Biol Psychiatry 10(1):101–104. Retrieved from http://www.ncbi.nlm.nih.gov/entrez/query.fcgi?cmd=Retrieve&db=PubMed&dopt=Citation&list_uids=1120172

Eguiluz VM, Chialvo DR, Cecchi GA, Baliki M, Apkarian AV (2005) Scale-free brain functional networks. Phys Rev Lett 94(1):018102. Retrieved from http://www.ncbi.nlm.nih.gov/entrez/query.fcgi?cmd=Retrieve&db=PubMed&dopt=Citation&list_uids=15698136

Eichhammer P, Langguth B, Marienhagen J, Kleinjung T, Hajak G (2003a) Neuronavigated repetitive transcranial magnetic stimulation in patients with tinnitus: a short case series. Biol Psychiatry 54(8):862–865. Retrieved from http://www.ncbi.nlm.nih.gov/entrez/query.fcgi?cmd=Retrieve&db=PubMed&dopt=Citation&list_uids=14550687

Eichhammer P, Langguth B, Marienhagen J, Kleinjung T, Hajak G (2003b) Neuronavigated repetitive transcranial magnetic stimulation in patients with tinnitus: a short case series. Biol Psychiatry 54:862–865

Elgoyhen AB, Langguth B, De Ridder D, Vanneste S (2015) Tinnitus: perspectives from human neuroimaging. Nat Rev Neurosci 16(10):632–642. https://doi.org/10.1038/nrn4003

Engineer ND, Riley JR, Seale JD, Vrana WA, Shetake JA, Sudanagunta SP et al (2011) Reversing pathological neural activity using targeted plasticity. Nature 470(7332):101–104. https://doi.org/10.1038/nature09656

Fertonani A, Pirulli C, Miniussi C (2011) Random noise stimulation improves neuroplasticity in perceptual learning. J Neurosci 31(43):15416–15423. https://doi.org/10.1523/JNEUROSCI.2002-11.2011

Fertonani A, Ferrari C, Miniussi C (2015) What do you feel if I apply transcranial electric stimulation? Safety, sensations and secondary induced effects. Clin Neurophysiol. https://doi.org/10.1016/j.clinph.2015.03.015

Fields H (2004) State-dependent opioid control of pain. Nat Rev Neurosci 5(7):565–575. https://doi.org/10.1038/nrn1431

Fleming MK, Sorinola IO, Newham DJ, Roberts-Lewis SF, Bergmann JH (2012) The effect of coil type and navigation on the reliability of transcranial magnetic stimulation. IEEE Trans Neural Syst Rehabil Eng 20(5):617–625. https://doi.org/10.1109/TNSRE.2012.2202692

Fodstad H, Hariz M (2007) Electricity in the treatment of nervous system disease. Acta Neurochir Suppl 97(Pt 1):11–19. Retrieved from http://www.ncbi.nlm.nih.gov/entrez/query.fcgi?cmd=Retrieve&db=PubMed&dopt=Citation&list_uids=17691352

Fornito A, Bullmore ET (2014) Connectomics: a new paradigm for understanding brain disease. Eur Neuropsychopharmacol. https://doi.org/10.1016/j.euroneuro.2014.02.011

Fornito A, Zalesky A, Breakspear M (2015) The connectomics of brain disorders. Nat Rev Neurosci 16(3):159–172. https://doi.org/10.1038/nrn3901
Fox MD, Halko MA, Eldaief MC, Pascual-Leone A (2012) Measuring and manipulating brain connectivity with resting state functional connectivity magnetic resonance imaging (fcMRI) and transcranial magnetic stimulation (TMS). Neuroimage 62(4):2232–2243. https://doi.org/10.1016/j.neuroimage.2012.03.035
Fox MD, Buckner RL, Liu H, Chakravarty MM, Lozano AM, Pascual-Leone A (2014) Resting-state networks link invasive and noninvasive brain stimulation across diverse psychiatric and neurological diseases. Proc Natl Acad Sci U S A 111(41):E4367–E4375. https://doi.org/10.1073/pnas.1405003111
Frank E, Schecklmann M, Landgrebe M, Burger J, Kreuzer P, Poeppl TB et al (2011) Treatment of chronic tinnitus with repeated sessions of prefrontal transcranial direct current stimulation: outcomes from an open-label pilot study. J Neurol. https://doi.org/10.1007/s00415-011-6189-4
Frank E, Schecklmann M, Landgrebe M, Burger J, Kreuzer P, Poeppl TB et al (2012) Treatment of chronic tinnitus with repeated sessions of prefrontal transcranial direct current stimulation: outcomes from an open-label pilot study. J Neurol 259(2):327–333. Retrieved from PM:21808984
Freeman WJ (2003) The wave packet: an action potential for the 21st century. J Integr Neurosci 2(1): 3–30. Retrieved from http://www.ncbi.nlm.nih.gov/entrez/query.fcgi?cmd=Retrieve&db=PubMed&dopt=Citation&list_uids=15011274
Freeman WJ (2005) A field-theoretic approach to understanding scale-free neocortical dynamics. Biol Cybern 92(6):350–359. https://doi.org/10.1007/s00422-005-0563-1
Freeman WJ, Rogers LJ (2002) Fine temporal resolution of analytic phase reveals episodic synchronization by state transitions in gamma EEGs. J Neurophysiol 87(2):937–945. Retrieved from http://www.ncbi.nlm.nih.gov/entrez/query.fcgi?cmd=Retrieve&db=PubMed&dopt=Citation&list_uids=11826058
Freeman WJ, Kozma R, Werbos PJ (2001) Biocomplexity: adaptive behavior in complex stochastic dynamical systems. Biosystems 59(2):109–123. Retrieved from http://www.ncbi.nlm.nih.gov/pubmed/11267739
Fregni F, Nitsche MA, Loo CK, Brunoni AR, Marangolo P, Leite J et al (2015) Regulatory considerations for the clinical and research use of transcranial direct current stimulation (tDCS): review and recommendations from an expert panel. Clin Res Regul Aff 32(1):22–35. https://doi.org/10.3109/10601333.2015.980944
Friedland DR, Gaggl W, Runge-Samuelson C, Ulmer JL, Kopell BH (2007) Feasibility of auditory cortical stimulation for the treatment of tinnitus. Otol Neurotol 28(8):1005–1012. Retrieved from http://www.ncbi.nlm.nih.gov/entrez/query.fcgi?cmd=Retrieve&db=PubMed&dopt=Citation&list_uids=18043428
Friston KJ (2011) Functional and effective connectivity: a review. Brain Connect 1(1):13–36. https://doi.org/10.1089/brain.2011.0008
Fuchs E, Flugge G (2014) Adult neuroplasticity: more than 40 years of research. Neural Plast 2014:541870. https://doi.org/10.1155/2014/541870
Fuggetta G, Pavone EF, Fiaschi A, Manganotti P (2008) Acute modulation of cortical oscillatory activities during short trains of high-frequency repetitive transcranial magnetic stimulation of the human motor cortex: a combined EEG and TMS study. Hum Brain Mapp 29(1):1–13. https://doi.org/10.1002/hbm.20371
Fujiyama H, Hyde J, Hinder MR, Kim SJ, McCormack GH, Vickers JC, Summers JJ (2014) Delayed plastic responses to anodal tDCS in older adults. Front Aging Neurosci 6:115. https://doi.org/10.3389/fnagi.2014.00115
Galvani L (1791) De viribus electricitatis in motu muscolari animalium. Bologna
Galvani L (1797) Memorie sull'elettricita' animale
Goebel G, Hiller W (1994) The tinnitus questionnaire. A standard instrument for grading the degree of tinnitus. Results of a multicenter study with the tinnitus questionnaire. HNO 42(3):166–172.

Retrieved from http://www.ncbi.nlm.nih.gov/entrez/query.fcgi?cmd=Retrieve&db=PubMed&dopt=Citation&list_uids=8175381

Goerigk S, Cretaz E, Sampaio-Junior B, Vieira ELM, Gattaz W, Klein I et al (2021) Effects of tDCS on neuroplasticity and inflammatory biomarkers in bipolar depression: results from a sham-controlled study. Prog Neuropsychopharmacol Biol Psychiatry 105:110119. https://doi.org/10.1016/j.pnpbp.2020.110119

Goetz SM, Kozyrkov IC, Luber B, Lisanby SH, Murphy DLK, Grill WM, Peterchev AV (2019) Accuracy of robotic coil positioning during transcranial magnetic stimulation. J Neural Eng 16(5):054003. https://doi.org/10.1088/1741-2552/ab2953

Grefkes C, Nowak DA, Wang LE, Dafotakis M, Eickhoff SB, Fink GR (2010) Modulating cortical connectivity in stroke patients by rTMS assessed with fMRI and dynamic causal modeling. Neuroimage 50(1):233–242. https://doi.org/10.1016/j.neuroimage.2009.12.029

Gribkoff VK, Kaczmarek LK (2017) The need for new approaches in CNS drug discovery: why drugs have failed, and what can be done to improve outcomes. Neuropharmacology 120:11–19. https://doi.org/10.1016/j.neuropharm.2016.03.021

Guleyupoglu B, Schestatsky P, Edwards D, Fregni F, Bikson M (2013) Classification of methods in transcranial electrical stimulation (tES) and evolving strategy from historical approaches to contemporary innovations. J Neurosci Methods 219(2):297–311. https://doi.org/10.1016/j.jneumeth.2013.07.016

Hagmann P, Cammoun L, Gigandet X, Meuli R, Honey CJ, Wedeen VJ, Sporns O (2008) Mapping the structural core of human cerebral cortex. PLoS Biol 6(7):e159. https://doi.org/10.1371/journal.pbio.0060159

Hayward G, Mehta MA, Harmer C, Spinks TJ, Grasby PM, Goodwin GM (2007) Exploring the physiological effects of double-cone coil TMS over the medial frontal cortex on the anterior cingulate cortex: an H2(15)O PET study. Eur J Neurosci 25(7):2224–2233. Retrieved from http://www.ncbi.nlm.nih.gov/entrez/query.fcgi?cmd=Retrieve&db=PubMed&dopt=Citation&list_uids=17439499

Hebb D (1949) The organization of behavior; a neuropsychological theory. Wiley-Interscience, New York

Hebel T, Abdelnaim M, Deppe M, Langguth B, Schecklmann M (2020) Attenuation of antidepressive effects of transcranial magnetic stimulation in patients whose medication includes drugs for psychosis. J Psychopharmacol 34(10):1119–1124

Heimrath K, Breitling C, Krauel K, Heinze HJ, Zaehle T (2015) Modulation of pre-attentive spectro-temporal feature processing in the human auditory system by HD-tDCS. Eur J Neurosci 41(12):1580–1586. https://doi.org/10.1111/ejn.12908

Heller AJ (2003) Classification and epidemiology of tinnitus. Otolaryngol Clin North Am 36(2):239–248. Retrieved from http://www.ncbi.nlm.nih.gov/entrez/query.fcgi?cmd=Retrieve&db=PubMed&dopt=Citation&list_uids=12856294

Herpich F, Melnick M, Huxlin K, Tadin D, Agosta S, Battelli L (2015) Transcranial random noise stimulation enhances visual learning in healthy adults. J Vis 15(12):40. https://doi.org/10.1167/15.12.40

Holland J (2014) Complexity. Oxford University Press, Oxford

Hunter AM, Minzenberg MJ, Cook IA, Krantz DE, Levitt JG, Rotstein NM et al (2019) Concomitant medication use and clinical outcome of repetitive transcranial magnetic stimulation (rTMS) treatment of major depressive disorder. Brain Behav 9(5):e01275. https://doi.org/10.1002/brb3.1275

Iyer MB, Mattu U, Grafman J, Lomarev M, Sato S, Wassermann EM (2005) Safety and cognitive effect of frontal DC brain polarization in healthy individuals. Neurology 64(5):872–875. Retrieved from PM:15753425

Jacquemin L, Shekhawat GS, Van de Heyning P, Mertens G, Fransen E, Van Rompaey V et al (2018) Effects of electrical stimulation in tinnitus patients: conventional versus high-definition tDCS. Neurorehabil Neural Repair 32(8):714–723. https://doi.org/10.1177/1545968318787916

Jastreboff PJ (1990) Phantom auditory perception (tinnitus): mechanisms of generation and perception. Neurosci Res 8(4):221–254. Retrieved from http://www.ncbi.nlm.nih.gov/entrez/query.fcgi?cmd=Retrieve&db=PubMed&dopt=Citation&list_uids=2175858

Joos K, De Ridder D, Vanneste S (2015) The differential effect of low- versus high-frequency random noise stimulation in the treatment of tinnitus. Exp Brain Res 233(5):1433–1440. https://doi.org/10.1007/s00221-015-4217-9

Kanai R, Chaieb L, Antal A, Walsh V, Paulus W (2008) Frequency-dependent electrical stimulation of the visual cortex. Curr Biol 18(23):1839–1843. https://doi.org/10.1016/j.cub.2008.10.027

Karuza EA, Thompson-Schill SL, Bassett DS (2016) Local patterns to global architectures: influences of network topology on human learning. Trends Cogn Sci 20(8):629–640. https://doi.org/10.1016/j.tics.2016.06.003

Keck ME, Welt T, Post A, Muller MB, Toschi N, Wigger A et al (2001) Neuroendocrine and behavioral effects of repetitive transcranial magnetic stimulation in a psychopathological animal model are suggestive of antidepressant-like effects. Neuropsychopharmacology 24(4):337–349. https://doi.org/10.1016/S0893-133X(00)00191-3

Keck ME, Welt T, Muller MB, Erhardt A, Ohl F, Toschi N et al (2002) Repetitive transcranial magnetic stimulation increases the release of dopamine in the mesolimbic and mesostriatal system. Neuropharmacology 43(1):101–109. Retrieved from http://www.ncbi.nlm.nih.gov/entrez/query.fcgi?cmd=Retrieve&db=PubMed&dopt=Citation&list_uids=12213264

Keeser D, Meindl T, Bor J, Palm U, Pogarell O, Mulert C et al (2011a) Prefrontal transcranial direct current stimulation changes connectivity of resting-state networks during fMRI. J Neurosci 31(43):15284–15293. https://doi.org/10.1523/JNEUROSCI.0542-11.2011

Keeser D, Padberg F, Reisinger E, Pogarell O, Kirsch V, Palm U et al (2011b) Prefrontal direct current stimulation modulates resting EEG and event-related potentials in healthy subjects: a standardized low resolution tomography (sLORETA) study. Neuroimage 55(2):644–657. https://doi.org/10.1016/j.neuroimage.2010.12.004

Kelloway P (1946) The part played by the electrical fish in the early history of bioelectricity and electrotherapy. Bull Hist Med 20:112–137

Keshner MS (1982) 1/f Noise. Proc IEEE 70(3):212–218

Kesselheim AS, Hwang TJ, Franklin JM (2015) Two decades of new drug development for central nervous system disorders. Nat Rev Drug Discov 14(12):815–816. https://doi.org/10.1038/nrd4793

Kimbrell TA, Little JT, Dunn RT, Frye MA, Greenberg BD, Wassermann EM et al (1999) Frequency dependence of antidepressant response to left prefrontal repetitive transcranial magnetic stimulation (rTMS) as a function of baseline cerebral glucose metabolism. Biol Psychiatry 46(12):1603–1613. Retrieved from http://www.ncbi.nlm.nih.gov/entrez/query.fcgi?cmd=Retrieve&db=PubMed&dopt=Citation&list_uids=10624541

Kimbrell TA, Dunn RT, George MS, Danielson AL, Willis MW, Repella JD et al (2002) Left prefrontal-repetitive transcranial magnetic stimulation (rTMS) and regional cerebral glucose metabolism in normal volunteers. Psychiatry Res 115(3):101–113. Retrieved from http://www.ncbi.nlm.nih.gov/entrez/query.fcgi?cmd=Retrieve&db=PubMed&dopt=Citation&list_uids=12208488

Kleinjung T, Steffens T, Sand P, Murthum T, Hajak G, Strutz J et al (2007) Which tinnitus patients benefit from transcranial magnetic stimulation? Otolaryngol Head Neck Surg 137(4):589–595. https://doi.org/10.1016/j.otohns.2006.12.007

Kong J, Loggia ML, Zyloney C, Tu P, Laviolette P, Gollub RL (2010) Exploring the brain in pain: activations, deactivations and their relation. Pain 148(2):257–267. https://doi.org/10.1016/j.pain.2009.11.008

Kreuzer PM, Landgrebe M, Schecklmann M, Poeppl TB, Vielsmeier V, Hajak G et al (2011) Can temporal repetitive transcranial magnetic stimulation be enhanced by targeting affective components of tinnitus with frontal rTMS? A randomized controlled pilot trial. Front Syst Neurosci 5:88. https://doi.org/10.3389/fnsys.2011.00088

Kreuzer PM, Landgrebe M, Husser O, Resch M, Schecklmann M, Geisreiter F et al (2012) Transcutaneous vagus nerve stimulation: retrospective assessment of cardiac safety in a pilot study. Front Psych 3:70. https://doi.org/10.3389/fpsyt.2012.00070

Kreuzer PM, Landgrebe M, Resch M, Husser O, Schecklmann M, Geisreiter F et al (2014) Feasibility, safety and efficacy of transcutaneous vagus nerve stimulation in chronic tinnitus: an open pilot study. Brain Stimul 7(5):740–747. https://doi.org/10.1016/j.brs.2014.05.003

Kreuzer PM, Poeppl TB, Rupprecht R, Vielsmeier V, Lehner A, Langguth B, Schecklmann M (2017) Individualized repetitive transcranial magnetic stimulation treatment in chronic tinnitus? Front Neurol 8:126. https://doi.org/10.3389/fneur.2017.00126

Kreuzer PM, Poeppl TB, Rupprecht R, Vielsmeier V, Lehner A, Langguth B, Schecklmann M (2019) Daily high-frequency transcranial random noise stimulation of bilateral temporal cortex in chronic tinnitus – a pilot study. Sci Rep 9(1):12274. https://doi.org/10.1038/s41598-019-48686-0

Kuo HI, Bikson M, Datta A, Minhas P, Paulus W, Kuo MF, Nitsche MA (2013) Comparing cortical plasticity induced by conventional and high-definition 4 x 1 ring tDCS: a neurophysiological study. Brain Stimul 6(4):644–648. https://doi.org/10.1016/j.brs.2012.09.010

Lang N, Siebner HR, Ward NS, Lee L, Nitsche MA, Paulus W et al (2005) How does transcranial DC stimulation of the primary motor cortex alter regional neuronal activity in the human brain? Eur J Neurosci 22(2):495–504. https://doi.org/10.1111/j.1460-9568.2005.04233.x

Langguth B, Eichhammer P, Kreutzer A, Maenner P, Marienhagen J, Kleinjung T et al (2006) The impact of auditory cortex activity on characterizing and treating patients with chronic tinnitus – first results from a PET study. Acta Otolaryngol (556):84–88. https://doi.org/10.1080/03655230600895317

Langguth B, Kreuzer PM, Kleinjung T, De Ridder D (2013) Tinnitus: causes and clinical management. Lancet Neurol 12(9):920–930. https://doi.org/10.1016/S1474-4422(13)70160-1

Larson PS, Cheung SW (2011) Deep brain stimulation in area LC controllably triggers auditory phantom percepts. Neurosurgery. https://doi.org/10.1227/NEU.0b013e3182320ab5

Larson PS, Cheung SW (2013) A stroke of silence: tinnitus suppression following placement of a deep brain stimulation electrode with infarction in area LC. J Neurosurg 118(1):192–194. https://doi.org/10.3171/2012.9.JNS12594

Laureys S, Faymonville ME, Degueldre C, Fiore GD, Damas P, Lambermont B et al (2000) Auditory processing in the vegetative state. Brain 123(Pt 8):1589–1601. Retrieved from http://www.ncbi.nlm.nih.gov/entrez/query.fcgi?cmd=Retrieve&db=PubMed&dopt=Citation&list_uids=10908189

Laureys S, Faymonville ME, Peigneux P, Damas P, Lambermont B, Del Fiore G et al (2002) Cortical processing of noxious somatosensory stimuli in the persistent vegetative state. Neuroimage 17(2):732–741

Leaver AM, Renier L, Chevillet MA, Morgan S, Kim HJ, Rauschecker JP (2011) Dysregulation of limbic and auditory networks in tinnitus. Neuron 69(1):33–43. https://doi.org/10.1016/j.neuron.2010.12.002

Lee HY, Choi MS, Chang DS, Cho CS (2017) Combined bifrontal transcranial direct current stimulation and tailor-made notched music training in chronic tinnitus. J Audiol Otol 21(1):22–27. https://doi.org/10.7874/jao.2017.21.1.22

Lefaucheur JP, Andre-Obadia N, Antal A, Ayache SS, Baeken C, Benninger DH et al (2014) Evidence-based guidelines on the therapeutic use of repetitive transcranial magnetic stimulation (rTMS). Clin Neurophysiol 125(11):2150–2206. https://doi.org/10.1016/j.clinph.2014.05.021

Lefaucheur JP, Antal A, Ayache SS, Benninger DH, Brunelin J, Cogiamanian F et al (2017) Evidence-based guidelines on the therapeutic use of transcranial direct current stimulation (tDCS). Clin Neurophysiol 128(1):56–92. https://doi.org/10.1016/j.clinph.2016.10.087

Lefaucheur JP, Aleman A, Baeken C, Benninger DH, Brunelin J, Di Lazzaro V et al (2020) Evidence-based guidelines on the therapeutic use of repetitive transcranial magnetic stimulation (rTMS): an update (2014-2018). Clin Neurophysiol 131(2):474–528. https://doi.org/10.1016/j.clinph.2019.11.002

Lefebvre-Demers M, Doyon N, Fecteau S (2020) Non-invasive neuromodulation for tinnitus: a meta-analysis and modeling studies. Brain Stimul 14(1):113–128. https://doi.org/10.1016/j.brs.2020.11.014

Lehtimaki J, Hyvarinen P, Ylikoski M, Bergholm M, Makela JP, Aarnisalo A et al (2012) Transcutaneous vagus nerve stimulation in tinnitus: a pilot study. Acta Otolaryngol. https://doi.org/10.3109/00016489.2012.750736

Leong SL, Glue P, Manning P, Vanneste S, Lim LJ, Mohan A, De Ridder D (2020) Anterior cingulate cortex implants for alcohol addiction: a feasibility study. Neurotherapeutics. https://doi.org/10.1007/s13311-020-00851-4

Lerner AJ, Wassermann EM, Tamir DI (2019) Seizures from transcranial magnetic stimulation 2012-2016: results of a survey of active laboratories and clinics. Clin Neurophysiol 130(8):1409–1416. https://doi.org/10.1016/j.clinph.2019.03.016

Levine RA (1999) Somatic (craniocervical) tinnitus and the dorsal cochlear nucleus hypothesis. Am J Otolaryngol 20(6):351–362. Retrieved from http://www.ncbi.nlm.nih.gov/entrez/query.fcgi?cmd=Retrieve&db=PubMed&dopt=Citation&list_uids=10609479

Lewis CM, Baldassarre A, Committeri G, Romani GL, Corbetta M (2009) Learning sculpts the spontaneous activity of the resting human brain. Proc Natl Acad Sci U S A 106(41):17558–17563. https://doi.org/10.1073/pnas.0902455106

Liang Z, Yang H, Cheng G, Huang L, Zhang T, Jia H (2020) Repetitive transcranial magnetic stimulation on chronic tinnitus: a systematic review and meta-analysis. BMC Psychiatry 20(1):547. https://doi.org/10.1186/s12888-020-02947-9

Liebetanz D, Koch R, Mayenfels S, Konig F, Paulus W, Nitsche MA (2009) Safety limits of cathodal transcranial direct current stimulation in rats. Clin Neurophysiol 120(6):1161–1167. https://doi.org/10.1016/j.clinph.2009.01.022

Lippold OC, Redfearn JW (1964) Mental changes resulting from the passage of small direct currents through the human brain. Br J Psychiatry 110:768–772. Retrieved from http://www.ncbi.nlm.nih.gov/entrez/query.fcgi?cmd=Retrieve&db=PubMed&dopt=Citation&list_uids=14211693

Litre CF, Theret E, Tran H, Leveque M, Portefaix C, Gierski F et al (2009) Surgical treatment by electrical stimulation of the auditory cortex for intractable tinnitus. Brain Stimul 2(3):132–137. https://doi.org/10.1016/j.brs.2008.11.001

Litre CF, Giersky F, Theret E, Leveque M, Peruzzi P, Rousseaux P (2010) Feasibility of auditory cortical stimulation for the treatment of tinnitus. Three case reports. Neurochirurgie 56(4):303–308. https://doi.org/10.1016/j.neuchi.2010.03.004

Llinas RR, Steriade M (2006) Bursting of thalamic neurons and states of vigilance. J Neurophysiol 95(6):3297–3308. https://doi.org/10.1152/jn.00166.2006

Llinas RR, Ribary U, Jeanmonod D, Kronberg E, Mitra PP (1999) Thalamocortical dysrhythmia: a neurological and neuropsychiatric syndrome characterized by magnetoencephalography. Proc Natl Acad Sci U S A 96(26):15222–15227. Retrieved from http://www.ncbi.nlm.nih.gov/entrez/query.fcgi?cmd=Retrieve&db=PubMed&dopt=Citation&list_uids=10611366

Llinas R, Urbano FJ, Leznik E, Ramirez RR, van Marle HJ (2005) Rhythmic and dysrhythmic thalamocortical dynamics: GABA systems and the edge effect. Trends Neurosci 28(6):325–333. Retrieved from http://www.ncbi.nlm.nih.gov/entrez/query.fcgi?cmd=Retrieve&db=PubMed&dopt=Citation&list_uids=15927689

Londero A, Bonfils P, Lefaucheur JP (2018) Transcranial magnetic stimulation and subjective tinnitus. A review of the literature, 2014-2016. Eur Ann Otorhinolaryngol Head Neck Dis 135(1):51–58. https://doi.org/10.1016/j.anorl.2017.12.001

Lorenz I, Muller N, Schlee W, Hartmann T, Weisz N (2009) Loss of alpha power is related to increased gamma synchronization-A marker of reduced inhibition in tinnitus? Neurosci Lett 453(3):225–228. https://doi.org/10.1016/j.neulet.2009.02.028

Luft CD, Pereda E, Banissy MJ, Bhattacharya J (2014) Best of both worlds: promise of combining brain stimulation and brain connectome. Front Syst Neurosci 8:132. https://doi.org/10.3389/fnsys.2014.00132

Malaguti A, Rossini D, Lucca A, Magri L, Lorenzi C, Pirovano A et al (2011) Role of COMT, 5-HT (1A) , and SERT genetic polymorphisms on antidepressant response to transcranial magnetic stimulation. Depress Anxiety 28(7):568–573. https://doi.org/10.1002/da.20815

Marks KL, Martel DT, Wu C, Basura GJ, Roberts LE, Schvartz-Leyzac KC, Shore SE (2018) Auditory-somatosensory bimodal stimulation desynchronizes brain circuitry to reduce tinnitus in guinea pigs and humans. Sci Transl Med 10(422). https://doi.org/10.1126/scitranslmed.aal3175

Marshall L, Helgadottir H, Molle M, Born J (2006) Boosting slow oscillations during sleep potentiates memory. Nature 444(7119):610–613. https://doi.org/10.1038/nature05278

Massimini M, Ferrarelli F, Huber R, Esser SK, Singh H, Tononi G (2005) Breakdown of cortical effective connectivity during sleep. Science 309(5744):2228–2232. https://doi.org/10.1126/science.1117256

Matsunaga K, Nitsche MA, Tsuji S, Rothwell JC (2004) Effect of transcranial DC sensorimotor cortex stimulation on somatosensory evoked potentials in humans. Clin Neurophysiol 115(2): 456–460. Retrieved from http://www.ncbi.nlm.nih.gov/pubmed/14744588

Mayberg HS, Lozano AM, Voon V, McNeely HE, Seminowicz D, Hamani C et al (2005) Deep brain stimulation for treatment-resistant depression. Neuron 45(5):651–660. Retrieved from http://www.ncbi.nlm.nih.gov/entrez/query.fcgi?cmd=Retrieve&db=PubMed&dopt=Citation&list_uids=15748841

Medeiros LF, de Souza IC, Vidor LP, de Souza A, Deitos A, Volz MS et al (2012) Neurobiological effects of transcranial direct current stimulation: a review. Front Psych 3:110. https://doi.org/10.3389/fpsyt.2012.00110

Meinzer M, Antonenko D, Lindenberg R, Hetzer S, Ulm L, Avirame K et al (2012) Electrical brain stimulation improves cognitive performance by modulating functional connectivity and task-specific activation. J Neurosci 32(5):1859–1866. https://doi.org/10.1523/JNEUROSCI.4812-11.2012

Meinzer M, Lindenberg R, Antonenko D, Flaisch T, Floel A (2013) Anodal transcranial direct current stimulation temporarily reverses age-associated cognitive decline and functional brain activity changes. J Neurosci 33(30):12470–12478. https://doi.org/10.1523/JNEUROSCI.5743-12.2013

Melloni L, Molina C, Pena M, Torres D, Singer W, Rodriguez E (2007) Synchronization of neural activity across cortical areas correlates with conscious perception. J Neurosci 27(11): 2858–2865. Retrieved from http://www.ncbi.nlm.nih.gov/entrez/query.fcgi?cmd=Retrieve&db=PubMed&dopt=Citation&list_uids=17360907

Mishima T, Nagai T, Yahagi K, Akther S, Oe Y, Monai H et al (2019) Transcranial direct current stimulation (tDCS) induces adrenergic receptor-dependent microglial morphological changes in mice. eNeuro 6(5). https://doi.org/10.1523/ENEURO.0204-19.2019

Mohan A, De Ridder D, Vanneste S (2017) Robustness and dynamicity of functional networks in phantom sound. Neuroimage 146:171–187. https://doi.org/10.1016/j.neuroimage.2016.04.033

Mohan A, Alexandra SJ, Johnson CV, De Ridder D, Vanneste S (2018a) Effect of distress on transient network dynamics and topological equilibrium in phantom sound perception. Prog Neuropsychopharmacol Biol Psychiatry 84(Pt A):79–92. https://doi.org/10.1016/j.pnpbp.2018.01.025

Mohan A, De Ridder D, Idiculla R, C DS, Vanneste S (2018b) Distress-dependent temporal variability of regions encoding domain-specific and domain-general behavioral manifestations of phantom percepts. Eur J Neurosci 48(2):1743–1764. https://doi.org/10.1111/ejn.13988

Mohsen S, Mahmoudian S, Talebian S, Pourbakht A (2018) Prefrontal and auditory tRNS in sequence for treating chronic tinnitus: a modified multisite protocol. Brain Stimul 11(5): 1177–1179. https://doi.org/10.1016/j.brs.2018.04.018

Mohsen S, Mahmoudian S, Talebian S, Pourbakht A (2019) Multisite transcranial random noise stimulation (tRNS) modulates the distress network activity and oscillatory powers in subjects with chronic tinnitus. J Clin Neurosci 67:178–184. https://doi.org/10.1016/j.jocn.2019.06.033

Moliadze V, Antal A, Paulus W (2010) Boosting brain excitability by transcranial high frequency stimulation in the ripple range. J Physiol 588(Pt 24):4891–4904. https://doi.org/10.1113/jphysiol.2010.196998

Moliadze V, Atalay D, Antal A, Paulus W (2012) Close to threshold transcranial electrical stimulation preferentially activates inhibitory networks before switching to excitation with higher intensities. Brain Stimul 5(4):505–511. https://doi.org/10.1016/j.brs.2011.11.004

Momi D, Ozdemir RA, Tadayon E, Boucher P, Shafi MM, Pascual-Leone A, Santarnecchi E (2020) Network-level macroscale structural connectivity predicts propagation of transcranial magnetic stimulation. Neuroimage 229:117698. https://doi.org/10.1016/j.neuroimage.2020.117698

Monai H, Hirase H (2018) Astrocytes as a target of transcranial direct current stimulation (tDCS) to treat depression. Neurosci Res 126:15–21. https://doi.org/10.1016/j.neures.2017.08.012

Moreno-Duarte I, Gebodh N, Schestatsky P, Guleyupoglu B, Reato D, Bikson M, Fregni F (2014) Transcranial electrical stimulation: transcranial direct current stimulation (tDCS), transcranial alternating current stimulation (tACS), transcranial pulsed current stimulation (tPCS), and transcranial random noise stimulation (tRNS). In: Kadosh RC (ed) The stimulated brain: cognitive enhancement using non-invasive brain stimulation. Academia Press, London

Mulquiney PG, Hoy KE, Daskalakis ZJ, Fitzgerald PB (2011) Improving working memory: exploring the effect of transcranial random noise stimulation and transcranial direct current stimulation on the dorsolateral prefrontal cortex. Clin Neurophysiol 122(12):2384–2389. https://doi.org/10.1016/j.clinph.2011.05.009

Nikolin S, Loo CK, Bai S, Dokos S, Martin DM (2015) Focalised stimulation using high definition transcranial direct current stimulation (HD-tDCS) to investigate declarative verbal learning and memory functioning. Neuroimage 117:11–19. https://doi.org/10.1016/j.neuroimage.2015.05.019

Nitsche MA, Paulus W (2001) Sustained excitability elevations induced by transcranial DC motor cortex stimulation in humans. Neurology 57(10):1899–1901. Retrieved from http://www.ncbi.nlm.nih.gov/pubmed/11723286

Nitsche MA, Cohen LG, Wassermann EM, Priori A, Lang N, Antal A et al (2008) Transcranial direct current stimulation: state of the art 2008. Brain Stimul 1(3):206–223. https://doi.org/10.1016/j.brs.2008.06.004

Nitsche MA, Kuo M, Paulus W, Antal A (2015) Transcranial direct current stimulation: protocols and physiological mechanisms of action. In: Knotkova H, Rasche D (eds) Textbook of neuromodulation: principles, methods and clinical application. Springer, New York

Nowogrodzki A (2018) The world's strongest MRI machines are pushing human imaging to new limits. Nature 563(7729):24–26. https://doi.org/10.1038/d41586-018-07182-7

Pascual-Leone A, Grafman J, Hallett M (1994) Modulation of cortical motor output maps during development of implicit and explicit knowledge. Science 263(5151):1287–1289. Retrieved from http://www.ncbi.nlm.nih.gov/entrez/query.fcgi?cmd=Retrieve&db=PubMed&dopt=Citation&list_uids=8122113

Paulus W (2011) Transcranial electrical stimulation (tES – tDCS; tRNS, tACS) methods. Neuropsychol Rehabil 21(5):602–617. https://doi.org/10.1080/09602011.2011.557292

Paus T, Sipila PK, Strafella AP (2001) Synchronization of neuronal activity in the human primary motor cortex by transcranial magnetic stimulation: an EEG study. J Neurophysiol 86(4):1983–1990. Retrieved from http://www.ncbi.nlm.nih.gov/entrez/query.fcgi?cmd=Retrieve&db=PubMed&dopt=Citation&list_uids=11600655

Pena-Gomez C, Sala-Lonch R, Junque C, Clemente IC, Vidal D, Bargallo N et al (2012) Modulation of large-scale brain networks by transcranial direct current stimulation evidenced by resting-state functional MRI. Brain Stimul 5(3):252–263. https://doi.org/10.1016/j.brs.2011.08.006

Plewnia C, Bartels M, Gerloff C (2003) Transient suppression of tinnitus by transcranial magnetic stimulation. Ann Neurol 53(2):263–266. Retrieved from http://www.ncbi.nlm.nih.gov/entrez/query.fcgi?cmd=Retrieve&db=PubMed&dopt=Citation&list_uids=12557296

Plewnia C, Reimold M, Najib A, Brehm B, Reischl G, Plontke SK, Gerloff C (2007) Dose-dependent attenuation of auditory phantom perception (tinnitus) by PET-guided repetitive transcranial magnetic stimulation. Hum Brain Mapp 28(3):238–246. Retrieved from http://www.ncbi.nlm.nih.gov/entrez/query.fcgi?cmd=Retrieve&db=PubMed&dopt=Citation&list_uids=16773635

Polania R, Nitsche MA, Paulus W (2011) Modulating functional connectivity patterns and topological functional organization of the human brain with transcranial direct current stimulation. Hum Brain Mapp 32(8):1236–1249. https://doi.org/10.1002/hbm.21104

Polania R, Nitsche MA, Korman C, Batsikadze G, Paulus W (2012a) The importance of timing in segregated theta phase-coupling for cognitive performance. Curr Biol 22(14):1314–1318. https://doi.org/10.1016/j.cub.2012.05.021

Polania R, Paulus W, Nitsche MA (2012b) Modulating cortico-striatal and thalamo-cortical functional connectivity with transcranial direct current stimulation. Hum Brain Mapp 33(10): 2499–2508. https://doi.org/10.1002/hbm.21380

Poreisz C, Boros K, Antal A, Paulus W (2007) Safety aspects of transcranial direct current stimulation concerning healthy subjects and patients. Brain Res Bull 72(4-6):208–214. https://doi.org/10.1016/j.brainresbull.2007.01.004

Pushpakom S, Iorio F, Eyers PA, Escott KJ, Hopper S, Wells A et al (2019) Drug repurposing: progress, challenges and recommendations. Nat Rev Drug Discov 18(1):41–58. https://doi.org/10.1038/nrd.2018.168

Rabenstein M, Unverricht-Yeboah M, Keuters MH, Pikhovych A, Hucklenbroich J, Vay SU et al (2019) Transcranial current stimulation alters the expression of immune-mediating genes. Front Cell Neurosci 13:461. https://doi.org/10.3389/fncel.2019.00461

Radman T, Ramos RL, Brumberg JC, Bikson M (2009) Role of cortical cell type and morphology in subthreshold and suprathreshold uniform electric field stimulation in vitro. Brain Stimul 2(4): 215–228., 228.e211–213. https://doi.org/10.1016/j.brs.2009.03.007

Ramirez RR, Kopell BH, Butson CR, Gaggl W, Friedland DR, Baillet S (2009) Neuromagnetic source imaging of abnormal spontaneous activity in tinnitus patient modulated by electrical cortical stimulation. Conf Proc IEEE Eng Med Biol Soc 1:1940–1944. https://doi.org/10.1109/IEMBS.2009.5333457

Ranck JB Jr (1975) Which elements are excited in electrical stimulation of mammalian central nervous system: a review. Brain Res 98(3):417–440. doi:0006-8993(75)90364-9 [pii]

Rauschecker JP, Leaver AM, Muhlau M (2010) Tuning out the noise: limbic-auditory interactions in tinnitus. Neuron 66(6):819–826. https://doi.org/10.1016/j.neuron.2010.04.032

Reckow J, Rahman-Filipiak A, Garcia S, Schlaefflin S, Calhoun O, DaSilva AF et al (2018) Tolerability and blinding of 4x1 high-definition transcranial direct current stimulation (HD-tDCS) at two and three milliamps. Brain Stimul 11(5):991–997. https://doi.org/10.1016/j.brs.2018.04.022

Redfearn JW, Lippold OC, Costain R (1964) A preliminary account of the clinical effects of polarizing the brain in certain psychiatric disorders. Br J Psychiatry 110:773–785. Retrieved from http://www.ncbi.nlm.nih.gov/pubmed/14211694

Ridding MC, Rothwell JC (2007) Is there a future for therapeutic use of transcranial magnetic stimulation? Nat Rev Neurosci 8(7):559–567. https://doi.org/10.1038/nrn2169

Romanska A, Rezlescu C, Susilo T, Duchaine B, Banissy MJ (2015) High-frequency transcranial random noise stimulation enhances perception of facial identity. Cereb Cortex 25(11): 4334–4340. https://doi.org/10.1093/cercor/bhv016

Rossini PM, Burke D, Chen R, Cohen LG, Daskalakis Z, Di Iorio R et al (2015) Non-invasive electrical and magnetic stimulation of the brain, spinal cord, roots and peripheral nerves: basic principles and procedures for routine clinical and research application. An updated report from an I.F.C.N. Committee. Clin Neurophysiol 126(6):1071–1107. https://doi.org/10.1016/j.clinph.2015.02.001

Rueger MA, Keuters MH, Walberer M, Braun R, Klein R, Sparing R et al (2012) Multi-session transcranial direct current stimulation (tDCS) elicits inflammatory and regenerative processes in the rat brain. PLoS One 7(8):e43776. https://doi.org/10.1371/journal.pone.0043776

Ruffini G, Fox MD, Ripolles O, Miranda PC, Pascual-Leone A (2014) Optimization of multifocal transcranial current stimulation for weighted cortical pattern targeting from realistic modeling of electric fields. Neuroimage 89:216–225. https://doi.org/10.1016/j.neuroimage.2013.12.002

Ruohonen J, Karhu J (2012) tDCS possibly stimulates glial cells. Clin Neurophysiol 123(10): 2006–2009. https://doi.org/10.1016/j.clinph.2012.02.082

Rush S, Driscoll DA (1968) Current distribution in the brain from surface electrodes. Anesth Analg 47(6):717–723. Retrieved from http://www.ncbi.nlm.nih.gov/entrez/query.fcgi?cmd=Retrieve&db=PubMed&dopt=Citation&list_uids=4972743

Santarnecchi E, Polizzotto NR, Godone M, Giovannelli F, Feurra M, Matzen L et al (2013) Frequency-dependent enhancement of fluid intelligence induced by transcranial oscillatory potentials. Curr Biol 23(15):1449–1453. https://doi.org/10.1016/j.cub.2013.06.022

Schlee W, Hartmann T, Langguth B, Weisz N (2009a) Abnormal resting-state cortical coupling in chronic tinnitus. BMC Neurosci 10:11. https://doi.org/10.1186/1471-2202-10-11

Schlee W, Mueller N, Hartmann T, Keil J, Lorenz I, Weisz N (2009b) Mapping cortical hubs in tinnitus. BMC Biol 7:80. https://doi.org/10.1186/1741-7007-7-80

Schoisswohl S, Agrawal K, Simoes J, Neff P, Schlee W, Langguth B, Schecklmann M (2019) RTMS parameters in tinnitus trials: a systematic review. Sci Rep 9(1):12190. https://doi.org/10.1038/s41598-019-48750-9

Schonfeldt-Lecuona C, Thielscher A, Freudenmann RW, Kron M, Spitzer M, Herwig U (2005) Accuracy of stereotaxic positioning of transcranial magnetic stimulation. Brain Topogr 17(4): 253–259. https://doi.org/10.1007/s10548-005-6033-1

Seidman MD, Ridder DD, Elisevich K, Bowyer SM, Darrat I, Dria J et al (2008) Direct electrical stimulation of Heschl's gyrus for tinnitus treatment. Laryngoscope 118(3):491–500. Retrieved from http://www.ncbi.nlm.nih.gov/entrez/query.fcgi?cmd=Retrieve&db=PubMed&dopt=Citation&list_uids=18094653

Shekhawat GS, Vanneste S (2018) Optimization of transcranial direct current stimulation of dorsolateral prefrontal cortex for tinnitus: a non-linear dose-response effect. Sci Rep 8(1): 8311. https://doi.org/10.1038/s41598-018-26665-1

Shekhawat GS, Searchfield GD, Stinear CM (2014) Randomized trial of transcranial direct current stimulation and hearing aids for tinnitus management. Neurorehabil Neural Repair 28(5): 410–419. https://doi.org/10.1177/1545968313508655

Shekhawat GS, Sundram F, Bikson M, Truong D, De Ridder D, Stinear CM et al (2015) Intensity, duration, and location of high-definition transcranial direct current stimulation for tinnitus relief. Neurorehabil Neural Repair. https://doi.org/10.1177/1545968315595286

Shen K, Hutchison RM, Bezgin G, Everling S, McIntosh AR (2015) Network structure shapes spontaneous functional connectivity dynamics. J Neurosci 35(14):5579–5588. https://doi.org/10.1523/JNEUROSCI.4903-14.2015

Shim HJ, Kwak MY, An YH, Kim DH, Kim YJ, Kim HJ (2015) Feasibility and safety of transcutaneous vagus nerve stimulation paired with notched music therapy for the treatment of chronic tinnitus. J Audiol Otol 19(3):159–167. https://doi.org/10.7874/jao.2015.19.3.159

Shore SE (2005) Multisensory integration in the dorsal cochlear nucleus: unit responses to acoustic and trigeminal ganglion stimulation. Eur J Neurosci 21(12):3334–3348. Retrieved from http://www.ncbi.nlm.nih.gov/entrez/query.fcgi?cmd=Retrieve&db=PubMed&dopt=Citation&list_uids=16026471

Shore SE, Koehler S, Oldakowski M, Hughes LF, Syed S (2008) Dorsal cochlear nucleus responses to somatosensory stimulation are enhanced after noise-induced hearing loss. Eur J Neurosci 27(1): 155–168. https://doi.org/10.1111/j.1460-9568.2007.05983.x

Shore SE, Roberts LE, Langguth B (2016) Maladaptive plasticity in tinnitus – triggers, mechanisms and treatment. Nat Rev Neurol 12(3):150–160. https://doi.org/10.1038/nrneurol.2016.12

Soleimani R, Jalali MM, Hasandokht T (2016) Therapeutic impact of repetitive transcranial magnetic stimulation (rTMS) on tinnitus: a systematic review and meta-analysis. Eur Arch Otorhinolaryngol 273(7):1663–1675. https://doi.org/10.1007/s00405-015-3642-5

Song JJ, Vanneste S, Van de Heyning P, De Ridder D (2012) Transcranial direct current stimulation in tinnitus patients: a systemic review and meta-analysis. ScientificWorldJournal 2012:427941. https://doi.org/10.1100/2012/427941

Speer AM, Kimbrell TA, Wassermann EM, J DR, Willis MW, Herscovitch P, Post RM (2000) Opposite effects of high and low frequency rTMS on regional brain activity in depressed patients. Biol Psychiatry 48(12):1133–1141. Retrieved from http://www.ncbi.nlm.nih.gov/entrez/query.fcgi?cmd=Retrieve&db=PubMed&dopt=Citation&list_uids=11137053

Sporns O, Chialvo DR, Kaiser M, Hilgetag CC (2004) Organization, development and function of complex brain networks. Trends Cogn Sci 8(9):418–425. https://doi.org/10.1016/j.tics.2004.07.008

Stagg CJ, Lin RL, Mezue M, Segerdahl A, Kong Y, Xie J, Tracey I (2013) Widespread modulation of cerebral perfusion induced during and after transcranial direct current stimulation applied to the left dorsolateral prefrontal cortex. J Neurosci 33(28):11425–11431. https://doi.org/10.1523/JNEUROSCI.3887-12.2013

Stefanescu RA, Koehler SD, Shore SE (2015) Stimulus-timing-dependent modifications of rate-level functions in animals with and without tinnitus. J Neurophysiol 113(3):956–970. https://doi.org/10.1152/jn.00457.2014

Steinmetz H, Furst G, Freund HJ (1990) Variation of perisylvian and calcarine anatomic landmarks within stereotaxic proportional coordinates. AJNR Am J Neuroradiol 11(6):1123–1130. Retrieved from http://www.ncbi.nlm.nih.gov/entrez/query.fcgi?cmd=Retrieve&db=PubMed&dopt=Citation&list_uids=2124038

Stramaccia DF, Penolazzi B, Sartori G, Braga M, Mondini S, Galfano G (2015) Assessing the effects of tDCS over a delayed response inhibition task by targeting the right inferior frontal gyrus and right dorsolateral prefrontal cortex. Exp Brain Res 233(8):2283–2290. https://doi.org/10.1007/s00221-015-4297-6

Suk WC, Kim SJ, Chang DS, Lee HY (2018) Characteristics of stimulus intensity in transcutaneous vagus nerve stimulation for chronic tinnitus. J Int Adv Otol 14(2):267–272. https://doi.org/10.5152/iao.2018.3977

Teismann H, Wollbrink A, Okamoto H, Schlaug G, Rudack C, Pantev C (2014) Combining transcranial direct current stimulation and tailor-made notched music training to decrease tinnitus-related distress – a pilot study. PLoS One 9(2):e89904. https://doi.org/10.1371/journal.pone.0089904

Terney D, Chaieb L, Moliadze V, Antal A, Paulus W (2008) Increasing human brain excitability by transcranial high-frequency random noise stimulation. J Neurosci 28(52):14147–14155. https://doi.org/10.1523/JNEUROSCI.4248-08.2008

Thut G, Schyns PG, Gross J (2011) Entrainment of perceptually relevant brain oscillations by non-invasive rhythmic stimulation of the human brain. Front Psychol 2:170. https://doi.org/10.3389/fpsyg.2011.00170

To WT, Ost J, Hart J Jr, De Ridder D, Vanneste S (2017) The added value of auditory cortex transcranial random noise stimulation (tRNS) after bifrontal transcranial direct current stimulation (tDCS) for tinnitus. J Neural Transm (Vienna) 124(1):79–88. https://doi.org/10.1007/s00702-016-1634-2

Tsanov M (2015) Septo-hippocampal signal processing: breaking the code. Prog Brain Res 219:103–120. https://doi.org/10.1016/bs.pbr.2015.04.002

Tyler S, Conto F, Battelli L (2015) Rapid effect of high-frequency tRNS over the parietal lobe during a temporal perceptual learning task. J Vis 15(12):393. https://doi.org/10.1167/15.12.393

Tyler R, Cacace A, Stocking C, Tarver B, Engineer N, Martin J et al (2017) Vagus nerve stimulation paired with tones for the treatment of tinnitus: a prospective randomized double-blind controlled pilot study in humans. Sci Rep 7(1):11960. https://doi.org/10.1038/s41598-017-12178-w

Vanneste S, De Ridder D (2011) Bifrontal transcranial direct current stimulation modulates tinnitus intensity and tinnitus-distress-related brain activity. Eur J Neurosci 34(4):605–614. https://doi.org/10.1111/j.1460-9568.2011.07778.x

Vanneste S, De Ridder D (2012a) The involvement of the left ventrolateral prefrontal cortex in tinnitus: a TMS study. Exp Brain Res 221(3):345–350. https://doi.org/10.1007/s00221-012-3177-6

Vanneste S, De Ridder D (2012b) Differences between a single session and repeated sessions of 1 Hz TMS by double-cone coil prefrontal stimulation for the improvement of tinnitus. Brain Stimul. https://doi.org/10.1016/j.brs.2012.03.019

Vanneste S, De Ridder D (2015) Deafferentation-based pathophysiological differences in phantom sound: Tinnitus with and without hearing loss. Neuroimage 129:80–94. https://doi.org/10.1016/j.neuroimage.2015.12.002

Vanneste S, Plazier M, Ost J, van der Loo E, Van de Heyning P, De Ridder D (2010) Bilateral dorsolateral prefrontal cortex modulation for tinnitus by transcranial direct current stimulation: a preliminary clinical study. Exp Brain Res 202(4):779–785. Retrieved from PM:20186404

Vanneste S, Focquaert F, Van de Heyning P, De Ridder D (2011) Different resting state brain activity and functional connectivity in patients who respond and not respond to bifrontal tDCS for tinnitus suppression. Exp Brain Res 210(2):217–227. https://doi.org/10.1007/s00221-011-2617-z

Vanneste S, van der Loo E, Plazier M, De Ridder D (2012) Parietal double-cone coil stimulation in tinnitus. Exp Brain Res 221(3):337–343. https://doi.org/10.1007/s00221-012-3176-7

Vanneste S, Fregni F, De Ridder D (2013a) Head-to-head comparison of transcranial random noise stimulation, transcranial AC stimulation, and transcranial DC stimulation for tinnitus. Front Psych 4:158. https://doi.org/10.3389/fpsyt.2013.00158

Vanneste S, Walsh V, Van De Heyning P, De Ridder D (2013b) Comparing immediate transient tinnitus suppression using tACS and tDCS: a placebo-controlled study. Exp Brain Res 226(1):25–31. https://doi.org/10.1007/s00221-013-3406-7

Vanneste S, Joos K, Langguth B, To WT, De Ridder D (2014) Neuronal correlates of maladaptive coping: an EEG-study in tinnitus patients. PLoS One 9(2):e88253. https://doi.org/10.1371/journal.pone.0088253

Vanneste S, Faber M, Langguth B, De Ridder D (2016) The neural correlates of cognitive dysfunction in phantom sounds. Brain Res. https://doi.org/10.1016/j.brainres.2016.03.016

Vanneste S, Alsalman O, De Ridder D (2019) Top-down and bottom-up regulated auditory phantom perception. J Neurosci 39(2):364–378. https://doi.org/10.1523/JNEUROSCI.0966-18.2018

Villamar MF, Volz MS, Bikson M, Datta A, Dasilva AF, Fregni F (2013) Technique and considerations in the use of 4x1 ring high-definition transcranial direct current stimulation (HD-tDCS). J Vis Exp 77:e50309. https://doi.org/10.3791/50309

Voroslakos M, Takeuchi Y, Brinyiczki K, Zombori T, Oliva A, Fernandez-Ruiz A et al (2018) Direct effects of transcranial electric stimulation on brain circuits in rats and humans. Nat Commun 9(1):483. https://doi.org/10.1038/s41467-018-02928-3

Vossen A, Gross J, Thut G (2015) Alpha power increase after transcranial alternating current stimulation at alpha frequency (alpha-tACS) reflects plastic changes rather than entrainment. Brain Stimul 8(3):499–508. https://doi.org/10.1016/j.brs.2014.12.004

Wagner T, Fregni F, Fecteau S, Grodzinsky A, Zahn M, Pascual-Leone A (2007) Transcranial direct current stimulation: a computer-based human model study. Neuroimage 35(3):1113–1124. https://doi.org/10.1016/j.neuroimage.2007.01.027

Walsh V, Rushworth M (1999) A primer of magnetic stimulation as a tool for neuropsychology. Neuropsychologia 37(2):125–135. Retrieved from http://www.ncbi.nlm.nih.gov/entrez/query.fcgi?cmd=Retrieve&db=PubMed&dopt=Citation&list_uids=10080370

Wang H, Wang X, Scheich H (1996) LTD and LTP induced by transcranial magnetic stimulation in auditory cortex. Neuroreport 7(2):521–525. Retrieved from http://www.ncbi.nlm.nih.gov/entrez/query.fcgi?cmd=Retrieve&db=PubMed&dopt=Citation&list_uids=8730820

Wang H, Wang X, Wetzel W, Scheich H (1999) Rapid-rate transcranial magnetic stimulation in auditory cortex induces LTP and LTD and impairs discrimination learning of frequency-modulated tones. Electroencephalogr Clin Neurophysiol Suppl 51:361–367. Retrieved from http://www.ncbi.nlm.nih.gov/pubmed/10590969

Wang TC, Tyler RS, Chang TY, Chen JC, Lin CD, Chung HK, Tsou YA (2018a) Effect of transcranial direct current stimulation in patients with tinnitus: a meta-analysis and systematic review. Ann Otol Rhinol Laryngol 127(2):79–88. https://doi.org/10.1177/0003489417744317

Wang Y, Zhang JN, Hu W, Li JJ, Zhou JX, Zhang JP et al (2018b) The characteristics of cognitive impairment in subjective chronic tinnitus. Brain Behav 8(3):e00918. https://doi.org/10.1002/brb3.918

Weber MJ, Messing SB, Rao H, Detre JA, Thompson-Schill SL (2014) Prefrontal transcranial direct current stimulation alters activation and connectivity in cortical and subcortical reward systems: a tDCS-fMRI study. Hum Brain Mapp 35(8):3673–3686. https://doi.org/10.1002/hbm.22429

Weisz N, Hartmann T, Dohrmann K, Schlee W, Norena A (2006) High-frequency tinnitus without hearing loss does not mean absence of deafferentation. Hear Res 222(1-2):108–114. https://doi.org/10.1016/j.heares.2006.09.003

Weisz N, Muller S, Schlee W, Dohrmann K, Hartmann T, Elbert T (2007) The neural code of auditory phantom perception. J Neurosci 27(6):1479–1484. Retrieved from http://www.ncbi.nlm.nih.gov/entrez/query.fcgi?cmd=Retrieve&db=PubMed&dopt=Citation&list_uids=17287523

Whittington MA, Traub RD, Adams NE (2018) A future for neuronal oscillation research. Brain Neurosci Adv 2:2398212818794827. https://doi.org/10.1177/2398212818794827

Wichova H, Alvi SA, Shew M, Lin J, Sale K, Larsen C, Staecker H (2018) Tinnitus perception in patients after vagal nerve stimulator implantation for epilepsy. Am J Otolaryngol 39(5):599–602. https://doi.org/10.1016/j.amjoto.2018.07.009

Witkowski M, Garcia-Cossio E, Chander BS, Braun C, Birbaumer N, Robinson SE, Soekadar SR (2015) Mapping entrained brain oscillations during transcranial alternating current stimulation (tACS). Neuroimage. https://doi.org/10.1016/j.neuroimage.2015.10.024

Ylikoski J, Lehtimaki J, Pirvola U, Makitie A, Aarnisalo A, Hyvarinen P, Ylikoski M (2017) Non-invasive vagus nerve stimulation reduces sympathetic preponderance in patients with tinnitus. Acta Otolaryngol 137(4):426–431. https://doi.org/10.1080/00016489.2016.1269197

Yokley BH, Hartman M, Slusher BS (2017) Role of academic drug discovery in the quest for new CNS therapeutics. ACS Chem Nerosci 8(3):429–431. https://doi.org/10.1021/acschemneuro.7b00040

Yuan T, Yadollahpour A, Salgado-Ramirez J, Robles-Camarillo D, Ortega-Palacios R (2018) Transcranial direct current stimulation for the treatment of tinnitus: a review of clinical trials and mechanisms of action. BMC Neurosci 19(1):66. https://doi.org/10.1186/s12868-018-0467-3

Yue L, Xiao-lin H, Tao S (2009) The effects of chronic repetitive transcranial magnetic stimulation on glutamate and gamma-aminobutyric acid in rat brain. Brain Res 1260:94–99. https://doi.org/10.1016/j.brainres.2009.01.009

Zaehle T, Rach S, Herrmann CS (2010) Transcranial alternating current stimulation enhances individual alpha activity in human EEG. PLoS One 5(11):e13766. https://doi.org/10.1371/journal.pone.0013766

Zaehle T, Beretta M, Jancke L, Herrmann CS, Sandmann P (2011) Excitability changes induced in the human auditory cortex by transcranial direct current stimulation: direct electrophysiological evidence. Exp Brain Res 215(2):135–140. https://doi.org/10.1007/s00221-011-2879-5

Zaghi S, Acar M, Hultgren B, Boggio PS, Fregni F (2010a) Noninvasive brain stimulation with low-intensity electrical currents: putative mechanisms of action for direct and alternating current stimulation. Neuroscientist 16(3):285–307. https://doi.org/10.1177/1073858409336227

Zaghi S, de Freitas Rezende L, de Oliveira LM, El-Nazer R, Menning S, Tadini L, Fregni F (2010b) Inhibition of motor cortex excitability with 15Hz transcranial alternating current stimulation (tACS). Neurosci Lett 479(3):211–214. https://doi.org/10.1016/j.neulet.2010.05.060

Zago S, Ferrucci R, Fregni F, Priori A (2008) Bartholow, Sciamanna, Alberti: pioneers in the electrical stimulation of the exposed human cerebral cortex. Neuroscientist 14(5):521–528. https://doi.org/10.1177/1073858407311101

Zeman F, Koller M, Figueiredo R, Aazevedo A, Rates M, Coelho C et al (2011) Tinnitus handicap inventory for evaluating treatment effects: which changes are clinically relevant? Otolaryngol Head Neck Surg 145(2):282–287. https://doi.org/10.1177/0194599811403882

Zhang K, Guo L, Zhang J, Rui G, An G, Zhou Y et al (2020) tDCS accelerates the rehabilitation of MCAO-induced motor function deficits via neurogenesis modulated by the notch1 signaling pathway. Neurorehabil Neural Repair 34(7):640–651. https://doi.org/10.1177/1545968320925474

Bimodal Auditory Electrical Stimulation for the Treatment of Tinnitus: Preclinical and Clinical Studies

Travis L. Riffle, David T. Martel, Gerilyn R. Jones, and Susan E. Shore

Contents

Abstract Tinnitus, or the phantom perception of sound, arises from pathological neural activity. Neurophysiological research has shown increased spontaneous firing rates and synchronization along the auditory pathway correlate strongly with behavioral measures of tinnitus. Auditory neurons are plastic, enabling external stimuli to be utilized to elicit long-term changes to spontaneous firing and synchrony. Pathological plasticity can thus be reversed using bimodal auditory plus nonauditory stimulation to reduce tinnitus. This chapter discusses preclinical and clinical evidence for efficacy of bimodal stimulation treatments of tinnitus, with highlights on sham-controlled, double-blinded clinical trials. The results from these studies have shown some efficacy in reducing the severity of tinnitus, based on subjective and objective outcome measures including tinnitus questionnaires and psychophysical tinnitus measurements. While results of some studies have been positive, the degree

T. L. Riffle, D. T. Martel, and G. R. Jones
Department of Otolaryngology, University of Michigan, Ann Arbor, MI, USA

S. E. Shore (✉)
Department of Otolaryngology, University of Michigan, Ann Arbor, MI, USA

Department of Biomedical Engineering, University of Michigan, Ann Arbor, MI, USA

Departments of Molecular and Integrative Physiology, University of Michigan, Ann Arbor, MI, USA

Kresge Hearing Research Institute, Ann Arbor, MI, USA
e-mail: sushore@umich.edu

Curr Topics Behav Neurosci (2021) 51: 295–324
https://doi.org/10.1007/7854_2020_180
Published Online: 21 October 2020

of benefit and the populations that respond to treatment vary across the studies. Directions and implications of future studies are discussed.

Keywords Auditory-somatosensory · Bimodal · Clinical trial · Plasticity · Tinnitus

1 Introduction

Tinnitus is the phantom perception of sound commonly referred to as "ringing" in the ears. Nearly 50 million people in the United States alone have tinnitus (Shargorodsky et al. 2010), which can be devastating to many sufferers. While commonly associated with and triggered by hearing loss (Eggermont and Roberts 2004), tinnitus can occur in participants with clinically normal audiograms (Schaette and McAlpine 2011; Gu et al. 2012), suggesting that the disorder arises from pathological neural activity in the central auditory pathway (Shore et al. 2016). Mounting evidence, from many labs, implicates pathologically altered homeostatic plasticity in tinnitus generation (Eggermont and Roberts 2004; Roberts et al. 2010; Wu et al. 2016b).

This chapter first discusses the role of neural plasticity in generating tinnitus as demonstrated in animal model preclinical studies. We will then highlight peer-reviewed, translational studies emanating from these rigorous preclinical studies to demonstrate how counteracting pathological plasticity can reduce and, in some cases, eliminate tinnitus. Detailed studies of spike-timing-dependent plasticity (STDP) in the cochlear nucleus (CN) and altered firing patterns in the auditory cortex (AC) have led to novel treatment paradigms based on reversing the pathological neural plasticity discovered in animal models. Each approach combines auditory and nonauditory stimulation. In the CN plasticity-derived treatment, auditory stimulation is combined with electrical stimulation of the somatosensory system in a precise temporal relationship to reduce aberrant STDP. In the AC plasticity-derived treatment, auditory stimulation is paired with vagal nerve stimulation (VNS) to normalize rearranged cortical maps. Other types of bimodal stimulation are also briefly discussed.

1.1 Animal Model Development of Tinnitus Treatments

1.1.1 Dorsal Cochlear Nucleus Plasticity as a Basis for Tinnitus Treatment

Dorsal Cochlear Nucleus Fusiform Cells Integrate Multisensory Information

The CN is the first nucleus in the auditory pathway and integrates auditory nerve input from the cochlea (Fig. 1) with somatosensory input from the dorsal column and trigeminal systems (Itoh et al. 1987; Kanold and Young 2001; Shore et al. 2003; Zhou and Shore 2006). Fusiform cells, the principle output neurons of the dorsal cochlear nucleus (DCN), receive auditory nerve input on their basal dendrites and

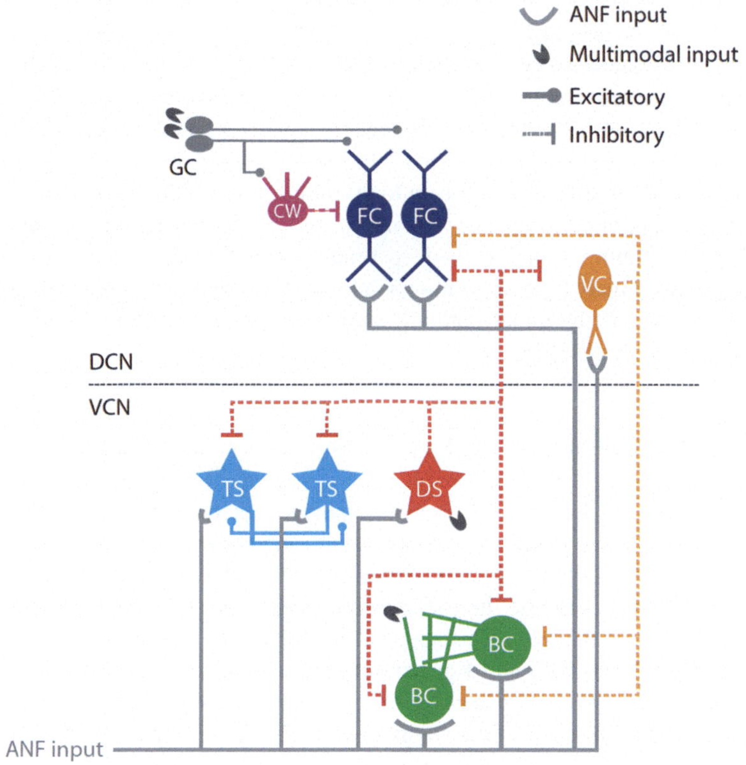

Fig. 1 Schematic of cochlear nucleus circuitry involved in tinnitus. Auditory nerve fibers (ANF) from the cochlea synapse on the cell bodies of bushy cells (BCs), dendrites of T-stellate (TS) cells and vertical cells (VCs), cell bodies and dendrites of D-stellate (DS) cells, and basal dendrites of fusiform cells (FCs). Multimodal auditory and somatosensory projections synapse on the dendritic fields of DS cells and BCs in the ventral cochlear nucleus (VCN) and granule cells (GCs). The axons of GCs terminate on the apical dendrites of FCs and inhibitory cartwheel cells (CW). DS cells provide wideband inhibition, and VCs provide narrowband inhibition to the output neurons of the VCN and DCN (from Shore and Wu 2019)

indirectly receive somatosensory input via granule cell axons on their apical dendrites (Osen et al. 1995; Fujino and Oertel 2003). The apical dendrite synapses are plastic so that their strength and ability to excite fusiform cells can sum over time, while the basal dendrites are not (Fujino and Oertel 2003).

The fusiform-cell circuit is an example of a cerebellar-like circuit, exhibiting structural and functional similarity to neurons in the cerebellum (Oertel and Young 2004; Tzounopoulos et al. 2004). Integration of auditory and somatosensory input by fusiform cells involves spike-timing-dependent plasticity (STDP), so that the order and time interval between the auditory and somatosensory stimuli determine whether increases or decreases in fusiform-cell excitability will occur (Tzounopoulos et al. 2004; Koehler and Shore 2013b). When the temporal relationship between auditory and somatosensory stimulation increases fusiform-cell excitability, the process is referred to as long-term potentiation (LTP), while decreases in fusiform-cell excitability are termed long-term depression (LTD). To determine whether STDP is primarily comprised of LTP or LTD, the intervals between auditory and somatosensory stimulations are varied, and fusiform-cell firing rate is measured as a percent change in spikes from pre- to post-auditory-somatosensory, or bimodal, stimulation. The percent change in spikes plotted versus the bimodal interval is referred to as a "timing rule." In the guinea pig, evoked LTP or LTD in the fusiform-cell circuit can last for up to 90 min (Shore 2005; Dehmel et al. 2012). Bimodal auditory-somatosensory activation of the fusiform-cell circuit can be elicited by combining sound with deep brain stimulation of somatosensory pathways, or noninvasively by combining sound with transdermal stimulation applied to the face (activating trigeminal pathways) or neck (activating dorsal column pathways) (Wu et al. 2015) (Fig. 2).

Tinnitus-Linked Neural Plasticity First Arises in the Cochlear Nucleus

Noise overexposure leading to cochlear damage elicits homeostatic changes to neural circuitry of the central auditory system, beginning in the CN (Zhang and Kaltenbach 1998; Kaltenbach and Afman 2000) and extending throughout the auditory pathway, including in the inferior colliculus (IC) (Bauer et al. 2008; Sturm et al. 2017), thalamus (Kalappa et al. 2014) and auditory cortex (Norena and Eggermont 2003; Seki and Eggermont 2003; Engineer et al. 2011). However, not all noise overexposure or hearing loss produces tinnitus (Roberts et al. 2010). Behavioral tests to assess tinnitus status in animals are required to understand how altered neural plasticity and electrophysiology produce tinnitus. The most widely used paradigm relies on gap-prepulse inhibition of the acoustic startle reflex (GPIAS) (Turner et al. 2006). GPIAS tests rely on two behaviors. First, animals will startle when presented with a loud sound pulse embedded in a background noise. Second, in normal-hearing animals, this startle is inhibited when a silent gap in the background noise closely precedes the startle pulse. However, in animals with tinnitus that matches the background noise, the tinnitus reduces detectability of the gap, resulting in the animals startling as if the gap were not present. The GPIAS

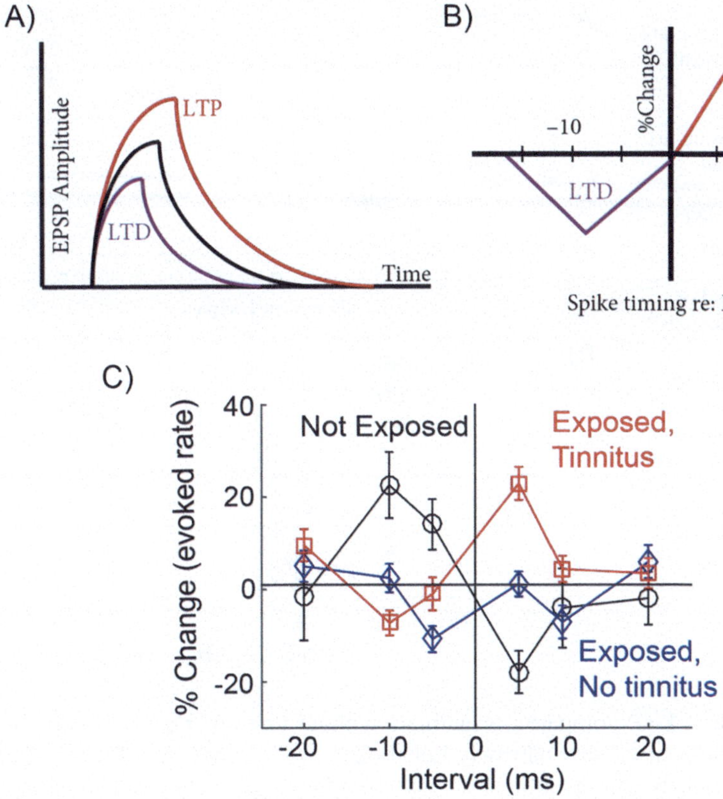

Fig. 2 Spike-timing rules in fusiform cells. (**a**) When apical dendrites in fusiform cells are activated, they can enhance the amplitude of excitatory postsynaptic potentials (EPSPs), termed long-term potentiation (LTP), represented by the red line. Conversely, when the spikes occur prior to apical dendritic activation, this reduces the EPSP amplitude, known as long-term depression (LTD), represented by the purple line. (**b**) Percent change in EPSP amplitude relative to EPSP spike timing is known as a timing rule. Positive values on the *x*-axis represent EPSPs before the spike activation, and negative values represent EPSPs after the spike activation. The *y*-axis represents % change in EPSPs (from Shore and Martel 2019). (**c**) Percent change in fusiform-cell firing rate after auditory-somatosensory bimodal stimulation as a function of bimodal interval. Normal and noise-overexposed guinea pigs with and without tinnitus are compared. Values on the *x*-axis represent the bimodal intervals; negative intervals indicate auditory before somatosensory (transdermal electrical) stimulation, and positive intervals indicate auditory after somatosensory stimulation. Nonexposed control animals (black) had equivalent LTP and LTD; exposed, tinnitus animals (red) had more LTP than LTD, indicating enhanced firing in the neural circuits; exposed, non-tinnitus animals (blue) had more LTD than LTP (from Koehler and Shore 2013a)

paradigm thus allows for tinnitus detection as well as providing a measurement of the animal's tinnitus spectrum.

DCN STDP is significantly altered in animals with noise-induced tinnitus, shifting STDP timing rules toward LTP (Koehler and Shore 2013a). Fusiform-cell timing rules in animals with tinnitus are also broader, meaning that more bimodal

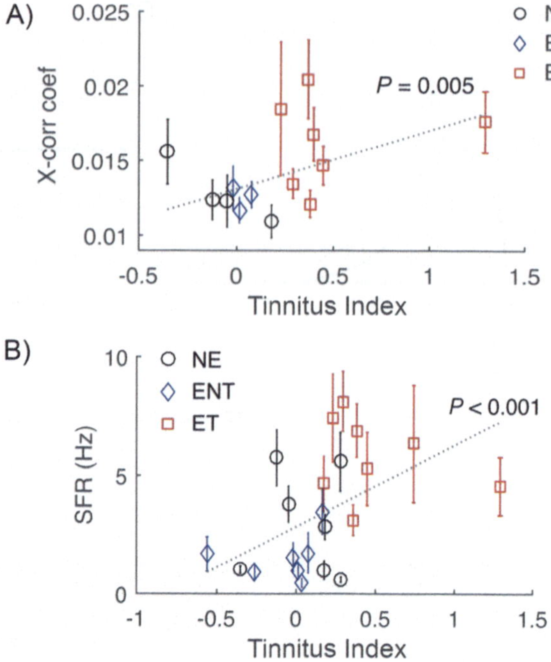

Fig. 3 Spontaneous synchrony and firing rate are increased in animals with tinnitus compared to animals without. Spontaneous synchrony (**a**) and firing rates (**b**) are increased in animals with tinnitus (ET = exposed tinnitus) compared to animals without (NE = noise-exposed or nonexposed sham animals, ENT = exposed non-tinnitus). Tinnitus index is a comparison of pre- and postexposure gap detection: larger tinnitus index values represent more impaired gap detection abilities after noise exposure, i.e., more tinnitus (from Wu et al. 2016a)

stimulus pairs elicit LTP compared to animals without tinnitus. Fusiform cells in animals with tinnitus also show narrowly tuned increases in spontaneous firing rate (SFR) and cross-unit synchrony at neurons with best frequencies close to behavioral tinnitus frequencies (Wu et al. 2016a) (Fig. 3). In contrast, animals without behavioral evidence of tinnitus, as well as nonexposed control animals, do not show increases in SFR or cross-unit synchrony. The fusiform-cell firing patterns in the animals with tinnitus are consistent with human psychophysical tinnitus measurements, in which tinnitus is characterized as tone-like or narrowly tuned bands (Roberts et al. 2006, 2008). Importantly, fusiform-cell increased spontaneous activity can arise as soon as 1 h after noise exposure (Kaltenbach and Afman 2000), suggesting that tinnitus-related increases in spontaneous activity in the auditory pathway first occur in the DCN (Kaltenbach and Zhang 2007; Finlayson and Kaltenbach 2009; Manzoor et al. 2013a), which then propagate through the more central pathways.

Toward a Tinnitus Treatment

As demonstrated in several studies, whether auditory-somatosensory bimodal stimulation elicits primarily LTP or LTD depends on order and the time interval between the auditory and the somatosensory stimulation (Fig. 4) (Koehler and Shore 2013a). The specific auditory-somatosensory order and timing that reduce fusiform-cell

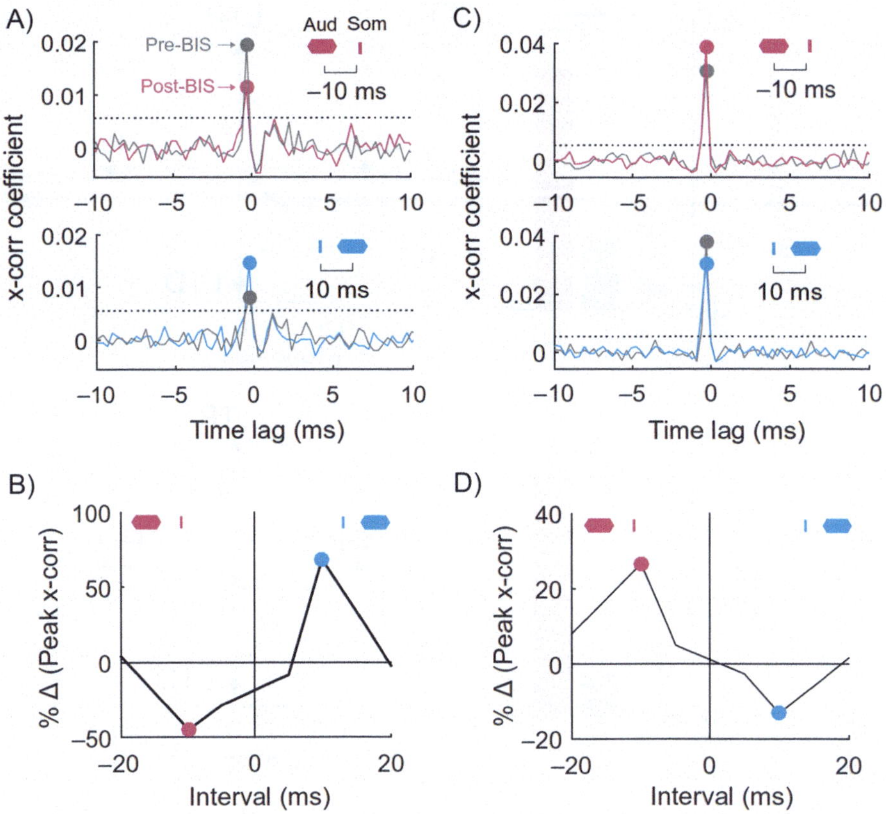

Fig. 4 Bimodal auditory-somatosensory stimulation can induce LTP and LTD in the fusiform-cell circuit. In a representative fusiform-cell pair, the peak cross-correlation coefficient decreased after auditory stimulation occurred 10 ms before somatosensory stimulation (**a** top) but increased when stimuli were presented in the opposite order (**a** bottom). This unit pair exhibited a Hebbian-like learning rule (**b**) where activation of fibers increased neural synchrony. However, in a different fusiform-cell pair (**c** and **d**), the stimulation resulted in opposite anti-Hebbian-like learning rule where the activation weakened neural synchrony (from Wu et al. 2016a, b)

excitability in the DCN through LTD create the opportunity for a potential tinnitus treatment. By reducing fusiform-cell excitability, SFR and cross-unit synchrony will decrease with corresponding reductions in tinnitus behavior.

Marks et al. (2018) used the bimodal auditory-somatosensory stimulation paradigm to develop a treatment for tinnitus based on STDP. The specific temporal sequence chosen to induce LTD was derived from the in vivo recordings in the DCN of guinea pigs. To test this hypothesis, Marks et al. (2018) noise overexposed guinea pigs with a 7 kHz-centered half-octave noise band presented at 97 dB SPL for 2 h, resulting in 72.7% of the guinea pigs having chronic tinnitus after 8 weeks. Animals with tinnitus had significantly increased synchrony ($p < 0.001$) and spontaneous activity ($p < 0.001$) in fusiform cells compared to the normal-hearing animals and

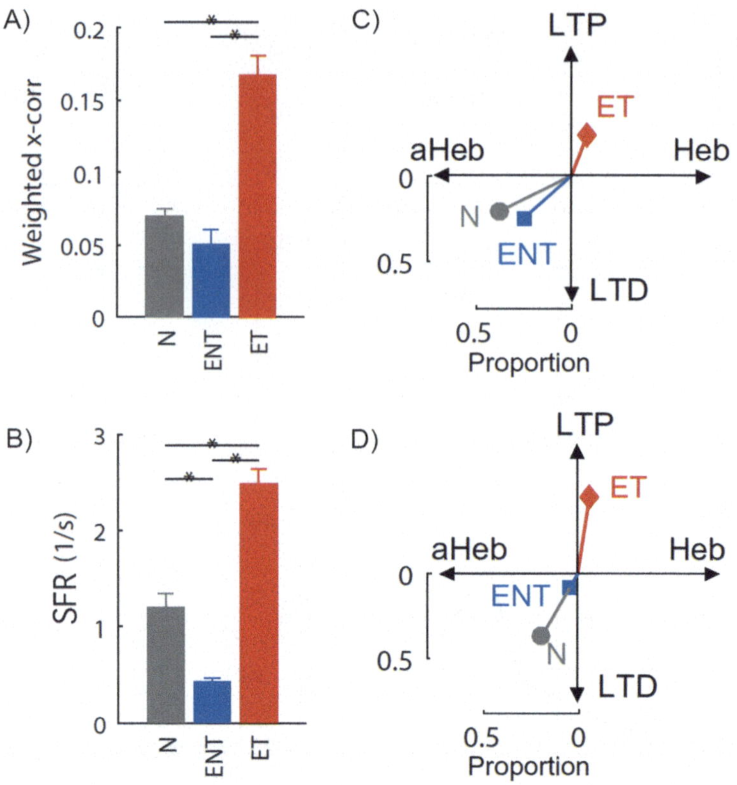

Fig. 5 Tinnitus animals show increased synchrony, SFR, and LTP-only learning rules. In Fig. **a**, following noise overexposure, animals with tinnitus (ET = exposed tinnitus) show increased synchrony via mean cross-correlation coefficient, weighted by the proportion of synchronous unit pairs compared to normal-hearing (N) and exposed-but-no-tinnitus (ENT) animals. Figure **b** shows increased mean spontaneous firing rate (SFR) for ET animals compared to N and ENT animals. Figures **c** and **d** show a shift in proportion of learning rules for ET animals toward Hebbian-like (Heb; *x*-axis) and long-term potentiation (LTP; *y*-axis) for synchrony (**c**) and SFR (**d**). Normal-hearing and ENT animals demonstrated anti-Hebbian-like (aHeb) and long-term depression (LTD) characteristics (from Marks et al. 2018)

noise-exposed non-tinnitus animals (Fig. 5). Additionally, the tinnitus group showed a significantly higher proportion of unit pairs that exhibited LTP-only learning rules compared to the normal-hearing and non-tinnitus animals that displayed LTD-only learning rules ($p = 0.001$) (Fig. 5). These results demonstrated that the neural pathway in animals with tinnitus was prone to exhibiting increased synchrony, spontaneous activity, and LTP compared to control and non-tinnitus animals.

Based on those findings, Marks et al. (2018) then hypothesized that inducing LTD should result in the opposite pattern of neural activity, i.e., decreased SFR, synchrony, and resulting tinnitus. Bimodal auditory-somatosensory stimulation was tested to determine which interval provided the greatest LTD. Transcutaneous electrical stimuli consisted of three 100 μs, biphasic pulses of 2–5 mA, paired with 10 ms pure tones presented at 40 dB SL. The stimuli were randomly tested at six different intervals (±5, ±10, and ±20 ms) with electrical-first or auditory-first ordering. Results showed that auditory stimulus presented 5 ms before electrical stimulation produced the most instances of LTD across DCN fusiform cells. GPIAS behavioral testing revealed that 8 kHz was the most prevalent tinnitus frequency observed in the noise-overexposed guinea pigs. Thus, the stimuli chosen for the treatment consisted of 8 kHz tone bursts and the −5 ms auditory electrical stimulation interval that induced the greatest LTD.

To evaluate this treatment paradigm, tinnitus was induced in guinea pigs by noise overexposure, followed by 25 days of the bimodal treatment for 20 min per day. Three control treatment groups were assessed: unimodal sound stimulation, unimodal somatosensory stimulation, and sedative only. Guinea pigs receiving the specific bimodal auditory-somatosensory stimulation treatment designed to induce LTD showed a significant decrease in tinnitus index scores (differences in gap-startle responses pre- and posttreatment) at the treatment sound frequency. Furthermore, the bimodal treatment group also showed significant reductions in fusiform-cell SFR and cross-unit synchrony associated with tinnitus whereas the sham group and unimodal sound group did not show any decrease in tinnitus index ($p < 0.01$). There were no significant changes in tinnitus in the unimodal auditory stimulation or sedative-only groups, while the animals receiving unimodal electrical stimulation showed increases in tinnitus behavior at some frequencies. Moreover, animals receiving unimodal auditory stimulation and sedative-only groups did not show changes in fusiform-cell SFR and cross-unit synchrony, consistent with a lack of LTP induction (Marks et al. 2018). The unimodal electrical stimulation showed some decreases, but also increases in SFR and synchrony reflecting LTP induction consistent with previous studies showing LTP induction by parallel fiber stimulation (Koehler et al. 2011). These results indicate that the bimodal auditory-somatosensory stimulation paradigm in animals shows promising capability of reducing physiological and behavioral indications of tinnitus.

Noise Overexposure Alters Excitability Throughout the Auditory Pathway

Physiological changes in the auditory pathway are passed on to more central auditory stations culminating in the auditory cortex. Studies have shown that reduced input from the auditory nerve results in homeostatic decreases in inhibitory neurotransmitters (Wang et al. 2009; Middleton et al. 2011) as well as increases in excitatory neurotransmitters (Wang et al. 2009; Barker et al. 2012; Zeng et al. 2012; Heeringa et al. 2018). The increased SFR generated by the CN has been observed in the IC (Manzoor et al. 2013b; Kalappa et al. 2014; Sturm et al. 2017), but tinnitus-specific behavioral and physiological measures are lacking in most studies in the IC. It is likely that changes in the IC related to tinnitus are a result of receiving altered input from the CN's primary output neurons. Since there are no physiologically or morphologically identified cells in the IC that respond to defined inputs and outputs, it is probable that IC cells receiving input predominantly from the CN would mimic the tinnitus phenotype that is generated in the CN (Shore and Wu 2019).

1.1.2 Cortical Plasticity as a Basis for Tinnitus Treatment

Tinnitus has also been proposed to arise as the result of aberrant cortical plasticity due to reduction of auditory input following cochlear damage (Eggermont and Roberts 2004). This theory is akin to the decrease in cortical input associated with phantom limb pain after peripheral amputation (Flor et al. 1995). The resulting cortical changes after cochlear damage include frequency-map reorganization, as well as increases in neural synchrony, spontaneous activity, excitability, and receptive field size (Engineer et al. 2013). Other studies have shown that increased SFR and synchrony are also observed in the medial geniculate body (Sametsky et al. 2015), which correlate with tinnitus behavior, suggesting that the cortex receives the already generated tinnitus-specific input from thalamocortical neurons (Llinás et al. 1999; De Ridder et al. 2015). It is likely that the thalamus itself is receiving already-processed tinnitus neural activity generated in the DCN. Since the thalamocortical output to the cortex is both excitatory and inhibitory, the increased activity from the thalamus likely alters the normal excitatory/inhibitory balance in the cortex (Rauschecker et al. 2010; Hamilton et al. 2013; Natan et al. 2017). These findings suggest that increased activity from subcortical neurons is the driving force behind cortical changes, rather than purely cortically driven homeostatic plasticity arising from decreased input (Shore and Wu 2019).

Treating Tinnitus by Reversing Pathological Cortical Plasticity in Animals

If pathological cortical plasticity contributes to tinnitus, then reversing the plasticity should reduce tinnitus. Previous studies showed that pairing auditory and electrical stimulation of subcortical structures could alter tuning in the auditory cortex through release of acetylcholine from the nucleus basalis (Kilgard and Merzenich 1998) or norepinephrine from the locus coeruleus (Edeline et al. 2011). However, direct stimulation of these structures to drive targeted cortical plasticity is highly invasive, making treatments of this type less practical. Instead, Engineer et al. (2011) developed a somewhat less invasive procedure that utilized vagus nerve stimulation (VNS) paired with sound. In animal studies using rats, a cuff electrode was placed on the left cervical vagus nerve to eliminate cardiac complications. Stimulation parameters were derived from previous VNS studies in rats and in humans suffering from epilepsy, with values set to approximately 1% of the approved levels used to treat depression and seizures (Engineer et al. 2013). In the first experiment, 0.8 mA pulses were delivered via the cuff electrode at a rate of 30 Hz with a pulse duration of 100 μs and were paired simultaneously with tones at 9 kHz or 19 kHz. This VNS-tone combination was presented in 500 ms bursts once approximately every 30 s for 2.5 h (totaling 300 stimulations per day), for 20 consecutive days. Cortical map recordings from the rats showed that the incidence of the primary auditory cortex (A1) neurons with best frequency in the frequency region of the tones was increased 70–79% compared to control animals 24 h after the paired VNS-tone stimulation. This finding demonstrated that paired VNS-tone pairs could be used to drive long-lasting plasticity in the auditory cortex.

A follow-up experiment aimed to examine whether this same mechanism could also be utilized to alleviate tinnitus by reversing pathological changes in the cortex after noise overexposure. A prevailing cortico-centric theory in tinnitus generation is that following cochlear damage, there is an overrepresentation of auditory cortical neurons that are tuned to the tinnitus frequency, which contributes to increased neural synchronization and spontaneous activity that correlates with the perception of tinnitus (Eggermont 2006). Accordingly, if VNS-tone pairs can expand the frequency map in the cortex outside of the tinnitus frequencies by pairing VNS with tones outside of the tinnitus frequency, this would decrease the overrepresented region corresponding to the tinnitus frequency. To test this theory, Engineer et al. (2011) exposed rats to 115 dB SPL noise at 16 kHz and demonstrated a significant increase in spontaneous activity and in synchrony in A1 neurons. GPIAS testing revealed that all noise-overexposed animals showed evidence of tinnitus in at least one frequency band, demonstrating both physiological and behavioral correlates of tinnitus.

The authors then paired VNS with multiple tones outside of the tinnitus frequency range. For the experimental group, these multiple VNS-tone pairs were randomly interleaved and presented 300 times/day using the same parameters from part 1 of the experiment. The controls were animals that received tone-only stimulation, VNS-only stimulation, or no stimulation. After 10 days of therapy, behavioral test results showed improved gap detection at the putative tinnitus frequency for rats receiving the active treatment but did not for rats receiving the sham treatment, indicating a reduction in tinnitus behavior (Fig. 6). These improvements were still observed 3 weeks after the end of the treatment. Furthermore, cortical excitability and synchronization both decreased to control levels, but spontaneous activity was unaffected. Collectively, these results demonstrated that VNS-tone pair therapy could reverse some of the pathophysiological changes in the auditory cortex associated with tinnitus in rats. These animal studies established a foundation in basic neuroscience investigating mechanisms of neuronal plasticity involved in tinnitus. The findings of these studies then informed the development of potential therapeutic treatments for tinnitus in humans.

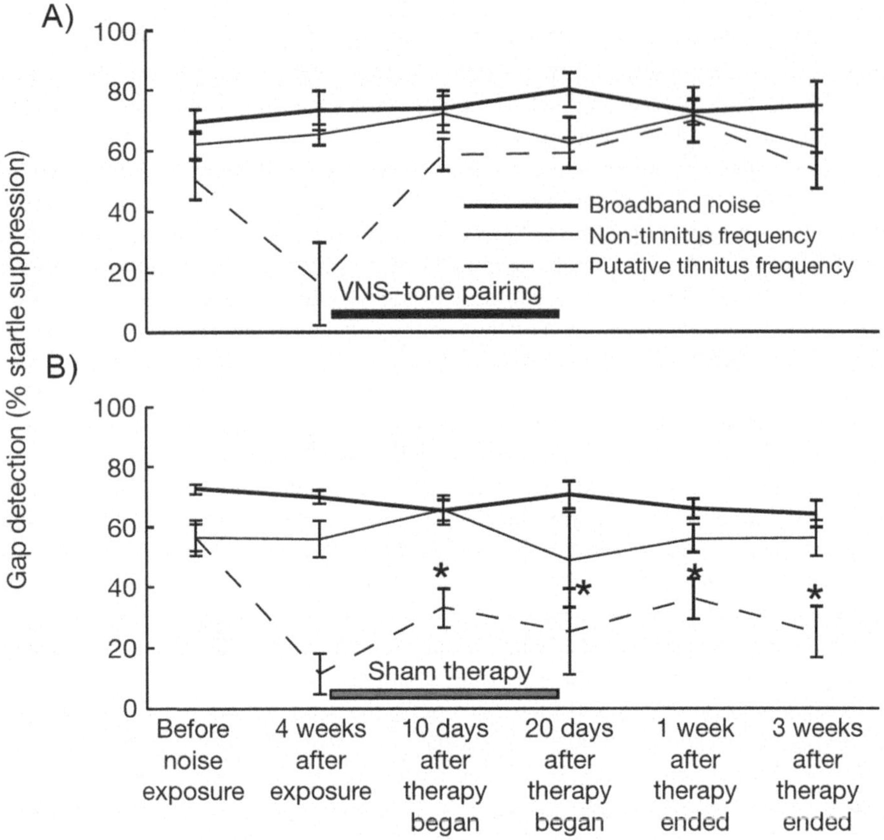

Fig. 6 Animals receiving bimodal VNS-sound stimulation, but not sound alone, showed improvements in tinnitus behavioral measures. Four weeks after noise overexposure, animals in both groups were less likely to detect the gaps at the putative tinnitus frequency compared to gaps in broadband noise or at non-tinnitus frequencies. After 10 days of paired tone-VNS treatment, the experimental group (**a**) gap detection at the putative tinnitus frequency improved to pre-noise exposure levels, whereas the gap detection for the sham group (**b**) did not improve after the sham treatment (from Engineer et al. 2011)

1.2 Translating Animal Studies to Humans

1.2.1 Bimodal Auditory-Somatosensory Stimulation to Reduce Tinnitus in Humans

The improvements in both behavioral and physiological correlates of tinnitus in animal studies were used as the basis for clinical bimodal auditory electrical stimulation experiments in humans. Using the precisely timed stimulation paradigm developed in the preclinical studies, a clinical study was performed in humans (Marks et al. 2018). Since the auditory portion of the bimodal stimulation in guinea

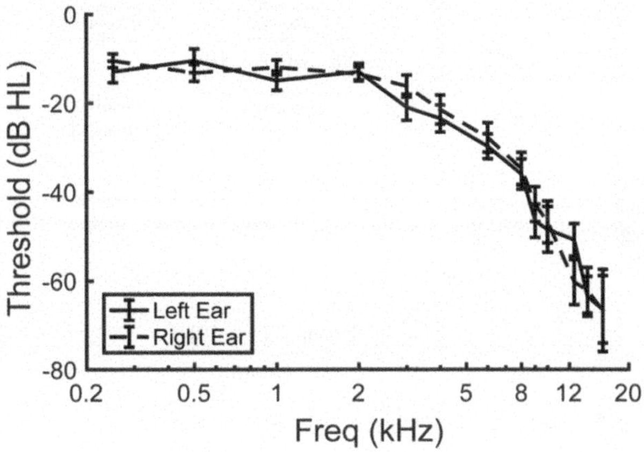

Fig. 7 Average audiogram from subjects in Marks et al. (2018)

pigs was a tone burst centered at the tinnitus frequency, for the human study, a complex tone burst derived from the participant's tinnitus spectrum (Roberts et al. 2008) served as the auditory portion of the bimodal stimulus. The sound was paired with transdermal electrical stimulation using the bimodal interval and order identified in the animal studies to produce maximum LTD in DCN fusiform cells. The sound was presented at 40 dB SL (sensation level) and was limited to no higher than 90 dB SPL to prevent hearing damage. Sound stimuli were delivered through a calibrated earphone, and somatosensory stimulation was provided via electrodes placed on either the cheek or cervical spine region of the neck. The electrical stimulation levels were just above the participants' threshold and well below levels that evoked muscle contractions. The device usage and function was monitored through device software to identify open and short circuits while receiving treatment. Software-controlled treatment was paused if electrode placement was altered.

Twenty participants were recruited into a double-blind, sham-controlled, crossover designed study. All participants had constant bothersome tinnitus, clinically normal-hearing thresholds with no hearing loss >40 dB HL up to 8 kHz (Fig. 7), and the ability to modulate their tinnitus with a somatic maneuver. Half of the participants were randomly assigned to the sham group ($n = 10$), which first received unimodal auditory treatment, and the other half were assigned to the active bimodal treatment group ($n = 10$) and had used no new tinnitus treatments 4 weeks prior to starting the study. Participants crossed over to the other treatment after 8 weeks (Fig. 7), so that each participant received both treatments and served as their own controls. Double-blinding and participant compliance, defined as daily use of the device for the full treatment period, was enabled by device software.

Participants were evaluated weekly for changes in their tinnitus spectra using interactive computer program (TinnTester) (Roberts et al. 2008) and tinnitus life impact using the Tinnitus Functional Index (TFI) (Meikle et al. 2012). Given that previous studies have shown that unimodal auditory stimulation does not elicit STDP (Koehler and Shore 2013a, b) and thus have not been expected to have a

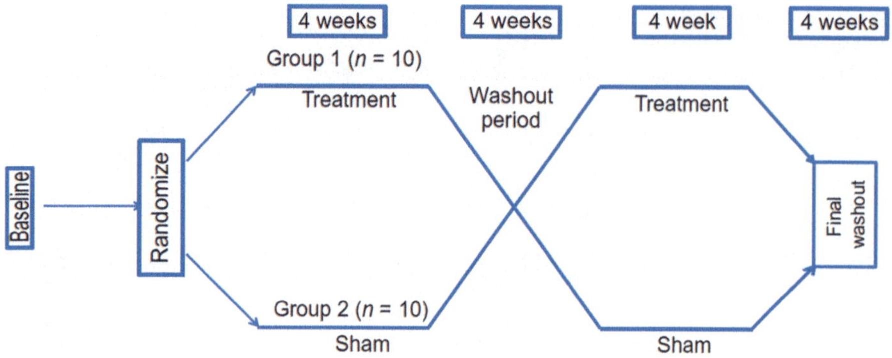

Fig. 8 Study design schematic from Marks et al. (2018)

long-lasting influence on tinnitus perception (Terry et al. 1983; Henry and Meikle 2000), the unimodal auditory stimulation was used as the sham treatment.

The participants receiving active treatment were given the bimodal auditory-somatosensory stimulation for 30 min per day for 28 consecutive days, while participants receiving the sham treatment were given the auditory stimulation for the same amount of time (Fig. 8). After 28 days, both treatment sessions were followed by a 4-week washout period to evaluate lasting impacts. Using the TinnTester software (Roberts et al. 2008), participants' tinnitus loudness and tinnitus spectrum data were collected throughout the entirety of the study.

Participants receiving the active bimodal treatment showed cumulative, statistically significant decrements in tinnitus loudness with a peak reduction of 13 dB over the 4 weeks of the study. Two participants reported complete elimination of their tinnitus. In comparison, the sham participants showed a nonsignificant reduction of 3 dB in tinnitus loudness. Absence of change in tinnitus loudness with unimodal auditory stimulation is consistent with the results from the animal studies, which demonstrated that unimodal sound stimulation was ineffective at eliciting fusiform-cell plasticity (Fujino and Oertel 2003; Dehmel et al. 2012; Wu et al. 2015), as would be expected by activation of fusiform-cell basal dendrites. The tinnitus life impact measured through TFI also showed cumulative, statistically significant reductions in overall scores during the active treatment (7.51 points) and active washout (6.71 points) compared to the sham periods ($p < 0.001$). A correlation showed there was a significant moderate relationship between the change in TFI scores and the change in tinnitus loudness (both relative to baseline) during the active treatment period, $r(78) = 0.31$, $p = 0.015$. Thus, a reduction in tinnitus loudness during the active treatment was related to improved quality of life (Figs. 9 and 10). Clinically significant reductions in TFI (a decrease of >13 points) (Meikle et al. 2012) were seen in 10/20 participants during the active treatment. Moreover, it was also found that the TFI improvements outlasted the reduction in tinnitus loudness following active treatment, with the most considerable improvements seen in sleep quality and reductions in tinnitus intrusiveness. The prolonged improvements in TFI scores could be due to alterations in the hippocampus, a part of the limbic system, which

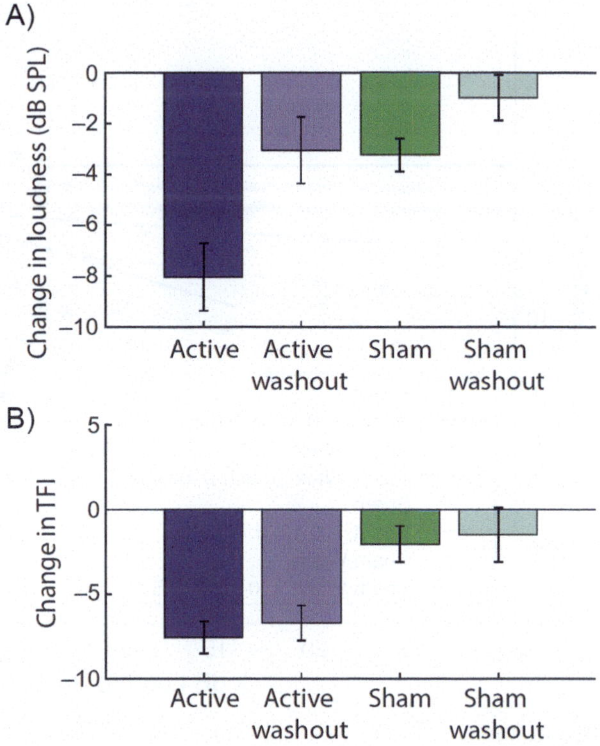

Fig. 9 Participants receiving active treatment, and not sham, show significant reductions in measured tinnitus loudness and TFI during bimodal auditory-somatosensory treatment. Participants receiving active treatment show significant reductions in measured tinnitus loudness (dB SPL, **a**) and TFI (points, **b**) during the active bimodal auditory-somatosensory treatment phase, but not during the sham treatment phase (from Marks et al. 2018)

could result in a reduced emotional response to tinnitus. Evidence for this is provided by tinnitus-specific physiological changes that occur in the hippocampus of guinea pigs after noise overexposure (Zhang et al. 2019).

1.2.2 Bimodal Auditory VNS to Treat Tinnitus in Humans

Based on preclinical VNS findings (Engineer et al. 2011), Tyler et al. (2017) conducted a prospective randomized double-blind controlled pilot study using paired VNS-tone treatment in humans. Thirty adults with chronic tinnitus that were unsuccessful with at least one other form of tinnitus treatment were selected for participation in the study. Participants were implanted with a VNS cuff electrode on their left cervical vagus nerve along with an internal pulse generator that was placed in the left pectoral region. Audiological and tinnitus assessments were performed to determine hearing levels and tinnitus pitch matching. Participants had average hearing thresholds ranging from normal mild to moderate sensorineural hearing loss from 500 to 8,000 Hz. Electrical stimulation of 0.8 mA, 100 μs pulses were delivered at a rate of 30 Hz; these were presented in pulse trains of 500 ms (totaling 15 stimulations per train). Tones ranging in frequency from 170 to 16,000 Hz were presented via circumaural headphones at a level based on each

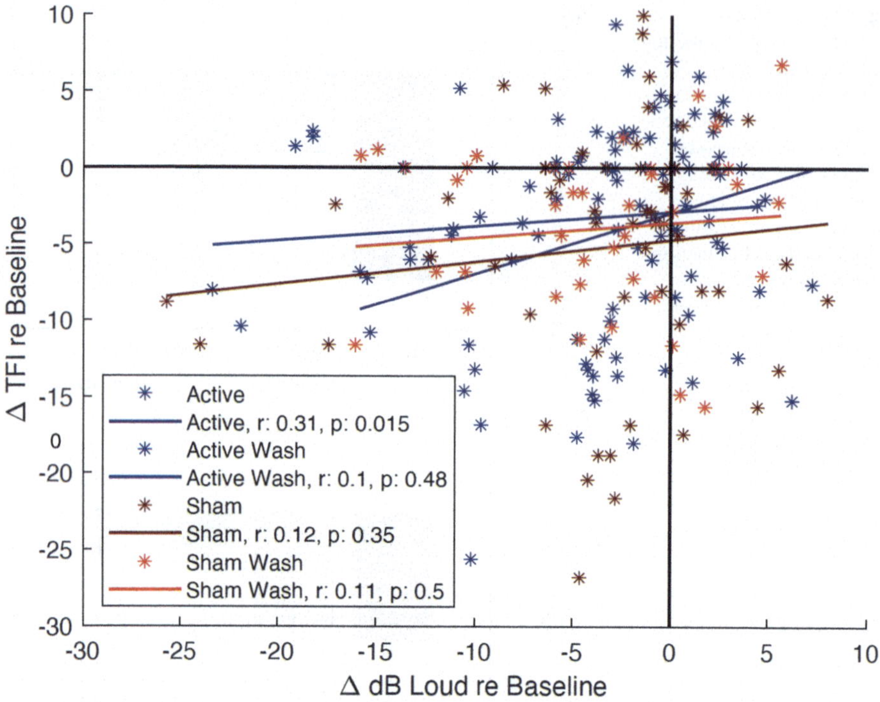

Fig. 10 Correlations for changes in TFI scores and changes in tinnitus loudness. Relative to baseline, changes in participants' TFI scores were compared to their changes in tinnitus loudness measures (dB) for each arm of the study. There was a significant positive correlation for reductions in TFI scores and decreases in tinnitus loudness during the active treatment arm of the study; as scores on the TFI decreased (improved), the loudness of tinnitus also decreased (improved). There were no other significant correlations between the two measures for the active washout, sham, or sham washout arms of the study (adapted from Marks et al. 2018)

participant's hearing thresholds and comfort (Fig. 11). The electrical pulses were paired with tones that were randomly selected from the frequency range excluding tones that were within a half-octave band of each participant's tinnitus frequency. The paired VNS-tones were presented approximately every 30 s for 2.5 h, totaling 300 stimulations per treatment session.

Participants were divided into two groups – a paired VNS-tone treatment group ($n = 16$) and a control group ($n = 14$) – and were randomly assigned to one of the groups in a double-blind crossover design. For the first 6-week arm of the study, participants in the treatment group received paired VNS-tone stimulation, and the control group received VNS treatment that was not paired with tones. After this period of experimental or sham treatment, both groups were unblinded, and the control group was crossed over to receive paired VNS-tone treatment, while the experimental group continued to receive active treatment for an additional 6 weeks. After the 12-week crossover trial, participants were allowed to continue using the device under less strictly controlled parameters if they wished, i.e., keeping the

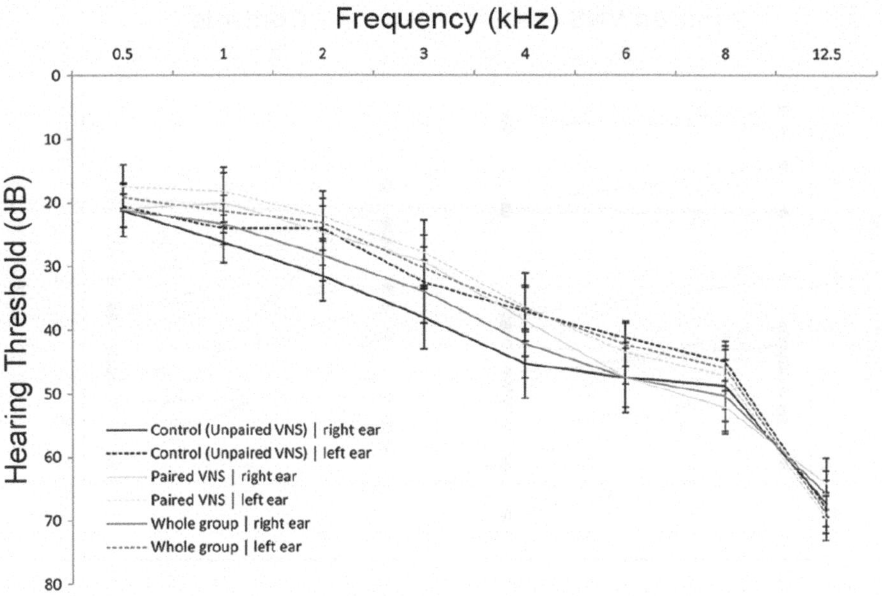

Fig. 11 Average audiogram from subjects in Tyler et al. (2017)

devices implanted and using them for shorter periods of time for fewer days per week for up to 1 year after the baseline visit. Outcome measures were assessed at baseline, 6 weeks, 12 weeks, 6 months, and 1 year.

The participants were assessed using subjective tinnitus life impact questionnaires and a loudness severity rating, as well as minimum masking level and tinnitus loudness matching. At the end of the 6-week blinded interval, results showed that the experimental group had a significant reduction of 17.7% in mean THI from baseline ($p = 0.001$), whereas the control group had a nonsignificant reduction (mean reduction of 7.3%, $p = 0.156$) (Figs. 12 and 13). There were no other significant differences in any outcome measurements from baseline scores to the week 6 scores for either group.

For the 12-week analysis after the control group had been unblinded and crossed over, results showed a significant reduction in scores on the questionnaire measures at the 6- and 12-week measurements (Figs. 12 and 13). However, there were no significant differences between groups, nor were there any interactions for these measures. There were also no significant results for any of the psychophysical measures. At the end of 6 weeks, 50% of the experimental group and 28% of the control group had clinically meaningful reductions in their THI scores although this was not significantly different ($p = 0.23$). At 12 weeks, the experimental group rate was 56%, and the control group improved to 43%. Lastly, the authors observed that treatment appeared to be more effective for participants that did not have hissing and/or blast-induced tinnitus, although no rationale was provided (Tyler et al. 2017). After removing 11 such participants, separate analyses were run on a subgroup of

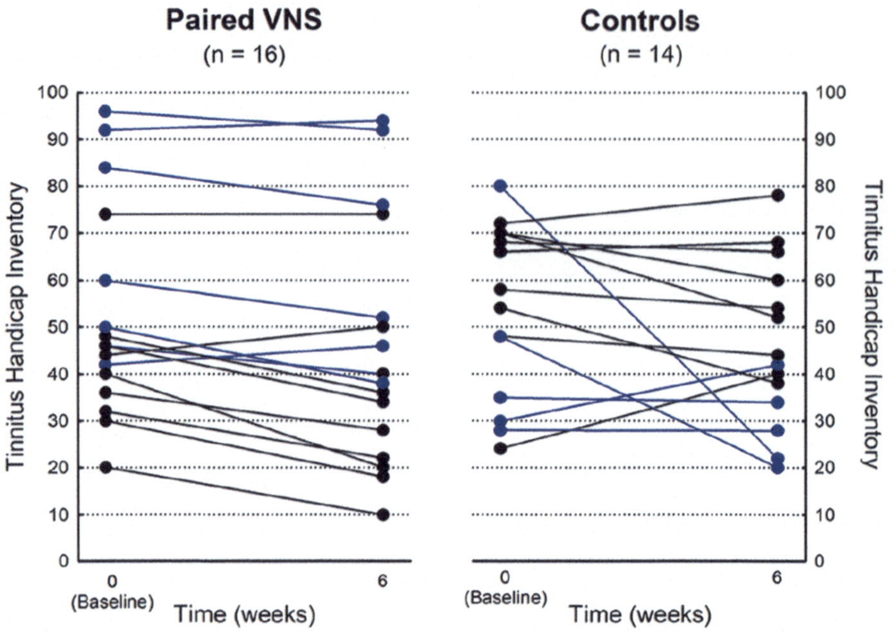

Fig. 12 Individual participant changes in THI pre- to post-VNS treatment. Individual participant changes in THI scores pre- to post-VNS treatment. Blue lines indicate participants with blast-induced and/or tonal tinnitus (paired-VNS group = 6, control group = 5) (from Tyler et al. 2017)

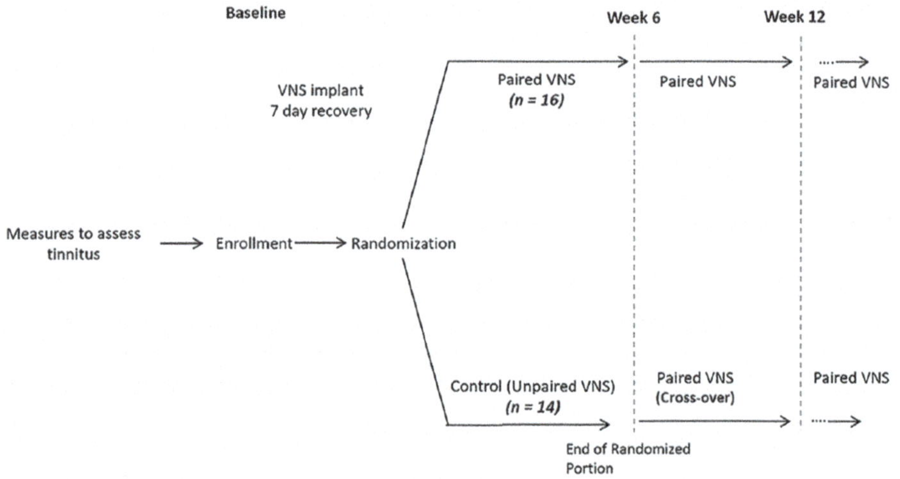

Fig. 13 Study design schematic from Tyler et al. (2017)

participants with tonal tinnitus. These results showed that the experimental group had statistically significant improvements of 24.3% at 6 weeks and 34% at 12 weeks compared to the 2% improvement in the control group at both 6 and 12 weeks ($p = 0.05$ and $p = 0.004$, respectively), although these analyses were likely exploratory in nature (Tyler et al. 2017). This study by Tyler et al. (2017) showed reductions in subjective tinnitus scores but had minimal effects on more objective psychophysical measures. Approximately 50% of participants showed clinically meaningful decreases on questionnaire scores after paired VNS-tone treatment, consistent with the findings by De Ridder et al. (2014). The treatment appeared to be more efficacious on a subset of the participants who had tonal tinnitus compared to those with hissing or blast-induced tinnitus (Tyler et al. (2017).

Based on previous studies showing that tinnitus patients display increased gamma band activity and reduced alpha band activity in the temporal cortex (Weisz et al. 2007; van der Loo et al. 2009), Vanneste et al. (2017) further examined the effects of paired VNS-tone treatment with electroencephalography (EEG) measures. EEG analysis showed a significant posttreatment reduction of gamma band activity, indicating a decrease in synchronized activity in the left auditory cortex, although no significant effects were observed in delta, theta, alpha, or beta frequency bands (Vanneste et al. 2017).

1.2.3 Open-Label Pilot Studies Using Bimodal Auditory Electrical Stimulation

Several open-label pilot studies have been performed using various methods of bimodal auditory electrical stimulation as potential tinnitus treatments. These studies have used noninvasive transcutaneous VNS (tVNS) paired with sound. Two studies utilized either a small electrode placed on the tragus (Lehtimaki et al. 2013) or concha of the external ear (Shim et al. 2015) to stimulate an afferent branch of the vagus nerve known as Arnold's nerve. Lehtimaki et al. (2013) tested ten adults using an electrode providing electrical impulses of 25 Hz at 0.8 mA paired with classical music presented in free field with a spectral notch corresponding to the participants' tinnitus frequency. Treatments were administered for 45–60 min/day for seven treatment sessions over 10 days. Posttreatment results showed statistically significant improvements on life impact questionnaires, as well as an overall reduction in cortical N1m amplitude measured by magnetoencephalography (MEG). The N1m is the MEG counterpart to the N1 wave that is commonly used as an electrophysiological measurement of cortical response to sound, which has been shown to be elevated in some patients with tinnitus (Lehtimaki et al. 2013). This finding suggests that tVNS could achieve some level of cortical neuromodulation via the afferent branch of the vagus nerve.

Shim et al. (2015) tested 30 adults using similar stimulation parameters as Lehtimaki et al. (2013): 25 Hz electrical stimulation at 1–10 mA and tailored notched classical music presented in free field. Treatments were comprised of 30 min/session for ten sessions that were spaced 1–4 days apart depending on the participants' condition, although no rationale was given for these differences. Posttreatment results showed that 50% of participants reported at least some improvement measured on a global improvement questionnaire, and mean overall tinnitus loudness ratings decreased by 18%, but THI scores were unaffected by the treatment. Both tVNS pilot studies were open-label non-randomized designs with no sham treatment or control groups used. Additionally, both studies implemented a short treatment session of only 7–10 treatments with no washout periods to examine any lasting effects. Therefore, the true effectiveness of the treatment method is still unclear, but tVNS could be investigated further as a noninvasive alternative to implanted VNS tinnitus treatment.

Referencing earlier studies that have indicated the involvement of the somatosensory system in tinnitus (Koehler et al. 2011; Engineer et al. 2011; Koehler and Shore 2013a, b; Wu et al. 2015) and a previous bimodal treatment in humans (De Ridder et al. 2014), another method of bimodal auditory-somatosensory stimulation has been investigated as a potential tinnitus treatment involving electrical stimulation of the tongue paired with sound. Hamilton et al. (2016) performed an open-label pilot study that utilized broad-spectrum noise with temporal fluctuations paired with electrical impulses delivered through a tongue stimulator. The tongue was chosen as a site of trigeminal nerve stimulation, although there were no referenced preclinical studies investigating this method of somatosensory activation.

Fifty-four adults with chronic tinnitus and sensorineural hearing loss of >25 dB HL in at least one ear and one frequency up to and including 8,000 Hz were recruited for the study, and 44 participants were used in the final analysis. The treatment device was an electrical stimulator consisting of an array of 32 transcutaneous electrodes placed on the antero-dorsal surface of the tongue, used in combination with headphones. Electrical stimuli were biphasic pulses of 17.5 μs duration and variable amplitude. The electrical stimulation was created by temporally and spectrally transposing the auditory stimuli to have a tonotopic organization spread across the 32 electrodes. Auditory stimuli consisted of broadband noise that was mixed with recorded rainfall and classical music, digitally processed to compensate for the participants' hearing loss and presented via high-fidelity headphones. The electrical and auditory stimulations were presented simultaneously during the treatment, and participants could adjust both levels based on comfort. The participants were instructed to use the device daily for 30–60 min for 10 weeks.

Outcome measures were the THI, tinnitus loudness matching, and minimum masking level (MML); these were measured at baseline and every 2 weeks throughout the duration of the study. Tested participants showed a significant reduction of 8.6 points on the THI and 8.1 dB on MML measurements (both $ps < 0.001$). There was a nonsignificant reduction of 5.4 dB on the tinnitus loudness measure. Clinically meaningful reductions on the THI (≥7 points) and MML (≥5.3 dB) were observed in 45% and 64% of participants, respectively. The authors also analyzed treatment compliance using a lenient exploratory criterion of 66% total device usage compared to 80% that is typically used in pharmaceutical studies (Hamilton et al. 2016). Based on this criterion, 30 participants were compliant with the recommended device usage and 14 were not. When analyzing the results for these groups separately, the compliant group showed significant reductions of 11.7 points on the THI, 7.5 dB on tinnitus loudness matching, and 9.8 dB on MML ($p < 0.001$). In contrast, the noncompliant group showed nonsignificant reductions of 1.9 points, 0.9 dB, and 4.7 dB, respectively. When the percentage of participants showing a clinically meaningful reduction of TFI and MML were further analyzed based on compliancy, the compliant group showed a greater percentage of responders for the THI (57%) and MML (73%) compared to the noncompliant group (21% and 43%, respectively).

Results of the study demonstrated the feasibility of the treatment, particularly for those who were compliant with the instructions. However, like other open-label pilot studies, all participants knowingly received active bimodal treatment, and there was no control group, sham treatment, or crossover period, so positive results for tongue sensory stimulation in conjunction with auditory stimuli for tinnitus treatment have not been well-established. The study by Hamilton et al. (2016) was designed to serve as pilot data for a future randomized, blinded clinical trial aimed to further investigate ideal parameter settings as well as the specific versus nonspecific therapeutic effects and their permanency. D'Arcy et al. (2017) and Conlon et al. (2019) have published experimental protocols and rationale for clinical trials utilizing this treatment device, but to date there have been no published supporting data.

2 Discussion

The methods of bimodal auditory electrical stimulation for tinnitus treatment that have been used in randomized controlled double-blinded clinical studies have solid foundations in preclinical basic science data. Prior studies studying the physiology of the auditory pathway showed that apical dendrites of DCN fusiform cells displayed STDP based on the timing of auditory and somatosensory stimulation. Guinea pigs that showed behavioral evidence of tinnitus displayed LTP patterns of neural activity associated with increased spontaneous firing rate and synchrony and that LTD neural patterns associated with a decrease in spontaneous firing rate and synchrony could be achieved by presenting auditory stimuli with a precise separation from the electrical stimulation (Koehler and Shore 2013a; Marks et al. 2018). In another approach, Engineer et al. (2011) found that stimulation of the vagus nerve paired with synchronous auditory stimulation could alter cortical plasticity in the auditory cortex and reverse cortical physiology patterns that are associated with tinnitus. Other studies have used the principles of STDP to investigate unimodal acoustic coordinated reset tones to desynchronize neural networks for tinnitus treatment (Tass et al. 2012), as well as bimodal acoustic transcranial magnetic stimulation of the auditory cortex to alter cortical excitability in normal hearing participants without tinnitus (Frank et al. 2012). However, these studies do not fit the criteria of bimodal stimulation for tinnitus treatment and therefore are beyond the scope of this chapter.

These studies used a variety of study designs to administer and investigate the efficacy of the treatments. The Marks et al. (2018) study used a randomized, double-blind crossover design in which all participants received active and sham treatments, and washout periods followed both treatment periods. In the Tyler et al. (2017) study, half of the participants received active treatment, while the other half received the sham treatment for the initial period. After the initial period, both groups were unblinded, and the sham group began receiving active treatment, and the active group continued to receive active treatment. Participants that initially received active treatment did not receive the sham treatment at any point in the study, and there was no washout period for either group. Lastly, the Tyler et al. (2017) study allowed participants to continue using their devices for up to 1 year after the initial 12-week investigation. In contrast, several open-label pilot studies did not use blinded, sham-controlled experimental paradigms (Lehtimaki et al. 2013; De Ridder et al. 2014; Shim et al. 2015; Hamilton et al. 2016). Additionally, the treatment durations among these studies varied considerably (see Table 1). Therefore, the ideal treatment time and the length of any lasting benefits from the treatments are still being investigated, and future studies may help answer these questions.

Studying tinnitus in humans is complicated by the wide variability in symptoms including onset, duration, frequency range, tonal quality, hearing levels, and etiology. The treatments in the auditory-somatosensory clinical study (Marks et al. 2018) as well as the paired VNS-tone clinical studies (Tyler et al. 2017; De Ridder et al. 2014; Vanneste et al. 2017) were derived from preclinical animal studies. Electrical stimuli in both auditory-somatosensory and VNS-tone studies utilized low-current electrical impulses, but auditory stimuli differed considerably. Marks et al. (2018) used auditory stimuli representative of the participants' tinnitus spectrum with precise timing relative to electrical stimuli, to induce LTD in those DCN fusiform cells with tinnitus-related hyperactivity. The paired tone-VNS studies used pure tones outside of the participant's tinnitus frequency. These stimuli were chosen to restructure a cortical map that had an overrepresentation of cortical activity corresponding to the tinnitus frequency. Thus, the auditory-somatosensory stimuli aimed to reduce activity in the putative tinnitus generation site, whereas the paired tone-VNS stimuli aimed to increase the activity of areas surrounding the tinnitus site to normalize the overall cortical map. A future avenue of research could investigate stimuli parameters to optimize individual settings for participants with tinnitus characteristics that did not respond to the treatments tested previously.

The improvements observed in the paired tone-VNS studies (Tyler et al. 2017; Vanneste et al. 2017) were revealed only through subjective questionnaires and not by more objective tinnitus loudness measurements. Thus, it is not clear if these findings reflect changes in emotional state or tinnitus (Deklerck et al. 2020). Psychophysical loudness measures and perceived loudness ratings are often poorly correlated, with ratings typically reflecting the person's reaction to the tinnitus rather than the actual loudness (Henry 2016). Thus, future studies would ideally incorporate outcome measures including subjective questionnaires, psychoacoustic measures, and physiological measures to assess any improvements that may be observed by tinnitus treatments. Non-paired VNS has been typically used for the treatment of depression and epilepsy (Engineer et al. 2013). Therefore, it is possible that paired tone-VNS treatment could primarily be influencing depression and not tinnitus per

Table 1 Comparison of human bimodal stimulation pilot study and clinical trial parameters, structure, and outcome measures

	Paired-notched music with transcutaneous VNS (Shim et al. 2015)	Auditory-somatosensory treatment (Hamilton et al. 2016)	Paired tone-VNS implanted device (Tyler et al. 2017)	Auditory-somatosensory treatment (Marks et al. 2018)
Study design	Open-label pilot study	Open-label pilot study	Double-blind, sham-controlled partial crossover	Double-blind, sham-controlled crossover
Device placement	Transcutaneous electrode pad on auricular concha of left ear, free-field speaker	Transcutaneous electrodes on antero-dorsal tongue surface, high-fidelity headphones	Implanted cuff electrode on cervical vagus nerve, circumaural headphones	Transcutaneous electrodes on cheek or neck, ear bud headphone
Participants	30 adults	54 adults (44 used for analysis)	30 adults	20 adults
Treatment timeframe	10 sessions, spread over 10–40 days	10 weeks of active tx	6 weeks of active or sham tx, followed by 6 weeks of active tx, 12 weeks total	4 weeks of active or sham tx followed by 4 weeks of washout (2×), 16 weeks total
Treatment usage	30 min/session	30–60 min/day	180 min/day	30 min/day
Electrical stimuli	200 μs pulse width, 1–10 mA amplitude, 25 Hz pulse rate	Biphasic anodic pulses 17.5 μs duration, variable amplitude, temporal-spectral transformation of auditory stimuli	100 μs pulse width, 0.8 mA amplitude, 30 Hz pulse rate, pulse train = 0.5 s	Biphasic pulses 100 μs duration, 2–5 mA amplitude
Auditory stimuli	Tailored notched classical music with 1/2 octave nearest tinnitus frequency removed	Wideband noise with high rate of temporal fluctuations (recorded rainfall) mixed with classical music, presented at comfortable level	Tones ranging from 170 to 16,000 Hz, excluding 1/2 octave around most prominent tinnitus frequency, presented at comfortable level, modified to have 3D spatial location, 500 ms duration	Matched tinnitus spectrum presented at 40 dB SL, 10 ms duration with 1 ms linear rise/fall time
Bimodal stimuli presentation	Audio presented simultaneously with electrical stim	Audio presented simultaneously with electrical stim	Audio presented simultaneously with electrical stim	Audio presented 5 ms prior to electrical stim

(continued)

Table 1 (continued)

	Paired-notched music with transcutaneous VNS (Shim et al. 2015)	Auditory-somatosensory treatment (Hamilton et al. 2016)	Paired tone-VNS implanted device (Tyler et al. 2017)	Auditory-somatosensory treatment (Marks et al. 2018)
Outcome measures	THI, tinnitus loudness scale, tinnitus awareness scale, global improvement scale	THI, tinnitus loudness matching, MML	THI, THQ, TFI, tinnitus loudness rating, tinnitus loudness matching, MML	TFI, tinnitus loudness matching (via TinnTester)
Results for blinded portion (if applicable)	50% reported relief in global improvement, significant reduction in loudness and awareness, no change in THI	Significant reductions in THI and MML. Responder rate = 45% for THI, 64% for MML	Significant reduction in THI scores for active group but not sham. No other significant findings. Responder rate = 50%	Significant reductions in loudness measure and TFI scores during active tx but not sham. Responder rate = 50%

se. In order to understand the effectiveness of VNS for tinnitus, future studies should aim to separate therapeutic effects on tinnitus loudness from emotional components (Deklerck et al. 2020).

A major difference between the double-blinded clinical studies examining auditory-somatosensory and paired tone-VNS therapy is the degree of invasiveness of the treatment devices. Bimodal auditory-somatosensory treatment is noninvasive and only requires superficial placement of electrodes on the skin, thus posing minimal risk to participants. Paired tone-VNS treatment devices require an invasive surgical procedure to implant the electrode around the cervical vagal nerve and the subcutaneous stimulator. Implantation is a safe and relatively common procedure that has been performed on as many as 100,000 people worldwide for treatment of depression and epilepsy; however, there are still associated risks (e.g., hoarseness, soreness after implantation, general surgical risks, cost of procedure) as there are with any surgical procedure. Thus, tinnitus patients seeking this therapy must weigh the risks over potential benefits from undergoing the procedure (Deklerck et al. 2020). Additionally, noninvasive tVNS may present another alternative to a surgical intervention for tinnitus, but more new research on this placement with more rigorous study designs must first be undertaken.

Both blinded clinical studies utilized a sham treatment consisting of unimodal auditory stimulation. Given that the electrical impulses might be perceptible, one weakness of the bimodal stimulation studies is the possibility that participants can determine which treatment is the sham (Hesse 2016). This issue was addressed in the auditory-somatosensory study of Marks et al. (2018) by intentionally setting the electrical stimuli at a level where perception was barely noticeable (just above threshold) and adapted over time, increasing the likelihood of maintaining blinding.

For the best study results, steps must be taken to minimize this phenomenon as much as possible, and analyses of sham treatments should be considered accordingly.

All bimodal treatments discussed here utilized auditory stimuli along with electrical stimulation. Thus, the hearing ability of the participants is an important factor in each treatment, requiring that participants were able to hear the stimuli. Both clinical studies utilized participants with similar average hearing profiles, i.e., normal sloping to mild or moderate sensorineural hearing loss, steady chronic tinnitus, and no history of retrocochlear pathology or Meniere's disease. While participants with severe hearing loss were excluded from the studies, the efficacy of treatments for those with more severe hearing loss should also be investigated in the future with modifications in the auditory stimulation paradigms. In some studies, reorganization of the auditory cortex was most pronounced in participants with severe hearing loss, but not tinnitus, supporting previous findings that found no cortical reorganization in participants with tinnitus and near-normal hearing, suggesting that the observed cortical reorganization is a result of hearing loss and not tinnitus (Langers et al. 2012; Koops et al. 2020). Therefore, it is likely that tonotopic reorganization in the cortex in the described studies is due to hearing loss and not tinnitus per se. Future therapeutic treatments should aim to target plasticity originating in subcortical pathways that are independent of hearing loss (Koops et al. 2020).

Bimodal auditory-electrical stimulation continues to be an exciting avenue of research for tinnitus alleviation. While the treatments discussed in this chapter differ in several ways, the bimodal treatments use a combination of electrical and auditory stimulation to modulate plasticity changes along the auditory pathway that target the underlying pathophysiology of tinnitus. As with any developing technology targeted for therapeutic use in humans, a strong basic science framework from animal studies should be used to inform feasibility and safety *for* pilot studies in a clinical population. Once these stipulations have been met, it is important that rigorous large-scale double-blinded sham-controlled clinical trials be performed to truly investigate the efficacy of each treatment and rule out any placebo effects. Tinnitus continues to prove to be a challenging area to research, especially in humans given the subjectivity and variability of the symptoms, as well as the subjective nature of many methods of capturing data (e.g., questionnaires and rating scales). Determining the most appropriate types of supporting electrophysiological data that could be used in conjunction with the subjective data would help quantify outcome measures. Future research should investigate the effects of treatments for different tinnitus subtypes, enabling a broader scope of treatment to be established to benefit the largest possible number of patients.

Acknowledgments This work was supported by National Institutes of Health Grants R01-DC004825 (SES), RF1 MH114244 01 (SES), T32-DC00011 (DTM), and P30-DC05188 and a grant from Michigan Institute for Clinical Health Research.

Competing Interests DTM and SES are co-inventors on US Patent 9,682,232, *Personalized auditory-somatosensory stimulation to treat tinnitus*. DTM and SES are co-founders of Auricle, Inc.

References

Barker M, Solinski HJ, Hashimoto H, Tagoe T, Pilati N, Hamann M (2012) Acoustic overexposure increases the expression of VGLUT-2 mediated projections from the lateral vestibular nucleus to the dorsal cochlear nucleus. PLoS One 7:e35955

Bauer CA, Turner JG, Caspary DM, Myers KS, Brozoski TJ (2008) Tinnitus and inferior colliculus activity in chinchillas related to three distinct patterns of cochlear trauma. J Neurosci Res 86:2564–2578

Conlon B, Hamilton C, Hughes S, Meade E, Hall DA, Vanneste S, Langguth B, Lim HH (2019) Noninvasive bimodal neuromodulation for the treatment of tinnitus: protocol for a second large-scale double-blind randomized clinical trial to optimize stimulation parameters. JMIR Res Protoc 8:e13176

D'Arcy S, Hamilton C, Hughes S, Hall DA, Vanneste S, Langguth B, Conlon B (2017) Bi-modal stimulation in the treatment of tinnitus: a study protocol for an exploratory trial to optimise stimulation parameters and patient subtyping. BMJ Open 7:e018465

De Ridder D, Vanneste S, Engineer ND, Kilgard MP (2014) Safety and efficacy of vagus nerve stimulation paired with tones for the treatment of tinnitus: a case series. Neuromodulation 17:170–179

De Ridder D, Vanneste S, Langguth B, Llinás RR (2015) Thalamocortical dysrhythmia: a theoretical update in tinnitus. Front Neurol 6:124

Dehmel S, Pradhan S, Koehler S, Bledsoe S, Shore S (2012) Noise overexposure alters long-term somatosensory-auditory processing in the dorsal cochlear nucleus--possible basis for tinnitus-related hyperactivity? J Neurosci 32:1660–1671

Deklerck AN, Marechal C, Pérez Fernández AM, Keppler H, Van Roost D, Dhooge IJM (2020) Invasive neuromodulation as a treatment for tinnitus: a systematic review. Neuromodul Technol Neural Interface 23:451–462 https://doi.org/10.1111/ner.13042

Edeline JM, Manunta Y, Hennevin E (2011) Induction of selective plasticity in the frequency tuning of auditory cortex and auditory thalamus neurons by locus coeruleus stimulation. Hear Res 274:75–84

Eggermont JJ (2006) Cortical tonotopic map reorganization and its implications for treatment of tinnitus. Acta Otolaryngol Suppl:9–12

Eggermont JJ, Roberts LE (2004) The neuroscience of tinnitus. Trends Neurosci 27:676–682

Engineer ND, Riley JR, Seale JD, Vrana WA, Shetake JA, Sudanagunta SP, Borland MS, Kilgard MP (2011) Reversing pathological neural activity using targeted plasticity. Nature 470:101–104

Engineer ND, Moller AR, Kilgard MP (2013) Directing neural plasticity to understand and treat tinnitus. Hear Res 295:58–66

Finlayson PG, Kaltenbach JA (2009) Alterations in the spontaneous discharge patterns of single units in the dorsal cochlear nucleus following intense sound exposure. Hear Res 256:104–117

Flor H, Elbert T, Knecht S, Wienbruch C, Pantev C, Birbaumer N, Larbig W, Taub E (1995) Phantom-limb pain as a perceptual correlate of cortical reorganization following arm amputation. Nature 375:482–484

Frank E, Schecklmann M, Landgrebe M, Burger J, Kreuzer P, Poeppl TB, Kleinjung T, Hajak G, Langguth B (2012) Treatment of chronic tinnitus with repeated sessions of prefrontal transcranial direct current stimulation: outcomes from an open-label pilot study. J Neurol 259:327–333

Fujino K, Oertel D (2003) Bidirectional synaptic plasticity in the cerebellum-like mammalian dorsal cochlear nucleus. Proc Natl Acad Sci U S A 100:265–270

Gu JW, Herrmann BS, Levine RA, Melcher JR (2012) Brainstem auditory evoked potentials suggest a role for the ventral cochlear nucleus in tinnitus. J Assoc Res Otolaryngol 13:819–833
Hamilton LS, Sohl-Dickstein J, Huth AG, Carels VM, Deisseroth K, Bao S (2013) Optogenetic activation of an inhibitory network enhances feedforward functional connectivity in auditory cortex. Neuron 80:1066–1076
Hamilton C, D'Arcy S, Pearlmutter BA, Crispino G, Lalor EC, Conlon BJ (2016) An investigation of feasibility and safety of bi-modal stimulation for the treatment of tinnitus: an open-label pilot study. Neuromodulation 19:832–837
Heeringa AN, Wu C, Chung C, West M, Martel D, Liberman L, Liberman MC, Shore SE (2018) Glutamatergic projections to the cochlear nucleus are redistributed in tinnitus. Neuroscience 391:91–103
Henry JA (2016) "Measurement" of tinnitus. Otol Neurotol 37:e276–e285
Henry JA, Meikle MB (2000) Psychoacoustic measures of tinnitus. J Am Acad Audiol 11:138–155
Hesse G (2016) Evidence and evidence gaps in tinnitus therapy. GMS Curr Top Otorhinolaryngol Head Neck Surg 15:Doc04
Itoh K, Kamiya H, Mitani A, Yasui Y, Takada M, Mizuno N (1987) Direct projections from the dorsal column nuclei and the spinal trigeminal nuclei to the cochlear nuclei in the cat. Brain Res 400:145–150
Kalappa BI, Brozoski TJ, Turner JG, Caspary DM (2014) Single unit hyperactivity and bursting in the auditory thalamus of awake rats directly correlates with behavioural evidence of tinnitus. J Physiol 592:5065–5078
Kaltenbach JA, Afman CE (2000) Hyperactivity in the dorsal cochlear nucleus after intense sound exposure and its resemblance to tone-evoked activity: a physiological model for tinnitus. Hear Res 140:165–172
Kaltenbach JA, Zhang J (2007) Intense sound-induced plasticity in the dorsal cochlear nucleus of rats: evidence for cholinergic receptor upregulation. Hear Res 226:232–243
Kanold PO, Young ED (2001) Proprioceptive information from the pinna provides somatosensory input to cat dorsal cochlear nucleus. J Neurosci 21:7848–7858
Kilgard MP, Merzenich MM (1998) Cortical map reorganization enabled by nucleus basalis activity. Science 279:1714–1718
Koehler SD, Shore SE (2013a) Stimulus timing-dependent plasticity in dorsal cochlear nucleus is altered in tinnitus. J Neurosci 33(50):19647–19656
Koehler SD, Shore SE (2013b) Stimulus-timing dependent multisensory plasticity in the guinea pig dorsal cochlear nucleus. PLoS One 8:e59828
Koehler SD, Pradhan S, Manis PB, Shore SE (2011) Somatosensory inputs modify auditory spike timing in dorsal cochlear nucleus principal cells. Eur J Neurosci 33(3):409–420
Koops EA, Renken RJ, Lanting CP, van Dijk P (2020) Cortical tonotopic map changes in humans are larger in hearing loss than in additional tinnitus. J Neurosci 40:3178–3185
Langers DR, de Kleine E, van Dijk P (2012) Tinnitus does not require macroscopic tonotopic map reorganization. Front Syst Neurosci 6:2
Lehtimaki J, Hyvarinen P, Ylikoski M, Bergholm M, Makela JP, Aarnisalo A, Pirvola U, Makitie A, Ylikoski J (2013) Transcutaneous vagus nerve stimulation in tinnitus: a pilot study. Acta Otolaryngol 133:378–382
Llinás RR, Ribary U, Jeanmonod D, Kronberg E, Mitra PP (1999) Thalamocortical dysrhythmia: a neurological and neuropsychiatric syndrome characterized by magnetoencephalography. Proc Natl Acad Sci U S A 96(26):15222–15227
Manzoor NF, Chen G, Kaltenbach JA (2013a) Suppression of noise-induced hyperactivity in the dorsal cochlear nucleus following application of the cholinergic agonist, carbachol. Brain Res 1523:28–36
Manzoor NF, Gao Y, Licari F, Kaltenbach JA (2013b) Comparison and contrast of noise-induced hyperactivity in the dorsal cochlear nucleus and inferior colliculus. Hear Res 295:114–123

Marks KL, Martel DT, Wu C, Basura GJ, Roberts LE, Schvartz-Leyzac KC, Shore SE (2018) Auditory-somatosensory bimodal stimulation desynchronizes brain circuitry to reduce tinnitus in guinea pigs and humans. Sci Transl Med 10:eaal3175

Meikle MB, Henry JA, Griest SE, Stewart BJ, Abrams HB, McArdle R, Myers PJ, Newman CW, Sandridge S, Turk DC (2012) The tinnitus functional index: development of a new clinical measure for chronic, intrusive tinnitus. Ear Hear 33:153–176

Middleton JW, Kiritani T, Pedersen C, Turner JG, Shepherd GM, Tzounopoulos T (2011) Mice with behavioral evidence of tinnitus exhibit dorsal cochlear nucleus hyperactivity because of decreased GABAergic inhibition. Proc Natl Acad Sci U S A 108:7601–7606

Natan RG, Rao W, Geffen MN (2017) Cortical interneurons differentially shape frequency tuning following adaptation. Cell Rep 21:878–890

Norena AJ, Eggermont JJ (2003) Changes in spontaneous neural activity immediately after an acoustic trauma: implications for neural correlates of tinnitus. Hear Res 183:137–153

Oertel D, Young ED (2004) What's a cerebellar circuit doing in the auditory system? Trends Neurosci 27:104–110

Osen KK, Storm-Mathisen J, Ottersen OP, Dihle B (1995) Glutamate is concentrated in and released from parallel fiber terminals in the dorsal cochlear nucleus: a quantitative immunocytochemical analysis in Guinea pig. J Comp Neurol 357:482–500

Rauschecker JP, Leaver AM, Muhlau M (2010) Tuning out the noise: limbic-auditory interactions in tinnitus. Neuron 66:819–826

Roberts LE, Moffat G, Bosnyak DJ (2006) Residual inhibition functions in relation to tinnitus spectra and auditory threshold shift. Acta Otolaryngol Suppl:27–33

Roberts LE, Moffat G, Baumann M, Ward LM, Bosnyak DJ (2008) Residual inhibition functions overlap tinnitus spectra and the region of auditory threshold shift. J Assoc Res Otolaryngol 9:417–435

Roberts LE, Eggermont JJ, Caspary DM, Shore SE, Melcher JR, Kaltenbach JA (2010) Ringing ears: the neuroscience of tinnitus. J Neurosci 30:14972–14979

Sametsky EA, Turner JG, Larsen D, Ling L, Caspary DM (2015) Enhanced GABAA-mediated tonic inhibition in auditory thalamus of rats with behavioral evidence of tinnitus. J Neurosci 35:9369–9380

Schaette R, McAlpine D (2011) Tinnitus with a normal audiogram: physiological evidence for hidden hearing loss and computational model. J Neurosci 31:13452–13457

Seki S, Eggermont JJ (2003) Changes in spontaneous firing rate and neural synchrony in cat primary auditory cortex after localized tone-induced hearing loss. Hear Res 180:28–38

Shargorodsky J, Curhan GC, Farwell WR (2010) Prevalence and characteristics of tinnitus among US adults. Am J Med 123:711–718

Shim HJ, Kwak MY, An YH, Kim DH, Kim YJ, Kim HJ (2015) Feasibility and safety of transcutaneous vagus nerve stimulation paired with notched music therapy for the treatment of chronic tinnitus. J Audiol Otol 19:159–167

Shore SE (2005) Multisensory integration in the dorsal cochlear nucleus: unit responses to acoustic and trigeminal ganglion stimulation. Eur J Neurosci 21:3334–3348

Shore SE, Martel DT (2019) Multimodal inputs to the cochlear nucleus and their role in the generation of tinnitus. In: Kandler K (ed) The Oxford handbook of the auditory brainstem, vol 226. Oxford University Press, Oxford

Shore SE, Wu C (2019) Mechanisms of noise-induced tinnitus: insights from cellular studies. Neuron 103:8–20

Shore SE, El Kashlan H, Lu J (2003) Effects of trigeminal ganglion stimulation on unit activity of ventral cochlear nucleus neurons. Neuroscience 119:1085–1101

Shore SE, Roberts LE, Langguth B (2016) Maladaptive plasticity in tinnitus--triggers, mechanisms and treatment. Nat Rev Neurol 12:150–160

Sturm JJ, Zhang-Hooks YX, Roos H, Nguyen T, Kandler K (2017) Noise trauma-induced behavioral gap detection deficits correlate with reorganization of excitatory and inhibitory local circuits in the inferior colliculus and are prevented by acoustic enrichment. J Neurosci 37:6314–6330

Tass PA, Adamchic I, Freund HJ, von Stackelberg T, Hauptmann C (2012) Counteracting tinnitus by acoustic coordinated reset neuromodulation. Restor Neurol Neurosci 30:137–159

Terry AM, Jones DM, Davis BR, Slater R (1983) Parametric studies of tinnitus masking and residual inhibition. Br J Audiol 17:245–256

Turner JG, Brozoski TJ, Bauer CA, Parrish JL, Myers K, Hughes LF, Caspary DM (2006) Gap detection deficits in rats with tinnitus: a potential novel screening tool. Behav Neurosci 120:188–195

Tyler R, Cacace A, Stocking C, Tarver B, Engineer N, Martin J, Deshpande A, Stecker N, Pereira M, Kilgard M, Burress C, Pierce D, Rennaker R, Vanneste S (2017) Vagus nerve stimulation paired with tones for the treatment of tinnitus: a prospective randomized double-blind controlled pilot study in humans. Sci Rep 7:11960

Tzounopoulos T, Kim Y, Oertel D, Trussell LO (2004) Cell-specific, spike timing-dependent plasticities in the dorsal cochlear nucleus. Nat Neurosci 7:719–725

van der Loo E, Gais S, Congedo M, Vanneste S, Plazier M, Menovsky T, Van de Heyning P, De Ridder D (2009) Tinnitus intensity dependent gamma oscillations of the contralateral auditory cortex. PLoS One 4:e7396

Vanneste S, Martin J, Rennaker RL 2nd, Kilgard MP (2017) Pairing sound with vagus nerve stimulation modulates cortical synchrony and phase coherence in tinnitus: an exploratory retrospective study. Sci Rep 7:17345

Wang H, Brozoski TJ, Turner JG, Ling L, Parrish JL, Hughes LF, Caspary DM (2009) Plasticity at glycinergic synapses in dorsal cochlear nucleus of rats with behavioral evidence of tinnitus. Neuroscience 164:747–759

Weisz N, Muller S, Schlee W, Dohrmann K, Hartmann T, Elbert T (2007) The neural code of auditory phantom perception. J Neurosci 27:1479–1484

Wu C, Martel DT, Shore SE (2015) Transcutaneous induction of stimulus-timing-dependent plasticity in dorsal cochlear nucleus. Front Syst Neurosci 9:116

Wu C, Martel DT, Shore SE (2016a) Increased synchrony and bursting of dorsal cochlear nucleus fusiform cells correlate with tinnitus. J Neurosci 36:2068–2073

Wu C, Stefanescu RA, Martel DT, Shore SE (2016b) Tinnitus: maladaptive auditory-somatosensory plasticity. Hear Res 334:20–29

Zeng C, Yang Z, Shreve L, Bledsoe S, Shore S (2012) Somatosensory projections to cochlear nucleus are upregulated after unilateral deafness. J Neurosci 32:15791–15801

Zhang JS, Kaltenbach JA (1998) Increases in spontaneous activity in the dorsal cochlear nucleus of the rat following exposure to high-intensity sound. Neurosci Lett 250:197–200

Zhang L, Wu C, Martel DT, West M, Sutton MA, Shore SE (2019) Remodeling of cholinergic input to the hippocampus after noise exposure and tinnitus induction in guinea pigs. Hippocampus 29:669–682

Zhou J, Shore S (2006) Convergence of spinal trigeminal and cochlear nucleus projections in the inferior colliculus of the Guinea pig. J Comp Neurol 495:100–112

Part V
Cognitive and Affective Aspects of Tinnitus

Neurobiology of Stress-Induced Tinnitus

Agnieszka J. Szczepek and Birgit Mazurek

Contents

Abstract Emotional stress has accompanied humans since the dawn of time and has played an essential role not only in positive selection and adaptation to an ever-changing environment, but also in the acceleration or even initiation of many illnesses. The three main somatic mechanisms induced by stress are the hypothalamus-pituitary-adrenal axis (HPA axis), the sympathetic-adreno-medullar (SAM) axis, and the immune axis. In this chapter, the stress-induced mechanisms that can affect cochlear physiology are presented and discussed in the context of tinnitus generation and auditory neurobiology. It is concluded that all of the presented mechanisms need to be further investigated. It is advised that clinical

A. J. Szczepek (✉)
Department of Otorhinolaryngology, Head and Neck Surgery, Charité-Universitätsmedizin Berlin, Corporate Member of Freie Universität Berlin, Humboldt-Universität zu Berlin, and Berlin Institute of Health, Berlin, Germany

Faculty of Medicine and Health Sciences, University of Zielona Góra, Zielona Góra, Poland
e-mail: agnes.szczepek@charite.de

B. Mazurek
Tinnitus Center, Charité-Universitätsmedizin Berlin, Corporate Member of Freie Universität Berlin, Humboldt-Universität zu Berlin, and Berlin Institute of Health, Berlin, Germany

Curr Topics Behav Neurosci (2021) 51: 327–348
https://doi.org/10.1007/7854_2020_215
Published Online: 19 February 2021

practitioners ask patients about stressful events or chronic stress preceding the tinnitus onset and measure the vital signs. Finally, taking into account that tinnitus itself acts as a stressor, the implementation of anti-stress therapies for tinnitus treatment is recommended.

Keywords Catecholamines · Cytokines · Glucocorticoids · Glutamate excitotoxicity · HPA axis · Immune axis · SAM axis · The cochlea

1 Introduction

The association between emotional stress and tinnitus has long been known, appearing in medical journals for at least 200 years ago (Curtis 1841). The crosstalk between tinnitus and stress is still a subject of intense research (Szczepek and Mazurek 2017; Aydin and Searchfield 2019; Biehl et al. 2019; Brueggemann et al. 2019; Moossavi et al. 2019). An epidemiological study with 12,166 subjects demonstrated that the correlation between tinnitus incidence and stress is as strong as between tinnitus and noise exposure (Baigi et al. 2011). The authors of another study that involved 658 tinnitus patients demonstrated a direct effect of stress level on tinnitus loudness and tinnitus distress (Probst et al. 2016). However, the question of *how* stress affects the auditory pathway to induce tinnitus remains open.

Non-auditory health conditions strongly associated with stress include depression and anxiety (Craske and Stein 2016; Michaelides and Zis 2019). Interestingly, tinnitus patients frequently report having depressive and anxious symptoms (Zöger et al. 2006; Adoga et al. 2008; Zirke et al. 2013; Gomaa et al. 2014; Salviati et al. 2014; Conrad et al. 2015; Waechter and Brännström 2015; Bruggemann et al. 2016; Brueggemann et al. 2019). Also, tinnitus patients are significantly more likely to have symptoms of depression and anxiety when compared to age-matched control subjects (Danioth et al. 2020) and a higher incidence of anxiety (26.1%) and depressive symptoms (25.6%) as compared to age-matched persons without tinnitus (9.2% incidence of anxiety and 9.1% of depressive symptoms) (Bhatt et al. 2017). In agreement, current epidemiological studies suggest a direct correlation between tinnitus and anxiety or depression (Hébert et al. 2012).

Definitive scientific evidence demonstrating that stress causes tinnitus is still lacking. There are multiple reasons for this knowledge gap, the main being the patients' diffuse knowledge about the time of onset of the phantom sound and a lack of medical and psychological information from that period. The other reason is that the persons who developed tinnitus may not necessarily be bothered by it. What is not lacking is the abundant clinical data demonstrating that the individuals affected by tinnitus are more likely to experience a higher level of stress than this experienced by tinnitus-free patients (Betz et al. 2017; Biehl et al. 2019; Mazurek et al. 2019). Accordingly, various therapeutic methods such as cognitive-behavioral therapy (CBT), mindfulness-based cognitive therapy (MBCT), mindfulness-based stress reduction (MBSR), brief solution-focused therapy, narrative therapy, acceptance and commitment therapy (ACT), and eye movement desensitization and reprocessing (EMDR) have been used to reduce the tinnitus-induced burden (National Guideline 2020). The success of these types of therapeutic approaches in treating tinnitus points towards the essential role that stress plays in tinnitus pathobiology. However, the outstanding questions are *if* and *how* the stress-induced responses affect the auditory pathway to produce a sensory activation without an acoustic stimulus. Before an attempt to answer that question, the mechanism evoked by stress needs to be described.

2 Stressors

Two general types of stressors are recognized: psychological and physical stressors. Psychological stressors include mental stressors (concentration tasks, memory requirements, intelligence tests) (Kirschbaum et al. 1993), social stress situations, and stressors acting throughout life (posttraumatic stress disorder after deprivation or abuse), which produce changes in stress regulation patterns over a lifetime and induce a central hyper-responsiveness (Heim and Nemeroff 2009; Lupien et al. 2009; Slavich and Shields 2018). Environmental (physical) stressors include hypo- or hyperthermia, noise, over-illumination, or overcrowding.

The effects of some physical stressors – namely noise and temperature changes – were and still are studied extensively in the auditory system (Seifert et al. 1998; El Ganzoury et al. 2012; Sliwinska-Kowalska and Davis 2012; Lie et al. 2016; Le et al. 2017). Also, the effects of psychological stressors on the auditory system have been investigated, providing insights into the pathophysiology of tinnitus and hyperacusis (Horner 2003; Mazurek et al. 2010b; Hasson et al. 2013; Mazurek et al. 2015). Interestingly, tinnitus itself is considered to be a stressor.

3 Neurobiological Mechanisms Associated with Tinnitus Induction

Tinnitus is a symptom, and the conditions associating with tinnitus are discussed in detail elsewhere in this book. In this chapter, selected processes leading to tinnitus are listed, and later, their association with stress is demonstrated.

It is well accepted that tinnitus initiation is associated with damage to the auditory periphery, whereas tinnitus maintenance correlates with the progressive changes in the central auditory system (Eggermont 1990; Eggermont and Roberts 2015; Haider et al. 2018). The cochlear structures that could be damaged include outer and inner hair cells, supporting cells, and spiral ganglion neurons. On the molecular level, the injury can be induced by glutamate excitotoxicity (Puel et al. 2002; Ryan and

Bauer 2016; Kim et al. 2019), an excess of free radicals (Evans and Halliwell 1999; Huang et al. 2000; Rybak et al. 2019), and all processes leading to apoptosis (Op de Beeck et al. 2011; Gauvin et al. 2018). On the structural level, cochlear synaptopathy has been proposed to represent an important mechanism contributing to tinnitus onset (Liberman and Kujawa 2017; Altschuler et al. 2019). However, this mechanism has recently been questioned for humans (Guest et al. 2017) and animals (Pienkowski 2018).

Interestingly, accumulating evidence supports the view that the limbic system makes an essential contribution to the onset and maintenance of tinnitus percept (Jastreboff 1990; Lockwood et al. 1998; Mühlau et al. 2006; Landgrebe et al. 2009; Leaver et al. 2016; Ryan and Bauer 2016; Caspary and Llano 2017; Qu et al. 2019; Kapolowicz and Thompson 2020). The limbic system was proposed to provide negative feedback to the central auditory system and, thus, to turn off the perception of the tinnitus sound. However, the stress-affected limbic system no longer provides that negative feedback, leaving the phantom sound uncancelled (Rauschecker et al. 2010). Corroborating studies have demonstrated significant volume reduction of grey matter in the (left) parahippocampal cortex of tinnitus patients (Landgrebe et al. 2009; Besteher et al. 2019; Liu et al. 2019).

4 Stress-Induced Responses

The term *stress* has been in use for about a century. Physicists first introduced it in an attempt to describe a distribution of energy leading to tension. In the twenties of the last century, an American physiologist Walter Cannon used the term *stress* when describing *fight or flight* response (Cannon 1922). A few years later, a Hungarian-Canadian endocrinologist, Hans Seyle originated the research on emotional stress (Selye and Fortier 1949; Selye 1950). The experiments performed with animals led to the discovery of the hypothalamus-pituitary-adrenal axis (HPA axis) (Fortier and Selye 1949), which we will describe later in detail. He also introduced the concept of positive or negative stress and explored various states of stress reactions. The followers of Seyle's model of stress continue to conduct research trying to understand how the emotional status may influence the functioning of cells, tissues, and the entire organism, leading in some cases to somatic pathologies.

The factors inducing stress are termed *stressors* and can be divided into physical and psychological stressors. The physical stressors (such as pain, heat, or cold) can cause similar but not identical effects on the organism compared to emotional stressors (Hermann et al. 2019). Also, duration of stress is an essential factor, where the outcome of acute, short-time stress differs from the chronic exposure to stress (Bryant 2018). Therefore, the result of stress differs, depending on the type of stressor, age, gender, genetics, social status, and education of the affected person (Fig. 1) (Oyola and Handa 2017).

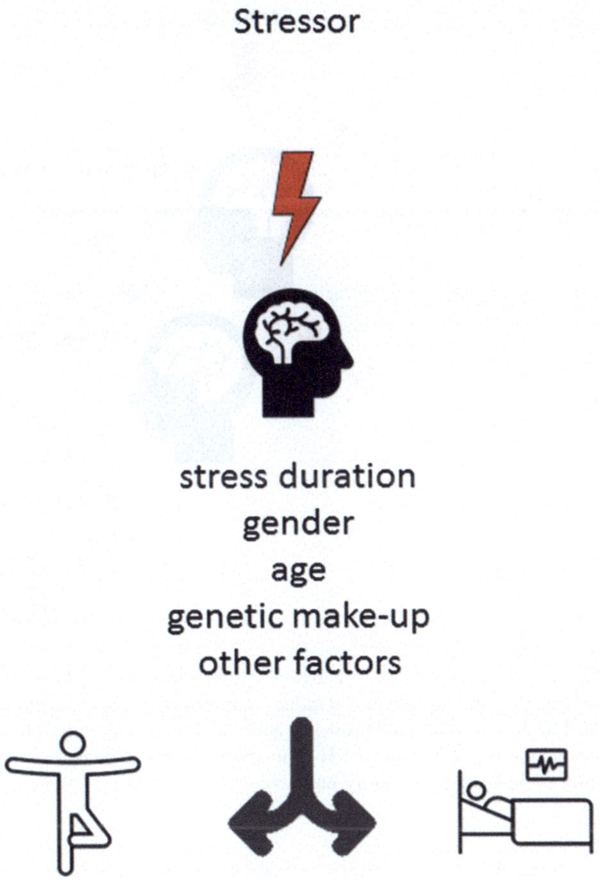

Fig. 1 Schematic illustration of the factors influencing the outcome of stress. The stressors involved in the stress event (physical, psychological, or both) induce effects, depending on age, gender, stress duration, and several other factors. The outcome of stress may range from staying healthy to acquiring a health condition

4.1 The Hypothalamus-Pituitary-Adrenal Axis (HPA Axis)

The pioneering work of Hans Seyle paved the way for understanding the somatic mechanisms induced by emotional stress (Selye 1937). The principal pathway caused by stress is the hypothalamic-pituitary-adrenal axis (HPA axis). HPA axis encompasses the following structural elements: the hypothalamus (paraventricular nucleus, PV), pituitary gland (the anterior lobe), and the adrenal cortex. The hypothalamic neurons in PV are capable of synthesizing vasopressin and corticotropin-releasing hormone (CRH). These two peptides are secreted upon stress and stimulate adrenocorticotropic hormone (ACTH) release from the pituitary gland. ACTH promotes the production and release of corticosteroids from the adrenal gland. The corticosteroids-driven negative feedback mechanism tightly regulates the HPA axis (Fig. 2). In humans, the principal corticosteroid produced is cortisol, whereas in rodents, it is corticosterone.

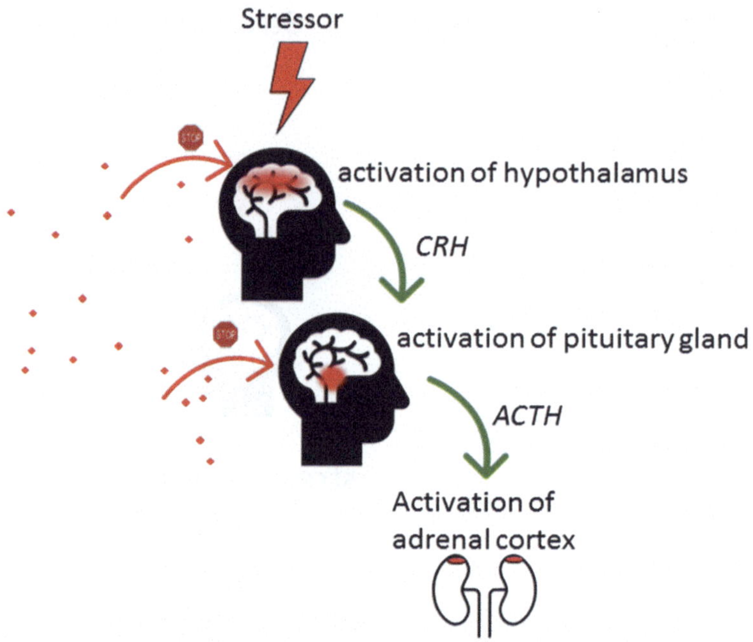

Fig. 2 The schematic HPA pathway. Upon stress, the CRH is released from the hypothalamus to activate the pituitary gland and induce adrenocorticotropic hormone (ACTH) release. In response to ACTH, the adrenal gland produces and releases glucocorticoid hormones, which in turn inhibit the production of ACTH and CRH. The green color indicates the stimulatory pathway, while the red color indicates the inhibitory pathway

In response to diurnal rhythm or stress, cortisol is released to the bloodstream to act on all tissues and cells of the body, influencing the metabolism and gene transcription regulation. The metabolic effect of cortisol is associated with the de novo production of glucose (gluconeogenesis) in the liver, kidney, intestine, muscle, and brain (Yip et al. 2016). In contrast, gene transcription regulation occurs almost in each somatic cell due to glucocorticoid receptors' ubiquitous presence (Fig. 3). The glucocorticoid-mediated transcriptional modulation is complex and comprises several types of processes involving direct binding of glucocorticoid-glucocorticoid receptor complex to specific sequences on the genomic DNA and activation/deactivation of several transcription factors through various mechanisms. Upon binding its receptor (GR) and translocation to the nucleus, GR inhibits or stimulates the expression of several genes – these genes belong to the so-called *glucocorticoid-responsive genes*. The glucocorticoid-responsive gene pattern differs depending on the cell type. For instance, in adipocytes, corticosteroids bind to 8,848 sites on the genomic DNA to upregulate the expression of 421 and downregulate the expression of 198 genes (Yu et al. 2010). In contrast, only 4,392 sites are bound by corticosteroids in A549 epithelial cell line carcinoma (Reddy et al. 2009). Similarly, in the neuronal cell line PC12, a unique,

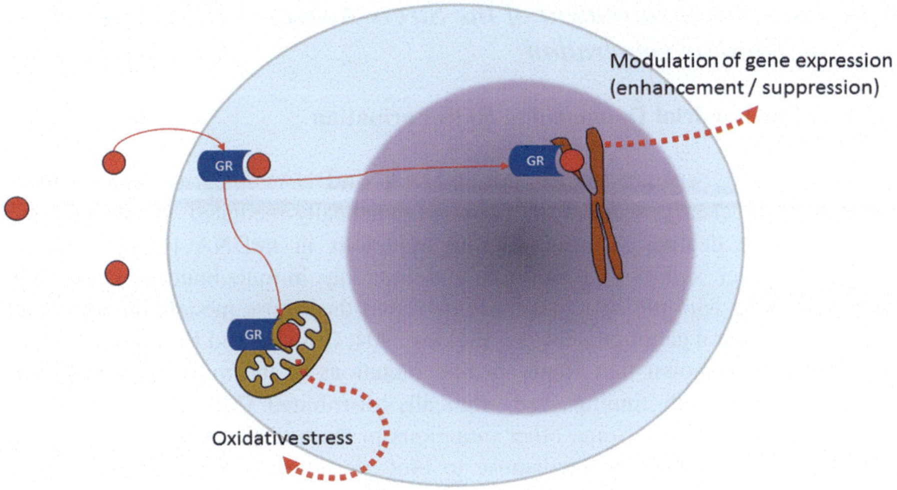

Fig. 3 The effect of corticosteroid on the cell on the molecular level. The corticosteroid attaches to the glucocorticoid receptor (GR) and translocates to the nucleus and the mitochondria, acting as a transcription factor. Upon binding to the chromosomal or mitochondrial DNA, GR influences gene transcription in both organelles, leading to altered transcription/translation (nucleus and mitochondria) and oxidative stress (mitochondria)

cell-type restricted GR specificity was described, demonstrating 1,183 genomic binding sites (Polman et al. 2012). The processes regulated by corticosteroids in the neuronal cells include neuron projection morphogenesis, neuron projection regeneration, synaptic transmission, and regulation of apoptosis, suggesting a strong influence of corticosteroids on neuronal plasticity.

The nuclear and mitochondrial DNA can be expressionally regulated by stress and glucocorticoids (Hunter et al. 2016), adding to the complexity of glucocorticoid effects in the cells, tissues, and entire body. Exposure to stress results in an inhibition of mitochondrial complex I activity and an increase in reactive oxygen species (ROS) production, damaging the affected cells (Fig. 3). The physiological significance of mitochondria affected by stress was demonstrated in the animal models of anxiety-related disorders and human anxiety disorders (Misiewicz et al. 2019).

Under ideal physiological circumstances, the HPA axis can be quickly activated by stress and promptly stopped by negative feedback via corticosteroids (cortisol in humans). However, under chronic stress, the inhibitory mechanisms are either no longer in place or aberrant. In response to experimental social stress (Trier Social Stress Task (Kirschbaum et al. 1993)), healthy subjects produce free cortisol detectable in saliva 30 min later (Hébert and Lupien 2007). Tinnitus patients were shown to have delayed reactions to the same stressor, suggesting anomalous HPA axis responses.

The potential effects of the HPA axis on the cochlea and its hypothetical contribution to the onset of tinnitus are discussed below.

4.2 Potential Involvement of the Stress-Activated HPA Axis in Tinnitus Generation

4.2.1 Mitochondrial Damage and ROS Formation

The association between mitochondrial DNA (mtDNA) integrity and cochlear physiology is clearly seen in the human genetically-mediated syndromic and non-syndromic deafness, caused by the mutations in mtDNA (Kokotas et al. 2007). Moreover, some of the isoforms and mutations in mitochondrial DNA have been associated with presbyacusis in humans and the mouse model and correlated with a loss of spiral ganglion neurons (Pickles 2004; Crawley and Keithley 2011). In addition, it was shown that some specific mutations in the mitochondrial DNA, which associate with tinnitus, are ethnically distributed (Mostafa et al. 2014; Lechowicz et al. 2018). Some other mutations in the mtDNA, which targeted the 12S rRNA gene known to predispose to ototoxicity, are also associated with a sudden tinnitus onset (Fischel-Ghodsian et al. 1997).

In the cochlea of humans and animals and the organ of Corti-derived cell line, ototoxic medications such as cisplatin or gentamicin were shown to induce overproduction of reactive oxygen species leading to mitochondrial damage and finally, hair cell death (Bertolaso et al. 2001; Poirrier et al. 2010; Sheth et al. 2017; Desa et al. 2018; O'reilly et al. 2019). Collectively, this evidence strongly implies the detrimental role of damaged mitochondria and the overproduction of reactive oxygen species in cochlear pathology. Even though no studies have examined the role of the HPA axis in the generation of ROS in the cochlea, this process might still play a role in the induction of cochlear hearing loss and tinnitus.

4.2.2 Glucocorticoid-Modified Expression of Genes

Glucocorticoid- (GR) and mineralocorticoid receptors (MR) are expressed in the cochlea by various cell types such as inner and outer hair cells, spiral ganglion cells, supporting cells (ten Cate et al. 1993; Zuo et al. 1995; Kil and Kalinec 2013). Stria vascularis expresses mainly MR, whereas fibrocytes type IV mainly GR (Kil and Kalinec 2013). In an animal model, the short-term acute restrain was used to study the effect of non-auditory stress on the auditory pathway. In the spiral ganglion neurons, the restrain has induced GR's nuclear translocation, whereas in the cochlea, it downregulated the expression of cochlear GR. The changes occurred 24 h after stress and indicated negative feedback mechanism (Tahera et al. 2006a). The GR translocation was associated with protection against noise-induced injury (Tahera et al. 2006b). In contrast to short-term stress, long-term stress was shown to be associated with an increased incidence of hearing loss and tinnitus in humans, indicating that the general dysregulation of the HPA axis might be detrimental to the auditory system (Canlon et al. 2013; Herr et al. 2018).

Many studies of gene expression in the cochlea following exposure to corticosteroid have been performed using synthetic steroid dexamethasone. These studies demonstrated, for instance, that dexamethasone modulates the expression of genes encoding apoptosis-relevant proteins in the cochlea (Hoang et al. 2009). The somewhat limited evidence obtained using the stress model is consistent with cochlear gene expression being modulated during or after stress. However, only a few genes were investigated, of which hypoxia-inducible factor 1 (Hif1) was downregulated in the cochlea of Wistar rats 7 days after mild, 24-h-long stress (Mazurek et al. 2010a).

To summarize, the influence of stress released corticosteroids on gene transcription in the cochlea should be further studied using a global approach (e.g., mRNA sequencing) and with a large-scale data.

4.2.3 Influence of HPA Axis on Glutamate Signaling

The role of cochlear glutamate-depending signaling in the generation of tinnitus has been suggested (Puel et al. 2002; Sahley et al. 2013). The model has been supported by a study using C57BL/6J mice that provided evidence of an imbalance between cochlear NMDA and AMPA receptors during a long-term administration of salicylate, associated with the induction of tinnitus (Cui et al. 2019). Consistent with this observation, salicylate-induced tinnitus could be inhibited by selective NMDA blocker memantine (Ralli et al. 2014). In primary hippocampal cultures, corticosterone was shown to increase the endocytosis of AMPAR, leading to its surface decrease (Martin et al. 2009). It remains to be determined if the HPA axis influences overexpression of NMDA or downregulation of AMPA receptors in the cochlea, therefore contributing to cochlear synaptic plasticity and eventually to a generation of tinnitus.

Stress (e.g., forced swim stress and restraint stress) has been shown to increase glutamate release in the medial prefrontal cortex, hippocampus, striatum, and nucleus accumbens in rats (Moghaddam 1993). Moreover, corticosterone application on rats hippocampal brain slices rapidly increased the glutamate release via MR (Karst et al. 2005). Although this phenomenon has not yet been investigated in the cochlea, it was demonstrated that spiral ganglion neurons (Furuta et al. 1994) and the inner hair cells express MR (Yao and Rarey 1996), making the glucocorticoid-mediated rapid glutamate release in the cochlea hypothetically possible. Such a quick release of glutamate could stimulate the auditory pathway without acoustic stimuli. Also, depending on a local concentration of freshly released glutamate, it could lead to excitotoxicity associated with peripheral deafferentation and tinnitus (Sahley and Nodar 2001).

4.3 The Sympathetic-Adreno-Medullar (SAM) Axis

The sympathetic-adreno-medullar axis is one of two stress axes, which alongside the HPA axis acts as a mediator for specific stress responses and adaptation to psychological and environmental stressors. The SAM axis mediates quick responses that

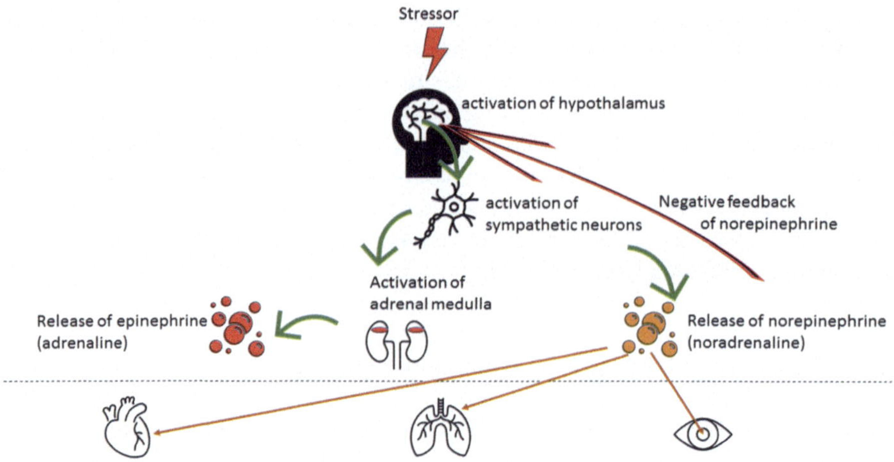

Fig. 4 The schematic presentation of the sympathetic-adreno-medullar (SAM) axis. Stress activates the hypothalamus, which in turn activates sympathetic neurons. The sympathetic neurons projecting to the adrenal medulla induce epinephrine (adrenaline) release into the bloodstream. Epinephrine increases the supply of glucose and oxygen to the brain and muscles (flight or fight) and suppresses the body's non-crisis functions (e.g., digestion). The sympathetic neurons themselves release the norepinephrine (noradrenaline) and, thus, activate all cells expressing adrenergic receptors. The effects of norepinephrine range from an increase in the blood volume pumped by the heart, expanding the respiratory pathway in the lungs, narrowing the blood vessels in non-essential organs, and pupil dilation

activate *fight or flight* reaction. During that reaction, several tissues and organs necessary for survival are activated (e.g., muscles, heart, and respiratory function), whereas tissues and organs with tasks that are non-essential for survival (e.g., digestion) are suppressed at the same time. The SAM axis comprises the hypothalamus, sympathetic neurons, and catecholamines (epinephrine and norepinephrine) (Fig. 4). Similar to the HPA axis, there is a negative feedback system mediated by epinephrine, which extinguishes the SAM activation.

4.4 Potential Involvement of the Stress-Activated SAM Axis in Tinnitus Generation

4.4.1 Arterial Hypertension

Arterial hypertension has often been considered a cause of tinnitus. This causative relationship's suggested mechanism is damage to cochlear microcirculation, induction of hearing loss, and deafferentation of the auditory periphery. Studies using spontaneously hypertensive (SH) and wild-type rats demonstrated differences in age-related hearing loss between the two rat strains (Borg and Viberg 1987). A progressive loss of the outer hair cells was observed in the SH but not wild-type rats already at the age of 3 months. A threshold shift in the high frequencies (16 and 24 kHz) was seen in the 21-month-old SH but not in wild-type rats.

Interestingly, the study scrutinizing the effect of noise on the cochlear vascular system in the SH and wild-type rats demonstrated dramatic differences between the two types of animals and suggested hypertension-induced cochlear vascular damage (Axelsson et al. 1983). In addition to vascular damage, arterial hypertension negatively affected the endocochlear potential (Mosnier et al. 2001). In agreement with that, several clinical studies have reported an association between arterial hypertension, hearing loss, and tinnitus (Figueiredo et al. 2015; Yang et al. 2015; Figueiredo et al. 2016). Also, a study with 80 tinnitus patients and 80 tinnitus-free subjects demonstrated that the nighttime systolic and diastolic blood pressure of tinnitus patients is higher than in the age-matched control subjects (Değirmenci et al. 2014), suggesting possible continuous upregulation of SAM axis.

4.4.2 Catecholamines

Catecholamines released during activation of the SAM axis mediate their effects through various receptors. One of those receptors is a G-protein coupled α2-adrenergic receptor, mediating vascular smooth muscle reaction, inhibiting the norepinephrine release, and platelets' activation. The presence of α2-adrenergic receptors in cochlear microvasculature was verified in an animal model (gerbils), and α2-adrenergic stimulation provided experimental evidence for catecholamine-induced cochlear vasoconstriction (Carrasco et al. 1990). Moreover, vasoconstriction is associated with hypoxia or ischemia. In the inner ear, experimentally induced hypoxic and ischemic events lead to hair cell loss in an animal model (Shirane and Harrison 1987; Mazurek et al. 2003), followed by threshold shift (Sawada et al. 2001) and likely tinnitus. The expression of α2-adrenergic receptors was also demonstrated on the outer and inner hair cells and the supporting cells, the spiral ganglion neurons, stria vascularis, and all five types of fibrocytes in the cochleae of developing rats (Cai et al. 2013).

Brimonidine, an α2 adrenergic agonist (activator), protected the auditory hair cells from gentamicin-induced toxicity (Cortada et al. 2017). Surprisingly, inhibition of the α2a adrenergic receptor with istradefylline (α2 adrenergic antagonist) also protected the hair cells from glutamate excitotoxicity (Han et al. 2019). However, the above experiments were performed on an isolated organ of Corti and might not reflect the real-life situation, where the interplay of hypoxic/ischemic and toxic

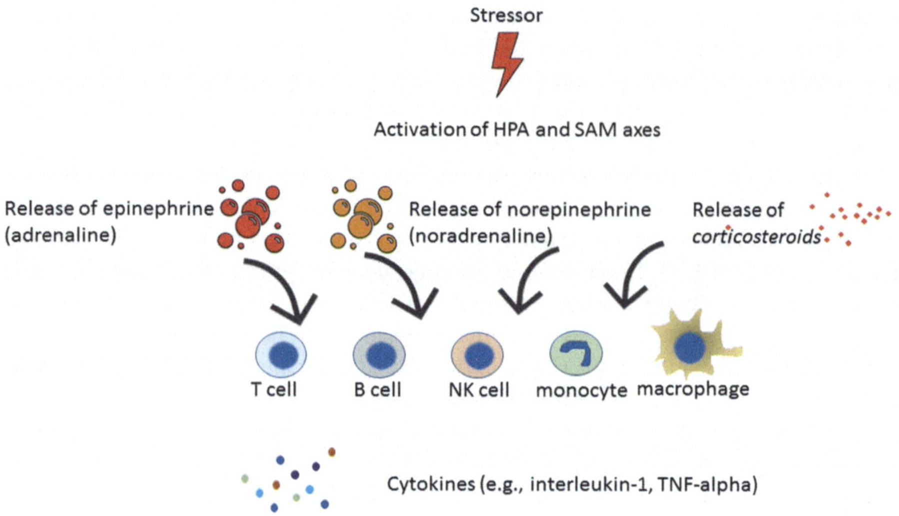

Fig. 5 Stress-induced immune axis

events and the protective mechanisms would yield an extrapolated role for α2-adrenergic receptors in the inner ear. Here too, more research should be performed to address the unanswered questions.

4.5 The Immune Axis

The immune system consists of specialized cells (neutrophils, eosinophils, basophils, lymphocytes, and monocytes) circulating in the blood, existing in primary lymphoid organs, or permanently residing in other tissues (e.g., resident macrophages). Under the steady-state condition, the immune system continually communicates with the endocrine and nervous systems, and there is a homeostatic balance between the three. During stress, the activation of the HPA and SAM axes leads to the release of mediators acting directly on the immune cells (Fig. 5), e.g., changing the number of circulating B cells (McGregor et al. 2016) or increasing the number of circulating T cells (Gupta et al. 2017). Also, chronic stress can change the activation status of immunocytes (Arranz et al. 2009) and modify cytokines' release (Jung et al. 2019).

The mediators set free by activation of the HPA and SAM axes (corticosteroids, norepinephrine, and epinephrine) act directly on the immune cells and modify their release from the primary lymphoid organs, affect their migration and alter the cytokine expression patterns.

4.6 Potential Involvement of the Stress-Activated Immune Axis in Tinnitus Generation

4.6.1 A Direct Influence of Stress on the Immune Cells in the Cochlea

Several types of immune cells have been found in a steady-state, healthy cochlea of humans and animals (Hu et al. 2018). The cochlear immunocytes include resident macrophages (Hu et al. 2018; Liu et al. 2018; Kishimoto et al. 2019), NK cells (Iguchi et al. 1997), and T cells (Liu and Rask-Andersen 2019). Of all the immune cell types, resident macrophages' cochlear function has been studied in the most detail. Macrophages phagocytose the damaged hair cells in Corti's organ, thus preventing cochlear inflammation and consecutive hearing loss (Hirose et al. 2017). In addition to phagocytic features, cochlear macrophages possess still not well-understood repair abilities that enable regeneration of ribbon synapses after noise exposure (Kaur et al. 2019). It is tempting to speculate that cochlear macrophages' activity might be impaired by the stress mediators, especially by corticosteroids. This impairment could be of a long duration during chronic stress and might result in reduced phagocytosis, migration, and a compromised spiral ganglion repair process upon acoustic injury. Macrophages also reside in the stria vascularis, where they regulate the cochlear intrastrial fluid–blood barrier. Here, long-term suppression of resident macrophages via stress hormones could result in modulation of endocochlear potential, which might induce hearing loss and tinnitus, similarly to what is observed during the aging process (Keithley 2020). Cochlear immunology is a rapidly developing field, and information is being published frequently, advancing our understanding of immune- and non-immune processes mediated by the cochlear immune cells.

4.6.2 Influence of Cytokines

Cytokines released by the immune and non-immune cells, such as interleukin 1-beta (IL-1beta), IL-6, and TNF-alpha, were shown to influence the plasticity in the brain and the peripheral nervous system (Aldskogius and Kozlova 1998; Levin and Godukhin 2017). The influence of the cytokines on the synaptic strength, plasticity, and integrity varies depending on the co-signaling molecules, the presence of other cells, and many other factors. Cytokines can act directly on neurons and different cell types (e.g., cochlear immunocytes), and the outcome ranges between cochlear regeneration and cell death (Barald et al. 2018). Studies using an animal model of noise induced hearing loss (NIHL) indicated that NIHL induced the expression of proinflammatory cytokine tumor necrosis factor-alpha (TNF-alpha) in the cochlea (Frye et al. 2019) and was associated with the development of tinnitus (Wang et al. 2019) and that pharmacological intervention or using genetically modified mice prevented this detrimental development. Further, in an animal model of salicylate-induced tinnitus, increased expression of genes encoding TNF-alpha and IL-1-beta was observed in the cochlea and correlated with behaviorally tested tinnitus (Hwang et al. 2011). In a sample of 30 patients with chronic tinnitus, a correlation was found between TNF-alpha in serum and tinnitus loudness, total perceived stress, tension, and depression (Szczepek et al. 2014). Another study investigating tinnitus in the elderly noted a negative association between IL-10 and tinnitus loudness and duration (Haider et al. 2020). It remains to be clarified if cytokines' systemic concentration reflects that in the cochlea and how aging, inflammation, and infections may affect this balance.

5 Summary and Conclusions

Tinnitus is a symptom that may arise as a result of various changes in the auditory system. Persons with disturbing tinnitus perceive it as an unpleasant, distressing signal that negatively affects life quality, associates with anxiety and depression, and may last a lifetime. The accepted view on the primary mechanism inducing tinnitus is that it is a consequence of a cochlear lesion, which could have been caused by noise, ototoxic medications, the physiological aging process, or other means. Here, the evidence was reviewed for the emotional stress-induced mechanisms, which could contribute to cochlear pathologies and, in consequence, to tinnitus (Fig. 6). These mechanisms involve HPA axis-induced corticosteroid action on MR and GR, possibly leading to glutamate excitotoxicity and altered gene expression in the cochlea. The immune axis of stress can also be affected by the HPA axis, leading to modulation of the resident cochlear macrophages' function. Lastly, corticosteroids released upon HPA activation could contribute to the NMDA/AMPA disbalance.

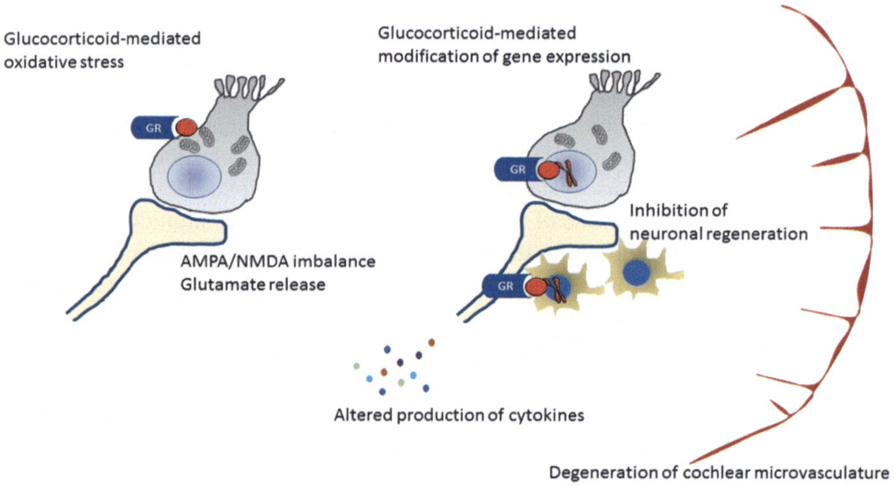

Fig. 6 Hypothetical model of some of the stress-induced events that could lead to cochlear pathologies and tinnitus

The SAM axis might increase blood pressure that induces degenerative changes in the cochlear microvasculature and changes in the organism's immune cells' repertoires, affecting cochlear immune cells.

Furthermore, SAM-induced vasoconstriction in the cochlea might likely cause hypoxia/ischemia, detrimental to the auditory hair cells and spiral ganglion neurons. It is concluded that all of the presented mechanisms need to be further investigated in the animal or ex vivo models. It is recommended for clinical practitioners to collect information about stressful events or chronic stress preceding the tinnitus onset. Furthermore, knowing the vital signs could add information to the stress-related status of a patient. Finally, taking into account that tinnitus itself acts as a stressor, the implementation of anti-stress therapies for tinnitus treatment is recommended.

References

Adoga AA, Adoga AS, Obindo JT (2008) Tinnitus and the prevalence of co-morbid psychological stress. Niger J Med 17:95–97

Aldskogius H, Kozlova EN (1998) Central neuron-glial and glial-glial interactions following axon injury. Prog Neurobiol 55:1–26

Altschuler RA, Halsey K, Kanicki A, Martin C, Prieskorn D, Deremer S, Dolan DF (2019) Small arms fire-like noise: effects on hearing loss, gap detection and the influence of preventive treatment. Neuroscience 407:32–40

Arranz L, De Vicente A, Muñoz M, De La Fuente M (2009) Impaired immune function in a homeless population with stress-related disorders. Neuroimmunomodulation 16:251–260

Axelsson A, Borg E, Hornstrand C (1983) Noise effects on the cochlear vasculature in normotensive and spontaneously hypertensive rats. Acta Otolaryngol 96:215–225

Aydin N, Searchfield GD (2019) Changes in tinnitus and physiological biomarkers of stress in response to short-term broadband noise and sounds of nature. Complement Ther Med 46:62–68

Baigi A, Oden A, Almlid-Larsen V, Barrenäs ML, Holgers KM (2011) Tinnitus in the general population with a focus on noise and stress: a public health study. Ear Hear 32:787–789

Barald KF, Shen YC, Bianchi LM (2018) Chemokines and cytokines on the neuroimmunoaxis: inner ear neurotrophic cytokines in development and disease. Prospects for repair? Exp Neurol 301:92–99

Bertolaso L, Martini A, Bindini D, Lanzoni I, Parmeggiani A, Vitali C, Kalinec G, Kalinec F, Capitani S, Previati M (2001) Apoptosis in the OC-k3 immortalized cell line treated with different agents. Audiology 40:327–335

Besteher B, Gaser C, Ivanšić D, Guntinas-Lichius O, Dobel C, Nenadić I (2019) Chronic tinnitus and the limbic system: reappraising brain structural effects of distress and affective symptoms. Neuroimage Clin 24:101976

Betz LT, Muhlberger A, Langguth B, Schecklmann M (2017) Stress reactivity in chronic tinnitus. Sci Rep 7:41521

Bhatt JM, Bhattacharyya N, Lin HW (2017) Relationships between tinnitus and the prevalence of anxiety and depression. Laryngoscope 127:466–469

Biehl R, Boecking B, Brueggemann P, Grosse R, Mazurek B (2019) Personality traits, perceived stress, and tinnitus-related distress in patients with chronic tinnitus: support for a vulnerability-stress model. Front Psychol 10:3093

Borg E, Viberg A (1987) Age-related hair cell loss in spontaneously hypertensive and normotensive rats. Hear Res 30:111–118

Brueggemann P, Seydel C, Schaefer C, Szczepek AJ, Amarjargal N, Boecking B, Rose M, Mazurek B (2019) ICD-10 symptom rating questionnaire for assessment of psychological comorbidities in patients with chronic tinnitus. HNO 67:46–50
Bruggemann P, Szczepek AJ, Rose M, Mckenna L, Olze H, Mazurek B (2016) Impact of multiple factors on the degree of tinnitus distress. Front Hum Neurosci 10:341
Bryant RA (2018) The current evidence for acute stress disorder. Curr Psychiatry Rep 20:111
Cai J, Li J, Mao Y, Bai X, Xu L, Wang H (2013) Immunohistochemical localization of α2-adrenergic receptors in the neonatal rat cochlea and the vestibular labyrinth. J Mol Neurosci 51:1010–1020
Canlon B, Theorell T, Hasson D (2013) Associations between stress and hearing problems in humans. Hear Res 295:9–15
Cannon WB (1922) Bodily changes in pain, hunger, fear and rage: an account of recent researches into the function of emotional excitement. D. Appleton
Carrasco VN, Prazma J, Faber JE, Triana RJ, Pillsbury HC (1990) Cochlear microcirculation. Effect of adrenergic agonists on arteriole diameter. Arch Otolaryngol Head Neck Surg 116:411–417
Caspary DM, Llano DA (2017) Auditory thalamic circuits and GABA(A) receptor function: putative mechanisms in tinnitus pathology. Hear Res 349:197–207
Conrad I, Kleinstauber M, Jasper K, Hiller W, Andersson G, Weise C (2015) The role of dysfunctional cognitions in patients with chronic tinnitus. Ear Hear 36
Cortada M, Levano S, Bodmer D (2017) Brimonidine protects auditory hair cells from in vitro-induced toxicity of gentamicin. Audiol Neurootol 22:125–134
Craske MG, Stein MB (2016) Anxiety. Lancet 388:3048–3059
Crawley BK, Keithley EM (2011) Effects of mitochondrial mutations on hearing and cochlear pathology with age. Hear Res 280:201–208
Cui W, Wang H, Cheng Y, Ma X, Lei Y, Ruan X, Shi L, Lv M (2019) Long-term treatment with salicylate enables NMDA receptors and impairs AMPA receptors in C57BL/6J mice inner hair cell ribbon synapse. Mol Med Rep 19:51–58
Curtis J (1841) Tinnitus aurium. Lancet 36:828–829
Danioth L, Brotschi G, Croy I, Friedrich H, Caversaccio MD, Negoias S (2020) Multisensory environmental sensitivity in patients with chronic tinnitus. J Psychosom Res 135:110155
Değirmenci H, Bakırcı EM, Salcan İ, Demirelli S, Duman H, Ceyhun G, Küçüksu Z (2014) Determination of correlation among heart rate variability, left atrium global strain, and nighttime blood pressure among patients with tinnitus. Med Sci Monit 20:1714–1719
Desa DE, Nichols MG, Smith HJ (2018) Aminoglycosides rapidly inhibit NAD(P)H metabolism increasing reactive oxygen species and cochlear cell demise. J Biomed Opt 24:1–14
Eggermont JJ (1990) On the pathophysiology of tinnitus; a review and a peripheral model. Hear Res 48:111–123
Eggermont JJ, Roberts LE (2015) Tinnitus: animal models and findings in humans. Cell Tissue Res 361:311–336
El Ganzoury MM, Kamel TB, Khalil LH, Seliem AM (2012) Cochlear dysfunction in children following cardiac bypass surgery. ISRN Pediatr 2012:375038
Evans P, Halliwell B (1999) Free radicals and hearing. Cause, consequence, and criteria. Ann N Y Acad Sci 884:19–40
Figueiredo RR, Azevedo AA, Penido NO (2016) Positive association between tinnitus and arterial hypertension. Front Neurol 7:171
Figueiredo RR, De Azevedo AA, Penido Nde O (2015) Tinnitus and arterial hypertension: a systematic review. Eur Arch Otorhinolaryngol 272:3089–3094
Fischel-Ghodsian N, Prezant TR, Chaltraw WE, Wendt KA, Nelson RA, Arnos KS, Falk RE (1997) Mitochondrial gene mutation is a significant predisposing factor in aminoglycoside ototoxicity. Am J Otolaryngol 18:173–178
Fortier C, Selye H (1949) Adrenocorticotrophic effect of stress after severance of the hypothalamo-hypophyseal pathways. Am J Physiol 159:433–439. illust

Frye MD, Ryan AF, Kurabi A (2019) Inflammation associated with noise-induced hearing loss. J Acoust Soc Am 146:4020
Furuta H, Mori N, Sato C, Hoshikawa H, Sakai S, Iwakura S, Doi K (1994) Mineralocorticoid type I receptor in the rat cochlea: mRNA identification by polymerase chain reaction (PCR) and in situ hybridization. Hear Res 78:175–180
Gauvin DV, Yoder J, Zimmermann ZJ, Tapp R (2018) Ototoxicity: the radical drum beat and rhythm of cochlear hair cell life and death. Int J Toxicol 37:195–206
Gomaa MA, Elmagd MH, Elbadry MM, Kader RM (2014) Depression, anxiety and stress scale in patients with tinnitus and hearing loss. Eur Arch Otorhinolaryngol 271:2177–2184
Guest H, Munro KJ, Prendergast G, Howe S, Plack CJ (2017) Tinnitus with a normal audiogram: relation to noise exposure but no evidence for cochlear synaptopathy. Hear Res 344:265–274
Gupta MA, Jarosz P, Gupta AK (2017) Posttraumatic stress disorder (PTSD) and the dermatology patient. Clin Dermatol 35:260–266
Haider HF, Bojić T, Ribeiro SF, Paço J, Hall DA, Szczepek AJ (2018) Pathophysiology of subjective tinnitus: triggers and maintenance. Front Neurosci 12
Haider HF, Ribeiro SF, Martins C, Ribeiro D, Trigueiros N, Szczepek AJ, Caria H, Hoare DJ, Paço J, Borrego LM (2020) Tinnitus, hearing loss and inflammatory processes in an older portuguese population. Int J Audiol 59:323–332
Han BR, Lin SC, Espinosa K, Thorne PR, Vlajkovic SM (2019) Inhibition of the adenosine A (2A) receptor mitigates excitotoxic injury in organotypic tissue cultures of the rat cochlea. Cell 8
Hasson D, Theorell T, Bergquist J, Canlon B (2013) Acute stress induces hyperacusis in women with high levels of emotional exhaustion. PLoS One 8:e52945
Hébert S, Canlon B, Hasson D, Magnusson Hanson LL, Westerlund H, Theorell T (2012) Tinnitus severity is reduced with reduction of depressive mood – a prospective population study in Sweden. PLoS One 7:e37733
Hébert S, Lupien SJ (2007) The sound of stress: blunted cortisol reactivity to psychosocial stress in tinnitus sufferers. Neurosci Lett 411:138–142
Heim C, Nemeroff CB (2009) Neurobiology of posttraumatic stress disorder. CNS Spectr 14:13–24
Hermann R, Biallas B, Predel HG, Petrowski K (2019) Physical versus psychosocial stress: effects on hormonal, autonomic, and psychological parameters in healthy young men. Stress 22:103–112
Herr RM, Bosch JA, Theorell T, Loerbroks A (2018) Bidirectional associations between psychological distress and hearing problems: an 18-year longitudinal analysis of the British household panel survey. Int J Audiol 57:816–824
Hirose K, Rutherford MA, Warchol ME (2017) Two cell populations participate in clearance of damaged hair cells from the sensory epithelia of the inner ear. Hear Res 352:70–81
Hoang KN, Dinh CT, Bas E, Chen S, Eshraghi AA, Van De Water TR (2009) Dexamethasone treatment of naïve organ of corti explants alters the expression pattern of apoptosis-related genes. Brain Res 1301:1–8
Horner K (2003) The emotional ear in stress. Neurosci Biobehav Rev 27:437–446
Hu BH, Zhang C, Frye MD (2018) Immune cells and non-immune cells with immune function in mammalian cochleae. Hear Res 362:14–24
Huang T, Cheng AG, Stupak H, Liu W, Kim A, Staecker H, Lefebvre PP, Malgrange B, Kopke R, Moonen G, Van De Water TR (2000) Oxidative stress-induced apoptosis of cochlear sensory cells: otoprotective strategies. Int J Dev Neurosci 18:259–270
Hunter RG, Seligsohn M, Rubin TG, Griffiths BB, Ozdemir Y, Pfaff DW, Datson NA, Mcewen BS (2016) Stress and corticosteroids regulate rat hippocampal mitochondrial DNA gene expression via the glucocorticoid receptor. Proc Natl Acad Sci U S A 113:9099–9104
Hwang JH, Chen JC, Yang SY, Wang MF, Chan YC (2011) Expression of tumor necrosis factor-α and interleukin-1β genes in the cochlea and inferior colliculus in salicylate-induced tinnitus. J Neuroinflammation 8:30
Iguchi H, Yamane H, Konishi K, Nakagawa T, Shibata S, Takayama M, Nishimura K, Sunami K, Nakai Y (1997) Asialo GM1-positive cells in mouse cochlea. Acta Otolaryngol Suppl 528:6–9

Jastreboff PJ (1990) Phantom auditory perception (tinnitus): mechanisms of generation and perception. Neurosci Res 8:221–254

Jung YH, Shin NY, Jang JH, Lee WJ, Lee D, Choi Y, Choi SH, Kang DH (2019) Relationships among stress, emotional intelligence, cognitive intelligence, and cytokines. Medicine (Baltimore) 98:e15345

Kapolowicz MR, Thompson LT (2020) Plasticity in limbic regions at early time points in experimental models of tinnitus. Front Syst Neurosci 13

Karst H, Berger S, Turiault M, Tronche F, Schütz G, Joëls M (2005) Mineralocorticoid receptors are indispensable for nongenomic modulation of hippocampal glutamate transmission by corticosterone. Proc Natl Acad Sci U S A 102:19204–19207

Kaur T, Clayman AC, Nash AJ, Schrader AD, Warchol ME, Ohlemiller KK (2019) Lack of fractalkine receptor on macrophages impairs spontaneous recovery of ribbon synapses after moderate noise trauma in C57BL/6 mice. Front Neurosci 13:620

Keithley EM (2020) Pathology and mechanisms of cochlear aging. J Neurosci Res 98(9):1674–1684. https://doi.org/10.1002/jnr.24439. Epub 2019 May 7

Kil SH, Kalinec F (2013) Expression and dexamethasone-induced nuclear translocation of glucocorticoid and mineralocorticoid receptors in guinea pig cochlear cells. Hear Res 299:63–78

Kim KX, Payne S, Yang-Hood A, Li SZ, Davis B, Carlquist J, V-Ghaffari B, Gantz JA, Kallogjeri D, Fitzpatrick JAJ, Ohlemiller KK, Hirose K, Rutherford MA (2019) Vesicular glutamatergic transmission in noise-induced loss and repair of cochlear ribbon synapses. J Neurosci 39:4434–4447

Kirschbaum C, Pirke KM, Hellhammer DH (1993) The 'Trier Social Stress Test' – a tool for investigating psychobiological stress responses in a laboratory setting. Neuropsychobiology 28:76–81

Kishimoto I, Okano T, Nishimura K, Motohashi T, Omori K (2019) Early development of resident macrophages in the mouse cochlea depends on Yolk Sac hematopoiesis. Front Neurol 10:1115

Kokotas H, Petersen MB, Willems PJ (2007) Mitochondrial deafness. Clin Genet 71:379–391

Landgrebe M, Langguth B, Rosengarth K, Braun S, Koch A, Kleinjung T, May A, De Ridder D, Hajak G (2009) Structural brain changes in tinnitus: grey matter decrease in auditory and non-auditory brain areas. Neuroimage 46:213–218

Le TN, Straatman LV, Lea J, Westerberg B (2017) Current insights in noise-induced hearing loss: a literature review of the underlying mechanism, pathophysiology, asymmetry, and management options. J Otolaryngol Head Neck Surg 46:41

Leaver AM, Seydell-Greenwald A, Rauschecker JP (2016) Auditory-limbic interactions in chronic tinnitus: challenges for neuroimaging research. Hear Res 334:49–57

Lechowicz U, Pollak A, Raj-Koziak D, Dziendziel B, Skarżyński PH, Skarżyński H, Ołdak M (2018) Tinnitus in patients with hearing loss due to mitochondrial DNA pathogenic variants. Eur Arch Otorhinolaryngol 275:1979–1985

Levin SG, Godukhin OV (2017) Modulating effect of cytokines on mechanisms of synaptic plasticity in the brain. Biochemistry (Mosc) 82:264–274

Liberman MC, Kujawa SG (2017) Cochlear synaptopathy in acquired sensorineural hearing loss: manifestations and mechanisms. Hear Res 349:138–147

Lie A, Skogstad M, Johannessen HA, Tynes T, Mehlum IS, Nordby KC, Engdahl B, Tambs K (2016) Occupational noise exposure and hearing: a systematic review. Int Arch Occup Environ Health 89:351–372

Liu W, Molnar M, Garnham C, Benav H, Rask-Andersen H (2018) Macrophages in the human cochlea: saviors or predators – a study using super-resolution immunohistochemistry. Front Immunol 9:223

Liu W, Rask-Andersen H (2019) Super-resolution immunohistochemistry study on CD4 and CD8 cells and the relation to macrophages in human cochlea. J Otol 14:1–5

Liu Y, Niu H, Zhu J, Zhao P, Yin H, Ding H, Gong S, Yang Z, Lv H, Wang Z (2019) Morphological neuroimaging biomarkers for tinnitus: evidence obtained by applying machine learning. Neural Plast 2019:1712342

Lockwood AH, Salvi RJ, Coad ML, Towsley ML, Wack DS, Murphy BW (1998) The functional neuroanatomy of tinnitus: evidence for limbic system links and neural plasticity. Neurology 50:114–120

Lupien SJ, Mcewen BS, Gunnar MR, Heim C (2009) Effects of stress throughout the lifespan on the brain, behaviour and cognition. Nat Rev Neurosci 10:434–445

Martin S, Henley JM, Holman D, Zhou M, Wiegert O, Van Spronsen M, Joëls M, Hoogenraad CC, Krugers HJ (2009) Corticosterone alters AMPAR mobility and facilitates bidirectional synaptic plasticity. PLoS One 4:e4714

Mazurek B, Boecking B, Brueggemann P (2019) Association between stress and tinnitus-new aspects. Otol Neurotol 40:e467–e473

Mazurek B, Haupt H, Joachim R, Klapp BF, Stöver T, Szczepek AJ (2010a) Stress induces transient auditory hypersensitivity in rats. Hear Res 259:55–63

Mazurek B, Stöver T, Haupt H, Klapp B, Adli M, Gross J, Szczepek A (2010b) The significance of stress: its role in the auditory system and the pathogenesis of tinnitus. HNO 58:162–172

Mazurek B, Szczepek A, Hebert S (2015) Stress and tinnitus. HNO 63:258–265

Mazurek B, Winter E, Fuchs J, Haupt H, Gross J (2003) Susceptibility of the hair cells of the newborn rat cochlea to hypoxia and ischemia. Hear Res 182:2–8

Mcgregor BA, Murphy KM, Albano DL, Ceballos RM (2016) Stress, cortisol, and B lymphocytes: a novel approach to understanding academic stress and immune function. Stress 19:185–191

Michaelides A, Zis P (2019) Depression, anxiety and acute pain: links and management challenges. Postgrad Med 131:438–444

Misiewicz Z, Iurato S, Kulesskaya N, Salminen L, Rodrigues L, Maccarrone G, Martins J, Czamara D, Laine MA, Sokolowska E, Trontti K, Rewerts C, Novak B, Volk N, Park DI, Jokitalo E, Paulin L, Auvinen P, Voikar V, Chen A, Erhardt A, Turck CW, Hovatta I (2019) Multi-omics analysis identifies mitochondrial pathways associated with anxiety-related behavior. PLoS Genet 15:e1008358

Moghaddam B (1993) Stress preferentially increases extraneuronal levels of excitatory amino acids in the prefrontal cortex: comparison to hippocampus and basal ganglia. J Neurochem 60:1650–1657

Moossavi A, Sadeghijam M, Akbari M (2019) The hypothetical relation between the degree of stress and auditory cortical evoked potentials in tinnitus sufferers. Med Hypotheses 130:109266

Mosnier I, Teixeira M, Loiseau A, Fernandes I, Sterkers O, Amiel C, Ferrary E (2001) Effects of acute and chronic hypertension on the labyrinthine barriers in rat. Hear Res 151(1–2):227–236. https://doi.org/10.1016/s0378-5955(00)00229-x

Mostafa H, Saad M, El-Attar A, Ahmed G, Berrettini S, Forli F, Siciliano G, Mancuso M (2014) Mitochondrial DNA (mtDNA) haplotypes and dysfunctions in presbyacusis. Acta Otorhinolaryngol Ital 34:54–61

Mühlau M, Rauschecker JP, Oestreicher E, Gaser C, Röttinger M, Wohlschläger AM, Simon F, Etgen T, Conrad B, Sander D (2006) Structural brain changes in tinnitus. Cereb Cortex 16:1283–1288

National Guideline (2020) Evidence review for psychological therapies: tinnitus: assessment and management: evidence review L. National Institute for Health and Care Excellence (UK), London. Copyright © NICE 2020

O'reilly M, Young L, Kirkwood NK, Richardson GP, Kros CJ, Moore AL (2019) Gentamicin affects the bioenergetics of isolated mitochondria and collapses the mitochondrial membrane potential in cochlear sensory hair cells. Front Cell Neurosci 13:416

Op De Beeck K, Schacht J, Van Camp G (2011) Apoptosis in acquired and genetic hearing impairment: the programmed death of the hair cell. Hear Res 281:18–27

Oyola MG, Handa RJ (2017) Hypothalamic-pituitary-adrenal and hypothalamic-pituitary-gonadal axes: sex differences in regulation of stress responsivity. Stress 20:476–494

Pickles JO (2004) Mutation in mitochondrial DNA as a cause of presbyacusis. Audiol Neurootol 9:23–33

Pienkowski M (2018) Prolonged exposure of CBA/Ca mice to moderately loud noise can cause cochlear synaptopathy but not tinnitus or hyperacusis as assessed with the acoustic startle reflex. Trends Hear 22:2331216518758109
Poirrier AL, Pincemail J, Van Den Ackerveken P, Lefebvre PP, Malgrange B (2010) Oxidative stress in the cochlea: an update. Curr Med Chem 17:3591–3604
Polman JAE, Welten JE, Bosch DS, De Jonge RT, Balog J, Van Der Maarel SM, De Kloet ER, Datson NA (2012) A genome-wide signature of glucocorticoid receptor binding in neuronal PC12 cells. BMC Neurosci 13:118
Probst T, Pryss R, Langguth B, Schlee W (2016) Emotional states as mediators between tinnitus loudness and tinnitus distress in daily life: results from the "TrackYourTinnitus" application. Sci Rep 6:20382
Puel JL, Ruel J, Guitton M, Wang J, Pujol R (2002) The inner hair cell synaptic complex: physiology, pharmacology and new therapeutic strategies. Audiol Neurootol 7:49–54
Qu T, Qi Y, Yu S, Du Z, Wei W, Cai A, Wang J, Nie B, Liu K, Gong S (2019) Dynamic changes of functional neuronal activities between the auditory pathway and limbic systems contribute to noise-induced tinnitus with a normal audiogram. Neuroscience 408:31–45
Ralli M, Troiani D, Podda MV, Paciello F, Eramo SL, De Corso E, Salvi R, Paludetti G, Fetoni AR (2014) The effect of the NMDA channel blocker memantine on salicylate-induced tinnitus in rats. Acta Otorhinolaryngol Ital 34:198–204
Rauschecker JP, Leaver AM, Mühlau M (2010) Tuning out the noise: limbic-auditory interactions in tinnitus. Neuron 66:819–826
Reddy TE, Pauli F, Sprouse RO, Neff NF, Newberry KM, Garabedian MJ, Myers RM (2009) Genomic determination of the glucocorticoid response reveals unexpected mechanisms of gene regulation. Genome Res 19:2163–2171
Ryan D, Bauer CA (2016) Neuroscience of tinnitus. Neuroimaging Clin N Am 26:187–196
Rybak LP, Mukherjea D, Ramkumar V (2019) Mechanisms of cisplatin-induced ototoxicity and prevention. Semin Hear 40:197–204
Sahley TL, Hammonds MD, Musiek FE (2013) Endogenous dynorphins, glutamate and N-methyl-d-aspartate (NMDA) receptors may participate in a stress-mediated type-I auditory neural exacerbation of tinnitus. Brain Res 1499:80–108
Sahley TL, Nodar RH (2001) A biochemical model of peripheral tinnitus. Hear Res 152:43–54
Salviati M, Bersani FS, Terlizzi S, Melcore C, Panico R, Romano GF, Valeriani G, Macri F, Altissimi G, Mazzei F, Testugini V, Latini L, Delle Chiaie R, Biondi M, Cianfrone G (2014) Tinnitus: clinical experience of the psychosomatic connection. Neuropsychiatr Dis Treat 10:267–275
Sawada S, Mori N, Mount RJ, Harrison RV (2001) Differential vulnerability of inner and outer hair cell systems to chronic mild hypoxia and glutamate ototoxicity: insights into the cause of auditory neuropathy. J Otolaryngol 30:106–114
Seifert E, Lamprecht-Dinnesen A, Asfour B, Rotering H, Bone HG, Scheld HH (1998) The influence of body temperature on transient evoked otoacoustic emissions. Br J Audiol 32:387–398
Selye H (1937) The significance of the adrenals for adaptation. Science 85:247–248
Selye H (1950) Stress and the general adaptation syndrome. Br Med J 1:1383–1392
Selye H, Fortier C (1949) Adaptive reactions to stress. Res Publ Assoc Res Nerv Ment Dis 29:3–18
Sheth S, Mukherjea D, Rybak LP, Ramkumar V (2017) Mechanisms of cisplatin-induced ototoxicity and otoprotection. Front Cell Neurosci 11:338
Shirane M, Harrison RV (1987) The effects of hypoxia on sensory cells of the cochlea in chinchilla. Scanning Microsc 1:1175–1183
Slavich GM, Shields GS (2018) Assessing lifetime stress exposure using the stress and adversity inventory for adults (adult STRAIN): an overview and initial validation. Psychosom Med 80:17–27
Sliwinska-Kowalska M, Davis A (2012) Noise-induced hearing loss. Noise Health 14:274–280
Szczepek A, Mazurek B (2017) Tinnitus and stress. Springer, Cham

Szczepek AJ, Haupt H, Klapp BF, Olze H, Mazurek B (2014) Biological correlates of tinnitus-related distress: an exploratory study. Hear Res 318:23–30

Tahera Y, Meltser I, Johansson P, Canlon B (2006a) Restraint stress modulates glucocorticoid receptors and nuclear factor kappa B in the cochlea. Neuroreport 17:879–882

Tahera Y, Meltser I, Johansson P, Hansson AC, Canlon B (2006b) Glucocorticoid receptor and nuclear factor-kappa B interactions in restraint stress-mediated protection against acoustic trauma. Endocrinology 147:4430–4437

Ten Cate WJ, Curtis LM, Small GM, Rarey KE (1993) Localization of glucocorticoid receptors and glucocorticoid receptor mRNAs in the rat cochlea. Laryngoscope 103:865–871

Waechter S, Brännström KJ (2015) The impact of tinnitus on cognitive performance in normal-hearing individuals. Int J Audiol 54(11):845–851

Wang W, Zhang LS, Zinsmaier AK, Patterson G, Leptich EJ, Shoemaker SL, Yatskievych TA, Gibboni R, Pace E, Luo H, Zhang J, Yang S, Bao S (2019) Neuroinflammation mediates noise-induced synaptic imbalance and tinnitus in rodent models. PLoS Biol 17:e3000307

Yang P, Ma W, Zheng Y, Yang H, Lin H (2015) A systematic review and meta-analysis on the association between hypertension and tinnitus. Int J Hypertens 2015:583493

Yao X, Rarey KE (1996) Localization of the mineralocorticoid receptor in rat cochlear tissue. Acta Otolaryngol 116:493–496

Yip J, Geng X, Shen J, Ding Y (2016) Cerebral gluconeogenesis and diseases. Front Pharmacol 7:521

Yu C-Y, Mayba O, Lee JV, Tran J, Harris C, Speed TP, Wang J-C (2010) Genome-wide analysis of glucocorticoid receptor binding regions in adipocytes reveal gene network involved in triglyceride homeostasis. PLoS One 5:e15188

Zirke N, Seydel C, Arsoy D, Klapp BF, Haupt H, Szczepek AJ, Olze H, Goebel G, Mazurek B (2013) Analysis of mental disorders in tinnitus patients performed with composite international diagnostic interview. Qual Life Res 22:2095–2104

Zöger S, Svedlund J, Holgers KM (2006) Relationship between tinnitus severity and psychiatric disorders. Psychosomatics 47:282–288

Zuo J, Curtis LM, Yao X, Ten Cate WJ, Bagger-Sjöbäck D, Hultcrantz M, Rarey KE (1995) Glucocorticoid receptor expression in the postnatal rat cochlea. Hear Res 87:220–227

Psychological Comorbidities of Tinnitus

Sylvie Hébert

Contents

Abstract In this chapter, I address the topic of tinnitus in the context of the patient's trajectory of care, with special attention to psychological comorbidities. Although most patients will cope with tinnitus and need no more than information and reassurance from professionals, a proportion of patients will need more supportive management. Assessment of psychological comorbidities is important to determine how urgent they should be seen in the clinic and their specific needs. The most frequent complaints are stress, depression, and anxiety. Although the direction of this relationship is still unclear (are comorbidities at the origin of tinnitus or are they a consequence of it), it is evident that the more serious comorbidities are at the onset of tinnitus, the worse the prognosis. Therefore, an assessment at the initial visit in the clinic is of utmost importance. There are valid and reliable psychometric tools to quickly draw a portrait of the psychological state of patients that can be used by audiologists, psychologists, or doctors. Therapeutic avenues can then be discussed with the patients to ensure them the best support possible.

Keywords Comorbidities · Depression · Psychological state · Stress · Tinnitus

S. Hébert (✉)
École d'orthophonie et d'audiologie, Faculty of Medicine, Université de Montréal, Montréal, QC, Canada
e-mail: sylvie.hebert@umontreal.ca

Curr Topics Behav Neurosci (2021) 51: 349–360
https://doi.org/10.1007/7854_2021_218
Published Online: 20 February 2021

1 Introduction

Individuals who have recently noticed, with a variable degree of alarm, the constant presence of an unwanted sound in their ears will usually turn to a health professional (most often a family doctor or ENT specialist, McFerran et al. 2018; Stockdale et al. 2017; Wu et al. 2018) to enquire about a possible formal medical *diagnosis*.

2 Diagnosis and Assessment of Tinnitus

A thorough audiological and psychological *assessment* of tinnitus (and patient's hearing status) is an important next step in tinnitus management. Assessment of the *psychoacoustic properties* of tinnitus (e.g., pitch and loudness, maskability, depth and duration of residual inhibition, location) and *audiological comorbidities* (e.g., hearing damage, hyperacusis) will help with selection and adjustment of appropriate sound therapy, if needed (Searchfield et al. 2017; Tyler et al. 2020). Assessment of the *psychological* comorbidities of tinnitus is also essential (Cima et al. 2019). Indeed, while the majority of individuals adapt to tinnitus without much problem (Bhatt et al. 2016), the experience of tinnitus as an unwelcome at best, and even threatening, sound can impair daily life significantly. Even diagnosis from a qualified health professional can be extremely distressing in itself, let alone learning that tinnitus will be chronic, that there is no universal cure to eliminate the sound, and that its time course is unpredictable. Understandably, the patients' ultimate preference would be for the sound to be removed in order to recover silence, or at least for a reduction in loudness (Pryce et al. 2018). Health professionals' goals are to bring patients to have realistic hopes in the face of available therapeutic options, that is, to increase knowledge about tinnitus, to dampen tinnitus awareness, and to relieve anxiety and stress (Husain et al. 2018). Recognition of the extent and severity of psychological comorbidities may help health professionals in the planning of the best possible management for individual patients. Tinnitus patients may have insights about the fact that their emotional reaction to the sound is somewhat distinct from the sound itself, i.e., that at times they have less tolerance towards their tinnitus and experience more distress, whereas at other times, they have the ability to cope with sound of a similar level (Colagrosso et al. 2019). Therefore, the management of psychological comorbidities provides a handle to decrease distress and improve quality of life.

3 Stress, Anxiety, and Depression Are Frequent Comorbidities of Tinnitus

Tinnitus onset shortly after the death of a loved one was reported long ago (Curtis 1841). Although obviously not all patients have stress at the inception of tinnitus, *emotional stress* is often reported as a contributor to the etiology of tinnitus (Khedr et al. 2010; Kreuzer et al. 2012; Probst et al. 2016b). A recent review on the association between stress and tinnitus onset estimated its prevalence at 13.5% to 28.3% in three studies including over 200 patients (Elarbed et al. 2020). However, tinnitus patients report different autobiographical memories than healthy controls, self-reporting alone may be biased, and therefore, such percentages might be either over- or under- estimate the stress-tinnitus axis (Andersson et al. 2013). Therefore, further studies are needed using precise definition of stress at recent tinnitus onset, a structured interview considering event recall, and independent corroboration by a third party.

A substantial proportion of patients subjectively report that emotional or mental stress makes their tinnitus worse (Pan et al. 2015). Although intuitive, this assumption is difficult to reproduce in the laboratory using controlled stressful tasks given the time constraints for measurement (e.g., Hébert and Lupien 2007, but see Betz et al. (2017) for a perceived increase in "tinnitus presence" concomitant with stress). Nonetheless, higher levels of self-reported tinnitus distress are associated with higher levels of self-reported stress (Biehl et al. 2019; Ciminelli et al. 2018; Hébert and Lupien 2009; Probst et al. 2016a) and with the occurrence of stress symptoms such as muscle tension, pain, and headaches (Scott and Lindberg 2000). Work-related stress has also been reported to be associated with worse tinnitus, as in orchestra musicians (Hasson et al. 2009), workers in large companies (Herr et al. 2016), and operators in call centers (Lin et al. 2009).

Large-scale population studies focusing on the relationship between stress and tinnitus have reported that stress is a significant risk factor for severe tinnitus. From the analysis of questionnaires returned by 9,756 working individuals (16–64 years old) enrolled in the Swedish Longitudinal Occupational Survey of Health, Hasson et al. (2011) reported clear and mostly linear correlations between hearing problems and work-related stressors, long-term illness, and several other health variables. In a sub-sample of the same study in which the focus was set more specifically on tinnitus (Hebert et al. 2012), specific factors – namely hearing loss, uncomfortable loudness levels, and long-term stress – were identified as significant predictors of tinnitus prevalence; moreover, long-term exposure to stressful conditions was highly correlated with tinnitus severity (Hebert et al. 2012). In yet another large-scale study, exposure to either stress or noise increased tinnitus prevalence about equally, whereas exposure to stress was the main determinant of transition from mild to severe tinnitus, in either sex (Baigi et al. 2011).

In smaller scale studies, psychosocial stress, depressive symptoms, and anxiety at tinnitus onset have all been identified as predictive factors of severe tinnitus (Holgers et al. 2000, 2005; Wallhausser-Franke et al. 2017). That is, the more severe these comorbidities are at the onset of tinnitus, the greater the subsequent tinnitus distress will be. In the context of catastrophic thinking – a tendency to overstate the impact of a condition and to expect negative consequences – tinnitus patients who misinterpret their tinnitus as catastrophic at onset have higher subsequent tinnitus-related distress scores and worsening quality of life (Cima et al. 2011; Weise et al. 2013). These data underline the importance of a thorough assessment as early as possible since the prognosis will differ according to the initial psychological state of the patients. Validated questionnaires that are used for assessment in this context are the Beck-Depression Inventory II (BDI-II (Beck et al. 1996), depressive symptomatology), Anxiety State and Trait Inventory (ASTA-STAI (Spielberger et al. 1970), trait and state anxiety), and Hospital Anxiety and Depression Scale (HADS (Zigmond and Snaith 1983), combined anxiety and depression). Validated stress questionnaires such as the Perceived Stress Questionnaire (Levenstein et al. 1993) and the Depression Anxiety Stress Scale (Lovibond and Lovibond 1995) are used less frequently but should also be considered for full assessment.

Using the objective measure of the stress hormone cortisol, studies have shown overall normal diurnal patterns in tinnitus (Hébert et al. 2004), but blunted cortisol responses to acute psychosocial stress (Hébert and Lupien 2007), which were interpreted as an indication of possible exhaustion of the hypothalamic-pituitary-adrenal axis (Simoens and Hébert 2012). One study (Kim et al. 2014) examining blood levels of norepinephrine (NE), epinephrine, a metabolite of serotonin (5-HIAA), and cortisol, did not find any differences in the global levels of these hormones between tinnitus and control groups matched for depression; however, a greater proportion of tinnitus was identified among individuals with elevated NE or 5-HIAA. In a pilot study looking at the response of young men to mental stress (one measure before and one after), Alsalman et al. (2016) did not find any differences in cortisol and neopterin hormones between tinnitus and control participants, but they did identify a blunted response of salivary α-amylase in the tinnitus group, reflecting an impaired sympathetic reactivity to stress. Likewise, Betz and colleagues reported a blunted heart rate response to stress, but without modification of heart rate variability (Betz et al. 2017). Heinecke and colleagues did not find any differences between tinnitus and controls in their stress responses when measuring skin conductance levels and muscle (EMG) activity (Heinecke et al. 2008). In general, the investigation of hormones and nervous system activity requires stringent inclusion and exclusion criteria, as well as several measurement time points and appropriate sample sizes. Not all studies conform to these conditions, and this can explain at least partly the conflicting data or lack of significant findings. In addition, tinnitus duration (recent versus chronic) and age at onset may be important factors to control (patients with later onset suffer more than those with an earlier onset in life, Schlee et al. 2011). Therefore, there is room for additional well-controlled and adequately powered research using stress biomarkers.

4 Depression or Severe Tinnitus?

Depression, either clinical or subclinical, is a frequent comorbidity of tinnitus. In clinical depression, defined by feelings of extreme sadness and hopelessness, reduced energy and decrease in activity lasting for at least 2 weeks, as well as rumination of the past (to distinguish from anxiety), depression has been identified for over 20 years as a comorbidity of tinnitus (Halford and Anderson 1991) and their association remains an important research topic (Husain 2020; Langguth et al. 2011; Weidt et al. 2016). Numerous studies have reported significant and rather strong correlations between questionnaires assessing depressive symptoms (sometimes along with anxiety symptoms) and those assessing tinnitus distress or handicap (Kehrle et al. 2016; Oishi et al. 2011). On the basis of some overlap between questions, it has been suggested that the correlations may have been artificially inflated (Ooms et al. 2011, 2012). However, it is likely that such overlap might be the direct consequence of the similarity of symptoms between distressful tinnitus and depression and that tinnitus questionnaires were developed to reliably reflect complaints of tinnitus patients (Langguth et al. 2011). In addition, cognitive (episodic/ autobiographical memory) patterns differ between patients with depression versus tinnitus (Andersson et al. 2013); this is in accordance with the notion that, whereas tinnitus and depression have some degree of phenomenological overlap, they display distinct psychopathological features. Longitudinal data suggest that tinnitus-related distress decreases concomitantly with depressive symptoms (Hébert et al. 2012). However there is conflicting evidence for the effectiveness of antidepressants in tinnitus (e.g., tricyclic antidepressants, selective serotonin reuptake inhibitor) (Baldo et al. 2012; Chang and Wu 2012; Oishi et al. 2011; Robinson et al. 2005) as the quality of evidence is limited by differences between studies with regard to the selection criteria (e.g., patient selection, assessment of depression and tinnitus-related distress, presence or absence of clinical depression), the treatment regimen (e.g., dosage), and the study design (e.g., presence of control groups).

5 Therapeutic Avenues for Decreasing Comorbidities of Tinnitus

A substantial number of meta-analyses deriving the highest level of evidence from clinical studies have reported that cognitive behavioral therapy (CBT) is efficient and cost-effective for the improvement of mood (i.e., depression and anxiety) and quality of life, as well as to decrease tinnitus-related distress as measured with questionnaires (R. F. Cima et al. 2012; Fuller et al. 2020; Grewal et al. 2014; Hesser et al. 2011; Landry et al. 2020; Maes et al. 2014). CBT is a clinically recommended solution for treating tinnitus according to the clinical practice guideline by the American Academy of Otolaryngology – Head and Neck Surgery (Tunkel et al. 2014). CBT was originally designed to treat depression and focuses on the development of personal coping strategies to change unhelpful patterns in cognition, behaviors, and emotional regulation. Face-to-face and internet-delivered CBT is also efficient to treat anxiety (Guo et al. 2020), another frequent comorbidity of tinnitus (Karaaslan et al. 2020; Pattyn et al. 2016).

When considering associations between stress and tinnitus, and despite uncertainties concerning the direction of the relationship (cause or consequence), it is natural that a therapeutic goal would be to reduce stress (Stattrop et al. 2013). However, it remains to be tested, with adequate stress questionnaires and measurements, whether CBT and any other forms of therapies aiming at specifically reducing stress might be efficient and for how long (see Elarbed et al. 2020, for the most recent and thorough scoping review on stress and tinnitus).

6 Final Remarks

Despite the established efficiency of CBT to improve tinnitus patients' quality of life, some individuals cannot or do not want to participate in this type of intervention, whereas others do not improve significantly. Therefore, for CBT as well as for other interventions, an interesting and new research avenue would be to develop objective measures to predict which patients will do better, why, and in how long, according to a personalized approach. For instance, what type of patients' initial symptoms (Uckelstam et al. 2019), what type of tinnitus, and socio-demographic profiles (Joutsenniemi et al. 2012), or what therapists' characteristics (Tschuschke et al. 2015) are the most important predictors for successful therapy in patients with different features? Are there indications in patients' language (Goodwin et al. 2019), facial expressions (Dibeklioglu et al. 2018), body movements (Kacem et al. 2018), or dyadic interactions (Scherer et al. 2014), which can predict outcome? Are there mandatory "stages," or a specific trajectory (De Smet et al. 2020), which patients must pass while evolving towards adaptation/acceptance stages at which tinnitus becomes increasingly tolerable? Answering these questions would undoubtedly help professionals to set goals for improved personalized therapy of their severely distressed tinnitus patients.

References

Alsalman OA, Tucker D, Vanneste S (2016) Salivary stress-related responses in tinnitus: a preliminary study in young male subjects with tinnitus. Front Neurosci 10:338. https://doi.org/10.3389/fnins.2016.00338

Andersson G, Hesser H, Cima RF, Weise C (2013) Autobiographical memory specificity in patients with tinnitus versus patients with depression and normal controls. Cogn Behav Ther 42 (2):116–126. https://doi.org/10.1080/16506073.2013.792101

Baigi A, Oden A, Almlid-Larsen V, Barrenas ML, Holgers KM (2011) Tinnitus in the general population with a focus on noise and stress: a public health study. Ear Hear 32(6):787–789. https://doi.org/10.1097/AUD.0b013e31822229bd

Baldo P, Doree C, Molin P, McFerran D, Cecco S (2012) Antidepressants for patients with tinnitus. Cochrane Database Syst Rev 9:CD003853

Beck AT, Steer RA, Brown GK (1996) BDI-II, beck depression inventory: manual, 2nd edn. Psychological Corporation, San Antonio

Betz LT, Muhlberger A, Langguth B, Schecklmann M (2017) Stress reactivity in chronic tinnitus. Sci Rep 7:41521. https://doi.org/10.1038/srep41521

Bhatt JM, Lin HW, Bhattacharyya N (2016) Prevalence, severity, exposures, and treatment patterns of tinnitus in the United States. JAMA Otolaryngol Head Neck Surg 142(10):959–965. https://doi.org/10.1001/jamaoto.2016.1700

Biehl R, Boecking B, Brueggemann P, Grosse R, Mazurek B (2019) Personality traits, perceived stress, and tinnitus-related distress in patients with chronic tinnitus: support for a vulnerability-stress model. Front Psychol 10:3093. https://doi.org/10.3389/fpsyg.2019.03093

Chang JP, Wu CC (2012) Serotonin-norepinephrine reuptake inhibitor treatment for tinnitus and depression. J Clin Psychopharmacol 32(5):729

Cima RF, Crombez G, Vlaeyen JW (2011) Catastrophizing and fear of tinnitus predict quality of life in patients with chronic tinnitus. Ear Hear 32(5):634–641. https://doi.org/10.1097/AUD.0b013e31821106dd

Cima RF, Maes IH, Joore MA, Scheyen DJ, El Refaie A, Baguley DM et al (2012) Specialised treatment based on cognitive behaviour therapy versus usual care for tinnitus: a randomised controlled trial. Lancet 379(9830):1951–1959. https://doi.org/10.1016/S0140-6736(12)60469-3

Cima R, Mazurek B, Haider H, Kikidis D, Lapira A, Norena A, Hoare DJ (2019) A multidisciplinary European guideline for tinnitus: diagnostics, assessment, and treatment. HNO 67(Suppl 1):10–42. https://doi.org/10.1007/s00106-019-0633-7

Ciminelli P, Machado S, Palmeira M, Carta MG, Beirith SC, Nigri ML et al (2018) Tinnitus: the sound of stress? Clin Pract Epidemiol Ment Health 14:264–269. https://doi.org/10.2174/1745017901814010264

Colagrosso EMG, Fournier P, Fitzpatrick EM, Hebert S (2019) A qualitative study on factors modulating tinnitus experience. Ear Hear 40(3):636–644. https://doi.org/10.1097/AUD.0000000000000642

Curtis HJ (1841) Tinnitus Aurium. Lancet 36(940):828–829

De Smet MM, Meganck R, Truijens F, De Geest R, Cornelis S, Norman UA, Desmet M (2020) Change processes underlying "good outcome": a qualitative study on recovered and improved patients' experiences in psychotherapy for major depression. Psychother Res 30(7):948–964. https://doi.org/10.1080/10503307.2020.1722329

Dibeklioglu H, Hammal Z, Cohn JF (2018) Dynamic multimodal measurement of depression severity using deep autoencoding. IEEE J Biomed Health Inform 22(2):525–536. https://doi.org/10.1109/JBHI.2017.2676878

Elarbed A, Fackrell K, Baguley DM, Hoare DJ (2020) Tinnitus and stress in adults: a scoping review. Int J Audiol:1–12. https://doi.org/10.1080/14992027.2020.1827306

Fuller T, Cima R, Langguth B, Mazurek B, Vlaeyen JW, Hoare DJ (2020) Cognitive behavioural therapy for tinnitus. Cochrane Database Syst Rev 1:CD012614. https://doi.org/10.1002/14651858.CD012614.pub2

Goodwin BJ, Constantino MJ, Westra HA, Button ML, Antony MM (2019) Patient motivational language in the prediction of symptom change, clinically significant response, and time to response in psychotherapy for generalized anxiety disorder. Psychotherapy (Chic) 56 (4):537–548. https://doi.org/10.1037/pst0000269

Grewal R, Spielmann PM, Jones SE, Hussain SS (2014) Clinical efficacy of tinnitus retraining therapy and cognitive behavioural therapy in the treatment of subjective tinnitus: a systematic review. J Laryngol Otol 128(12):1028–1033. https://doi.org/10.1017/S0022215114002849

Guo S, Deng W, Wang H, Liu J, Liu X, Yang X et al (2020) The efficacy of internet-based cognitive behavioural therapy for social anxiety disorder: a systematic review and meta-analysis. Clin Psychol Psychother. https://doi.org/10.1002/cpp.2528

Halford JB, Anderson SD (1991) Anxiety and depression in tinnitus sufferers. J Psychosom Res 35 (4–5):383–390. http://www.ncbi.nlm.nih.gov/entrez/query.fcgi?cmd=Retrieve&db=PubMed&dopt=Citation&list_uids=1920169

Hasson D, Theorell T, Liljeholm-Johansson Y, Canlon B (2009) Psychosocial and physiological correlates of self-reported hearing problems in male and female musicians in symphony orchestras. Int J Psychophysiol 74(2):93–100. https://doi.org/10.1016/j.ijpsycho.2009.07.009

Hasson D, Theorell T, Wallen MB, Leineweber C, Canlon B (2011) Stress and prevalence of hearing problems in the Swedish working population. BMC Public Health 11:130. https://doi.org/10.1186/1471-2458-11-130

Hébert S, Lupien SJ (2007) The sound of stress: blunted cortisol reactivity to psychosocial stress in tinnitus sufferers. Neurosci Lett 411(2):138–142. http://www.ncbi.nlm.nih.gov/entrez/query.fcgi?cmd=Retrieve&db=PubMed&dopt=Citation&list_uids=17084027

Hébert S, Lupien SJ (2009) Salivary cortisol levels, subjective stress, and tinnitus intensity in tinnitus sufferers during noise exposure in the laboratory. Int J Hyg Environ Health 212 (1):37–44. https://doi.org/10.1016/j.ijheh.2007.11.005

Hébert S, Paiement P, Lupien SJ (2004) A physiological correlate for the intolerance to both internal and external sounds. Hear Res 190(1–2):1–9. http://www.ncbi.nlm.nih.gov/entrez/query.fcgi?cmd=Retrieve&db=PubMed&dopt=Citation&list_uids=15051125

Hebert S, Canlon B, Hasson D (2012) Emotional exhaustion as a predictor of tinnitus. Psychother Psychosom 81(5):324–326

Hébert S, Canlon B, Hasson D, Magnusson Hanson LL, Westerlund H, Theorell T (2012) Tinnitus severity is reduced with reduction of depressive mood--a prospective population study in Sweden. PLoS One 7(5):e37733. https://doi.org/10.1371/journal.pone.0037733

Heinecke K, Weise C, Schwarz K, Rief W (2008) Physiological and psychological stress reactivity in chronic tinnitus. J Behav Med 31(3):179–188. https://doi.org/10.1007/s10865-007-9145-0

Herr RM, Loerbroks A, Bosch JA, Seegel M, Schneider M, Schmidt B (2016) Associations of organizational justice with tinnitus and the mediating role of depressive symptoms and burnout-findings from a cross-sectional study. Int J Behav Med 23(2):190–197. https://doi.org/10.1007/s12529-015-9505-z

Hesser H, Weise C, Westin VZ, Andersson G (2011) A systematic review and meta-analysis of randomized controlled trials of cognitive-behavioral therapy for tinnitus distress. Clin Psychol Rev 31(4):545–553. https://doi.org/10.1016/j.cpr.2010.12.006

Holgers KM, Erlandsson SI, Barrenas ML (2000) Predictive factors for the severity of tinnitus. Audiology 39(5):284–291. http://www.ncbi.nlm.nih.gov/entrez/query.fcgi?cmd=Retrieve&db=PubMed&dopt=Citation&list_uids=11093613

Holgers KM, Zoger S, Svedlund K (2005) Predictive factors for development of severe tinnitus suffering-further characterisation. Int J Audiol 44(10):584–592. http://www.ncbi.nlm.nih.gov/entrez/query.fcgi?cmd=Retrieve&db=PubMed&dopt=Citation&list_uids=16315449

Husain FT (2020) Perception of, and reaction to, tinnitus: the depression factor. Otolaryngol Clin N Am 53(4):555–561. https://doi.org/10.1016/j.otc.2020.03.005

Husain FT, Gander PE, Jansen JN, Shen S (2018) Expectations for tinnitus treatment and outcomes: a survey study of audiologists and patients. J Am Acad Audiol 29(4):313–336. https://doi.org/10.3766/jaaa.16154

Joutsenniemi K, Laaksonen MA, Knekt P, Haaramo P, Lindfors O (2012) Prediction of the outcome of short- and long-term psychotherapy based on socio-demographic factors. J Affect Disord 141 (2–3):331–342. https://doi.org/10.1016/j.jad.2012.03.027

Kacem A, Hammal Z, Daoudi M, Cohn J (2018) Detecting depression severity by interpretable representations of motion dynamics. In: Proceedings of international conf autom face gesture recognit, vol 2018. pp 739–745. https://doi.org/10.1109/FG.2018.00116

Karaaslan O, Kantekin Y, Hacimusalar Y, Dagistan H (2020) Anxiety sensitivities, anxiety and depression levels, and personality traits of patients with chronic subjective tinnitus: a case-control study. Int J Psychiatry Clin Pract:1–6. https://doi.org/10.1080/13651501.2020.1757117

Kehrle HM, Sampaio AL, Granjeiro RC, de Oliveira TS, Oliveira CA (2016) Tinnitus annoyance in normal-hearing individuals: correlation with depression and anxiety. Ann Otol Rhinol Laryngol 125(3):185–194. https://doi.org/10.1177/0003489415606445

Khedr, E. M., Ahmed, M. A., Shawky, O. A., Mohamed, E. S., El Attar, G. S., & Mohammad, K. A. (2010). Epidemiological study of chronic tinnitus in Assiut, Egypt. Neuroepidemiology, 35(1), 45–52. http://ovidsp.ovid.com/ovidweb.cgi?T=JS&CSC=Y&NEWS=N&PAGE=fulltext&D=medl&AN=20389124

Kim DK, Chung DY, Bae SC, Park KH, Yeo SW, Park SN (2014) Diagnostic value and clinical significance of stress hormones in patients with tinnitus. Eur Arch Otorhinolaryngol 271 (11):2915–2921. https://doi.org/10.1007/s00405-013-2785-5

Kreuzer PM, Landgrebe M, Schecklmann M, Staudinger S, Langguth B, Group, T. R. I. D. S (2012) Trauma-associated tinnitus: audiological, demographic and clinical characteristics. PLoS One 7 (9):e45599. https://doi.org/10.1371/journal.pone.0045599

Landry EC, Sandoval XCR, Simeone CN, Tidball G, Lea J, Westerberg BD (2020) Systematic review and network meta-analysis of cognitive and/or behavioral therapies (CBT) for tinnitus. Otol Neurotol 41(2):153–166. https://doi.org/10.1097/MAO.0000000000002472

Langguth B, Landgrebe M, Kleinjung T, Sand GP, Hajak G (2011) Tinnitus and depression. World J Biol Psychiatry. https://doi.org/10.3109/15622975.2011.575178

Levenstein S, Prantera C, Varvo V, Scribano ML, Berto E, Luzi C, Andreoli A (1993) Development of the perceived stress questionnaire: a new tool for psychosomatic research. J Psychosom Res 37(1):19–32. https://doi.org/10.1016/0022-3999(93)90120-5

Lin YH, Chen CY, Lu SY (2009) Physical discomfort and psychosocial job stress among male and female operators at telecommunication call centers in Taiwan. Appl Ergon 40(4):561–568. https://doi.org/10.1016/j.apergo.2008.02.024

Lovibond PF, Lovibond SH (1995) The structure of negative emotional states: comparison of the depression anxiety stress scales (DASS) with the beck depression and anxiety inventories. Behav Res Ther 33(3):335–343. https://doi.org/10.1016/0005-7967(94)00075-u

Maes IH, Cima RF, Anteunis LJ, Scheijen DJ, Baguley DM, El Refaie A et al (2014) Cost-effectiveness of specialized treatment based on cognitive behavioral therapy versus usual care for tinnitus. Otol Neurotol 35(5):787–795. https://doi.org/10.1097/MAO.0000000000000331

McFerran D, Hoare DJ, Carr S, Ray J, Stockdale D (2018) Tinnitus services in the United Kingdom: a survey of patient experiences. BMC Health Serv Res 18(1):110. https://doi.org/10.1186/s12913-018-2914-3

Oishi N, Shinden S, Kanzaki S, Saito H, Inoue Y, Ogawa K (2011) Influence of depressive symptoms, state anxiety, and pure-tone thresholds on the tinnitus handicap inventory in Japan. Int J Audiol 50(7):491–495. https://doi.org/10.3109/14992027.2011.560904

Ooms E, Meganck R, Vanheule S, Vinck B, Watelet JB, Dhooge I (2011) Tinnitus severity and the relation to depressive symptoms: a critical study. Otolaryngol Head Neck Surg 145(2):276–281. https://doi.org/10.1177/0194599811403381

Ooms E, Vanheule S, Meganck R, Vinck B, Watelet JB, Dhooge I (2012) Tinnitus severity and its association with cognitive and somatic anxiety: a critical study. Eur Arch Otorhinolaryngol 269 (11):2327–2333. https://doi.org/10.1007/s00405-011-1887-1

Pan T, Tyler RS, Ji H, Coelho C, Gogel SA (2015) Differences among patients that make their tinnitus worse or better. Am J Audiol 24(4):469–476. https://doi.org/10.1044/2015_AJA-15-0020

Pattyn T, Van Den Eede F, Vanneste S, Cassiers L, Veltman DJ, Van De Heyning P, Sabbe BC (2016) Tinnitus and anxiety disorders: a review. Hear Res 333:255–265. https://doi.org/10.1016/j.heares.2015.08.014

Probst T, Pryss R, Langguth B, Schlee W (2016a) Emotion dynamics and tinnitus: daily life data from the "TrackYourTinnitus" application. Sci Rep 6:31166. https://doi.org/10.1038/srep31166

Probst T, Pryss R, Langguth B, Schlee W (2016b) Emotional states as mediators between tinnitus loudness and tinnitus distress in daily life: results from the "TrackYourTinnitus" application. Sci Rep 6:20382. https://doi.org/10.1038/srep20382

Pryce H, Hall A, Shaw R, Culhane BA, Swift S, Straus J, Claesen B (2018) Patient preferences in tinnitus outcomes and treatments: a qualitative study. Int J Audiol 57(10):784–790. https://doi.org/10.1080/14992027.2018.1484184

Robinson SK, Viirre ES, Bailey KA, Gerke MA, Harris JP, Stein MB (2005) Randomized placebo-controlled trial of a selective serotonin reuptake inhibitor in the treatment of nondepressed tinnitus subjects. Psychosom Med 67(6):981–988

Scherer S, Hammal Z, Yang Y, Morency LP, Cohn JF (2014) Dyadic behavior analysis in depression severity assessment interviews. In: Proceedings of ACM international conf multimodal interact, vol 2014. pp 112–119. https://doi.org/10.1145/2663204.2663238

Schlee W, Kleinjung T, Hiller W, Goebel G, Kolassa IT, Langguth B (2011) Does tinnitus distress depend on age of onset? PLoS One 6(11):e27379. https://doi.org/10.1371/journal.pone.0027379

Scott B, Lindberg P (2000) Psychological profile and somatic complaints between help-seeking and non-help-seeking tinnitus subjects. Psychosomatics 41(4):347–352. http://www.ncbi.nlm.nih.gov/entrez/query.fcgi?cmd=Retrieve&db=PubMed&dopt=Citation&list_uids=10906357

Searchfield GD, Durai M, Linford T (2017) A state-of-the-art review: personalization of tinnitus sound therapy. Front Psychol 8:1599. https://doi.org/10.3389/fpsyg.2017.01599

Simoens VL, Hébert S (2012) Cortisol suppression and hearing thresholds in tinnitus after low-dose dexamethasone challenge. BMC Ear Nose Throat Disord 12:4. https://doi.org/10.1186/1472-6815-12-4

Spielberger CD, Gorsuch RL, Lushene RE (1970) Manual for the state-trait anxiety inventory. Consulting Psychologists, Palo Alto

Stattrop U, Goebel G, Voderholzer U (2013) Correspondence (letter to the editor): the therapeutic objective: reducing the stress caused by tinnitus. Dtsch Arztebl Int 110(35–36):599–600

Stockdale D, McFerran D, Brazier P, Pritchard C, Kay T, Dowrick C, Hoare DJ (2017) An economic evaluation of the healthcare cost of tinnitus management in the UK. BMC Health Serv Res 17(1):577. https://doi.org/10.1186/s12913-017-2527-2

Tschuschke V, Crameri A, Koehler M, Berglar J, Muth K, Staczan P et al (2015) The role of therapists' treatment adherence, professional experience, therapeutic alliance, and clients' severity of psychological problems: prediction of treatment outcome in eight different psychotherapy approaches. Preliminary results of a naturalistic study. Psychother Res 25(4):420–434. https://doi.org/10.1080/10503307.2014.896055

Tunkel DE, Bauer CA, Sun GH, Rosenfeld RM, Chandrasekhar SS, Cunningham ER Jr, Archer SM, Blakley BW, Carter JM, Granieri EC, Henry JA, Hollingsworth D, Khan FA, Mitchell S, Monfared A, Newman CW, Omole FS, Phillips CD, Robinson SK, Taw MB, Tyler RS, Waguespack R, Whamond EJ (2014) Clinical practice guideline: tinnitus. Otolaryngol Head Neck Surg 151(2 Suppl):S1–S40

Tyler RS, Perreau A, Powers T, Watts A, Owen R, Ji H, Mancini PC (2020) Tinnitus sound therapy trial shows effectiveness for those with tinnitus. J Am Acad Audiol 31(1):6–16. https://doi.org/10.3766/jaaa.18027

Uckelstam CJ, Philips B, Holmqvist R, Falkenstrom F (2019) Prediction of treatment outcome in psychotherapy by patient initial symptom distress profiles. J Couns Psychol 66(6):736–746. https://doi.org/10.1037/cou0000345

Wallhausser-Franke E, D'Amelio R, Glauner A, Delb W, Servais JJ, Hormann K, Repik I (2017) Transition from acute to chronic tinnitus: predictors for the development of chronic distressing tinnitus. Front Neurol 8:605. https://doi.org/10.3389/fneur.2017.00605

Weidt S, Delsignore A, Meyer M, Rufer M, Peter N, Drabe N, Kleinjung T (2016) Which tinnitus-related characteristics affect current health-related quality of life and depression? A cross-sectional cohort study. Psychiatry Res 237:114–121. https://doi.org/10.1016/j.psychres.2016.01.065

Weise C, Hesser H, Andersson G, Nyenhuis N, Zastrutzki S, Kroner-Herwig B, Jager B (2013) The role of catastrophizing in recent onset tinnitus: its nature and association with tinnitus distress and medical utilization. Int J Audiol 52(3):177–188. https://doi.org/10.3109/14992027.2012.752111

Wu V, Cooke B, Eitutis S, Simpson MTW, Beyea JA (2018) Prise en charge de l'acouphène. Can Family Phys/Le Médecin de famille canadien 64(July/juillet):e293–e298

Zigmond AS, Snaith RP (1983) The hospital anxiety and depression scale. Acta Psychiatr Scand 67 (6):361–370. https://www.ncbi.nlm.nih.gov/pubmed/6880820

Psychosocial Variables That Predict Chronic and Disabling Tinnitus: A Systematic Review

Maria Kleinstäuber and Cornelia Weise

Contents

Abstract To improve tinnitus management we have to gain more knowledge of factors that explain how a persistent distressing tinnitus develops. The central aim of this systematic review was to identify longitudinal studies that investigated psychosocial variables predicting the transition from an acute to a chronic, disabling tinnitus (i.e. tinnitus decompensation) or tinnitus outcomes in chronic tinnitus sufferers. We conducted a systematic literature search of electronic databases and searched manually reference lists. We identified 16 eligible studies: Four longitudinal studies targeted predictors of the transition from acute to chronic tinnitus and 12 longitudinal

Electronic Supplementary Material The online version of this chapter (https://doi.org/10.1007/7854_2020_213) contains supplementary material, which is available to authorized users.

M. Kleinstäuber (✉)
Department of Psychological Medicine, Otago Medical School – Dunedin Campus, University of Otago, Dunedin, New Zealand
e-mail: maria.kleinstaeuber@otago.ac.nz

C. Weise
Department of Clinical Psychology and Psychotherapy, Philipps-University Marburg, Marburg, Germany

Curr Topics Behav Neurosci (2021) 51: 361–380
https://doi.org/10.1007/7854_2020_213
Published Online: 2 February 2021

studies investigated predictors of tinnitus distress (k = 9 observational, longitudinal studies; k = 3 ecological momentary assessment [EMA] and diary studies). The results of this systematic review showed that tinnitus distress, general psychological distress, tendencies to somatize, tinnitus-related delay of sleep onset, certain health behaviors, general illness coping, and certain personality traits (e.g. neuroticism) predicted the transition from acute to chronic, disabling tinnitus. General psychological, mental disorders, tinnitus distress, tinnitus disability (e.g. in different domains of physical, emotional, and social functioning; sleep disturbances), certain health behaviors (e.g. physical exercise), the level of physical and social functioning, and the report of other somatic problems such as pain were predictors of tinnitus outcomes in chronic tinnitus patients at a later follow-up. Studies that examined psychosocial variables as predictors of tinnitus distress are rare and had substantial methodological shortcomings. Future research should focus on core outcome domains and use standardized outcome measures to improve the comparability of results from different studies. Numerous psychosocial variables that have already been investigated as correlates of tinnitus sufferers' functioning in cross-sectional studies are worth investigating with longitudinal designs in future research. Identified predictors of the transition from acute to chronic, disabling tinnitus have to be addressed by health care practitioners who commonly function as the first contact person of individuals with acute tinnitus in the healthcare system.

Keywords Disability · Distress · Ecological momentary assessment · EMA · Longitudinal · Predictor · Psychosocial · Tinnitus

1 Introduction

Tinnitus lasting for more than 5 min is a common phenomenon and affects approximately 34% of the general population (McCormack et al. 2016). However, not every individual who reports tinnitus perceives the ringing in the ear as debilitating and bothering. The prevalence of persistent tinnitus that has a moderate to severe effect on the ability to lead a normal life is approximately 1.0–2.8% (Davis 1989; Fujii et al. 2011; McCormack et al. 2016). Moreover, a significant proportion of tinnitus sufferers can be helped to improve their symptom management with low-intensity interventions (e.g. self-help interventions based on cognitive behavior therapy) (Hesser et al. 2011; Nyenhuis et al. 2013). Nonetheless, there is also a significant number of individuals with tinnitus who are at an increased risk of developing psychological distress and disability (Bhatt et al. 2017).

Health care systems worldwide lack efficient management plans to significantly reduce the number of individuals developing a persistent distressing tinnitus (Wise et al. 2015). To improve tinnitus management, more knowledge of factors is needed that explain how an acute tinnitus becomes chronic and of factors that predict the course of tinnitus outcomes. Researchers in the field of tinnitus have developed numerous theories to address tinnitus decompensation, which means the transition from an acute phantom sound to a disorder that is associated with severe suffering and disability (De Ridder et al. 2014a; Hallam et al. 1984; Jastreboff 1990; Searchfield et al. 2012; Sedley et al. 2016). For example, Hallam and colleagues (Hallam et al. 1984) postulated that individuals developed feelings of distress towards a ringing in their ear when they were not able to learn to tolerate it. Habituation towards tinnitus is described as a healthy process of adjustment that can be disturbed by individual factors (e.g. tinnitus characteristics, hearing loss, central nervous system pathology, certain information processing styles), sensory factors (e.g. masking level), or perceptual factors (e.g. competing attentional demands). Jastreboff (1990) emphasized in his neurophysiological model that tinnitus is a complex process that involves somatic, psychological, and social factors. The model assumes that the tinnitus perception is processed not only on each level of the auditory pathway but also in non-auditory systems. Jastreboff (1990) hypothesized that factors such as emotions, negative thoughts, and learning processes are involved in the process of developing a bothering ringing in the ear. Searchfield et al. (2012) postulate in their adaptation level theory that the tinnitus percept is a weighted product of the magnitude of the tinnitus stimulus itself, the magnitude of the background sound, and residual factors such as psychological factors (e.g. personality traits). Attention is considered in this model as a weighting factor. Sedley et al. (2016) and de Ridder et al. (2014b) propose a new framework that builds on a so-called predictive coding approach. According to this approach, our brain uses sensory inputs (from the environment or from spontaneous sensory activity) to update memory-based beliefs that help us to predict our world. The precision of the sensory input, the tinnitus precursor, can be increased by various factors, such as attention, stress, or emotional distress.

These models have commonly conceptualized that bothering and disabling tinnitus is not only a perception of a phantom sound but is a multidimensional phenomenon that involves – in addition to a sound sensational dimension – emotional, cognitive, and behavioral dimensions. In addition to audiological and tinnitus characteristics, cognitive, emotional, and behavioral variables are supposed to play an important role as predisposing and perpetuating factors of persistent and debilitating tinnitus. Evidence for these models mainly build on studies that implement cross-sectional designs. Psychosocial correlates of tinnitus distress, such as personality traits (e.g., Durai et al. 2017), cognitive processing styles (Conrad et al. 2015; Weise et al. 2013), anxiety sensitivity (e.g., Hesser and Andersson 2009), or fear avoidance (Cima et al. 2011; Kleinstäuber et al. 2013) have been explored. However, cross-sectional study designs do not reach conclusions about how well psychosocial variables predict tinnitus outcomes. For this purpose, longitudinal trial designs with repeated observations of the same variables within the same subjects would be needed.

The central aim of this systematic review is to identify and summarize findings of longitudinal studies that either investigated psychosocial variables predicting the transition from an acute to chronic and disabling tinnitus (i.e. tinnitus decompensation) or the course of tinnitus outcomes in chronic tinnitus sufferers. The review shall provide an insight into how previous research has recognized psychosocial factors as potential predisposing or perpetuating factors of chronic tinnitus distress in comparison to other variables, such as audiological or tinnitus characteristics.

2 Methods

2.1 Search Procedure

We searched the following electronic databases: Pubmed, Web of Science, Psyndex, PsycINFO, and Pubpsych (during January 2020). We updated our search of electronic databases twice: on 3 May 2020 and on 29 October 2020. We manually searched lists of references of relevant reviews. Date and language restrictions were not applied to the searches. The complete search strategy for each electronic database is available in Table S1 (Supplementary Material).

2.2 Study Selection

The literature search resulted in 8,921 hits in total. After eliminating duplicates, abstracts and titles of 5,414 records were screened. At this stage 5,342 references were excluded. Two reviewers (MK, CW) independently screened the full texts of the remaining 72 articles. Disagreements were resolved by consensus. Finally, 16 studies (23 references) were included in this systematic review. Figure 1 summarizes the process of study selection. Table S2 (Supplementary Material) summarizes references of all studies that were excluded after the review of full texts.

2.3 Eligibility Criteria

Studies were included if they met the following inclusion criteria:

(a) Participants suffered from subjective tinnitus.
(b) The study implemented a longitudinal study design with at least two data collection points.
(c) At least one psychosocial variable, tinnitus characteristic, or audiological characteristic was measured at minimum one data collection point. At least one tinnitus-related outcome (e.g. tinnitus severity, distress, disability, loudness, or sequelae [e.g., tinnitus-related emotional disturbances or sleep difficulties]) was assessed at minimum one data collection point. The measurement of the psychosocial variable/tinnitus characteristic/audiological characteristic (=predictor) had to be dated before the measurement of the tinnitus outcome.

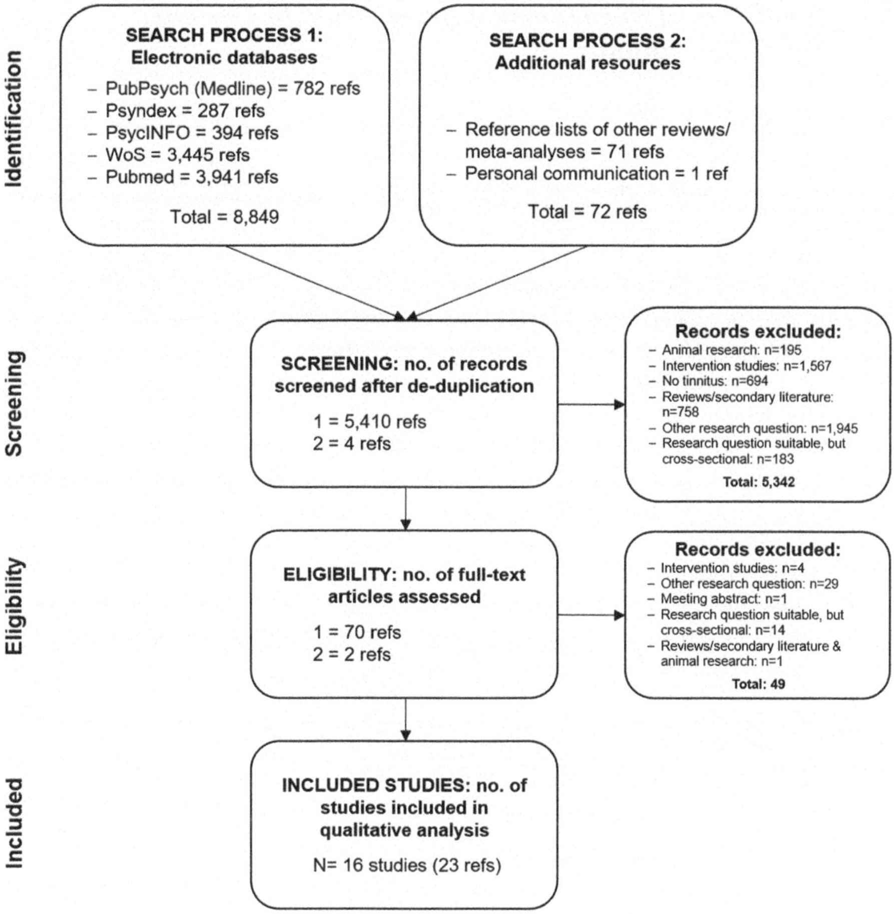

Fig. 1 PRISMA flow diagram

Animal research was excluded from this review. Repeated measures to evaluate a trial's control intervention (e.g. predictor assessed at the baseline of a placebo intervention or a waitlist control group predicting the outcome at the end of the control intervention) were not included to this review.

3 Results

Details of the results of all studies included to this review are summarized in Table S3 (Supplementary Material). An overview of variables that were studied as predictors of tinnitus decompensation are summarized in Table S4 (Supplementary Material).

3.1 Predictors of the Transition from Acute to Chronic, Disabling Tinnitus

Our literature search revealed four studies that examined individuals with an acute tinnitus (i.e. within 4 weeks after the phantom sound started) and followed them up to investigate tinnitus decompensation in the chronic phase, between 3 and 24 months post-tinnitus onset (Jäger et al. 2004a, b, 2006, 2005; Langenbach et al. 2005; Olderog et al. 2004; Vielsmeier et al. 2020; Wallhäusser-Franke et al. 2015, 2017; see Table 1). Only information about the study design was reported for one of the four studies (Jäger et al. 2004b), but results of this one study were reported without sufficient detail as part of conference abstracts (Jäger et al. 2004a, 2006, 2005). We therefore excluded this one study from our systematic review. Sample sizes in the remaining three included studies ranged between 44 and 49 participants. Participants' average age ranged between 41 and 47 years (Langenbach et al. 2005; Vielsmeier et al. 2020; Wallhäusser-Franke et al. 2017). Two studies recruited participants in an outpatient ENT settings (Langenbach et al. 2005; Vielsmeier et al. 2020), one study gained participants from an inpatient ENT clinic (Wallhäusser-Franke et al. 2017).

All three included studies (Langenbach et al. 2005; Olderog et al. 2004; Vielsmeier et al. 2020; Wallhäusser-Franke et al. 2015, 2017) examined different outcomes as indicators of tinnitus decompensation:

- tinnitus distress,
- tinnitus loudness (i.e. subjective perception),
- tinnitus disability in different domains of daily life, and,
- chronic manifestation of distressing tinnitus.

Psychosocial variables within the domains of:

- *tinnitus distress* (i.e. negative emotions and feelings of distress caused by tinnitus, e.g. tinnitus-related depressive symptoms or annoyance),
- *tinnitus sequelae* (including tinnitus disability in different domains physical, emotional, and social functioning, and sleep and concentration difficulties),
- *general psychological distress* (i.e. general psychological problems and symptoms of psychopathology that are not [necessarily] related to tinnitus, e.g. symptoms of depression or anxiety),
- *health behaviors* (e.g. alcohol use),
- *personality traits,*
- *illness coping,* and,
- *resilience* (i.e. a combination of serious risk experiences and a relatively positive psychological outcome despite those experiences (Rutter 2006)),

that were assessed in the acute phase of tinnitus and were explored as potential predictors of tinnitus decompensation in the chronic phase (see Table S4/Supplementary Material).

Table 1 Characteristics of included studies

Study/ Reference/ Country	Total $N_{analyzed\ sample}$ ($N_♀$)	Mean (SD) age [years]	Mean (SD) tinnitus length [months]	Setting
Predictors of transition from acute to chronic, disabling tinnitus				
Jäger et al. (2004)/ Jäger et al. (2004a, b, 2006, 2005)/ Germany	210 (108)	44.5 (13.6)	n.a.	Outpatient ENT practices
Langenbach et al. (2005)/ Langenbach et al. (2005)/ Olderog et al. (2004)/ Germany	44 (16)	47.3 (n.a.)	0.4 (0.3)	Inpatient ENT clinic
Vielsmeier et al. (2020)/ Vielsmeier et al. (2020)	49 (21)	46.0 (14)	0.3 (0.2)	Outpatient ENT clinic
Wallhäuser-Franke et al. (2017)/ Wallhäusser-Franke et al. (2015, 2017)/ Germany	47 (20)	41.4 (15.5)	0.5 (0.3)	Outpatient ENT department at university medical center, ENT practices
Predictors of tinnitus outcomes in individuals with chronic tinnitus: Longitudinal studies with ≥ 2 data collection points				
Bleich et al. (2001)/ Bleich et al. (2000, 2001)/ Germany	73 (n.a.)	54.3 (10.6)	n.a.	ENT practice
Clifford et al. (2019)/ Clifford et al. (2019)/ USA	2,600 (0)	22.3 (3.3)	n.a.	Active-duty Marine/Navy servicemen
Dawes et al. (2010)/ Dawes and Welch (2010)/ New Zealand	970 (n.a.)	32.0 (0.0)	n.a.	Dunedin Multidisciplinary Health and Development Study (longitudinal study of a birth cohort)
Erlandsson et al. (2006)/ Erlandsson and Persson (2009)/ Sweden	70 (25)	49.0 (n.a.)	8.0 (n.a.)	Specialist tinnitus clinic

(continued)

Table 1 (continued)

Study/ Reference/ Country	Total $N_{\text{analyzed sample}}$ ($N_♀$)	Mean (SD) age [years]	Mean (SD) tinnitus length [months]	Setting
Hallam et al. (1996)/ Hallam (1996)/ UK	26 (16)	54.7 (n.a.)	6.5 (n.a.)	Hospital neuro-otology department, follow-up assessment of therapy study of psychological intervention
Hébert et al. (2012)/ Hébert et al. (2012)/ Sweden	2,571 (1,223)	52.4 (11.2)	n.a.	Swedish Longitudinal Occupational Survey of Health (SLOSH)
Holgers et al. (2000)/ Holgers et al. (2000, 2005)/ Sweden	172 (54)	53.6 (13.9)	n.a.	Hospital department of audiology
Olsen et al. (2013)/ Olsen et al. (2013)/ Sweden	174 (64)	61.0 (12.7)	n.a.	ENT clinic
Westin et al. (2008)/ Westin et al. (2008)/ Sweden	47 (24)	50.6 (12.1)	75.6 (98.4)	Local audiology department
Predictors of tinnitus outcomes in individuals with chronic tinnitus: Ecological momentary assessment (EMA) and diary studies				
Andersson et al. (1997)/ Andersson et al. (1997)/ Sweden	20 (13)	50.6 (9.7)	63.9 (82.1)	ENT practice, Menière's disease patients
Goldberg et al. (2017)/ Goldberg et al. (2017)/ USA	40 (10)	60.0 (10.5)	Md = 180 (IQR = 7–600)	ENT/audiology clinics and local practices, American Tinnitus Association, university volunteers registers for health research
Probst et al. (2016)/ Probst et al. (2016)/ Germany	306 (87)	42.6 (17.6)	10.1 (12.0)	Users of the Track Your Tinnitus app

Md = Median, IQR = Interquartile range

Results demonstrated that *tinnitus distress* (e.g., tinnitus was experienced to be tedious or agonizing or was associated with tension, stress or suicidal thoughts) in the acute phase post-tinnitus onset was an important predictor of tinnitus decompensation 6 months past tinnitus onset (Langenbach et al. 2005; Olderog et al. 2004; Wallhäusser-Franke et al. 2015, 2017). *Tinnitus sequelae* in the acute phase were demonstrated to predict chronic tinnitus distress at a 6-month follow-up post-tinnitus onset. For example, individuals with increased levels of tinnitus-related disability in different domains of daily living in the acute phase were at risk of experiencing high levels of disability 6 months past tinnitus onset (Wallhäusser-Franke et al. 2015, 2017). Tinnitus-related delayed sleep onset, but not sleep disruption, in the acute phase predicted tinnitus decompensation in the chronic phase (Langenbach et al. 2005; Olderog et al. 2004). Concentration difficulties shortly after tinnitus onset did not predict tinnitus distress at a 6-month follow-up post-tinnitus onset (Wallhäusser-Franke et al. 2015, 2017). Interestingly the studies consistently demonstrated that *general psychological distress* (e.g., symptoms of depression or anxiety) and a general tendency to somatize in the acute phase were risk factors of developing a persistent disabling tinnitus (Langenbach et al. 2005; Olderog et al. 2004; Wallhäusser-Franke et al. 2015, 2017). Vielsmeier et al. (2020) identified *health behaviors*, alcohol use in particular, during the acute phase as a significant predictor of a chronic manifestation of distressing tinnitus 6 months post-tinnitus onset. Langenbach et al. (2005) investigated *personality traits* as predictors of tinnitus decompensation: Whereas a general satisfaction with life was identified to be a protective factor, individuals who were generally more emotionally-inclined and excitable (i.e. behavioral excitability, increased sensitivity to instigation behavior, and lack of self-control) and who reported somatic complaints more frequently, such as insomnia, headaches or gastrointestinal problems, for example, were at a higher risk of decompensating in the later course of their tinnitus (Langenbach et al. 2005). Vielsmeier et al. (2020) investigated general optimism and pessimism during the acute tinnitus phase and did not find any relationship to a chronic manifestation of distressing tinnitus 6 months post-symptom onset. Wallhäusser-Franke et al. (2017) investigated styles of general *illness coping* as predictors of tinnitus decompensation. Interestingly, individuals who coped with illness in an active and problem-oriented manner during the acute phase of their tinnitus, were at risk of developing increased tinnitus distress, loudness, and negative impact of tinnitus on family relationships 6 months post-tinnitus onset (Wallhäusser-Franke et al. 2017). Self-distraction and self-affirmation as illness coping strategies were related to increased tinnitus loudness and work-related disability 6 months post-tinnitus onset (Wallhäusser-Franke et al. 2017). Other illness coping styles such as religiousness, searching for meaning in events that happened in life or exercising trivialization measured during the acute phase tinnitus did not predict tinnitus decompensation 6 months post-tinnitus onset (Wallhäusser-Franke et al. 2017). Resilience towards stressful events during the acute phase of tinnitus was not related to tinnitus decompensation in the chronic phase (Wallhäusser-Franke et al. 2017).

Besides psychosocial variables, *tinnitus* and *audiological characteristics* as predictors of tinnitus decompensation were examined. The temporal pattern of the acute tinnitus (onset, persistence) was not related to tinnitus decompensation 6 months post-tinnitus onset (Langenbach et al. 2005; Olderog et al. 2004). Tinnitus localized in the right ear (Langenbach et al. 2005; Olderog et al. 2004) and tinnitus loudness (Wallhäusser-Franke et al. 2015, 2017) in the acute phase predicted higher levels of tinnitus distress and tinnitus loudness, respectively, in the chronic phase. Other tinnitus characteristics, such as tinnitus awareness, distractibility or discomfort measured within 4 weeks post-tinnitus onset were not related to a chronic tinnitus manifestation 6 months later (Vielsmeier et al. 2020). The most frequently studied *audiological characteristic* of individuals with tinnitus was hearing loss. Hearing loss in the acute phase predicted tinnitus loudness but not distress in the chronic phase (Wallhäusser-Franke et al. 2015, 2017). In the contrary, noise sensitivity measured shortly after the onset of tinnitus was related to more tinnitus distress (Vielsmeier et al. 2020; Wallhäusser-Franke et al. 2015, 2017) but not loudness (Wallhäusser-Franke et al. 2015, 2017) 6 months post-tinnitus onset. Individuals who experienced vertigo and ear pressure accompanying the tinnitus in the acute phase were at increased risk of developing chronic tinnitus distress 6 months post-tinnitus onset (Langenbach et al. 2005; Olderog et al. 2004; Vielsmeier et al. 2020).

The results reported so far were gained from bivariate analyses. Multivariate analyses confirmed *delayed sleep onset*, *symptoms of anxiety and depression*, *low general life satisfaction*, and *active, problem-solving illness coping* as robust predictors of chronic decompensation 6 months post-tinnitus onset.

3.2 Predictors of Tinnitus Outcomes in Individuals with Chronic Tinnitus: Longitudinal Studies with ≥2 Data Collection Points

Our literature search revealed nine studies that assessed individuals with chronic disabling tinnitus at several data collection points to investigate variables that predict tinnitus-related outcomes (see Table 1). Sample sizes varied across the studies, between $N = 26$ and $N = 2{,}571$. The average age of participants ranged between 49 and 61 years, except for participants in two studies (Clifford et al. 2019; Dawes and Welch 2010). Clifford et al. (2019) investigated a young cohort of individuals, aged 22 years on average. Dawes and Welch (2010) evaluated participants from one birth cohort (age at follow-up: 32 years) gained from the Dunedin Multidisciplinary Health and Development Study. Participants had suffered from their tinnitus on average between 7 months and 6 years. Of the nine eligible studies, six recruited in outpatient ENT departments or specialized tinnitus clinics (Bleich et al. 2001; Erlandsson and Persson 2009; Hallam 1996; Holgers et al. 2005; Olsen et al. 2013; Westin et al. 2008), two studies gained data from longitudinal population studies (Dawes and Welch 2010; Hébert et al. 2012), and one study recruited active Navy and Marine servicemen (Clifford et al. 2019). Three studies had short follow-up intervals lasting between 1 week and 7 months (Clifford et al. 2019; Hallam 1996;

Westin et al. 2008), the remaining six studies had longer follow-up periods between 1.5 and 17 years (see Table S3/Supplementary Material).

The studies examined the following outcomes:

- tinnitus severity,
- tinnitus distress and tinnitus-related symptoms of depression and anxiety,
- tinnitus loudness,
- tinnitus progression (i.e. worsening of symptoms),
- tinnitus persistence,
- tinnitus-related sequelae (i.e. tinnitus disability in different domains physical, emotional, and social functioning, and negative impact of tinnitus on individuals' quality of life), and,
- tinnitus acceptance (i.e. pursuit of life activities regardless of tinnitus [Westin et al. 2008]).

The following domains of psychosocial variables:

- *tinnitus sequelae* (i.e. disability or feeling of disturbance in different domains of physical, emotional, and social functioning due to tinnitus, reduced employment, sleep disturbances),
- *tinnitus acceptance,*
- *tinnitus distress,*
- *general psychological distress,*
- *mental disorders* (i.e. diagnosed mental health conditions, e.g. PTSD or personality disorders),
- *health behaviors* (e.g. diet, physical exercise, nicotine or alcohol use),
- *general physical, mental, and social functioning* (e.g. level of energy, physical mobility, or social isolation),
- *distressing and traumatic events* (e.g. traumatic events during deployment), and,
- *report of somatic symptoms and medical conditions* (e.g. TBI, pain).

were explored as potential predictors of tinnitus outcomes.

The predictive value of *tinnitus sequelae* and *tinnitus acceptance* was investigated in three studies (Bleich et al. 2001; Holgers et al. 2000; Westin et al. 2008). Different measures of tinnitus disability (e.g. social isolation, lack of energy) and sleep disturbances were associated with an increased tinnitus-related absence from work at an 18-month follow-up (Holgers et al. 2000). Reduced employment (due to tinnitus in 29% of the cases) resulted in lower quality of life but also higher tinnitus acceptance 7 months post-baseline assessment (Westin et al. 2008). How much individuals felt disturbed in their level of functioning by their tinnitus did not predict how much individuals felt distressed by their tinnitus at a 7-month or 5- to 10-year follow-up (Bleich et al. 2001; Westin et al. 2008). How well individuals were able to pursuit with life activities despite of their tinnitus, i.e. of *tinnitus acceptance,* was neither associated with symptom acceptance nor with tinnitus distress, nor with quality of life at a 7-month follow-up (Westin et al. 2008).

Tinnitus distress was investigated as predictive variable in two studies (Hallam 1996; Westin et al. 2008). Tinnitus distress predicted increased tinnitus distress, symptoms of depression and anxiety, reduced quality of life and less tinnitus acceptance at a 7-month follow-up (Westin et al. 2008). Tinnitus annoyance predicted sleep difficulties one week post-baseline assessment (Hallam 1996).

Health behaviors, general physical, mental, and social functioning and report of *somatic symptoms,* pain in particular, were examined as predictors of tinnitus outcomes in one study (Holgers et al. 2000). An increased body mass index and lack of regular physical exercise were associated with increased tinnitus disability at an 18-month follow-up. Neither alcohol use nor smoking were related to tinnitus disability at the 18-month follow-up. Individuals who reported pain symptoms were at increased risk of experiencing more tinnitus-related disability at the 18-month follow-up. Several indicators of general physical and psychosocial functioning were found to be predictors of tinnitus distress at the 18-month follow-up (Holgers et al. 2000): Individuals with chronic tinnitus who experienced less energy in general, whose physical mobility was restricted, and who lived socially more isolated were at risk of experiencing more disability due to their tinnitus at the follow-up assessment.

General psychological distress or pre-existing *mental disorders* predicted tinnitus-related outcomes in four studies (Clifford et al. 2019; Erlandsson and Persson 2009; Hébert et al. 2012; Holgers et al. 2000). Symptoms of depression were found to predict increased tinnitus severity at a 2-year follow-up (Hébert et al. 2012). Tinnitus sufferers with increased emotional disturbances were at risk of developing more tinnitus disability at an 18-month follow-up (Holgers et al. 2000). One study (Erlandsson and Persson 2009) investigated a subgroup of tinnitus sufferers with personality disorders. Individuals with one or more personality disorders showed increased symptoms of anxiety and depression at an 18-month follow-up compared to those tinnitus sufferers without personality disorder.

Seven studies examined *tinnitus* and *audiological characteristics* as predictors of tinnitus outcomes (Bleich et al. 2001; Clifford et al. 2019; Dawes and Welch 2010; Hébert et al. 2012; Holgers et al. 2000; Olsen et al. 2013; Westin et al. 2008). Three studies (Bleich et al. 2001; Hébert et al. 2012; Westin et al. 2008) investigated *tinnitus characteristics* as predictors of different tinnitus outcomes (including tinnitus severity, loudness, and duration). Tinnitus severity at a baseline assessment predicted tinnitus severity at a 2-year follow-up (Hébert et al. 2012). Neither tinnitus loudness nor tinnitus onset were associated with tinnitus distress at a 7-month or 5- to 10-year follow-up (Bleich et al. 2001; Westin et al. 2008). *Audiological characteristics* were investigated as predictors of different tinnitus outcomes in seven studies (Bleich et al. 2001; Clifford et al. 2019; Dawes and Welch 2010; Hébert et al. 2012; Holgers et al. 2000; Olsen et al. 2013; Westin et al. 2008). Results regarding the impact of hearing loss were contradictory and appeared to be moderated by additional factors. Two studies consistently showed that hearing loss was not associated with tinnitus distress 7 months or 5–10 years post-baseline assessment (Bleich et al. 2001; Westin et al. 2008). Hearing loss also did not predict tinnitus distress, symptoms of depression or anxiety, changes of quality of life, or tinnitus acceptance at a 7-month follow-up (Westin et al. 2008). One study showed hearing loss to be a risk factor of increased tinnitus severity 2 years post-baseline assessment

(Hébert et al. 2012), whereas another study showed no predictive association between both variables (Olsen et al. 2013). The relationship between hearing loss and tinnitus persistence or disability probably depends on additional factors. One study (Dawes and Welch 2010) demonstrated that individuals who suffered in their childhood from hearing loss and additionally from an otitis media were at particular risk of developing a more persistent tinnitus in adulthood, compared to individuals with childhood hearing loss only. Another study (Holgers et al. 2000) demonstrated that hearing loss in ranges of lower frequency (0.5–2 kHz) was associated with an increased risk of tinnitus disability at an 18-month follow-up. Other audiological abnormalities (e.g. tympanic membrane abnormality, deviations in acoustic reflex threshold or speech recognition) in childhood were not associated with tinnitus outcomes at later follow-up assessments (Olsen et al. 2013).

As mentioned earlier, the study by Clifford et al. (2019) examined a specific sample of tinnitus sufferers, active Navy and Marine servicemen. The trial investigated risk factors of tinnitus progression (which means worsening tinnitus status or maintaining worst tinnitus status) after deployment. The authors showed that *hearing loss, partial post-traumatic stress disorder (PTSD),* and *traumatic brain injury* before the deployment and *stressful, traumatic events* (i.e. combat intensity) during the deployment were risk factors for tinnitus progression 3 months after the end of deployment. Combat intensity is defined as combat-related circumstances such as firing a weapon, being fired on, being attacked, or witnessing an attack.

The results that have been reported so far have been based on bivariate analyses. Multivariate analyses were performed in two studies (Hébert et al. 2012; Holgers et al. 2000). One of these studies (Holgers et al. 2000) confirmed *emotional disturbances due to tinnitus, lack of regular physical exercise, physical immobility, hearing loss* in ranges of low frequency and *sleep disturbances* as robust predictors of work-related disability due to tinnitus at an 18-month follow-up. Another study (Hébert et al. 2012) demonstrated that *symptoms of depression, hearing loss,* and *tinnitus severity* were robust predictors of tinnitus severity at a 2-year follow-up in a multivariate analysis controlled for gender, age, and income. An overview of variables that were studied as predictors of tinnitus outcomes are summarized in Table S5 (Supplementary Material).

3.3 Predictors of Tinnitus Outcomes in Individuals with Chronic Tinnitus: Ecological Momentary Assessment (EMA) and Diary Studies

In three studies predictors of tinnitus-related outcomes were examined with a diary (Andersson et al. 1997) or ecological momentary assessment (Goldberg et al. 2017; Probst et al. 2016). Participants in these studies were on average between 43 and 60 years old and had suffered from their tinnitus over 10 months to 15 years. All three studies focused on stress or emotional states as predictors of tinnitus-related outcomes.

The diary study by Andersson et al. (1997) was published in the 1990s and is based on a small sample ($N = 20$) of individuals with Menière's disease. Data interpretation is therefore limited and cannot be generalized to the entire population of tinnitus sufferers. Amongst other variables, perceived stress (as predictor), and tinnitus severity (as outcome) were assessed once daily with a visual analogue scale (VAS) over 194 days on average. The results showed no relationship between tinnitus severity and perceived stress measured the day or 2 days before.

Goldberg et al. (2017) studied the relationship between tinnitus bother, loudness, the overall emotional feeling, level of stress, and noise environment in 40 individuals with chronic tinnitus with four daily EMA assessments over a 2-week period. The authors demonstrated that, on an individual level, tinnitus bother, loudness, and stress all vary together over time. That means when an individual experienced increased stress over time the person also experienced increased tinnitus bother and loudness. Analyses showed that these changes over time were all caused by the same underlying factor that varies over time. The authors also demonstrated that all five EMA variables varied across participants. This means in individuals who reported higher levels of stress or negative feelings tinnitus bother was likely to be higher, compared with individuals reporting lower levels of stress and negative feelings. The authors' model suggested the underlying factor that explained changes within subjects and across individuals was stress.

Probst et al. (2016) investigated 306 users of the Track Your Tinnitus app with five daily EMA assessments over 2 months on average. They examined the relationship between the individuals' emotional state and tinnitus distress or loudness. Tinnitus-related outcomes were measured with VAS. The quality and intensity of the emotional state were assessed with Self-Assessment Manikin scales. Probst et al. then created a score indicating the extent of intra-individual variability of affect intensity ("pulse" score) and affect quality ("spin" score). The authors demonstrated that the effect of tinnitus loudness on tinnitus distress was moderated by the pulse and spin score. This means that tinnitus loudness predicted more tinnitus distress in individuals who experienced more dynamics in the intensity and quality of their emotions. Spin and pulse scores both were not associated with the increase of tinnitus distress over time. The level of variability between different feelings was associated with tinnitus loudness. The level of variability of affect intensity was not.

4 Discussion

The central goal of this systematic literature review was to summarize previous research of psychosocial variables that predict tinnitus decompensation (i.e. the transition from an acute to a chronic disabling tinnitus) and tinnitus outcomes in individuals suffering from chronic tinnitus. Although epidemiologic studies showed that a significant proportion of individuals with tinnitus develop serious levels of distress and disability from their ringing in the ear (Davis 1989; Fujii et al. 2011;

McCormack et al. 2016), we have limited knowledge about factors that explain this transition from an acute to a chronic, debilitating tinnitus.

Our systematic literature search revealed only four longitudinal studies that examined tinnitus sufferers during the acute phase of their tinnitus and then monitored their transition to a chronic, decompensated status. Only three of these four studies reported sufficient results to become included to this systematic review. These three studies targeted both psychosocial and audiological or tinnitus characteristics as predictors of tinnitus decompensation. *General psychological distress* (i.e. symptoms of anxiety and depression), individuals' *report of other somatic symptoms*, and *tinnitus-related delay of the sleep onset* were identified as robust predictors of the development of a decompensated ringing in the ear. These results were in accordance with findings of cross-sectional studies of emotional well-being in individuals with chronic tinnitus compared with healthy individuals (Trevis et al. 2018). Symptoms of anxiety and depression, somatoform symptoms, and sleep difficulties were found to be consistently increased in tinnitus sufferers compared with healthy individuals (Trevis et al. 2018). One study (Wallhäusser-Franke et al. 2017) investigated illness coping as predictor. Interestingly individuals who applied *active, problem-solving coping* strategies during the acute stage were at risk of developing tinnitus decompensation during the chronic phase. Studies of other chronic medical conditions such as inflammatory pain conditions (e.g. rheumatoid arthritis) (Englbrecht et al. 2013) or HIV (Chan et al. 2006) revealed similar findings. Active and problem-solving coping was associated with less physical functioning and less emotional well-being. One way of explaining this result could be that active problem-solving does not well apply to manage persistent symptoms such as tinnitus. Individuals repeatedly report the experience that their attempts to cope fail. They feel frustrated and struggle to develop symptom acceptance. Several *personality traits* were identified to predict tinnitus decompensation in the chronic phase. Individuals who tended to be more emotional or excitable and who often reported somatic complaints were at higher risk of decompensating from their chronic ringing in the ear. Individuals who were in general more satisfied with their life were able to better compensate their tinnitus. The findings of longitudinal studies in this review are consistent with results of cross-sectional studies that have repeatedly demonstrated a relationship between neurotic, hysterical, and extraverted personality traits and tinnitus decompensation (Durai et al. 2017; Langguth et al. 2007; Trevis et al. 2018).

Psychosocial predictors of tinnitus outcomes in individuals with a chronic ringing in their ear were investigated in longitudinal observational as well as in EMA and diary studies. *Tinnitus distress* as well as *general psychological distress* and pre-existing *mental disorders* were demonstrated to be robust predictors of negative tinnitus outcomes across several studies. EMA studies showed that negative emotional states and stress experience predicted negative tinnitus outcomes in daily life of chronic tinnitus sufferers. Study findings that targeted the predictive value of psychosocial variables other than *tinnitus-related* or *general psychological distress* have to be interpreted with caution, because they were mostly examined in only one study. A study on *health behaviors* and the *level of functioning* for example (Holgers et al. 2000, 2005) showed that physical activity and physical mobility predicted better tinnitus outcomes at a later follow-up. The protective effect of physical exercise has already been demonstrated in numerous studies of other medical conditions (Baum and Posluszny 1999; Stewart and Yuen 2011). Several pathways are assumed to explain the protective impact of physical exercise, for example, it can buffer stress, it improves physical functioning and resilience (Baum and Posluszny 1999; Stewart and Yuen 2011). One study investigated the predictive value of *tinnitus acceptance* (Westin et al. 2008). Acceptance was not associated with tinnitus outcomes at a later follow-up. The results from longitudinal observational studies contradict findings from cross-sectional analyses that show acceptance to be correlated with less negative tinnitus impact (Trevis et al. 2018). Similar to the studies that examined predictors of tinnitus decompensation, the report of other somatic symptoms or medical conditions, such as *pain*, predicted negative tinnitus outcomes in chronic tinnitus sufferers.

The number of longitudinal observational studies (excluding EMA and diary studies) that examined *tinnitus characteristics* or *audiological characteristics* as predictors of tinnitus decompensation and tinnitus outcomes in individuals with chronic tinnitus was equal to the number of studies that examined psychosocial variables as predictors. The results on *tinnitus characteristics* or *audiological characteristics* as predictors of tinnitus outcomes were, however, less consistent than on psychosocial variables. For example, hearing loss was the most frequently investigated audiological variable (Bleich et al. 2001; Clifford et al. 2019; Dawes and Welch 2010; Hébert et al. 2012; Holgers et al. 2000; Olsen et al. 2013; Westin et al. 2008). It can be assumed that the relationship between hearing loss and tinnitus outcomes is more complex than expected and is determined by other factors such as the type of tinnitus outcome (tinnitus distress vs. loudness vs. severity), the frequency range in which the hearing loss is present or ENT conditions in childhood (Dawes and Welch 2010; Holgers et al. 2000, 2005).

This systematic review shows that there are still many gaps in the knowledge of predictors of tinnitus outcomes. Observational, longitudinal studies are rare, in particular those that examined the transition from an acute tinnitus to a chronic debilitating tinnitus. In this systematic review the quality of the included studies was low. Six out of the 16 included studies had a sample size smaller than $N = 50$ and loss to follow-up was increased in several studies (e.g., Erlandsson and Persson 2009; Hébert et al. 2012). Most of the included trials lacked power calculations to determine their optimal sample size. We assume that six of the included 16 studies were not primarily conceptualized as longitudinal studies of predictors of tinnitus outcomes (Bleich et al. 2001; Dawes and Welch 2010; Erlandsson and Persson 2009; Hébert et al. 2012; Holgers et al. 2005; Olsen et al. 2013). Another shortcoming is that several predicting variables were investigated in only one study and the results would need to be replicated before they can be interpreted appropriately. Most of the statistical analyses of predictors and outcomes were bivariate. Multivariate analyses were performed in only four studies (Hébert et al. 2012; Holgers et al. 2005; Langenbach et al. 2005; Wallhäusser-Franke et al. 2017), multilevel analyses in only 2 studies (Goldberg et al. 2017; Probst et al. 2016) out of 16 studies.

Furthermore, we have to consider critically that tinnitus outcomes varied substantially across the studies, as did the measures of the predictors of tinnitus outcomes. To increase the comparability between studies in the future it would be important to follow standards of core outcomes and measures (Landgrebe et al. 2010, 2012). Finally, most of the studies assessed tinnitus outcomes with self-report measures. We suggest that future research should apply also other modes of measures, such as neurophysiological or behavioral measures. For example, an outcome of interest could be psychophysiological response to stress, measured through cardiovascular outcomes (e.g. heart rate variability). Research on other persistent somatic symptoms, such as chronic pain for example, shows that the chronic, distressing experience of a somatic symptom is associated with cardiovascular changes (Tracy et al. 2016). In pain research behavioral, observational measures are used to assess the level of disability and limitations in mobility caused by the somatic symptom, such as the Back Performance Scale (Strand et al. 2002) for example. A similar approach could be applied to assess tinnitus disability.

In summary, this systematic review shows that general psychological distress, individuals' report of somatic complaints, tinnitus-related delay of sleep onset, active illness coping, and certain personality traits (e.g. neuroticism) predict the transition from an acute to a chronic, disabling tinnitus. General psychological and tinnitus-related distress, the report of other somatic problems such as pain, certain health behaviors and the level of physical and social functioning are predictors of tinnitus outcomes at a later follow-up in chronic tinnitus patients. The number of psychosocial predictors that have been investigated in longitudinal study designs is still limited. Results from cross-sectional studies showed that there are several psychosocial variables that correlate with tinnitus sufferers' level of functioning, such as cognitive processing styles (Weise et al. 2013), anxiety sensitivity (e.g., Hesser and Andersson 2009), symptom catastrophizing (Weise et al. 2013) or fear avoidance (Cima et al. 2011; Kleinstäuber et al. 2013). It would be important to examine these variables also with longitudinal study designs. Moreover future research should consider variables that correlated with outcomes in individuals with other chronic conditions, such as chronic pain for example. Examples of these variables are social support (e.g., Gallant 2003), health locus of control (e.g., Stewart et al. 2018), discomfort intolerance (e.g., Schmidt et al. 2006), dispositional mindfulness (e.g., Garland and Fredrickson 2019; Tomlinson et al. 2018), or dispositional optimism (e.g., Stanton et al. 2007). Future tinnitus research should examine these variables as potential predictors of the transition from acute to chronic, disabling tinnitus, and other outcomes in chronic tinnitus sufferers. Finally the results of our review have important implications for clinical practice and delivery of health services to individuals with acute tinnitus. Psychosocial variables that robustly predict a risk of developing a chronic, decompensating tinnitus should be assessed and addressed by clinicians who are commonly contacted by individuals with acute tinnitus. Depending on the structure of a national healthcare system, audiologists, general practitioners, or ENT specialists are commonly the first contact person for individuals with acute tinnitus. Besides the audiological and medical examination, characteristics such as pre-existing mental health problems, increased levels of distress associated with the acute tinnitus, sleep difficulties, an increased report of other somatic symptoms, and certain problematic health behaviors should be assessed.

References

Andersson G, Hägnebo C, Yardley L (1997) Stress and symptoms of Menière's disease: a time-series analysis. J Psychosom Res 43:595–603

Baum A, Posluszny DM (1999) Health psychology: mapping biobehavioral contributions to health and illness. Ann Rev Psychol 50:137–163

Bhatt JM, Bhattacharyya N, Lin HW (2017) Relationships between tinnitus and the prevalence of anxiety and depression. Laryngoscope 127:466–469

Bleich T, Lamprecht F, Lamm H et al (2000) Long-term course of chronic tinnitus aurium. Psychother Psych Med 50:99–99

Bleich T, Lamprecht F, Lamm H et al (2001) The long-term course of chronic tinnitus [Der Langzeitverlauf des chronischen tinnitus aurium]. Z Med Psychol 10:79–86

Chan I, Au A, Li P et al (2006) Illness-related factors, stress and coping strategies in relation to psychological distress in HIV-infected persons in Hong Kong. AIDS Care 18:977–982

Cima RFF, Crombez G, Vlaeyen JWS (2011) Catastrophizing and fear of tinnitus predict quality of life in patients with chronic tinnitus. Ear Hearing 32:634–641

Clifford RE, Baker D, Risbrough VB et al (2019) Impact of TBI, PTSD, and hearing loss on tinnitus progression in a US marine cohort. Mil Med 184:839–846

Conrad I, Kleinstäuber M, Jasper K et al (2015) The role of dysfunctional cognitions in patients with chronic tinnitus. Ear Hearing 36:e279–e289

Davis AC (1989) The prevalence of hearing impairment and reported hearing disability among adults in great-Britain. Int J Epidemiol 18:911–917

Dawes PJD, Welch D (2010) Childhood hearing and its relationship with tinnitus at thirty-two years of age. Ann Oto Rhinol Laryn 119:672–676

De Ridder D, Vanneste S, Freeman W (2014a) The Bayesian brain: phantom percepts resolve sensory uncertainty. Neurosci Biobehav Rev 44:4–15

De Ridder D, Vanneste S, Weisz N et al (2014b) An integrative model of auditory phantom perception: tinnitus as a unified percept of interacting separable subnetworks. Neurosci Biobehav Rev 44:16–32

Durai M, O'keeffe MG, Searchfield GD (2017) The personality profile of tinnitus sufferers and a nontinnitus control group. J Am Acad Audiol 28:271–282

Englbrecht M, Kruckow M, Araujo E et al (2013) The interaction of physical function and emotional Well-being in rheumatoid arthritis-what is the impact on disease activity and coping? Semin Arthritis Rheum 42:482–491

Erlandsson S, Persson V (2009) A longitudinal study investigating the contribution of mental illness in chronic tinnitus patients. Audiol Med 4:124–133

Fujii K, Nagata C, Nakamura K et al (2011) Prevalence of tinnitus in community-dwelling japanese adults. J Epidemiol 21:299–304

Gallant MP (2003) The influence of social support on chronic illness self-management: a review and directions for research. Health Educ Behav 30:170–195

Garland EL, Fredrickson BL (2019) Positive psychological states in the arc from mindfulness to self-transcendence: extensions of the mindfulness-to-meaning theory and applications to addiction and chronic pain treatment. Curr Opin Psychol 28:184–191

Goldberg RL, Piccirillo ML, Nicklaus J et al (2017) Evaluation of ecological momentary assessment for tinnitus severity. JAMA Otolaryngol-Head Neck Surg 143:700–706

Hallam RS (1996) Correlates of sleep disturbance in chronic distressing tinnitus. Scand Audiol 25:263–266

Hallam RS, Rachman S, Hinchcliffe R (1984) Psychological aspects of tinnitus. In: Rachman S (ed) Contributions of medical psychology. Pergamon, Oxford, pp 31–54

Hébert S, Canlon B, Hasson D et al (2012) Tinnitus severity is reduced with reduction of depressive mood - a prospective population study in Sweden. PLoS One 7:e37733

Hesser H, Andersson G (2009) The role of anxiety sensitivity and behavioral avoidance in tinnitus disability. Int J Audiol 48:295–299

Hesser H, Weise C, Westin VZ et al (2011) A systematic review and meta-analysis of randomized controlled trials of cognitive-behavioral therapy for tinnitus distress. Clin Psychol Rev 31:545–553

Holgers KM, Erlandsson SI, Barrenas ML (2000) Predictive factors for the severity of tinnitus. Audiology 39:284–291

Holgers KM, Zöger S, Svedlund K (2005) Predictive factors for development of severe tinnitus suffering-further characterisation. Int J Audiol 44:584–592

Jäger B, Malewski P, Koch T et al (2004a) Prediction of disease experience in tinnitus during one year - results of the Hanover tinnitus study [Prädiktion des Krankheitserlebens bei tinnitus über ein Jahr – Ergebnisse der Hannover-tinnitus-Verlaufsstudie]. Psychother Psych Med 54:95–96

Jäger B, Malewski P, Lamprecht F (2004b) Coping with tinnitus: results of the Hannover tinnitus course study [Krankheitsbewältigung bei tinnitus: erste Ergebnisse aus der Hannover-tinnitus-Verlaufsstudie]. In: Petermann F, Ehlebracht-König I (eds) Motivierung, Compliance und Krankheitsbewältigung. Roderer, Regensburg, pp 155–168

Jäger B, Malewski P, Schwab B et al (2005) Prediction of decompensation and treatment requirements in chronicized tinnitus - 2-year-catamnesis of the Hannover tinnitus development study [Prädiktion der Dekompensation und des Behandlungsbedarfs beim chronifizierten tinnitus – 2-Jahres-Katamnese der Hannover-tinnitus-Verlaufsstudie]. Psychother Psych Med 55:102–102

Jäger B, Malewski P, Schwab B et al (2006) Development and prognosis of acute subjective tinnitus over 2 years [Verlauf und Prognose des akuten subjektiven tinnitus über 2 Jahre]. Psychother Psych Med 56:91–91

Jastreboff PJ (1990) Phantom auditory perception (tinnitus): mechanisms of generation and perception. Neurosci Res 8:221–254

Kleinstäuber M, Jasper K, Schweda I et al (2013) The role of fear-avoidance cognitions and behaviors in patients with chronic tinnitus. Cogn Behav Ther 42:84–99

Landgrebe M, Zeman F, Koller M et al (2010) The tinnitus research initiative (TRI) database: a new approach for delineation of tinnitus subtypes and generation of predictors for treatment outcome. BMC Med Inform Decis Mak 10:42

Landgrebe M, Azevedo A, Baguley D et al (2012) Methodological aspects of clinical trials in tinnitus: a proposal for an international standard. J Psychosom Res 73:112–121

Langenbach M, Olderog M, Michel O et al (2005) Psychosocial and personality predictors of tinnitus-related distress. Gen Hosp Psychiat 27:73–77

Langguth B, Kleinjung T, Fischer B et al (2007) Tinnitus severity, depression, and the big five personality traits. In: Langguth B, Hajak G, Kleinjung T, Cacace A, Moller A (eds) Tinnitus: pathophysiology and treatment. Elsevier, München, pp 221–225

McCormack A, Edmondson-Jones M, Somerset S et al (2016) A systematic review of the reporting of tinnitus prevalence and severity. Hearing Res 337:70–79

Nyenhuis N, Zastrutzki S, Weise C et al (2013) The efficacy of minimal contact interventions for acute tinnitus: a randomised controlled study. Cogn Behav Ther 42:127–138

Olderog M, Langenbach M, Michel O et al (2004) Predictors and mechanisms of tinnitus distress - a longitudinal analysis. Laryngo-Rhino-Otologie 83:5–13

Olsen US, Lello E, Nagaraj VS et al (2013) Factors affecting severity of tinnitus - a follow-up study of tinnitus subjects at an ear nose throat clinic in Sweden. Int Tinnitus J 18:115–121

Probst T, Pryss R, Langguth B et al (2016) Emotion dynamics and tinnitus: daily life data from the "TrackYourTinnitus" application. Sci Rep 6:9

Rutter M (2006) Implications of resilience concepts for scientific understanding. Ann N Y Acad Sci 1094:1–12
Schmidt NB, Richey JA, Fitzpatrick KK (2006) Discomfort intolerance: development of a construct and measure relevant to panic disorder. J Anxiety Disord 20:263–280
Searchfield GD, Kobayashi K, Sanders M (2012) An adaptation level theory of tinnitus audibility. Front Syst Neurosci 6:46
Sedley W, Friston KJ, Gander PE et al (2016) An integrative tinnitus model based on sensory precision. Trends Neurosci 39:799–812
Stanton AL, Revenson TA, Tennen H (2007) Health psychology: psychological adjustment to chronic disease. Annu Rev Psychol 58:565–592
Stewart DE, Yuen T (2011) A systematic review of resilience in the physically ill. Psychosomatics 52:199–209
Stewart JA, Aebischer V, Egloff N et al (2018) The role of health locus of control in pain intensity outcome of conservatively and operatively treated hand surgery patients. Int J Behav Med 25:374–379
Strand LI, Moe-Nilssen R, Ljunggren AE (2002) Back performance scale for the assessment of mobility-related activities in people with back pain. Phys Ther 82:1213–1223
Tomlinson ER, Yousaf O, Vitterso AD et al (2018) Dispositional mindfulness and psychological health: a systematic review. Mindfulness (N Y) 9:23–43
Tracy LM, Ioannou L, Baker KS et al (2016) Meta-analytic evidence for decreased heart rate variability in chronic pain implicating parasympathetic nervous system dysregulation. Pain 157:7–29
Trevis KJ, Mclachlan NM, Wilson SJ (2018) A systematic review and meta-analysis of psychological functioning in chronic tinnitus. Clin Psychol Rev 60:62–86
Vielsmeier V, Stiel RS, Kwok P et al (2020) From acute to chronic tinnitus: pilot data on predictors and progression. Front Neurol 11:997
Wallhäusser-Franke E, Repik I, Delb W et al (2015) Long-term development of acute tinnitus [Langzeit-Entwicklung von akutem tinnitus]. Laryngo-Rhino-Otologie 94:759–769
Wallhäusser-Franke E, D'amelio R, Glauner A et al (2017) Transition from acute to chronic tinnitus: predictors for the development of chronic distressing tinnitus. Front Neurol 8:605
Weise C, Hesser H, Andersson G et al (2013) The role of catastrophizing in recent onset tinnitus: its nature and association with tinnitus distress and medical utilization. Int J Audiol 52:177–188
Westin V, Hayes SC, Andersson G (2008) Is it the sound or your relationship to it? The role of acceptance in predicting tinnitus impact. Behav Res Ther 46:1259–1265
Wise KJ, Bird PA, O'beirne GA (2015) Tinnitus healthcare in New Zealand. N Z Med J 128:20–23

Part VI
Clinical Assessment of Tinnitus

Momentary Analysis of Tinnitus: Considering the Patient

Brian C. Deutsch and Jay F. Piccirillo

Contents

Abstract Ecological momentary assessment is a valuable research technique meant to capture real-time data and contextualize disease. While more common in neuropsychiatric research, this methodology is exceptionally fit for tinnitus. Tinnitus has been shown to be affected by many patient-level and environment-specific factors. From an individual's baseline anxiety to the level of ambient noise in their environment, the level of bother experienced by those with tinnitus can vary widely. Only assessing tinnitus within a clinical environment can distort the true impact of the

B. C. Deutsch (✉)
Department of Otolaryngology, Head and Neck Surgery, Washington University School of Medicine in St. Louis, St. Louis, MO, USA

Department of Medical Education, Icahn School of Medicine at Mount Sinai, New York, NY, USA
e-mail: brian.deutsch@icahn.mssm.edu

J. F. Piccirillo
Department of Otolaryngology, Head and Neck Surgery, Washington University School of Medicine in St. Louis, St. Louis, MO, USA

Curr Topics Behav Neurosci (2021) 51: 383–402
https://doi.org/10.1007/7854_2020_176
Published Online: 18 August 2020

disease. Ecological data can minimize bias while generating an individualistic picture of the burden being experienced by the patient. Individual data can also compliment new research methods rooted in precision medicine, providing clearer, better-suited treatments for each patient on the tinnitus spectrum.

Keywords Application · Assessment · Ecological · EMA · Momentary · Smartphone · Tinnitus

1 Introduction: Momentary Assessment

Self-reported health information using out-of-office ecological assessments are common in biomedical research practice for well-known neuropsychiatric illnesses that display a range of fluctuating symptoms. Ecological momentary assessment (EMA) is a research tool that allows for self-reported data to be collected with context. Clinically, this translates to gathering relevant data while also assessing the surrounding factors and individual experience of those factors (Kubiak and Smyth 2019). A wide variety of sampling techniques and recording styles of self-reported data have been studied in a broad range of illnesses including insomnia (Buysse et al. 2007), headache (Kikuchi et al. 2006), schizophrenia (Granholm et al. 2008), and many others (Stone et al. 1998; Hufford et al. 2002; Wonderlich et al. 2007). Ecological assessment has evolved from pencil and paper diaries to utilizing modern technologies like pagers, preset watches, computers, and mobile phones.

With the turn of the millennium, the NIH endorsed self-reporting as an important consideration for the future of biomedical research (Stone et al. 2000). The ensuing decades of technical innovation brought along devices that make this kind of self-reporting simple and allowed for easy data abstraction. Smartphones can notify their users in real-time, record responses, and timestamp data and then store or send the information with increasing simplicity. This simplicity of collecting data and monitoring compliance addresses commonly cited shortcomings of EMA such as the feasibility of administering multiple assessments for each patients and whether the participant is completing the assessments at the appropriate time rather than backfilling the information. The built-in functionality with smartphone applications may open the floodgates to a new strata of research design and patient-centered clinical practice.

EMA is beneficial for providing a more longitudinal "real-world" picture of disease severity. It is also recognized as novel in its ability to discern specific symptom variability (Ross et al. 2007). The continuity of symptom reporting and patient-engagement paints a more precise landscape of the progression of diseases than can be obtained with scheduled assessment. Collinearity or sequentiality of symptom changes is more easily discerned with the granularity of data provided by EMA. The strength of this data is that it can be used to see reductionist principles as well as the bigger clinical picture. The methodology allows for multiple samplings to break down disease processes and the specific effects on patients, which delineates the progression for diagnoses with complex etiologies and stochastic progressions (Moskowitz and Young 2006).

The use of EMA for measurement in tinnitus is likely to avoid some of the systematic biases that are common in retrospective reporting tools such as the tinnitus questionnaire (TQ) (Hallam et al. 1988), Tinnitus Functional Index (TFI) (Meikle et al. 2012), and Tinnitus Handicap Inventory (THI) (Newman et al. 1996). The fluctuating nature of tinnitus symptoms and the retrospective nature of patient-reported outcome measures leave them vulnerable to recall bias. Human memory is imperfect, and the data captured by memory tends to be biased to an individual's personal heuristics; human recall is not like playing back a recorded movie but rather reconstructing events and experiences based on these rules and heuristics (Gorin and Stone 2001).

The structure of EMA is one which attempts to mitigate the effects of subjective bias by minimizing the need to rely on these heuristics. Common pitfalls of retrospective surveys are peak and recency bias, where the patient preferentially recalls extremes or the most recent of their symptoms. EMA provides a more accurate picture of a subject's experience by making them more aware of their symptoms throughout the period of study (Stone et al. 1997). When the subject completes multiple surveys in a day, recency and peak bias are spread across multiple data points, providing a higher resolution of symptomatic burden over time. How much resolution is possible can vary depending on what disease is being studied, how the data is collected, what questions are being asked, and how often the subject is reporting symptoms (Ross et al. 2007).

1.1 Designing an Ecological Assessment

EMA is a useful tool but does not exist as a singular protocol that will be inherently useful for all diseases. Formatting the EMA questionnaire and the dissemination for a specific research question are an important aspect of its design. EMA is a sampling technique, which requires thoughtful research design that accounts for questions, hypotheses, practicality of the intervention, and background knowledge of the disease being studied. Additionally, with this sampling method, investigators must consider the sampling schedule in the specific ecological context being assessed. Sampling can be event based, where the subject voluntarily inputs data during a pre-defined experience or symptom, or it can be time based. Time-based sampling is further divided into structured sampling where data collection is planned for specific times during the day, and it can also be randomly collected by using phone applications. There is no benefit to using one method versus the other outright; however, it is important to consider one's research question when making the decision about sampling methodology.

When considering the use of EMA to study tinnitus, there is no single "correct" methodology. Tinnitus can be considered a syndrome with varying levels of complexity (Han et al. 2009), which complicates straightforward study design. Since EMA can be constructed to handle event-based or time-based questions (Ross et al. 2007), depending on the outcome of interest, one must first define that outcome. If one were interested in something discrete, say the number of times a subject experiences ringing in their ears throughout the day, then it would be as simple as instructing the subject to record each instance of tinnitus in an event-based strategy. On the other hand, if we were interested in a patients' mood throughout the day, then a time-based collection would be more useful to gather a relevant distribution of temperament, which is known to fluctuate throughout the day (Stone et al. 1994). The benefit of using EMA comes from its ability to discern small changes in a subject's symptoms or affect throughout the day, which makes it keenly useful for therapeutic trials where fluctuation in these attributes is expected.

1.2 Precision Medicine and Ecological Data Sampling

Traditional research designs rooted in retrospective questionnaires and individual office visits may not be able to answer all questions on the origin, severity, and treatment effectiveness of a disease clearly or thoroughly. Depending on the research question, there can be a benefit to using a hybrid model that allows for event sampling with additional time-based surveys. In this hybrid method, the subject serves as their own control where the time-based surveys can give contrast to the event-based reporting of particular symptoms or sequelae of interest. This within-subject case-control method may be the most useful to consider for something as complex as tinnitus as it will allow for modification of therapy or reassessment of clinical triggers such as anxiety levels, loudness of ambient environment, and social stressors.

Reassessment of clinical triggers is a fundamental tenant of patient-centered medicine. EMA is integral to precision medicine and specifically precision clinical trials (PCT). The PCT framework aims to generate individualized treatments through various methods meant to target complex neurobehavioral and somatic disorders. These methods include enriching target populations with brief run-in periods (dubbed "treatment-targeted enrichment") designed to identify those susceptible to specific treatments or therapies and modifying treatments *during* the trial for optimal patient outcomes (Lenze et al. 2020). The PCT design is heavily reliant on very precise and reliable feedback about outcomes, compliance, and other predictors. This new patient-centered precision medicine is highly congruous with the capabilities of EMA and can inform the next generation of clinical trials. A review article by Torous and Keshavan suggests that PCTs are ideal for those with a high clinical risk for psychosis; using smartphone tools like digital phenotyping can offer affordable, scalable, and personalized intervention (Torous and Keshavan 2020).

2 Analytical Approaches to Ecological Data

The massive amounts of longitudinal data that are generated from an EMA study provide a unique opportunity to use statistical methods originally used for economic time-series data. The longitudinal data provides individual and contextual patterns for illnesses, which informs a personalized medicine approach. The statistical methodologies for EMA have been adapted by clinical researchers in psychology and psychiatry to assess causes over time for related variables. Vector autoregression (VAR) is one of these statistical methods (Lutkepohl 2007), which allows testing of how different variables relate within a single individual over time. For example, a patient could report on negative thinking, stress, and tinnitus bother repeatedly. VAR techniques can test which of these variables might plausibly cause the others over time within an individual. Interpretation of the data from PCTs can be difficult because each subject becomes a single study. A technique that looks at *both* individuals *and* the group as a whole is preferable. This preferable situation can be assessed with multilevel extension of dynamic structural equation modeling (ML-DSEM), where group and individual models can both be examined (Schwartz and Stone 1998; Shiffman 2014; Muthén and Muthén 2017).

ML-DSEM allows longitudinal data to be modeled with respect to subjects' level of tinnitus bother, cognitive symptoms, anxiety symptoms, and potentially other variables (e.g., psychological or medical symptoms; anything that can vary over time and that the individual can report). Using ML-DSEM, a researcher or clinician could examine whether cognitive symptoms or anxiety may prospectively predict tinnitus bother at a later time point for a single individual. The degree to which a person's tinnitus bother is predicted by anxiety symptoms can then become a variable to predict response to treatment and functional connectivity with key brain networks. In other words, the characteristics of patients' individual profiles can be used to predict both treatment response and imaging results.

3 Use and Utility of Momentary Assessment in Other Fields

When considering the spectrum of disease processes, EMA is particularly useful in research surrounding chronic, complicated, and not well-understood diagnoses. This is not a tool equipped to take the place of physicians in determining a general diagnosis but rather an adjuvant useful in specifying sub-characteristics. The diagnostic capacity of EMA lies less in its ability to inform which disease is likely but rather which subtype of a particular disorder may be present as well as how the disease changes over time. EMA helps to create sections of complicated diseases with nonuniform phenotypes, which can be helpful to identify which domains of a disease are affecting a patient's health. This ability to define domains of disease can be useful for diagnoses that do not fit in specific "boxes" and may require a more complicated treatment approach. This more complicated presentation is a common phenomenon in many psychiatric disorders. Commonly described barriers for progress related to mental disorders include the preferential focus on the diagnosis rather than the network of interrelated symptoms (Borsboom 2008; Borsboom and Cramer 2013; Cramer et al. 2010, 2012) as well as the predilection to focus on groups of patients within a single diagnosis rather than how symptoms persist as ongoing processes within an individual (idiographic) (Fisher and Boswell 2016; Fisher 2015; Molenaar 2013). These tendencies are equally relevant in tinnitus research, and it is possible that tinnitus can be better understood as a spectrum of experiences originating from different sources. Meaning, that for one person with tinnitus, the experience of anxiety may predominant, while for another person, cognitive symptoms may be most significant. Further, how strongly these symptoms are associated with tinnitus may vary widely among patients due to unknown factors.

While tinnitus is relatively new to the world of ecological assessment, there are many examples of EMA's capacity to better define the complexity of a diagnosis. For example, eating disorders are a complicated amalgam of behaviors and psychological principals that have generally been classified into restrictive and binge type (Bailey et al. 2014). But many experts agree that patients don't typically fit one set of criteria for any particular disease, and the present systems of classification don't reflect patients' clinical realities (Uher and Rutter 2012). Clinical realities for eating disorders are inherently ecological, and EMA has been a useful tool for characterizing this pathology more clearly. Specifically, studies have used EMA to assess symptoms that are commonly employed in models for disordered eating like emotional avoidance and cognitive behavioral theory (Haynos et al. 2015; Lavender et al. 2013). Emotional avoidance is the concept that an individual will displace negative feelings and affect with behaviors like food restriction (Wildes et al. 2010). One study found that for patients with anorexia nervosa, high levels of food restriction were not associated with avoidance of negative emotions, which is discordant with the emotional avoidance model as previously described (Haynos et al. 2015). Another study looking at the cognitive behavioral theory of "body checking," a manifestation of elevated concerns about one's body, weight, and eating, found the frequency of body checks to be directly associated with level of dietary restriction (Lavender et al. 2013). The ability of EMA to associate isolated symptomatology with specific actions, emotions, or settings allows for a more granular understanding of both the disease process and its variability. A similar methodology would be useful to assess the direct associations of tinnitus with negative ideation, anxiety levels, or catastrophizing.

4 Tinnitus

4.1 Tinnitus as an Ecological Entity

There are numerous sources that debate the etiology of tinnitus as well as a litany of neuroimaging correlations with behavioral measures and suggestions for treatment.

The consensus on tinnitus has evolved to the likely conclusion that there *is* no unifying consensus. Clinically, the spectrum of tinnitus severity ranges from mildly irritating to incapacitating (Han et al. 2009). Initially, the severity of disease was postulated to be due to a singular biological underpinning. For example, there are several regions of the brain that are associated with the symptoms of tinnitus including the auditory cortex (Burton et al. 2012; Kim et al. 2012; Landgrebe et al. 2009; Maudoux et al. 2012; Vanneste et al. 2011a, b), attention centers (Burton et al. 2012; Roberts et al. 2013), basal ganglia (Maudoux et al. 2012), prefrontal cortex (Kim et al. 2012; Maudoux et al. 2012; Schlee et al. 2009; Seydell-Greenwald et al. 2012), parahippocampal regions (Maudoux et al. 2012; Vanneste et al. 2011a; Schmidt et al. 2017), and insula (Burton et al. 2012; Vanneste et al. 2011b). Tinnitus has also been associated with the limbic system, and its severity fluctuates with a person's emotional state (Jastreboff and Hazell 1993). These associations suggest that tinnitus is a manifestation of heterogeneous disease processes with different biological manifestations. This heterogeneity is consistent with the existence of a spectrum of patient profiles. While it is possible the heterogeneity of etiology research can be explained by methodological and sampling inconsistencies (Hall et al. 2013; Husain and Schmidt 2014; Scott-Wittenborn et al. 2017), the existing findings are also consistent with a spectrum of patient profiles. Tinnitus is also the result of interactions between a patient's neuroplasticity, cognition, and surroundings. These connections between the patient and their environment are important mediators in the perception of tinnitus and indirectly of these patient profiles (Searchfield 2014). This spectrum of profiles within different researchers' sampling suggests different and not necessarily contradictory conclusions among different individual studies. The use of an ecological approach in tinnitus can provide a supplementary understanding to the neurophysiologic research.

Previous work on the ecology of tinnitus focused on the interaction of the interplay between social, environmental, and psychosocial factors (Fig. 1). The model itself stems from Adaptation Level Theory (ALT) which suggests that tinnitus varies constantly and is dependent on "subtle changes in cochlear outflow, emotion, context, and attention." The presentation of this model includes a mathematical relationship suggested by Heson (Helson 1964); however, Searchfield writes that the model is generally showing "the interplay between sound, tinnitus signal, individual psychology, and attention" (Searchfield 2014). Searchfield suggests two models to test the ecological model of tinnitus: (1) long-term follow-up using a variety of assessments and methods and (2) addition of ecological validity to the existing reductionist method, by allowing focus on individual differences within a group (Searchfield 2014). Both of these models can be readily interrogated using EMA as either a primary or adjuvant tool. The ability to deliver assessment tools to a smartphone simplifies the methodology for long-term follow-up while incorporating ecological context by being available consistently throughout the day.

The experience of sound is not the same for everyone; it is the product of personal neurobiological phenomenon in constant interaction with our environment. Upper-level auditory processing is also prone to change and adapt with changes in sound perception (Rabinowitz et al. 2011). These realities of the communication between

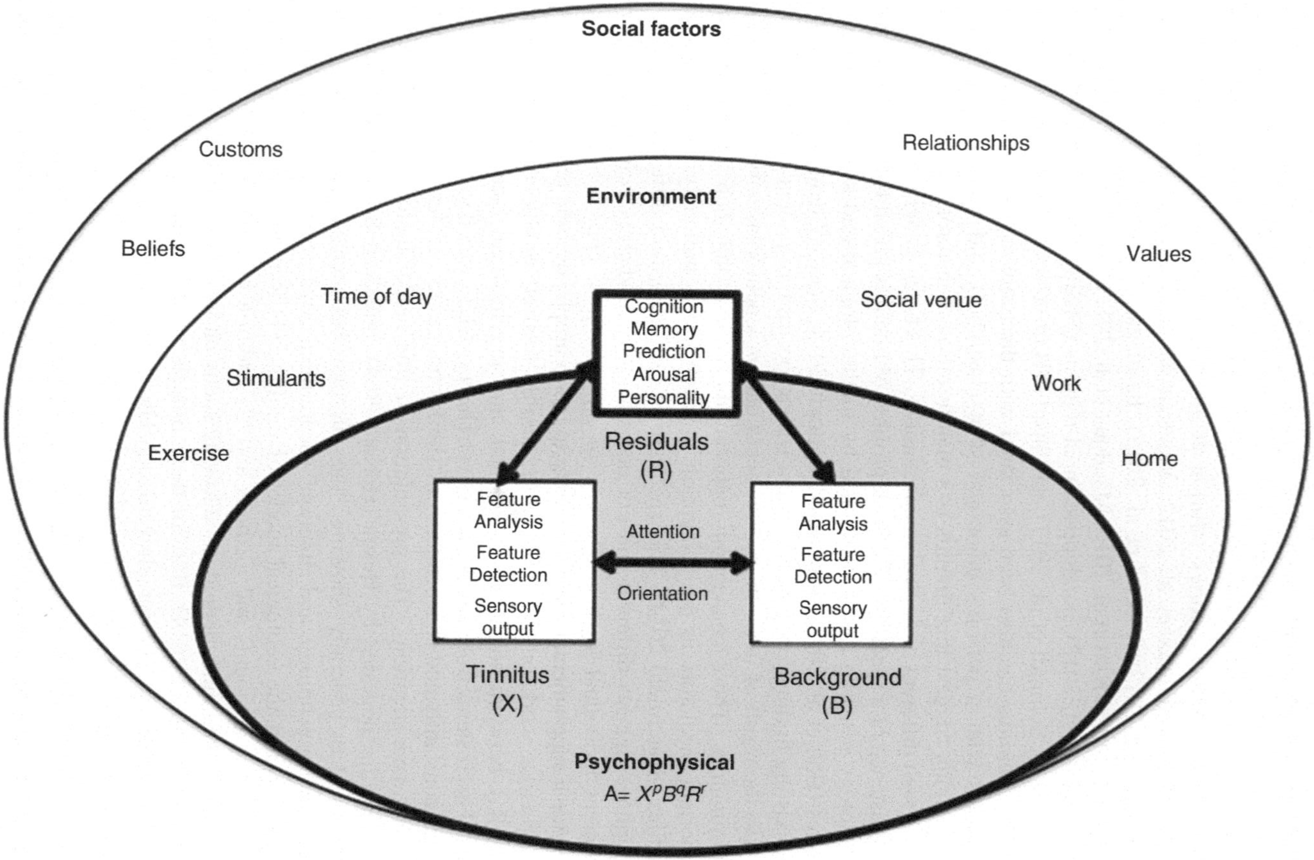

Fig. 1 A conceptual ecological model for tinnitus. The ecological model of tinnitus. This model consists of a psychophysical core described by adaptation level theory in which tinnitus and background sound perception are under influence of individual psychology factors classified in ALT as "residuals." These factors

are influenced by the environment and social context. The adaptation level is the weighted product of: $\overline{X}$, the intensity of tinnitus signal, B, intensity of background neural activity, and R, intensity of residual components (e.g., memory, arousal, and personality). The weighting coefficients p, q, and r determine the relative contributions of components to adaptation level and are considered to reflect attention and auditory scene analysis. Helson expressed this relationship mathematically: $A = \overline{X}^p B^q R^r$ (Helson 1964). Reprinted from "Tinnitus What and Where: An Ecological Framework" in *Frontiers in Neurology*, by Dr. Grant D. Searchfield, December 15, 2014, retrieved from https://www.ncbi.nlm.nih.gov/pmc/articles/PMC4266022/. Copyright 2014 by Grant Searchfield changing physiology and sound perception suggest a more complex story of tinnitus. A story with many chapters which require synthesizing to create a coherent narrative. The perceived impact of tinnitus is not a simple metric of intensity but an interplay between personal heuristics and contextual perception of sound; as a whole, the compound effect is a combination of volume, attention, and tinnitus severity (Schmidt et al. 2014). Variations in things like time of day (Probst et al. 2017), amount of ambient sound (Tyler and Baker 1983), and anxiety levels (Dauman et al. 2017) can alter an individual's tinnitus experience. The combination of varying external stimuli and individual psychosocial factors reinforces the idea that tinnitus is a highly personal affliction, the measurement of which dramatically changes over time. Things like anxiety, stress, and depression are linked to worsened perception of tinnitus symptoms (Pattyn et al. 2016; Reiss et al. 1986). Even personality has been suggested to affect response to sound, and those with tinnitus are more likely to be socially withdrawn, isolated, reactive to stress, and have poorer self-control (Welch and Dawes 2008). Comorbid health issues, coping strategies, and baseline acceptance are also contributors to tinnitus. Not only can physical health affect the presence of tinnitus directly, but comorbidities such as cardiovascular disease (Stobik et al. 2005) and arthritis (Nondahl et al. 2011) further exacerbate its severity. On top of physical health, acceptance and coping strategies are also implicated in the perception of tinnitus bother: specifically, avoiding catastrophizing, seeking out support systems, and attempting to avoid focusing on the perceived sound of tinnitus can decrease bother (Budd and Pugh 1996; Sullivan et al. 1994).

Moreover, while tinnitus and any associated hearing complaints may be the identifiable primary complaint for many of these patients, it is important to recognize the broader implications of this diagnosis. That is, tinnitus is not so much an auditory stimulus as it is a psychopathological reaction to a perceived auditory stimulus. Many problems for these patients aren't necessarily due to the direct effects of the tinnitus sounds but rather the downstream social and cognitive impairments. The sequelae of tinnitus can lead to social isolation and withdrawal due to the inability to communicate (Dauman et al. 2017). The inability to control the progression of hearing loss and misunderstanding from potential support networks contributes to learned helplessness and the loss of appropriate coping behaviors (Overmier 2002). The overall effect being the potential to adopt emotionally maladaptive mechanisms for living with tinnitus (Budd and Pugh 1996). Another avenue to utilize the infrastructure of EMA is being able to collect and analyze data quickly, giving providers additional information about the patient's experience that cannot be feasibly collected during a clinic visit. This is especially true when reporting psychosocial problems which may affect the patient in complex ways that are not easily, quickly, or clearly expressed directly to the physician during a visit. There are many and varied causes, effects, and psychosocial modulators of tinnitus. To capture a representative set of data about each of these considerations, we must be able to record data as they occur. We cannot expect a patient to recall every instance of feeling helpless or recognize when they are being socially isolative. These are the responsibility of the researcher to design the appropriate questionnaires and survey schedules to extract the factors intensifying these patients' tinnitus. The recognition of the multifactorial impacts on tinnitus has led to the development of questionnaires meant to capture this information. Tools such as the TFI (Meikle et al. 2012), the THI (Newman et al. 1996), the Fear of Tinnitus Questionnaire (FTQ), the Tinnitus Catastrophizing Scale (TCS) (Cima et al. 2011), and the Perceived Stress Questionnaire (PSQ-30) (Levenstein et al. 1993) are all validated and commonly used in tinnitus-related research. These tools are meant to incorporate the personal experience of tinnitus and capture individual characteristics that may exacerbate its impact. While a step in the direction of capturing a more holistic picture of tinnitus, the retrospective capturing of this data continues to pose the same reporting biases and eschews ecological validity. The development of these tools is an incredible asset and has provided the first starting point for ecological studies about tinnitus (Wilson et al. 2015; Goldberg et al. 2017; Henry et al. 2012). They have proved very adaptable and easy to implement (Fig. 2). The initial data from momentary assessment using established questionnaires helps to interrogate the ecological validity of these already existing, reductionist questions, as suggested by Searchfield (2014). Essentially, while the questionnaires capture a tiny puzzle piece of information about the disease, EMA is used to see if we can generate multiple pieces to see if they fit into a clearer picture.

4.2 *Current Assessment*

While there are no specific, physiological tests for tinnitus, it is worth comparing to a phenomenon which often accompanies tinnitus: hearing loss (Meikle and Taylor-Walsh 1984). Current clinical assessments of hearing also tend to lack real-world, or ecological, validity (Keidser 2016) and variably reflect self-reported hearing, with moderate correlations at best (Timmer et al. 2015). This is similar to the discordance experienced by tinnitus patients with equivocal measures of disease severity and no delineated treatment protocol. Tinnitus is notoriously complicated in its severity, frequency, onset, and progression (Han et al. 2009). Therefore, designing an EMA tool appropriate for tinnitus must incorporate many patient-centric factors leading to a more dynamic tool than is customary in current practice. Beyond intentional design, one must also consider the possibility of fatigue when it comes to constant sampling using EMA. Of the limited existing literature concerning ecological sampling techniques, patients with tinnitus showed very good compliance in using phone-based EMA (Wilson et al. 2015; Goldberg et al. 2017; Henry et al. 2012). Initial ecological studies of tinnitus are a proof-of-concept that using EMA is a feasible bridge to personalized treatments for our patients with tinnitus that can provide crucial data about its clinical heterogeneity.

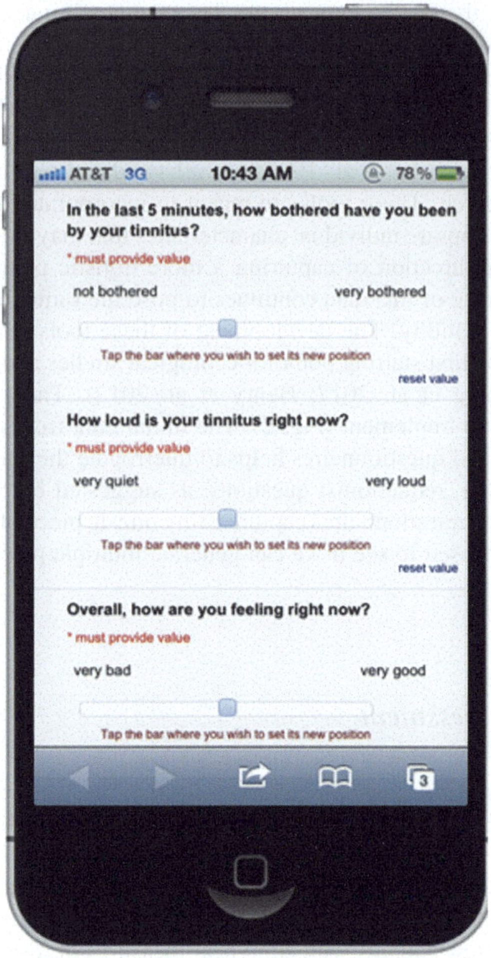

Fig. 2 A real-life example of the simple interface for integrating previous reductionist techniques within an ecologically valid platform. Using the smartphone interface, the researcher can assess when and where the patient is experiencing symptoms and the severity of the experience. Survey questions were accessible through smartphone's internet browser. Reprinted from "Ecological Momentary Assessment of Tinnitus Using Smartphone Technology: A Pilot Study" in, by Michael B. Wilson et al., vol 152, issue 5, pp. 897–903. Copyright © May 2015 by *Otolaryngology – Head & Neck Surgery*. Reprinted by Permission of SAGE Publications, Inc. retrieved from https://www.ncbi.nlm.nih.gov/pmc/articles/PMC4580970/. While every effort has been made to ensure that the contents of this publication are factually correct, neither the authors nor the publisher accepts, and they hereby expressly exclude to the fullest extent permissible under applicable law, any and all liability arising from the contents published in this Article, including without limitation, from any errors, omissions, and inaccuracies in original or following translation, or for any consequences arising therefrom. Nothing in this notice shall exclude liability which may not be excluded by law. Approved product information should be reviewed before prescribing any subject medications

4.3 Examples of EMA

The existing investigations of EMA in tinnitus have laid the foundation for this new technology, by studying compliance, risk to well-being, and effects of repeated measures. In a study of 20 participants with tinnitus, 79.4% (889/1120) surveys were completed over a 2-week time period, suggesting the broad feasibility of using smartphone EMA in a tinnitus sample (Wilson et al. 2015). In a pilot study of EMA use in tinnitus patients, Henry and colleagues showed that 24 subjects did not have a significant change in tinnitus distress over a 2-week period (Henry et al. 2012). In 2016, Schlee and colleagues performed a cohort analysis using a smartphone application called Track Your Tinnitus (TYT) to follow 857 subjects between April 2014 and February 2016. Similar to Henry's findings, regular testing through the application was shown to not have an influence on perceived tinnitus loudness or distress. They also reported a sub-analysis of groups with more than 1 month of data reporting ($n = 66$) compared to those with less than 1 month ($n = 134$) which also showed no difference. Little difference between the different lengths of surveys suggests that extended surveys can be used in this population without exacerbating tinnitus bother. This study also utilized a customizable administration of their daily questionnaires, allowing the user to accept random allocation or to customize when to receive their questions (Schlee et al. 2016).

TYT has also been used to assess how emotional stress impacts tinnitus. Probst et al. surveyed 658 subjects about their experience of tinnitus along with the self-assessment manikin which is used to assess emotional arousal and valence. Their findings showed that emotional states affect perception of tinnitus loudness and subsequent distress (Probst et al. 2016). Probst followed this work with another study using TYT, which found time of day affected both perceived loudness and distress in tinnitus patients. Probst and colleagues utilized multilevel modeling which allowed to control for nesting within the model (Probst et al. 2017). Multilevel modeling was also used in a 6-week observational study by Goldberg and colleagues to test a 2-level confirmatory factor analysis model assessing within-individual and between-individual responses. The findings suggest a significant association between perceived stress and tinnitus bother (Goldberg et al. 2017). Goldberg et al. utilized multilevel modeling strategies to show within-person and between-individual differences. The model explored the association between individual question responses from consecutive time points. It assessed that if one answer increases in severity whether other answers are also likely to change along with it for that single participant. Participants were also compared to each other such that if a participant reported a higher severity for an EMA item compared to another participant, were other answers likely to be more severely rated for the former participant, as well? These results can be used to quantify a latent factor which summarizes an individual's vulnerability to tinnitus bother (Fig. 3) (Goldberg et al. 2017).

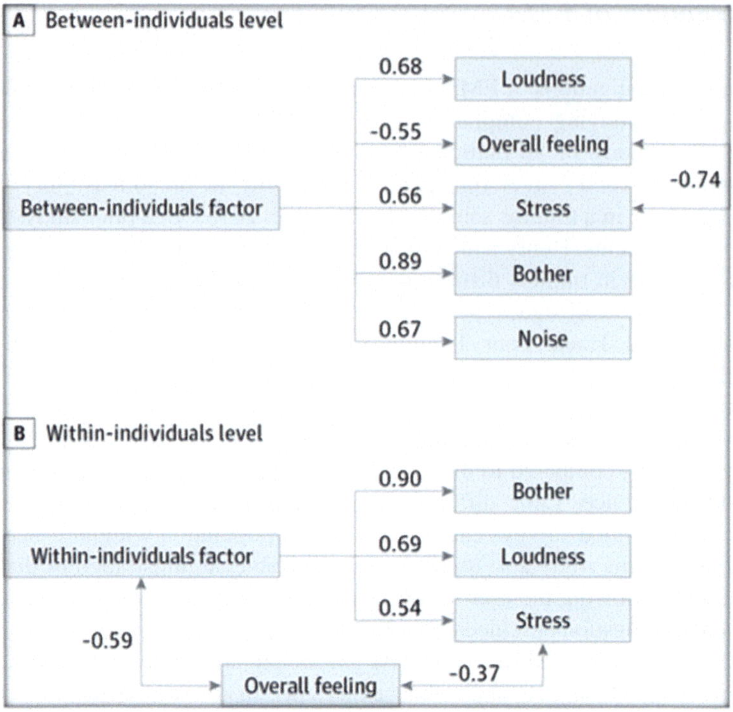

Fig. 3 The ability to assess different aspects of tinnitus bother within and among subjects is a key benefit of the analysis of EMA data. In the final model, noise was not included on the within-individuals factor. Single-headed arrows represent standardized factor loadings. Double-headed arrows represent correlations. All standardized estimates are significant at $P < 0.001$. Used with permission of *JAMA Otolaryngology – Head & Neck Surgery*, from "Evaluation of Ecological Momentary Assessment for Tinnitus Severity" in, by Rachel L. Goldberg et al., vol 143, issue 7, July 2017, retrieved from https://jamanetwork.com/journals/jamaotolaryngology/fullarticle/2618944

4.4 Where We Are

The examples of EMA use in tinnitus provide prerequisite proof-of-concept that tinnitus patients will respond to frequent, ecological questionnaires, and the use of EMA in longitudinal studies is feasible and appropriate. Further, many of the reports show that while tinnitus bother may increase at times, overall, the levels are consistent, suggesting there is no systemic bias introduced by the repeated measures design (Schlee et al. 2019). Furthermore, one study also reported that 80% of participants approved of using EMA to report tinnitus to a medical professional (Goldberg et al. 2017). Statistical techniques capable of properly assessing this kind of data are available. The multilevel modeling and factor analysis techniques serve as models for future studies. The use of customizable data entry methods by Schlee and colleagues is a feasible implementation of EMA for precision medicine methods, such as PCT (Lenze et al. 2020; Schlee et al. 2016).

4.5 *Limitations of Momentary Assessment*

EMA is not without limitations. Much of the literature on real-time data capturing such as EMA has focused on correcting the biases inherent in retrospective collection. The field does not yet know what biases lurk in this new format, and there is the possibility that we are swapping the known bias of retrospective data collection with the unknown biases of real-time data collection.

Furthermore, the lack of supervision by a member of the research team means there is no way to confirm the identity of who is filling out the survey nor is there a way to assure correct usage of the survey or device. Lastly, to expect all patients to have access to a smartphone and the Internet will bias these studies toward technology users and against those with low socioeconomic status.

5 Future Directions

Clinical research should adapt to new technologies. New tools will be developed to best utilize the benefits of these technologies. We are in a phase of constant transformation and growth as we begin to incorporate the new field of ecological data. Initial strategies are using ssmartphone applications to disseminate already-validated clinical questionnaires (Wilson et al. 2015; Goldberg et al. 2017; Henry et al. 2012). Future studies may incorporate tools more attuned to the inherent benefits of EMA. With the vast amount of data available, machine learning algorithms could be developed and trained to identify participants most likely to respond to different therapies (Chekroud et al. 2016).

Adoption of EMA in tinnitus requires investment of time and resources on the side of the researchers as well as the patients. To be effective tools, researchers cannot expect unreasonable amounts of detail from patients and must have realistic expectations of how data can be interpreted and used. It has been noted that even with modern data collection technologies, there can still be significant burden on both participants and the research team in EMA-based studies (Burke et al. 2017). Ideally, there will exist a future where most EMA can be automated via biosensors, and a few questionnaires will automatically capture data, which will be stored in a centralized data repository, and analyzed in real time.

A cursory search using the terms "Ecological Momentary Assessment" and "Tinnitus" using the NIH Research Portfolio Online Reporting Tool (RePORTER, version 7.41.0) (Research Portfolio Online Reporting Tools (RePORT) 2020) yields a single study funded from the beginning of 2019 through the time of this chapter's writing (May 2020). When the word "Tinnitus" is removed from that search criteria, the total number of grants is 374. Funding barriers in hearing research in general (and tinnitus specifically) have always existed due to conflicting evidence and lack of a clearly marketable product (McFerran et al. 2019). The good news is the trend for funding is upward, and with the accumulation of promising data to support the validity and effective use of EMA in tinnitus, we anticipate its broad acceptance.

6 Conclusion

EMA has made great strides as a relevant clinical research methodology. It has also gained acceptance in many fields, most notably in psychiatry and psychology. The initial proof-of-concept for EMA's use in tinnitus has been completed, proving this is a viable research methodology for the tinnitus patient population. The existence of tools like the TYT smartphone application make data collection easier and methodology like ML-DSEM allows that data to be interpreted clearly. The tools are ready to be used; we just need to use them.

Acknowledgments and Disclaimers Special thanks to Lauren Yaeger for helping identify key literature for citation in this chapter. Research reported in this publication was supported by the National Center for Advancing Translational Sciences of the National Institutes of Health under Award Number TL1TR002344. The content is solely the responsibility of the authors and does not necessarily represent the official views of the National Institutes of Health.

References

Bailey AP, Parker AG, Colautti LA, Hart LM, Liu P, Hetrick SE (2014) Mapping the evidence for the prevention and treatment of eating disorders in young people. J Eat Disord 2:5–5

Borsboom D (2008) Psychometric perspectives on diagnostic systems. J Clin Psychol 64 (9):1089–1108

Borsboom D, Cramer AO (2013) Network analysis: an integrative approach to the structure of psychopathology. Annu Rev Clin Psychol 9:91–121

Budd RJ, Pugh R (1996) Tinnitus coping style and its relationship to tinnitus severity and emotional distress. J Psychosom Res 41(4):327–335

Burke LE, Shiffman S, Music E et al (2017) Ecological momentary assessment in behavioral Research: addressing technological and human participant challenges. J Med Internet Res 19(3): e77–e77

Burton H, Wineland A, Bhattacharya M, Nicklaus J, Garcia KS, Piccirillo JF (2012) Altered networks in bothersome tinnitus: a functional connectivity study. BMC Neurosci 13:3

Buysse DJ, Thompson W, Scott J et al (2007) Daytime symptoms in primary insomnia: a prospective analysis using ecological momentary assessment. Sleep Med 8(3):198–208

Chekroud AM, Zotti RJ, Shehzad Z et al (2016) Cross-trial prediction of treatment outcome in depression: a machine learning approach. Lancet Psychiatry 3(3):243–250

Cima RF, Crombez G, Vlaeyen JW (2011) Catastrophizing and fear of tinnitus predict quality of life in patients with chronic tinnitus. Ear Hear 32(5):634–641

Cramer AO, Waldorp LJ, van der Maas HL, Borsboom D (2010) Comorbidity: a network perspective. Behav Brain Sci 33(2–3):137–150. discussion 150-193

Cramer AO, Kendler KS, Borsboom D (2012) A constructionist account of emotional disorders. Behav Brain Sci 35(3):146–147

Dauman N, Erlandsson SI, Albarracin D, Dauman R (2017) Exploring Tinnitus-induced disablement by persistent frustration in aging individuals: a grounded theory study. Front Aging Neurosci 9:272

Fisher AJ (2015) Toward a dynamic model of psychological assessment: implications for personalized care. J Consult Clin Psychol 83(4):825–836

Fisher AJ, Boswell JF (2016) Enhancing the personalization of psychotherapy with dynamic assessment and modeling. Assessment 23(4):496–506

Goldberg RL, Piccirillo ML, Nicklaus J et al (2017) Evaluation of ecological momentary assessment for tinnitus severity. JAMA Otolaryngol Head Neck Surg 143(7):700–706

Gorin A, Stone A (2001) Recall biases and cognitive errors in retrospective self reports: a call for momentary assessments

Granholm E, Loh C, Swendsen J (2008) Feasibility and validity of computerized ecological momentary assessment in schizophrenia. Schizophr Bull 34(3):507–514

Hall D, Mohamad N, Firkins L, Fenton M, Stockdale D (2013) Identifying and prioritizing unmet research questions for people with tinnitus: the James Lind Alliance tinnitus priority setting partnership. Clin Invest 3:21–28

Hallam RS, Jakes SC, Hinchcliffe R (1988) Cognitive variables in tinnitus annoyance. Br J Clin Psychol 27(3):213–222

Han BI, Lee HW, Kim TY, Lim JS, Shin KS (2009) Tinnitus: characteristics, causes, mechanisms, and treatments. J Clin Neurol 5(1):11–19

Haynos AF, Crosby RD, Engel SG et al (2015) Initial test of an emotional avoidance model of restriction in anorexia nervosa using ecological momentary assessment. J Psychiatr Res 68:134–139

Helson H (1964) Adaptation-level theory: an experimental and systematic approach to behavior. Harper and Row, New York

Henry JA, Galvez G, Turbin MB, Thielman EJ, McMillan GP, Istvan JA (2012) Pilot study to evaluate ecological momentary assessment of tinnitus. Ear Hear 33(2):179–290

Hufford MR, Shields AL, Shiffman S, Paty J, Balabanis M (2002) Reactivity to ecological momentary assessment: an example using undergraduate problem drinkers. Psychol Addict Behav 16(3):205–211

Husain FT, Schmidt SA (2014) Using resting state functional connectivity to unravel networks of tinnitus. Hear Res 307:153–162

Jastreboff PJ, Hazell JW (1993) A neurophysiological approach to tinnitus: clinical implications. Br J Audiol 27(1):7–17

Keidser G (2016) Introduction to special issue: towards ecologically valid protocols for the assessment of hearing and hearing devices. J Am Acad Audiol 27(7):502–503

Kikuchi H, Yoshiuchi K, Miyasaka N et al (2006) Reliability of recalled self-report on headache intensity: investigation using ecological momentary assessment technique. Cephalalgia 26 (11):1335–1343

Kim JY, Kim YH, Lee S et al (2012) Alteration of functional connectivity in tinnitus brain revealed by resting-state fMRI? A pilot study. Int J Audiol 51(5):413–417

Kubiak T, Smyth JM (2019) Connecting domains—ecological momentary assessment in a mobile sensing framework. In: Baumeister H, Montag C (eds) Digital phenotyping and mobile sensing: new developments in psychoinformatics. Springer, Cham, pp 201–207

Landgrebe M, Langguth B, Rosengarth K et al (2009) Structural brain changes in tinnitus: grey matter decrease in auditory and non-auditory brain areas. NeuroImage 46(1):213–218

Lavender JM, Wonderlich SA, Crosby RD et al (2013) A naturalistic examination of body checking and dietary restriction in women with anorexia nervosa. Behav Res Ther 51(8):507–511

Lenze EJ, Rodebaugh TL, Nicol GE (2020) A framework for advancing precision medicine in clinical trials for mental disorders. JAMA Psychiatry 77(7):663–664
Levenstein S, Prantera C, Varvo V et al (1993) Development of the perceived stress questionnaire: a new tool for psychosomatic research. J Psychosom Res 37(1):19–32
Lutkepohl H (2007) New introduction to multiple time series analysis. Springer, Cham
Maudoux A, Lefebvre P, Cabay JE et al (2012) Auditory resting-state network connectivity in tinnitus: a functional MRI study. PLoS One 7(5):e36222
McFerran DJ, Stockdale D, Holme R, Large CH, Baguley DM (2019) Why is there no cure for tinnitus? *Front Neurosci* 13:802
Meikle M, Taylor-Walsh E (1984) Characteristics of tinnitus and related observations in over 1800 tinnitus clinic patients. J Laryngol Otol Suppl 9:17–21
Meikle MB, Henry JA, Griest SE et al (2012) The tinnitus functional index: development of a new clinical measure for chronic, intrusive tinnitus. Ear Hear 33(2):153–176
Molenaar PCM (2013) On the necessity to use person-specific data analysis approaches in psychology. Eur J Dev Psychol 10(1):29–39
Moskowitz DS, Young SN (2006) Ecological momentary assessment: what it is and why it is a method of the future in clinical psychopharmacology. J Psychiatry Neurosci 31(1):13–20
Muthén LK, Muthén BO (2017) MPlus users guide, 8th edn. Los Angeles, CA
Newman CW, Jacobson GP, Spitzer JB (1996) Development of the tinnitus handicap inventory. Arch Otolaryngol Head Neck Surg 122(2):143–148
Nondahl DM, Cruickshanks KJ, Huang GH et al (2011) Tinnitus and its risk factors in the beaver dam offspring study. Int J Audiol 50(5):313–320
Overmier JB (2002) Sensitization, conditioning, and learning: can they help us understand somatization and disability? Scand J Psychol 43(2):105–112
Pattyn T, Van Den Eede F, Vanneste S et al (2016) Tinnitus and anxiety disorders: a review. Hear Res 333:255–265
Probst T, Pryss R, Langguth B, Schlee W (2016) Emotional states as mediators between tinnitus loudness and tinnitus distress in daily life: results from the "TrackYourTinnitus" application. Sci Rep 6(1):20382
Probst T, Pryss RC, Langguth B et al (2017) Does tinnitus depend on time-of-day? An ecological momentary assessment study with the "TrackYourTinnitus" application. *Front Aging Neurosci* 9:253
Rabinowitz NC, Willmore BD, Schnupp JW, King AJ (2011) Contrast gain control in auditory cortex. Neuron 70(6):1178–1191
Reiss S, Peterson RA, Gursky DM, McNally RJ (1986) Anxiety sensitivity, anxiety frequency and the prediction of fearfulness. Behav Res Ther 24(1):1–8
Research Portfolio Online Reporting Tools (RePORT) (2020) 7.41.0: https://www.nih.gov/. Accessed Apr 2020
Roberts LE, Husain FT, Eggermont JJ (2013) Role of attention in the generation and modulation of tinnitus. Neurosci Biobehav Rev 37(8):1754–1773
Ross J, Schreiber I, Atienza A, Nebeling L, Shiffman S, Stone A (2007) The science of real-time data capture : self-reports in health research. Oxford University Press, Cary
Schlee W, Hartmann T, Langguth B, Weisz N (2009) Abnormal resting-state cortical coupling in chronic tinnitus. BMC Neurosci 10:11
Schlee W, Pryss RC, Probst T et al (2016) Measuring the moment-to-moment variability of tinnitus: the trackyourtinnitus smart phone app. Front Aging Neurosci 8:294
Schlee W, Kraft R, Schobel J et al (2019) Momentary assessment of tinnitus—how smart mobile applications advance our understanding of tinnitus. Springer, Cham, pp 209–220
Schmidt CJ, Kerns RD, Griest S, Theodoroff SM, Pietrzak RH, Henry JA (2014) Toward development of a tinnitus magnitude index. Ear Hear 35(4):476–484
Schmidt SA, Carpenter-Thompson J, Husain FT (2017) Connectivity of precuneus to the default mode and dorsal attention networks: a possible invariant marker of long-term tinnitus. Neuroimage Clin 16:196–204

Schwartz JE, Stone AA (1998) Strategies for analyzing ecological momentary assessment data. Health Psychol 17(1):6–16

Scott-Wittenborn N, Karadaghy OA, Piccirillo JF, Peelle JE (2017) A methodological assessment of studies that use voxel-based morphometry to study neural changes in tinnitus patients. Hear Res 355:23–32

Searchfield GD (2014) Tinnitus what and where: an ecological framework. Front Neurol 5:271–271

Seydell-Greenwald A, Leaver AM, Turesky TK, Morgan S, Kim HJ, Rauschecker JP (2012) Functional MRI evidence for a role of ventral prefrontal cortex in tinnitus. Brain Res 1485:22–39

Shiffman S (2014) Conceptualizing analyses of ecological momentary assessment data. *Nicotine Tob Res* 16(Suppl 2):S76–S87

Stobik C, Weber RK, Munte TF, Walter M, Frommer J (2005) Evidence of psychosomatic influences in compensated and decompensated tinnitus. Int J Audiol 44(6):370–378

Stone AA, Broderick JE, Porter LS, Krupp L et al (1994) Fatigue and mood in chronic fatigue syndrome patients: results of a momentary assessment protocol examining fatigue and mood levels and diurnal patterns. Ann Behav Med 16(3):228–234

Stone AA, Broderick JE, Porter LS, Kaell AT (1997) The experience of rheumatoid arthritis pain and fatigue: examining momentary reports and correlates over one week. Arthritis Care Res 10 (3):185–193

Stone AA, Schwartz JE, Neale JM et al (1998) A comparison of coping assessed by ecological momentary assessment and retrospective recall. J Pers Soc Psychol 74(6):1670–1680

Stone AA, Turkkan JS, Bachrach CA, Jobe JB, Kurtzman HS, Cain VS (2000) The science of self-report: Implications for research and practice. In: The science of self-report: implications for research and practice. Lawrence Erlbaum Associates Publishers, Mahwah, p xi, 380

Sullivan M, Katon W, Russo J, Dobie R, Sakai C (1994) Coping and marital support as correlates of tinnitus disability. Gen Hosp Psychiatry 16(4):259–266

Timmer BH, Hickson L, Launer S (2015) Adults with mild hearing impairment: are we meeting the challenge? Int J Audiol 54(11):786–795

Torous J, Keshavan M (2020) Towards precision clinical trials and personalized prevention in CHR with smartphone digital phenotyping and personal sensing tools. Schizophr Res 20(S0920-9964)

Tyler RS, Baker LJ (1983) Difficulties experienced by tinnitus sufferers. J Speech Hear Disord 48 (2):150–154

Uher R, Rutter M (2012) Classification of feeding and eating disorders: review of evidence and proposals for ICD-11. World Psychiatry 11(2):80–92

Vanneste S, Heyning PV, Ridder DD (2011a) Contralateral parahippocampal gamma-band activity determines noise-like tinnitus laterality: a region of interest analysis. Neuroscience 199:481–490

Vanneste S, van de Heyning P, De Ridder D (2011b) The neural network of phantom sound changes over time: a comparison between recent-onset and chronic tinnitus patients. Eur J Neurosci 34(5):718–731

Welch D, Dawes PJ (2008) Personality and perception of tinnitus. Ear Hear 29(5):684–692

Wildes JE, Ringham RM, Marcus MD (2010) Emotion avoidance in patients with anorexia nervosa: initial test of a functional model. Int J Eat Disord 43(5):398–404

Wilson MB, Kallogjeri D, Joplin CN et al (2015) Ecological momentary assessment of tinnitus using smartphone technology: a pilot study. Otolaryngol Head Neck Surg 152(5):897–903

Wonderlich SA, Rosenfeldt S, Crosby RD et al (2007) The effects of childhood trauma on daily mood lability and comorbid psychopathology in bulimia nervosa. J Trauma Stress 20(1):77–87

Tinnitus Questionnaires for Research and Clinical Use

Sarah M. Theodoroff

Contents

Abstract The lack of an objective measure of tinnitus has led to self-report questionnaires as the best option to evaluate tinnitus symptoms and quantify the degree to which quality of life is negatively impacted. There are many tinnitus questionnaires to choose from and it can be difficult to decide which one is best. From an evidence-based perspective, knowing how the questionnaire is designed, including its intended purpose, can help determine if it is appropriate or not to use. For example, a questionnaire designed to screen for the presence or absence of tinnitus should not be used as an outcome measure to answer questions about treatment effectiveness. Often, using more than one questionnaire is preferable to relying on just one. This chapter will review important factors to consider when selecting a questionnaire for research purposes and/or routine clinical care.

S. M. Theodoroff (✉)
Veterans Affairs (VA) RR&D National Center for Rehabilitative Auditory Research (NCRAR), VA Portland Health Care System, Portland, OR, USA

Department of Otolaryngology, Head and Neck Surgery, Oregon Health and Science University, Portland, OR, USA
e-mail: sarah.theodoroff@va.gov

Curr Topics Behav Neurosci (2021) 51: 403–418
https://doi.org/10.1007/7854_2020_175
Published Online: 11 October 2020

Keywords Evidence-based practice · Needs assessment · Surveys and questionnaires · Tinnitus

1 Introduction

1.1 Characteristics of Questionnaires

In his chapter "Psychometrics in Clinical Settings," Dr. Hamilton Fairfax, a counseling psychologist, addresses two main principles underlying psychometric testing: "One principal use of psychometrics clinically is to establish if a particular diagnosis, problem or attribute is present in an individual at a given time. The second main aim [principle] is to establish to what degree the person has the quality being tested." (Fairfax 2017). The first principle, applied to the focus of this chapter, translates to diagnosing the presence or absence of tinnitus, including subtypes, and the second principle addresses establishing the degree of tinnitus-related distress. These principles highlight the rationale for why questionnaires are used in research and clinical practice. Specifically, questionnaires are measurement tools that can ascertain if tinnitus is present and if so, quantify its effects. Because of the reliance on questionnaires, it is essential to use valid and reliable ones so that researchers and clinicians have confidence in applying the results to answer scientific questions and help patients make informed decisions as to the best course of action.

1.2 Item Wording and Measurement Scales

How a questionnaire's items and instructions are worded influences a patient's response, as does the number and ordering of response options. Many tinnitus questionnaires use verbal rating scales. Consistent with a patient-centered approach, which emphasizes the need to focus on the patient's perspective, individuals select a rating associated with their level of distress for each question. There is often a trade-off between the number of response options and the precision of the measurement, such that fewer response options can result in a less precise measure. Also, if the instructions and responses are too complex, reading level becomes a bigger factor and there is an increased likelihood of the patient or research participant experiencing fatigue. These factors highlight why it is important for clinicians and researchers to verify their patients and/or participants understand the instructions and how they are to respond to each item on a questionnaire.

Table 1 shows examples of instructions and response options for two questionnaires that use verbal rating scales. The first example is from the World Health Organization Disability Assessment Schedule (WHODAS), a quality of life

Table 1 Example of verbal scale instructions and response options: The World Health Organization disability assessment schedule and hearing handicap for adults

Instructions	Response options
WHODAS 2.0	
This questionnaire asks about difficulties due to health conditions. Health conditions include diseases or illnesses, other health problems that may be short or long lasting, injuries, mental or emotional problems, and problems with alcohol or drugs Think back over the past 30 days and answer these questions, thinking about how much difficulty you had doing the following activities. For each question, please circle only one response	None Mild Moderate Severe Extreme, or cannot do
Hearing handicap inventory for adults	
The purpose of the scale is to identify the problems your hearing loss may be causing you. Check Yes, Sometimes, or No for each question. Do not skip a question if you avoid a situation because of your hearing problem	Yes Sometimes No

assessment tool that assesses six domains of functioning and has five response options (Üstün et al. 2010). The second example is a question from the Hearing Handicap Inventory for Adults, which, as its name suggests, assesses self-perceived hearing handicap and has three response options (Newman et al. 1991). These two non-tinnitus questionnaires were selected as examples so as not to show preference of one tinnitus questionnaire over another. They are provided to demonstrate how the wording of a questionnaire's instructions and response options influence decision making.

The three most commonly used rating scales are the verbal rating scale, numeric rating scale, and visual analog scale. Visual analogue scales and numeric rating scales are most commonly used to evaluate pain (Johnson 2005) and have been adapted to evaluate tinnitus symptoms (see Fig. 1a, b). These scales are used in a variety of ways, both as a means to gauge and document acute effects, such as immediately following a therapeutic session, and to evaluate if benefit sustains, such as a month after therapy has been completed. Meikle et al. (2008) reviewed the strengths and weaknesses of using rating scales for tinnitus. Their main strengths are that they are easy to administer and interpret; their main weakness is that they do not have established criteria as to how much of a change is needed pre- vs. post-intervention to be considered meaningful improvement (Meikle et al. 2008).

Standardized versions of visual analogue and numeric rating scales for tinnitus do not exist. These measures are a perfect example of how the instructions given and choice of anchors can influence the patient's response. Consider how responses might differ if asked to make a judgment based on "the last 3 days" compared to "at this moment." Another caveat to using these types of scales to rate aspects of tinnitus is the uncertainty as to what exactly these instruments are "measuring" in terms of tinnitus perceptions, such as loudness and pitch, versus reactions, meaning the negative impact or effects of tinnitus, i.e., tinnitus-related distress (Henry 2016; Theodoroff et al. 2017).

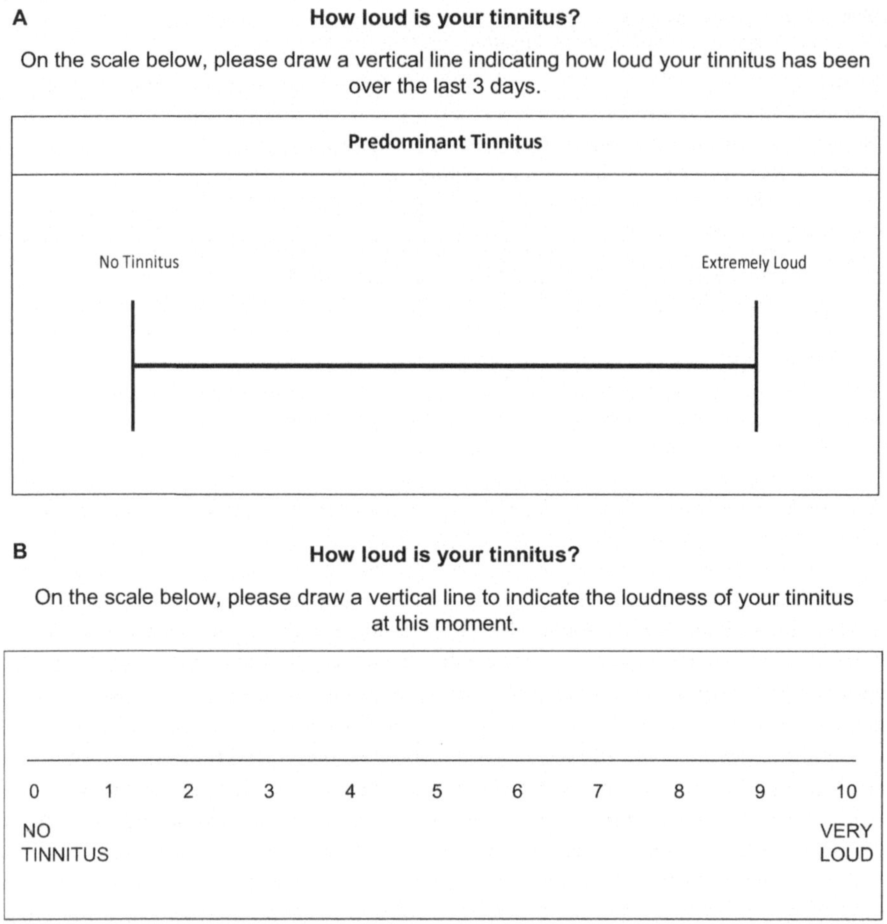

Fig. 1 (**a** and **b**) An example of a Visual Analogue Scale is provided in (**a**) and a Numeric Rating Scale in (**b**)

Using self-reported measures is good practice, but it is important to remember that bias is inherent in self-assessment, which influences judgments and responses. Due to the subjective nature of tinnitus, a valid and reliable questionnaire offers a means to determine from the patient's point of view, if symptoms have improved or worsened following an intervention. When evaluating treatment-related change, because of the diversity in patients' goals and expectations, it is essential to know not only how much of a change matters, but also what kind of change is considered important to the patient. It is imperative to have this knowledge prior to starting any type of intervention or management approach to minimize the chances of a disconnect between patients' and clinicians' treatment expectations, which unfortunately happens all too often (Husain et al. 2018).

2 Psychometric Properties

Psychometric properties such as validity, reliability, interpretability, acceptability, and responsiveness describe attributes of a measurement tool. Validity indicates how well an instrument, such as a questionnaire, truly measures what it purports to measure. There are multiple types of validity tests that all focus on how well an instrument can measure the underlying construct of interest (Lohr 2002; Bolarinwa 2015). Reliability addresses to what degree the instrument's result, such as the total score, is free from measurement error and can be replicated. Two types of reliability testing are (1) internal consistency and (2) test–retest reliability. Internal consistency addresses the degree to which the questionnaire's items all measure the same underlying construct. This is typically determined by examining correlations between the items. The most common test statistic used to evaluate this attribute is Cronbach's alpha, but other methods can be used to assess this type of reliability (for a review, see McNeish 2018). Test–retest reliability addresses how well a questionnaire's results are reproducible from one time point to another (Bolarinwa 2015). This type of reliability examines the stability of the scores from the same individuals across two time points; it evaluates how well a questionnaire yields the same outcome, one that is not expected to vary from time point 1 to time point 2.

Any field that employs questionnaires needs to be concerned with the psychometric characteristics of the instruments used for outcome assessments. Lohr (2002) recommends evaluating health status and quality of life instruments using a set of criteria recommended by the Medical Outcomes Trust Scientific Advisory Committee (SAC). The SAC developed these criteria to serve as a guideline to judge the quality of health status survey instruments, which includes tinnitus questionnaires. Table 2 lists the eight attributes recommended by the SAC, which are (1) conceptual and measurement model; (2) reliability; (3) validity; (4) responsiveness; (5) interpretability; (6) respondent and administrative burden; (7) alternative forms; and (8) cultural and language adaptations (translations). Lohr (2002) goes on to stress:

> An instrument that works well for one purpose or in one setting or population may not do so when applied for another purpose or in another setting or population. The relative importance of the eight attributes may differ depending on the intended uses and applications specified for the instrument. Instruments may, for instance, document the health status or attitudes of individuals at a point in time, distinguish between two or more groups, assess change over time among groups or individuals, predict future status, or some combinations of these. Hence, the weight placed on one or another set of criteria may differ according to the purposes claimed for the instrument. (Lohr 2002, p. 197).

Overall, these criteria can guide researchers and clinicians to find the appropriate questionnaire to accomplish their respective aims and goals, such as testing research hypotheses or conducting a needs assessment for a tinnitus patient. The following sections address how to select among questionnaires if the intended use is for evaluation and assessment or quantifying improvement in symptoms following an intervention (i.e., outcome).

Table 2 Attributes and criteria for reviewing questionnaires

Attribute	Review criteria
1. Conceptual and measurement model The rationale for and description of the concept and the populations that a measure is intended to assess and the relationship between these concepts	• Concept to be measured • Conceptual and empirical bases for item content and combinations • Target population involvement in content derivation • Information on dimensionality and distinctiveness of scales • Evidence of scale variability • Intended level of measurement • Rationale for deriving scale scores
2. Reliability The degree to which an instrument is free from random error *Internal consistency* The precision of a scale based on the homogeneity (intercorrelations) of the scale's items at one point in time *Reproducibility* Stability of an instrument over time (test–retest) and inter-rater agreement at one point in time	*Internal consistency* • Methods to collect reliability data • Reliability estimates and standard errors for all score elements (classical test) or standard error of the mean over the range of scale and marginal reliability of each scale (modern IRT) • Data to calculate reliability coefficients or actual calculations of reliability coefficients • Above data for each major population of interest, if necessary *Reproducibility* • Methods employed to collect reproducibility data • Well-argued rational to support the design of the study and the interval between first and subsequent administration to support the assumption that the population is stable • Information on test–retest reliability and inter-rater reliability based on intraclass correlation coefficients • Information on the comparability of the item parameter estimates and on measurement precision over repeated administrations
3. Validity The degree to which the instrument measures what it purports to measure *Content-related:* Evidence that the domain of an instrument is appropriate relative to its intended use *Construct-related:* Evidence that supports a proposed interpretation of scores based on theoretical implications associated with the constructs being measured *Criterion-related:* Evidence that shows the extent to which scores of the instrument are related to a criterion measure	• Rationale supporting the particular mix of evidence presented for the intended uses • Clear description of the methods employed to collect validity data • Composition of the sample used to examine validity (in detail) • Above data for each major population of interest • Hypothesis tested and data relating to the tests • Clear rationale and support for the choice of criteria measures
4. Responsiveness An instrument's ability to detect change over time	• Evidence on the changes in scores of the instrument • Longitudinal data that compare a group that is expected to change with a group that is expected to remain stable

(continued)

Table 2 (continued)

Attribute	Review criteria
	• Population(s) on which responsiveness has been tested, including the time intervals of assessment, the interventions or measures involved in evaluating change, and the populations assumed to be stable
5. Interpretability The degree to which one can assign easily understood meaning to an instrument's quantitative scores	• Rationale for selection of external criteria of populations for purposes of comparison and interpretability of data • Information regarding the ways in which data from the instrument should be reported and displayed • Meaningful "benchmarks" to facilitate interpretation of the scores
6. Burden The time, effort, and other demands placed on those to whom the instrument is administered (respondent burden) or on those who administer the instrument (administrative burden)	*Respondent burden* • Information on: (a) average and range of the time needed to complete the instrument, (b) reading and comprehension level, and (c) any special requirements or requests made of respondent • Evidence that the instrument places no undue physical or emotional strain on the respondent • When or under what circumstances the instrument is not suitable for respondents *Administrative burden* • Information about any resources required for administration of the instrument • Average time and range of time required of a trained interviewer to administer the instrument in face-to-face interviews, by telephone, or with computer-assisted formats • Amount of training and level of education or professional expertise and experience needed by administrative staff
7. Alternative modes of administration These include self-report, interviewer-administered, trained observer rating, computer-assisted interviewer-administered, performance-based measures	• Evidence on reliability, validity, responsiveness, interpretability, and burden for each mode of administration • Information on the comparability of alternative modes
8. Cultural and language adaptations or translations Involves two primary steps: 1. Assessment of conceptual and linguistic equivalence 2. Evaluation of measurement properties	• Methods to achieve conceptual equivalence • Methods to achieve linguistic equivalence • Any significant differences between the original and translated versions • How inconsistencies were reconciled

Lohr (2002) specifies that developers are expected to provide definitions, descriptions, explanations, and/or empirical information. Adapted with permission from Lohr, KN. Assessing health status and quality-of-life instruments: Attributes and review criteria, *Qual Life Res*, 2002, 11:193–205. Table 1, pp. 196–197

3 Assessment

3.1 Tinnitus Case History

When considering which instrument to use for an intake history, clinical practice guidelines recommend asking questions to determine what type of tinnitus the patient has. Again, there is no standardized questionnaire for this purpose, but there are published recommendations to assist the clinician and researcher in this endeavor (e.g., Tinnitus Sample Case History Questionnaire, see Langguth et al. 2007; Somatosensory Tinnitus, see Michiels et al. 2018). Identifying and characterizing tinnitus subtypes and symptoms provide the clinician or researcher essential information to develop an appropriate treatment plan, which many times warrants taking an interprofessional approach (Newman and Sandridge 2016).

For example, at this stage in the evaluation process, the clinician can learn what, if any, triggers exist that exacerbate or improve the patient's tinnitus. If patients report that jaw movements change their tinnitus, referral to a dentist would be appropriate to determine the possibility of comorbid temporomandibular joint dysfunction. Other examples that might lead to taking an interprofessional approach include reports of headaches exacerbating a patient's tinnitus, which might lead to a neurology consult and reports of stress or anxiety associated with tinnitus-distress, which could lead to engaging a mental health professional in the care of the patient. When taking an interprofessional approach, it is helpful to know which specific screening tools your health care collaborators prefer. Using these tools, such as to screen for depression or anxiety, will greatly assist in this process of knowing when it is appropriate to involve other health professionals in the care of tinnitus patients.

Many patients with tinnitus have comorbid hearing loss. In these instances, communication difficulties are often misattributed to be due solely because of the tinnitus rather than the hearing loss, e.g., "I can't hear you because of my tinnitus." Tinnitus-related problems are often intertwined with other auditory-related issues, making it difficult for the patient to disentangle tinnitus-related problems from hearing-related problems. Because of this, Henry et al. (2015) developed the Tinnitus and Hearing Survey. This brief and easy to administer questionnaire is validated to differentiate tinnitus-related problems from hearing-related problems. The Tinnitus and Hearing Survey is an excellent counseling tool that can be used as part of the needs assessment to evaluate and discuss with the patient whether tinnitus and/or hearing problems are impacting aspects of daily life.

3.2 Performing a Needs Assessment

The concept of performing a needs assessment to understand the patient's experience and satisfaction with treatment is not new and is routinely used in mental health and other medical fields (Davidson et al. 2004). A needs assessment provides valuable insight into the patient's perspective regarding how their tinnitus impacts their daily functioning as well as their motivation and readiness to accomplish the goals of treatment. Evaluating self-efficacy is often included as part of the needs assessment. Smith and Fagelson (2011) developed the Self-Efficacy for Tinnitus Management Questionnaire (SETMQ). The SETMQ is a valid and reliable questionnaire that can be used to determine how confident the patient or research participant is in managing their tinnitus in terms of: (1) routine tinnitus management; (2) emotional response; (3) internal thoughts and interactions with others; (4) tinnitus concepts; and (5) using devices.

Other questionnaires that are available for needs assessment include the Sound Therapy Option Profile (STOP; Newman and Sandridge 2006), the Tinnitus Activities Questionnaire (TAQ; Tyler et al. 2007), and the Client Oriented Scale of Improvement in Tinnitus (COSIT; Searchfield 2019). The COSIT fosters a collaborative relationship between the clinician and patient in that it is an open-ended questionnaire used pre-therapy to ascertain patient-specific priorities related to goals and treatment expectations. It is also used post-therapy to determine to what degree the patient's goals were met by having the patient answer these two questions: (1) "With the tinnitus therapy, my tinnitus is. . ." worse; no different; slightly better; better; much better; and (2) "I am annoyed by my tinnitus. . ." almost always; most of the time; half the time; occasionally; hardly ever.

3.3 *Screening Tools and Outcome Measures*

An outcome measure is distinguished from a screening tool in that it was designed to identify treatment-related change associated with an intervention, which is the psychometric property known as responsiveness. Hall et al. make the distinction between an outcome domain and outcome measure by stating: "First, the outcome domain refers to any aspect of that condition which matters most to patients and clinicians, such as tinnitus intrusiveness, sense of control, or impact on work. Second, the outcome instrument refers to how that domain is to be measured." (Hall et al. 2018, p. 2). When selecting an outcome instrument, it is important to keep these concepts in mind.

In the last two to three decades, numerous tinnitus questionnaires have been developed for screening and/or outcome purposes in various countries and languages. Widely used ones include the Tinnitus Questionnaire (TQ; Hallam et al. 1984), Tinnitus Handicap Questionnaire (THQ; Kuk et al. 1990), Tinnitus Reaction Questionnaire (TRQ; Wilson et al. 1991), Tinnitus Handicap Inventory (THI; Newman et al. 1996), Tinnitus Functional Index (TFI; Meikle et al. 2012), and the Tinnitus Primary Function Questionnaire (TPFQ; Tyler et al. 2014). Each of these questionnaires has strengths and weaknesses, like all self-reported instruments. To determine which questionnaire to use for screening purposes or outcome assessment, it is recommended to become familiar with how the questionnaire was developed, paying close attention to if its results will be meaningful for your patient population and intended purpose (i.e., screening tool vs. outcome measure).

To aid in the decision-making process, Greenhalgh et al. (1998) created a checklist to facilitate comparing different questionnaires. The checklist provides a systematic method to evaluate and compare instruments in the following seven areas: (1) purpose; (2) background; (3) description; (4) user-centeredness, which is the extent that an instrument captures the desired outcomes from the viewpoint of multiple stakeholders; (5) psychometrics; (6) feasibility; and (7) utility. Checklists for the common tinnitus questionnaires mentioned above are found in Tables 3, 4, 5, 6, 7, and 8. References are included so that additional information, such as the population used to develop the questionnaire, can be found.

To determine how well results on a questionnaire will generalize to your patient population, it is important to know the population that was sampled from to establish its validity and reliability. This information is typically found in the Methods section of the publication about the questionnaire's development. How were the individuals recruited? Were they selected from a group of treatment-seeking patients with tinnitus? Were they native speakers of the language used in the questionnaire? Was the sample drawn from native speakers of the language used in the questionnaire? When a questionnaire is translated into different languages, it is important to verify that appropriate cultural and language adaptations are made. This point is

Table 3 Greenhalgh et al. (1998) checklist – Tinnitus Questionnaire (TQ)

Purpose	Meet the need for a tinnitus-specific psychological measure; developed to evaluate tinnitus-related reactions, specifically annoyance and emotional distress
Description	52-item questionnaire; only 41 of the 52 are included in the scoring Total score depends on how many items included; could be scored 0–84 or 0–104 Response options: True, Partly True, or Not True; point value 0, 1, 2 is determined by each question and certain items are reversed scored; higher value indicates presence of complaint often indicated by a response of "true" Five domains: (1) Emotional distress; (2) Auditory perceptual difficulties; (3) Intrusiveness; (4) Sleep disturbance; (5) Somatic complaints
User-centeredness	Developed over the course of a few years at the Royal National Throat Nose and Ear Hospital in London to evaluate psychological interventions for tinnitus (e.g., Cognitive Behavioral Therapy); many of the questions designed to ask about patients' beliefs and emotional distress
Psychometrics	The psychometric properties of the TQ have been extensively tested and show the TQ is a valid and reliable instrument
Feasibility	Self-administered Takes approximately 15–20 min to complete
Clinical utility	The TQ is used in a wide variety of research and clinical settings Some questions are reversed scored
Other comments	Recommended scoring is based on clinical populations Shortened TQ (33-items) available for clinical purposes, not recommended for research purposes
Sources	Hallam et al. (1984, 1988); Henry and Wilson (1998); Jacquemin et al. (2019)

Table 4 Greenhalgh et al. (1998) checklist – Tinnitus Handicap Questionnaire (THQ)

Purpose	Meet the need for an instrument to reliably measure a patient's perceived handicap attributed to tinnitus
Description	27-item questionnaire Total score calculated as average score based on all items; higher score reflects increased tinnitus handicap Response options: 0 (disagree) to 100 (strongly agree) Three domains/factors: (1a) Physical health; (1b) Emotional status; (1c) Social consequences of tinnitus; (2) Hearing difficulties associated with tinnitus; (3) Patient's views related to tinnitus
User-centeredness	Developed in two phases: Phase 1 involved item development Phase 2 determines psychometric properties of questionnaire items developed during phase 1 in a clinical population
Psychometrics	The psychometric properties of the THQ reported in Kuk et al. (1990) show it to be a valid and reliable instrument Only Factors 1 and 2 were established as reliable for comparison with normative data
Feasibility	Self-administered
Clinical utility	The THQ is used in research and clinical settings
Other comments	Recommended scoring is based on clinical populations Shortened version available for clinical purposes, not recommended for research purposes
Source	Kuk et al. (1990)

Table 5 Greenhalgh et al. (1998) checklist – Tinnitus Reaction Questionnaire (TRQ)

Purpose	Meet the need to evaluate psychological distress associated with tinnitus
Description	26-item questionnaire Total score 0–104 Response options: 5-point scale (scored 0–4) "not at all"; "a little of the time"; "some of the time"; "a good deal of the time"; "almost all of the time"
User-centeredness	Developed using three clinical populations Factor analysis grouped items into the following domains: (1) General distress; (2) Interference; (3) Severity; and (4) Avoidance; 4-factor analysis revealed small degree of variance; majority of variance accounted for by factor 1 (General distress)
Psychometrics	The psychometric properties of the TRQ reported in Wilson et al. (1991) show it to be a valid and reliable instrument with moderate to high correlations with measures of depression and anxiety
Feasibility	Self-administered
Clinical utility	The TRQ is used in both research and clinical settings Some questions are reversed scored
Other comments	Recommended scoring is based on clinical populations Shortened version available for clinical purposes, not recommended for research purposes
Source	Wilson et al. (1991)

Table 6 Greenhalgh et al. (1998) checklist – Tinnitus Handicap Inventory (THI)

Purpose	Meet the need to have a quick, easy to administer and interpret self-reported measure of tinnitus handicap
Description	25-item questionnaire Total score 0–100 Three response options: No (0); Sometimes (2); Yes (4) Three domains: (1) Functional; (2) Emotional; (3) Catastrophic
User-centeredness	Developed in two Phases: Phase I: Item development Phase 2: Tested the validity and reliability of the THI using two different tinnitus patient populations, civilian tinnitus patients and United States Veterans with tinnitus
Psychometrics	The psychometric properties of the THI have been extensively tested and show the THI is a valid and reliable instrument
Feasibility	Self-administered Takes approximately 10–15 min to complete The questions are easy to understand
Clinical utility	The THI is used in a wide variety of research and clinical settings
Other comments	Recommended scoring is based on clinical populations Provides an overall score of tinnitus-related distress 12-item screening version available
Sources	Newman et al. (1996, 2008); Kamalski et al. (2010)

emphasized in #8 of Table 2, which notes this process involves two primary steps: (1) assessment of conceptual and linguistic equivalence and (2) evaluation of measurement properties. When reading articles about questionnaires that were translated into another language, the methods should address steps taken to achieve conceptual equivalence, linguistic equivalence, and how inconsistencies for any significant differences between the original and translated versions were reconciled (Lohr 2002).

There is no consensus on which questionnaire(s) to use to evaluate tinnitus treatment effectiveness. Lack of agreement makes comparison of the same treatment across clinics and/or research groups challenging. The need for standardized outcomes has been discussed for over a decade, but due to the complexity of the issue, widespread agreement does not exist as to which questionnaires to use for research studies or clinical purposes, nor whether or not new questionnaires should be developed to address this need (Meikle et al. 2008; Williamson et al. 2012; Hall et al. 2018).

Regardless of which questionnaires are used, as mentioned earlier, there are multiple sources of bias inherent in self-reported outcomes. Consequently, results can be influenced by a variety of factors including non-tinnitus health concerns such as hearing loss, insomnia, post-traumatic stress disorder, memory issues, and so on. Therefore, when making decisions based on questionnaire data for a patient, it is helpful to keep these influential factors and other sources of measurement error in mind.

Table 7 Greenhalgh et al. (1998) checklist – Tinnitus Functional Index (TFI)

Purpose	Meet the need to have a tinnitus questionnaire sensitive to treatment-related change (i.e., responsiveness)
Description	25-item questionnaire Total score 0–100 Likert-type scale for each item (0–10) Eight subscales/domains: (1) Intrusiveness; (2) Sense of Control; (3) Cognitive; (4) Sleep; (5) Auditory; (6) Relaxation; (7) Quality of life; (8) Emotional
User-centeredness	Phase 1: Panel of 17 experts formed to review nine widely used questionnaires; panel members identified domains of tinnitus distress and items most likely to be responsive to treatment effects Phase 2: Clinical evaluation of questionnaire (Prototype 1) tested at five clinics to determine construct validity of instrument; patients provided follow-up data at 3- and 6-months post-treatment Phase 3: A priori criteria were used to construct next version of TFI (Prototype 2), which was tested at four clinics, including patients who provided follow-up data at 3- and 6-months post-treatment Phase 4: Results synthesized and used to create final version of TFI
Psychometrics	The psychometric properties of the TFI have been extensively tested and show the TFI is a valid and reliable instrument to evaluate treatment responsiveness; provides a measure of clinically important change
Feasibility	Self-administered Takes approximately 10–15 min to complete The questions are easy to understand
Clinical utility	The TFI is used in a wide variety of research and clinical settings
Other comments	Recommended scoring is based on clinical populations Provides an overall score of tinnitus-related distress and individual subscale scores Research-based guidelines established to determine bothersome tinnitus vs. non-bothersome as well as degree of tinnitus impact: "not a problem"; "small problem"; "moderate problem"; "big problem"; "very big problem"
Sources	Meikle et al. (2012); Henry et al. (2016); Jacquemin et al. (2019)

3.4 Permission for Use

Prior to using a questionnaire for clinical or research purposes, it is necessary to contact whomever holds the questionnaire's license and obtain permission for its use. The license grants you the right to use the print or electronic version and will specify if the permission is for commercial or non-commercial uses, as well as for clinical or research purposes. It is important to note that obtaining a license to use a questionnaire in one language is not transferable to the same questionnaire in a different language, which usually requires a separate agreement.

Table 8 Greenhalgh et al. (1998) checklist – Tinnitus Primary Function Questionnaire (TPFQ)

Purpose	Meet the need to evaluate primary activities of a patient's life impacted by tinnitus
Description	12-item questionnaire Total score 0–100; calculated from average of each subscale summed and then divided by 4 Response options: 0 (completely disagree) to 100 (completely agree) Four domains: (1) Emotion; (2) Hearing; (3) Sleep; (4) Concentration
User-centeredness	Developed in two phases: Phase 1 first version "Tinnitus Activities Questionnaire" 20-items Phase 2 shortened to 12-items
Psychometrics	The psychometric properties of the TPFQ reported in Tyler et al. (2014) show it to be a valid and reliable instrument
Feasibility	Self-administered Takes approximately 5–10 min to complete
Clinical utility	The TPFQ is used in research and clinical settings
Other comments	Recommended score based on clinical population Can use 20-item or 12-item version (subset of 20-items)
Source	Tyler et al. (2014)

4 Summary

Questionnaires are useful tools to evaluate patients' needs and engage them to actively participate in the decision-making process regarding their health care (Theodoroff and Saunders 2019). A critical first step, regardless if the intended use is for research or clinical purposes, is to verify the appropriateness of the selected questionnaire for its intended purpose. This chapter has outlined suggested criteria that can be used to guide researchers and clinicians in this endeavor. Additional help to compare commonly used tinnitus questionnaires is provided in the form of checklists following the recommendations of Greenhalgh et al. (1998).

Due to the heterogeneity of tinnitus and the fact that individual differences in tinnitus attributes and severity greatly impact what is considered "bothersome," interpreting tinnitus questionnaire scores is not always straightforward. Administering more than one questionnaire can help gain insight into what domains of functioning are negatively impacted and what health services are most appropriate to pursue to best meet the needs of patient care or the research objectives.

Acknowledgments This material is the result of work supported with resources and the use of facilities at the VA Rehabilitation Research and Development, National Center for Rehabilitative Auditory Research (Center Award #C9230C) at the VA Portland Health Care System in Portland, Oregon. These contents are solely the responsibility of the author and do not necessarily represent the official views of the U.S. Department of Veterans Affairs, Department of Defense, or the United States Government.

References

Bolarinwa OA (2015) Principles and methods of validity and reliability testing of questionnaires used in social and health science researches. Niger Postgrad Med J 22:195–201

Davidson P, Cockburn J, Daly J, Fisher RS (2004) Patient-centered needs assessment: rationale for a psychometric measure for assessing needs in heart failure. J Cardiovasc Nurs 19(3):164–171

Fairfax H (2017) Psychometrics in clinical settings. In: Cripps B (ed) Psychometric testing: critical perspectives. Wiley, Chichester, pp 175–184

Greenhalgh J, Long AF, Brettle AJ, Grant MJ (1998) Reviewing and selecting outcome measures for use in routine practice. J Eval Clin Pract 4(4):339–350

Hall DA, Smith H, Heffernan E et al (2018) Recruiting and retaining participants in e-Delphi surveys for core outcome set development: evaluating the COMiT'ID study. PLoS One 13(7): e0201378

Hallam RS, Rachman S, Hinchcliffe R (1984) Psychological aspects of tinnitus. In: Rachman S (ed) Contributions to medical psychology, vol 3. Pergamon Press, Oxford

Hallam RS, Jakes SC, Hinchcliffe R (1988) Cognitive variables in tinnitus annoyance. Br J Clin Psychol 27:213–222

Henry JA (2016) "Measurement" of tinnitus. Otol Neurotol 37(8):e276–e285

Henry JL, Wilson PH (1998) The psychometric properties of two measures of tinnitus complaint and handicap. Int Tinnitus J 4(2):114–121

Henry JA, Griest S, Zaugg TL et al (2015) Tinnitus and hearing survey: a screening tool to differentiate bothersome tinnitus from hearing difficulties. Am J Audiol 24(1):66–77

Henry JA, Griest S, Thielman E et al (2016) Tinnitus functional index: development, validation, outcomes research, and clinical application. Hear Res 334:58–64

Husain FT, Gander PE, Jansen JN, Shen S (2018) Expectations for tinnitus treatment and outcomes: a survey study of audiologists and patients. J Am Acad Audiol 29:313–336

Jacquemin L, Mertens G, Van de Heyning P et al (2019) Sensitivity to change and convergent validity of the tinnitus functional index (TFI) and the tinnitus questionnaire (TQ): clinical and research perspectives. Hear Res 382:107796

Johnson C (2005) Measuring pain. Visual analog scale versus numeric pain scale: what is the difference? Chiropr Med 4(1):43–44

Kamalski DM, Hoekstra CE, van Zanten BG et al (2010) Measuring disease-specific health-related quality of life to evaluate treatment outcomes in tinnitus patients: a systematic review. Otolaryngol Head Neck Surg 143(2):181–185

Kuk FK, Tyler RS, Russell D, Jordan H (1990) The psychometric properties of a tinnitus handicap questionnaire. Ear Hear 11:434–445

Langguth B, Goodey R, Azevedo A et al (2007) Consensus for tinnitus patient assessment and treatment outcome measurement: tinnitus research initiative meeting, Regensburg, July 2006. Prog Brain Res 2007(166):525–536

Lohr KN (2002) Assessing health status and quality-of-life instruments: attributes and review criteria. Qual Life Res 11:193–205

McNeish D (2018) Thanks coefficient alpha, we'll take it from here. Psychol Methods 23 (3):412–433

Meikle MB, Stewart BJ, Griest SE, Henry JA (2008) Tinnitus outcomes assessment. Trends Amplif 12(3):223–235

Meikle MB, Henry JA, Griest SE et al (2012) The tinnitus functional index: development of a new clinical measure for chronic, intrusive tinnitus. Ear Hear 33(2):153–176

Michiels S, Ganz Sanchez T, Oron Y et al (2018) Diagnostic criteria for somatosensory tinnitus: a Delphi process and face-to-face meeting to establish consensus. Trends Hear 22:1–10

Newman CW, Sandridge SA (2006) Sound therapy option profile (STOP): a tool for selecting devices used in tinnitus treatment. Audiol Online:1–6

Newman CW, Sandridge SA (2016) Care path for patients with tinnitus: an interprofessional collaborative model. Perspectives ASHA Special Interest Groups (SIG 7) 1:24–42

Newman CW, Weinstein BE, Jacobson GP, Hug GA (1991) Test-retest reliability of the hearing handicap inventory for adults. Ear Hear 12(5):355–357

Newman CW, Jacobson GP, Spitzer JB (1996) Development of the tinnitus handicap inventory. Arch Otolaryngol Head Neck Surg 122(2):143–148

Newman CW, Sandridge SA, Bolek L (2008) Development and psychometric adequacy of the screening version of the tinnitus handicap inventory. Otol Neurotol 29(3):276–281

Searchfield GD (2019) A client oriented scale of improvement in tinnitus for therapy goal planning and assessing outcomes. J Am Acad Audiol 30(4):327–337

Smith SL, Fagelson M (2011) Development of the self-efficacy for tinnitus management questionnaire. J Am Acad Audiol 22(7):424–440

Theodoroff SM, Saunders GH (2019) Key findings from tinnitus research and clinical implications. Am J Audiol 28(1S):239–240

Theodoroff SM, McMillan GP, Zaugg TL et al (2017) Randomized controlled trial of a novel device for tinnitus sound therapy during sleep. Am J Audiol 26(4):543–554

Tyler RS, Gogel SA, Gehringer AK (2007) Tinnitus activities treatment. Prog Brain Res 166:425–434. https://doi.org/10.1016/S0079-6123(07)66041-5

Tyler R, Ji H, Perreau A et al (2014) Development and validation of the tinnitus primary function questionnaire. Am J Audiol 23(3):260–272

Üstün TB, Chatterji S, Kostanjsek N et al (2010) Developing the World Health Organization disability assessment schedule 2.0. Bull World Health Organ 88:815–823

Williamson PR, Altman DG, Blazeby JM et al (2012) Developing core outcome sets for clinical trials: issues to consider. Trials 13:132

Wilson PH, Henry J, Bowen M, Haralambous G (1991) Tinnitus reaction questionnaire: psychometric properties of a measure of distress associated with tinnitus. J Speech Hear Res 34:197–201

Principles and Methods for Psychoacoustic Evaluation of Tinnitus

Dunja Vajsakovic, Michael Maslin, and Grant D. Searchfield

Contents

D. Vajsakovic and G. D. Searchfield (✉)
Section of Audiology, The University of Auckland, Auckland, New Zealand

Eisdell Moore Centre, The University of Auckland, Auckland, New Zealand

Centre for Brain Research, The University of Auckland, Auckland, New Zealand

Brain Research New Zealand, A Centre of Research Excellence, Auckland, New Zealand
e-mail: g.searchfield@auckland.ac.nz

M. Maslin
Eisdell Moore Centre, The University of Auckland, Auckland, New Zealand

School of Psychology, Speech and Hearing, The University of Canterbury, Canterbury, New Zealand

Curr Topics Behav Neurosci (2021) 51: 419–460
https://doi.org/10.1007/7854_2020_211
Published Online: 7 February 2021

Abstract Tinnitus, the perception of sound in the absence of a physical sound in the environment, is highly heterogeneous. It varies in its etiology, characteristics, and impact on an individual's life. The sound is commonly described as "ringing," "buzzing," "crickets," "hissing," "humming." Tinnitus can be acute or chronic, mild or disabling. It can be perceived unilaterally or, more commonly, bilaterally. The sound and its location differ from person to person and fluctuate in the same individual over a certain period of time. This heterogeneity in characterization has important implications for research and clinical practice. Identifying patterns in how tinnitus sounds and its relationship to hearing may aid in identifying different forms of tinnitus and revealing their underlying mechanisms. However, the subjective nature of characterizing tinnitus makes it difficult to reliably define and measure. This chapter will focus on reviewing the psychoacoustic assessment of tinnitus, its relationship to cognitive and behavioral aspects of tinnitus, and its neuropathophysiology. In particular, it will describe the heterogeneity of tinnitus and tinnitus matching, and how individual variability in measures may be used to guide treatment and as a prognostic factor.

Keywords Assessment · Loudness · Masking · Pitch · Psychoacoustic · Tinnitus

1 Introduction

1.1 Overview of Clinical Characteristics of Tinnitus

It is important to evaluate and quantify tinnitus so as to understand its mechanisms and possible means of treatment (Henry et al. 2004). Psychoacoustic measurements have been obtained primarily for two reasons: (1) to define the auditory features related to how it is perceived by the patient and (2) to define to what degree external sounds have an effect on the tinnitus (Henry 2016). There is now increasing interest in how psychoacoustic measures can inform treatment and serve as prognostic tools. Pitch and loudness measurements aid in determining the auditory attributes of tinnitus, while masking and residual inhibition effects reveal how a patient's tinnitus behaves following the application of an external sound (Henry 2016). In addition to characterizing the auditory features of tinnitus and its response to auditory stimuli, conducting psychoacoustic measurements is important for a number of secondary reasons: (1) to understand the underlying etiology and mechanisms of tinnitus, (2) to be able to replicate the patient's tinnitus and demonstrate its characteristics and features to both the patient and their family, (3) to reassure the patient and their family that the tinnitus is real, (4) to explore management options and select the treatment most beneficial to an individual patient, (5) to monitor any changes in the tinnitus and its perception, (6) to determine if a patient is likely to benefit from a certain type of treatment, (7) to determine the effects of treatment on tinnitus and its perception, (8) to reproduce the patients tinnitus for use in treatments such as sound

Table 1 Patients' descriptions of their tinnitus percept. Responses were obtained from the OHSU Tinnitus Clinic Database 1981–1994 ($n = 1{,}625$) and Stouffer and Tyler (1990) ($n = 528$)

Sound	OHSU tinnitus clinic 1981–1994 % of responses	Stouffer and Tyler (1990) % of responses
Ringing	56.9	37.5
Hissing	19.5	7.8
Clear tone	17.3	–
High tension wire	14.2	–
Buzzing	12.0	11.2
Sizzling	7.6	–
Transformer noise	7.1	–
Crickets, insects	6.2	8.5
Pulsating	6.0	3.8
Whistle	6.0	6.6
Hum	6.0	5.3
Roaring	4.8	4.5
Pounding	1.1	0.4
Clicking	0.5	0.6
Music/musical note	0.4	4.2
Other	21.9	9.8

OHSU = Oregon Health and Science University. OHSU data reflects all patients regardless of number of predominant sounds reported and amounts to a total value greater than 100% as patients were permitted to provide multiple responses (Meikle et al. 2004)

therapy, (9) to aid in the exploration and characterization of different tinnitus subcategories, and (10) to assist in legal issues (Henry et al. 2004; Henry et al. 2013; Manning et al. 2019; Moore 2014; Nageris et al. 2010; Ristovska et al. 2019; Sandlin and Olsson 1999; Schechter and Henry 2002; Switalski and Sanchez 2019; Tyler 2000; Vernon 2000).

Tinnitus has traditionally been classified as being one of the two general types: (1) the rare objective type (somatosounds) generated as an acoustic signal located in the head or neck as a result of muscular, skeletal, respiratory, or vascular irregularities/disturbances or (2) the more common subjective type (true tinnitus) thought to arise within the auditory and non-auditory pathways as a result of lesion-induced reactive neural plasticity (Douek 1981; Henry 2016). This review is focused on subjective tinnitus.

Patients describe their tinnitus percept in a number of ways; most commonly as a ringing or hissing sensation, but in some cases as more complex sounds such as crickets and even music (Table 1). It is not uncommon for patients to report hearing more than one sound (together or separately) and/or complex sounds (Douek 1981; Meikle and Taylor-Walsh 1984; Tyler 2000). For example, some individuals might experience a high-pitched whistle and low-pitched "ocean waves" in the same ear, with perhaps one of the sounds fluctuating while the other is continuous (Tyler 2000), or may describe their tinnitus as sounding like music (Vanneste et al. 2013). Further, tinnitus can have multiple and/or variable localizations, being perceived in the ear (s) or frequently in various locations in/or around the head (Meikle and Taylor-Walsh 1984; Tyler 2000).

1.2 The Historic Foundations of Tinnitus Matching

Psychoacoustics is the study of the perception of sound and can trace its roots to Aristotle and later Leonardo De Vinci. However our understanding of psychoacoustics has accelerated with the development of telecommunications (Yost 2015). The history of tinnitus assessment has followed a similar trajectory (Stephens 1984). In the early nineteenth century, French physician Jean Marie Itard outlined early attempts at describing tinnitus masking, including piercing a tiny hole in a water-filled vase and letting the water trickle into a large copper bowl of the same capacity, burning greener or slightly damp firewood, and even advising tinnitus sufferers to take up residence near a water mill (J.M.G. Itard, quoted in Stephens 1984). As early as 1903, Spaulding acknowledged the advantages of matching a patient's tinnitus to musical notes of the same pitch. His approach included playing musical scales on his violin and instructing patients to indicate when they felt a note best-matched the pitch of their perceived tinnitus (Spaulding 1903). Spaulding was one of the first along with Josephson (1931) to identify and shed light on the temporary suppression of tinnitus following sound, known as residual inhibition (Vernon and Fenwick 1984). Building on Spaulding's work, Josephson (1931) designed a method for the measurement of both tinnitus pitch and "intensity." For the evaluation of tinnitus pitch, he employed a pure-tone pitch-matching method in which the pitch of the tinnitus was estimated by stimulation over a range of frequencies. Throughout his masking experiments, Josephson noted that tinnitus did not appear to behave as an external sound. If it did, a summation of the tinnitus and pitch-matched sound loudness would have been observed, instead masking occurred (Josephson 1931). According to his methods, the "intensity" of tinnitus could be measured by defining the difference between the threshold of the superimposed sound as compared to the normal hearing threshold at that frequency (Josephson 1931). The pure-tone method was also implemented by Wegel (1931), who used measurements of pitch and loudness to construct masking curves of tinnitus. The work of these researchers both set the scene and laid the foundation for the more detailed investigations that were to follow. Pivotal contributions to tinnitus measurement were made mid-last century by Edmund Prince Fowler, the first researcher to determine tinnitus loudness in dB sensation level (dB SL) by balancing the tinnitus loudness in one ear with the loudness of a tone in the contralateral ear (Fowler 1936, 1937). Fowler emphasized the importance of psychoacoustic measurements in defining the spectral characteristics of tinnitus. He also considered the possibility that features such as tinnitus tone, frequency, intermittency (or conversely constancy), and reaction to masking and environmental noise might suggest different etiologies and site of lesion (Fowler 1940). Fowler described methods of pitch and loudness matching that are comparable to what is being used in clinics and research today. He also identified many of the issues and limitations of psychoacoustic measures that clinicians and researchers still remain mindful of when assessing patients.

Goodhill (1952) believed that the validity with which patients described their tinnitus depended to a great degree on their knowledge of musical terms, having observed that a number of his patients had difficulties with defining their tinnitus. Reed (1960) conducted the first large-scale tinnitus study, using audiometric testing to evaluate the spectral features of tinnitus in 200 patients. Instead of applying the tinnitus measurement techniques available (such as those described by Fowler and others), he chose to develop his own method for matching tinnitus "frequency, content, and loudness," commenting on the inability to replicate the methods of others due to the lack of sufficient detail provided. In addition to defining the localization, intermittency (or conversely constancy), degree of hearing loss, description of tinnitus sound, frequency, and loudness, Reed (1960) also noted that the severity of a patient's tinnitus did not appear to be correlated with any of the psychoacoustic measures (central frequency, bandwidth, or loudness). Furthermore, he advised that tinnitus without a degree of hearing loss is rare, occurring in only 7.5% of tinnitus cases. Exploring tinnitus and hearing loss in more depth, Graham and Newby (1962) assessed the characteristics of tinnitus in four different groups of patients; those with normal hearing sensitivity, sensorineural hearing loss, conductive hearing loss, and mixed sensorineural-conductive hearing loss. The study found that the pitch matches for patients with conductive hearing loss were significantly lower in frequency than for patients in the other two hearing loss groups. Graham and Newby (1962) suggested that this might reflect an underlying mechanism for the generation of tinnitus in patients with conductive hearing loss that is distinct from that experienced by patients with other types of hearing loss.

The first systematic evaluation of tinnitus masking was conducted by Feldmann (1971), who investigated the effect that sounds of certain frequencies had on the tinnitus sensation of 200 patients. Feldmann (1971) noted that sound-on-sound masking generates five different masking patterns (see *The Minimum Masking Level* section): one similar to sound-on-sound masking in which the closer the frequency of the masker is to the frequency of the tone being masked, the lower the intensity required to mask it. The others did not follow conventional sound-on-sound masking principles (Feldmann 1971).

Forty years ago efforts were made at a Ciba Foundation Conference in London to promote the standardization of tinnitus characterization procedures, advocating for the routine measurement of four features of tinnitus deemed crucial to its evaluation, namely tinnitus pitch and loudness, the maskability of tinnitus (ability of an external sound to conceal the tinnitus), and residual inhibition (reduction or complete elimination of the perception of tinnitus following auditory stimulation) (Evered and Lawrenson 1981). Working closely with the Ciba Symposium, Vernon and Meikle (1981) designed a protocol detailing the methods for conducting these four measurements. However, despite these efforts, standardized methods have not been adopted universally (Henry et al. 2004). The absence of standardization presents severe limitations to the obtainment of valid and reliable psychoacoustic measurements. It also prevents the collation, comparison, and interpretation of data across clinics and research laboratories on a global scale, in turn impeding progress in the field of tinnitus research.

2 Psychoacoustic Measures: Pitch

Pitch is the perceptual equivalent of the frequency of sound. Pitch matching of tinnitus is a fundamental psychoacoustic measure in most clinical or research assessment protocols. This section will consider the values and shortcomings of different measures of tinnitus pitch. Pitch measures:

- Aid in characterizing the perceived tinnitus and can be used as a reference point (useful for monitoring changes in tinnitus, especially during treatment).
- Help/support the clinician in determining an optimal route of treatment for a particular patient.
- Aid in the selection and fitting process for acoustic instrumentation and sound therapy.
- Form a critical component for establishing and implementing therapeutic masking for tinnitus.
- Contribute to our understanding of tinnitus including its origin and etiology by allowing interindividual comparisons to be made, focusing on patients who experience different tinnitus frequencies, have hearing loss in addition to their tinnitus, or an altogether different comorbidity (Kim et al. 2017; Nageris et al. 2010; Switalski and Sanchez 2019).

2.1 Methodologies of Tinnitus Pitch Matching

2.1.1 Test Ear

The ear chosen for the test stimuli is important as it can influence the results of pitch matching and in turn the final tinnitus-matched frequency (Tyler 2000). If the tone is presented to the same ear where the tinnitus is heard (i.e. ipsilaterally), then the tone could affect the tinnitus perception in some way. Alternatively, if it is presented in the ear opposite to that where the tinnitus is heard (i.e. contralaterally), there is a risk of binaural diplacusis occurring, giving rise to inaccurate pitch matching (Tyler 2000). If the tester/investigator chooses to present the tone binaurally, the patient might experience confounding by both of these effects, in turn finding it hard to not only define their perceived tinnitus, but do so accurately and reliably from trial to trial.

A number of researchers suggest ipsilateral sound presentation, primarily to avoid any effects of binaural diplacusis. Though the majority of patients have bilateral tinnitus (Meikle and Taylor-Walsh 1984), recommendations have been made to conduct monaural ipsilateral testing in each ear separately (if they are found to differ), making sure to perform at least seven replications, with the examiner noting down the test ear each time (Tyler and Conrad-Armes 1983a; Vernon and Fenwick 1984). Conversely, there are also those who stand in support of contralateral test tone presentation, maintaining that it is less confusing for patients during pitch-matching procedures to have the tone presented in the ear that is free from distracting sound sensations (Evered and Lawrenson 1981; Sandlin and Olsson 1999).

In most cases it simply does not matter which ear the sound is presented to (Baguley et al. 2013; Vernon and Fenwick 1984). The best procedure is to play a test tone to the patient separately in each ear and allow them to decide for themselves which ear they would like to have the tones presented (see Vernon and Fenwick 1984). In the case that they feel more comfortable with the tones being presented to the contralateral ear, it is recommended that the examiner repeats the final pitch-match tone on the ipsilateral side to account for possible diplacusis-related complications and ensure that the tinnitus-matched frequency established in the contralateral ear also holds for the ipsilateral ear (Vernon and Fenwick 1984). An alternative approach, advocated when there are hearing asymmetries, is to choose the better hearing ear; this is likely to have least disruption of tonotopicity, frequency resolution, and less complications related to cochlear recruitment.

2.1.2 Matching Method

Numerous methods of pitch matching have been developed and tested, with varying levels of success. These methods include, but are not limited to, the adjustment, limits, and adaptive methods (Tyler and Conrad-Armes 1983a), 2-alternative forced-choice (2AFC) method (Vernon and Fenwick 1984), binary-2AFC (Henry et al. 2001), forced-choice double-staircase (FCDS) technique (Penner and Bilger 1992; Penner and Klafter 1992), heptatonic scale (Ohsaki et al. 1990), subject-guided procedure (Henry et al. 2004), and tinnitus likeness ratings (also known as tinnitus spectrum measurements) (Norena et al. 2002; Roberts et al. 2006) (see Table 2).

Tyler and Conrad-Armes (1983a) were among the first to investigate psychoacoustic measures, developing three methods (adjustment, limits, and adaptive) and evaluating their ability to define the pitch and loudness of tinnitus for ten participants. The adaptive and limits methods share a similar protocol, while the method of adjustment differs primarily in that it is patient-guided; the patient "adjusts" the main frequency dial of a pulsed-tone oscillator to localize the frequency most representative of their tinnitus, before making further adjustments using a fine-control dial to more precisely define and finalize their tinnitus frequency (Tyler and Conrad-Armes 1983a). Although the methods varied in their respective protocols, no significant

Table 2 Summary of methods used to characterize and evaluate tinnitus pitch

Method	Studies using this method, number of participants (n)	Protocol	Test ear (to which sound is presented to)	General comments
Conventional single-tone pitch matching (2AFC)	Vernon and Fenwick (1984), $n =$ review	A tinnitus synthesizer is used to present two loudness-matched tones separated by 1,000 Hz in an alternating manner so that each tone is heard four to five times. The subject is instructed to choose which of the two tones is most like their tinnitus. This is done using a bracketing approach (in which the patient's decision dictates the subsequent frequencies presented) until the frequency of the tinnitus has been established. Once the tinnitus frequency has been defined, it is verified in the ipsilateral ear to avoid binaural diplacusis	Ipsilateral and contralateral (determined by patient preference)	N/A
	Ohsaki et al. (1990), $n = 55$		Not disclosed	Reproducibility of pitch matching is not as good as for the heptachord method
Binary-2AFC	Henry et al. (2001), $n = 20$	The 2AFC procedure is followed; however, binary bracketing is applied, narrowing the testing frequencies down until tinnitus frequency is reached. First, the subject is presented with two frequencies and is instructed to decide which is closer in pitch to their tinnitus. This initial frequency	Contralateral	Pitch matches could be obtained within 20–25 min, with response reliability being good for some subjects but not others
	Henry et al. (2004), $n = 42$		Contralateral	Excellent response reliability for about half of the subjects. Defining the range of pitch matches might be more appropriate than identifying single pitch matches

(continued)

Table 2 (continued)

Method	Studies using this method, number of participants (n)	Protocol	Test ear (to which sound is presented to)	General comments
		choice results in binary bracketing, either to the lower or upper frequency range. Movement to new frequencies progresses in octave steps, and the computer further brackets the pitch match to within an octave. Following this, matches are made in $\frac{1}{3}$ octave steps		
Recursive 2-interval forced-choice (RIFT)	Korth et al. (2020), $n = 117$	The RIFT procedure is similar to the 2AFC. 17 tones are presented in the range from 1 to 16 kHz in ¼ octave steps. First, the level of the tones is manually adjusted by the participant using a scrolling volume slider until the tinnitus and presented tone is matched in loudness. Following this, the stimulus frequency is then limited to the highest frequency that is audible to the participant. Two tone pairs (2 octaves apart) are presented and the participant is instructed to indicate the frequency most representative of their tinnitus. Based on their decision, further tone pairs are	Ipsilateral (unilateral tinnitus) OR Ear with more dominant tinnitus perception/less average hearing loss (bilateral tinnitus)	Reliable estimations of tinnitus pitch can be obtained as long as initial and redundant sessions, and participants with poor pitch-match performance are excluded

(continued)

Table 2 (continued)

Method	Studies using this method, number of participants (n)	Protocol	Test ear (to which sound is presented to)	General comments
		presented, reducing the octave step each time until a decision on the smallest step of a twelfth octave is reached and one final frequency is chosen. If the subject's choices are contradictory twice in a row, the test is cancelled and restarted		
Tinnitus likeness ratings	Norena et al. (2002), $n = 10$	A pure tone with a pseudo-randomly selected frequency is presented and subjects are asked to match the intensity of the tone to the loudness of their tinnitus. If the subject indicates that the pitch of the pure tone corresponds to a component of their tinnitus sensation, they are to rate on a ten-point scale the degree to which this pitch contributes to their overall tinnitus percept	Ipsilateral	In most cases, the "internal tinnitus spectra" demonstrated a broad peak sitting within the range of hearing loss
	Roberts et al. (2006), $n = 32$	Tones are chosen from a set of three stimuli depending on the subjects tinnitus (tonal, ringing, or hissing). Subjects are then instructed to rate the pitch of each presented sound for similarity to their perceived tinnitus using a Borg CR100	Not disclosed	Results are in agreement with Norena et al. (2002); however, it is important to consider the effect hearing loss has on the perception of sounds used to measure the tinnitus spectrum

(continued)

Table 2 (continued)

Method	Studies using this method, number of participants (n)	Protocol	Test ear (to which sound is presented to)	General comments
		scale (0 = not at all similar, 30 = not very similar, 50 = somewhat similar, 70 = very similar, and 100 = identical)		
	Hoare et al. (2014), $n = 28$	11 tonal or narrow-band sound clips are presented to subjects, with each sound being played three times in a random order. Subjects are asked to rate the similarity of the pitch of each of the sounds presented to their tinnitus on a 100-point scale	Not disclosed	Tinnitus likeness rating across a range of frequencies is highly variable and unpredictable
FCDS	Penner and Bilger (1992), $n = 11$	In the double-staircase, the experimenter selects two starting points for two sequences of trials; one with the comparison stimuli clearly above the patients tinnitus pitch, and the other clearly below it. In the forced-choice procedure, the experimenter presents the tinnitus ("standard") stimulus, accompanied by a 750 ms flash of light, followed by an external comparison tone also marked with a flash of light. The subject is instructed to choose which of the two	Ipsilateral	Within-session variability of pitch matching to the tinnitus pitch for the FCDS procedure is consistent and greater than for a method of adjustment (bracketing technique similar to the method of adaptation – not to be confused with the adjustment method developed by Tyler and Conrad-Armes (1983a))

(continued)

Table 2 (continued)

Method	Studies using this method, number of participants (n)	Protocol	Test ear (to which sound is presented to)	General comments
		stimuli is "lower" or "higher" in frequency (depending on the experimenter's instruction). 100-Hz step sizes are employed for obtaining pitch matches		
	Penner and Klafter (1992), $n = 7$	The FCDS procedure is followed, but step sizes are reduced to 0.2%	Ipsilateral OR Ear in which tinnitus is louder	The step sizes used in the original FCDS procedure (Penner and Bilger 1992) were substantially larger than the frequency difference limen for normal-hearing subjects. Following step size reduction, the frequency difference limen for tinnitus matches is comparable to that obtained for external stimuli
Subject-guided	Henry et al. (2004), $n = 42$	Loudness-matched test frequency is presented at random (from a selection of 17 frequencies) using an automated computer system. The subject then has the option to respond and alter the tone so that it becomes "higher," "much higher," "lower," or "much lower" in frequency until a tone is presented that is equal in pitch to the subjects tinnitus	Contralateral	Excellent response reliability for about half of the subjects. Defining the range of pitch matches might be more appropriate than identifying single pitch matches

(continued)

Table 2 (continued)

Method	Studies using this method, number of participants (*n*)	Protocol	Test ear (to which sound is presented to)	General comments
Heptatonic scale	Ohsaki et al. (1990), $n = 55$	A heptachord generator is used to present 56 tones in eight octaves. Presentation of the tones is done in ascending runs. A tone is considered representative of the patients tinnitus frequency when it has been obtained twice continuously in increasing runs	Not disclosed	The heptachord method is more reliable and accurate than the conventional "one-octave-interval" method, with coincidence ratios as high as 100%, 94.8%, and 88.9%, respectively, for intra-daily variation, intra-weekly variation, and eight consecutive measurements in one test
Method of limits	Tyler and Conrad-Armes (1983a), $n = 10$	The subject is instructed to state whether their tinnitus is "higher" or "lower" in pitch than the tone presented by the tester. The tester then presents a sequence of pulsed tones and allows the subject to make a decision. The tones are presented in ascending and descending runs, with an ascending run always being accompanied by a descending run (order of presentation is randomized). The last "higher" and "lower" responses (in the respective ascending and descending runs) are averaged to	Ipsilateral	No significant differences were found in the group means or standard deviations for the tinnitus frequencies obtained using these three methods. For clinical use, it is recommended that the adaptive or adjustment method be used due to their time efficiency (only 1–2 min per pitch match). Ipsilateral presentation is recommended to avoid binaural diplacusis. Seven to nine pitch matches are recommended per subject due to large variability in pitch reproducibility

(continued)

Table 2 (continued)

Method	Studies using this method, number of participants (n)	Protocol	Test ear (to which sound is presented to)	General comments
		give the overall tinnitus pitch		
Adaptive method		The subject is instructed as per the "method of limits," with the tester presenting a series of pulsed tones and allowing the subject to make a "higher" or "lower" decision. The first stage includes locating the subjects tinnitus pitch to within a 1-octave band. In the second stage, tones are presented whose frequencies are within the 1-octave range determined in stage 1, with the final pitch being located to within a $\frac{1}{6}$-octave range		
Adjustment method		The subject is instructed to adjust the pitch of the pulsing tone using a dial on the pitch-matching apparatus, first making wide sweeps with the dial and then gradually narrowing down to the pitch most representative of their tinnitus		

differences were observed in either the group means or standard deviations for the pitch matches. Tyler and Conrad-Armes (1983a) suggested that the adaptive and adjustment methods were superior for use in the clinical setting due to their ability to obtain single pitch matches within 1–2 min (compared with 4–5 min for the method of limits). A second recommendation was to conduct a minimum of seven pitch-match replicates for each patient to account for the large variability in a patient's ability to accurately reproduce their tinnitus pitch (Tyler and Conrad-Armes 1983a).

The 2AFC procedure has received the most attention and gained wide acceptance as the conventional method for tinnitus pitch assessment due to its simplicity for patients and relatively short completion time (Kim et al. 2017). Development of the 2AFC method began as an attempt by Vernon and Fenwick (1984) to provide a standardized measure for tinnitus characterization. Over the years, progressive advances in technology have enabled 2AFC to be applied using different platforms: manual and computer-automated (Henry et al. 2004), web-based (Mahboubi et al. 2012), through portable media players (Wunderlich et al. 2015), and iPods (Korth et al. 2020). Korth et al. (2020) used an adaptation of the 2AFC, known as the recursive 2-interval forced-choice test (RIFT), in an iPod-based automated tinnitus pitch-matching procedure. The study found that recursive matching resulted in reliable tinnitus pitch matching in patients with tonal tinnitus once initial and redundant sessions, and patients with poor pitch-matching performance were excluded (Korth et al. 2020).

Penner and Bilger (1992) explored the FCDS procedure (Jesteadt 1980) as a psychoacoustic measure for tinnitus pitch and loudness, believing it had a number of advantages over measures based on bracketing and sequential presentation of tones. For one, methods which present matching tones in a monotonic series (such as those in which the experimenter adjusts the frequency in equal steps according to a subject's request to raise or lower it) are subject to response bias as a result of sequential effects. FCDS avoids response bias by not presenting successive tones in a predictable sequence, instead forcing the subject to judge each stimulus independent to the judgment of previous stimuli as the stimuli bear no relation to each other (Penner and Bilger 1992). Further, the FCDS enables the subject to classify comparison stimuli with respect to their tinnitus rather than simply matching pure tones to the percept, as is the protocol applied in most pitch (and loudness) matching measures. Evaluating the pitch-match reliability of the FCDS relative to that obtained using a "method of adjustment,[1]" the investigators observed a lower within-session variability with the FCDS procedure. Although the method has been reported as being capable of producing reliable pitch matches, it is rarely used in the clinical setting due to issues regarding comprehension of the testing concept and the lengthy completion time involved (Henry et al. 2013; Kim et al. 2017).

Most methods available today share a similar basis for pitch matching (namely presenting a series of tones at varying frequencies and adjusting them according to

[1]The "method of adjustment" as referred to by Penner and Bilger (1992) is simply a bracketing method in which the subject attempts to match their tinnitus to tones presented by the experimenter (for more detailed protocol, see Penner and Bilger (1992)). It is not to be confused with Tyler and Conrad-Armes (1984) "method of adjustment."

the subject's response until the tinnitus frequency is achieved). They differ only in the finer details such as the instructions given to the subject, sequence in which the tones are presented, or perhaps the frequency differences of test tones. Novel approaches have been trialed including matching using the standard musical "do-re-mi" scale (Ohsaki et al. 1990). The usefulness of this method is limited by the nature of musical tonality which begins to break down above 4,000 Hz. As will be discussed shortly, many patients perceive their tinnitus pitch above this frequency; hence, it becomes difficult to apply methods using music intervals for testing and pitch matching.

Perhaps the most complete pitch-matching procedures are the tinnitus likeness rating methods developed by Norena et al. (2002) and Roberts et al. (2006). Instead of participants' choosing between test tones in an "all-or-nothing" manner, they were instructed to rate each presented tone for the degree to which the particular tone contributed to the overall tinnitus sensation. What is generated as a result is an "internal tinnitus spectrum" which shows the frequency components of an individual's tinnitus, highlighting dominant frequencies. Roberts et al. (2006) used tinnitus likeness software to evaluate psychoacoustic properties of tinnitus and residual inhibition in 32 tinnitus patients. After identifying the quality of sound of their tinnitus ("tonal," "ringing," or "hissing"), the patients were instructed to match the loudness of 11 "tonal," "ringing," or "hissing" sounds depending on their initial selection (each with increasing center frequency) to the loudness of their tinnitus. Once the loudness was established, the patients were then replayed with each of the 11 sounds at this level, rating the sounds based on their likeness (similarity) to the tinnitus, in turn generating a spectrum of tinnitus frequency components. The most prominent component was replayed to the patients, and they were asked to rate the sound based on its similarity to the tinnitus percept using a Borg CR100 scale approach where 0 = "not at all," 30 = "not very similar," 50 = "somewhat similar," 70 = "very similar," and 100 = "identical." The study revealed a tendency for the tinnitus spectra to span the region of hearing loss in agreement with the results of Norena et al. (2002) who also found that the majority of their participants displayed a broad peak sitting within the range of hearing loss. Likeness ratings offer more complete account of participants' tinnitus, but the method is time-consuming, limiting its clinical applicability compared to the 2AFC method.

2.2 *Tinnitus Pitch and Hearing Loss*

It is not uncommon for patients to complain about difficulties in hearing as a result of their tinnitus; however, it appears that this is primarily a consequence of an underlying hearing loss rather than the tinnitus itself (Ratnayake et al. 2009). Auditory processing mechanisms do appear to be disrupted by tinnitus in a pitch specific manner; specifically, auditory streaming of tones is disrupted at tinnitus pitch (Durai et al. 2019).

As demonstrated by the tinnitus likeness rating measurements, there is a clear relationship between tinnitus pitch and hearing loss. Individuals who suffer from tinnitus are highly likely to have some degree of hearing loss (Axelsson and Ringdahl 1989; Henry 2016; Josephson 1931; Moore 2012; Moore and Vinay 2010; Norena et al. 2002; Ristovska et al. 2019; Roberts et al. 2008; Schechter and Henry 2002; Vernon 1977; Ward and Baumann 2009; Wegel 1931). While the exact pitch perceived varies from patient to patient, most patients tend to match their tinnitus to a high-frequency tone at or above 3,000 Hz (Meikle and Taylor-Walsh 1984; Mitchell et al. 1984; Reed 1960; Roeser and Price 1980; Sandlin and Olsson 1999; Stouffer and Tyler 1990; Tyler 2000; Vernon 2000). The high-frequency locus of tinnitus has, for some time, been thought to be linked to the idea that tinnitus is closely related to hearing loss, in particular, high-frequency hearing loss, in many cases due to noise induced trauma/exposure (Henry 2016). Conducting the first large-scale tinnitus study, Reed (1960) found that in 38% of cases the tinnitus was pitch matched to a pure tone within the 3,000–5,000 Hz range; several subsequent studies have observed similar trends (Meikle and Taylor-Walsh 1984; Roeser and Price 1980; Sandlin and Olsson 1999; Stouffer and Tyler 1990; Tyler 2000; Vernon 2000). Mitchell et al. (1984) confirmed that most patients suffer from a high-pitch tinnitus, however noted a broader range of reported pitch-match frequencies (1,000–8,000 Hz). The high-frequency nature of tinnitus is not only evident from the results of psychoacoustic testing, but has also been demonstrated by patient complaints (Stouffer and Tyler 1990; Tyler 2000).

Although the pitch of tinnitus commonly falls in the high-frequency range, this is not always the case. A study by Pan et al. (2009) identified various trends in pitch matching among 195 tinnitus patients; those who described a tone-like tinnitus reported a higher pitch (mean = 5,385 Hz) relative to those experiencing a noise-like sensation (mean = 3,266 Hz). Further, patients with a flat audiogram demonstrated a higher chance of describing a tinnitus that was noise-like, unilateral, and had a pitch-match frequency $< 2{,}000$ Hz. In addition, those with a notched audiogram often identified a pitch $\leq 8{,}000$ Hz, while those with normal hearing up to 8,000 Hz often matched a pitch $\geq 8{,}000$ Hz (Pan et al. 2009).

Alongside the fairly universal acceptance of a general relationship between the frequency of hearing loss and tinnitus pitch, specific theories relating the degree of hearing loss to tinnitus have been proposed to explain tonal pitch matches. The two most widely recognized theories are the "edge frequency" and "region of maximal hearing loss" theories. The edge frequency theory posits that the pitch of a patient's tinnitus corresponds to the "edge frequency" of the audiogram; more specifically, the frequency at which hearing loss transitions from normal to abnormal hearing relatively abruptly (Josephson 1931; Moore 2014; Moore and Vinay 2010). Moore and Vinay (2010) found that patients' final pitch matches were generally at the lower end of the spectrum (1,630 Hz) with a strong correlation ($r = 0.94$) between the matches and edge frequency of the audiogram. The edge theory is concordant with the tonotopic reorganization model of tinnitus; when a certain frequency region is affected as a result of hearing loss, there is a lack of inhibition from neurons that were once tuned to that particular region. This then leads to a downstream release of lateral inhibition and resultant increase in neural activity in adjacent regions where there is less or no hearing loss; the consequence is tinnitus with a dominant frequency that corresponds to the audiometric edge (Moore and Vinay 2010).

Several psychoacoustic studies have failed to find strong support for the edge frequency theory (Pan et al. 2009; Ristovska et al. 2019; Sereda et al. 2011). Ristovska et al. (2019) reported a clear relationship between tinnitus pitch and hearing loss, however found no relationship between tinnitus pitch and the edge frequency of the audiogram. The tinnitus pitch corresponded to the edge frequency in only 16.5% of patients; comparatively, the tinnitus frequency corresponded to the frequency range of hearing loss and greatest region of hearing loss in 70.8% and 37.3% of cases, respectively. Pan et al. (2009) and Sereda et al. (2011) found a subset of participants in which the tinnitus pitch was associated with the audiometric edge. Sereda et al. (2011) reported that these participants exhibited a narrow tinnitus bandwidth. It is possible that a relationship between tinnitus pitch and the audiogram edge exists, but perhaps only in certain subgroups, possibly alluding to different underlying mechanisms involved (Pan et al. 2009).

A number of researchers believe that tinnitus pitch corresponds to a frequency range where the hearing loss was greatest, giving rise to the "region of maximal hearing loss" theory (Moore 2012; Sandlin and Olsson 1999). This theory is supported by the homeostatic plasticity hypothesis which posits that tinnitus is the result of homeostatic mechanisms acting to compensate for the reduced sensory input that occurs in hearing loss by reducing inhibitory and/or increasing facilitatory mechanisms (Schecklmann et al. 2012). Changes in neuronal activity take place in the frequency ranges where there is sensory deprivation, in turn leading to ongoing increased neuronal activity and/or synchrony in the affected central auditory pathways. This increase in central gain and resultant neuronal hyperactivity are thought to represent a neural correlate of tinnitus, with the frequency of tinnitus corresponding to the frequency of hearing loss (Norena 2011; Schaette and Kempter 2009; Schecklmann et al. 2012).

Patients with hearing loss generally face challenges when performing tinnitus pitch matches as their ability to hear and discriminate between the matching tones presented is limited to frequency regions of normal hearing (Ward and Baumann 2009). Considering the fact that the tinnitus frequency is often found in regions of hearing loss, it is important to accommodate for this by presenting the matching sound at a level that is safe but audible to the patient (Mitchell et al. 1984). It has also been noted that patients with significant hearing loss can experience difficulties in pitch matching due to the testing tone not having a clear pitch. This is often the case when the frequency of the sound used for pitch matching leads to maximum basilar membrane vibration in the region of the cochlea where the number of functioning inner hair cells and/or neurons is scarce or even nonexistent; this is known as the cochlear dead region (Moore 2014). Even with training, these patients are still limited in their ability to make appropriate pitch matches (Henry et al. 2001).

Although there is a clear link between tinnitus pitch and hearing loss, it does not explain the tinnitus experienced by individuals clear of any hearing difficulties. Recent findings have demonstrated that even individuals with a normal/healthy audiogram are likely to have some small degree of hearing injury ("hidden hearing loss") which might in turn give rise to the tinnitus (Plack et al. 2014).

2.3 Reliability of Pitch-Matching Measures

Pitch-matching measures are complicated by the fact that pitch matching reliability varies widely across patients. As such, it becomes difficult to discern whether the reliability of a measure has been compromised by the method itself or is simply a reflection of the heterogenous nature of the patient's tinnitus.

The reliability of pitch matching is also complicated due to the oversimplification of a patient's tinnitus through the use of single tones during psychoacoustic testing. Even in cases where tinnitus is described as "tonal," it is often comprised of a spectrum of frequencies, in turn intrinsically limiting pitch-match reliability (Hébert 2018). Hébert (2018) investigated individual test–retest reliability of the 2AFC and tinnitus likeness rating methods in 31 patients over a one-month period. The study reported a superior test–retest reliability for the tinnitus likeness rating relative to the 2AFC protocol, with at least one of three dominant tinnitus frequencies being reproducibly identified at the second session by the majority of participants (>80%), and two dominant frequencies being reproducibly identified by half of the patients. Only 13% of patients could reproducibly identify as many as three dominant tinnitus frequencies; this is similar to the proportion of patients in whom the final tinnitus frequency could be determined at the same ear using the 2AFC method (Hébert 2018). Though the tinnitus likeness rating protocol has been praised for its ability to offer a more complete view of an individual's tinnitus, it is not immune to problems and complications; in particular relating to its methodology. Unlike in the case of conventional methods where the complexity of tinnitus can be severely underestimated by use of single-tone pitch matches, tinnitus likeness ratings risk the tinnitus sensation being described as having a broad spectral pattern, when in fact it may be narrow (Norena et al. 2002). This could occur due to patient laxity or perhaps misunderstanding of the protocol. For instance, a broad spectral pattern would arise if a patient is not strict enough in their likeness-rating criteria, in turn simply rating the overall similarity between the presented tone and their tinnitus sensation rather than the degree to which the presented tone contributed to their overall tinnitus percept. Neff et al. (2019) found that the 2AFC and tinnitus likeness rating (as well as the adjustment method) all had good reliability, with participants being less satisfied with the 2AFC method, and the likeness rating protocol being more time-consuming.

A source of pitch-match inaccuracy and complication comes from octave confusion, where patients find it difficult to differentiate frequencies one octave apart from each other, considering them to be identical (Graham and Newby 1962; Kim et al. 2017). Though the effect is widely recognized and has been identified in a number of studies (Graham and Newby 1962; Kim et al. 2017; Ristovska et al. 2019), there are researchers who have failed to observe octave confusion among patients (Penner

1983; Tyler and Conrad-Armes 1983a) or have identified the effect in only a small subset of patients (Ristovska et al. 2019). Even so, testing for octave confusion has been recommended as an integral part of a standard tinnitus evaluation battery and should be performed to ensure patients are generating reliable matches true to their tinnitus pitch (Evered and Lawrenson 1981). The protocol involves presenting the matching sound one octave above and below the initial match and allowing the patient to determine which – if either – of the tones appears to be a better match to their tinnitus relative to the initial frequency chosen (Hazell and Wood 1981).

In addition to test reliability, the heterogeneous nature of tinnitus – and its perceived pitch – must also be considered. Conducting an "in-depth" tinnitus characterization study on 528 patients, Stouffer and Tyler (1990) noted approximately 36% of patients reporting regular fluctuation in pitch of their tinnitus varying from day to day (Stouffer and Tyler 1990). It was not uncommon for patients to report noticing a change in their tinnitus since its onset; while 76% reported they experienced no change, 19% of patients found their tinnitus pitch increased, and 5% observed a decrease in pitch Stouffer and Tyler (1990). The majority of patients (73%) reported that their tinnitus has always been constant, while those who indicated that their tinnitus changed to a completely different sound noted that this change either occurred suddenly (16%) or gradually (11%) (Stouffer and Tyler 1990).

The 2AFC method is widely used due to its easy-to-follow instructions and efficient time to complete; however, its pitch-matching reliability has been questioned (Hébert 2018; Neff et al. 2019). The FCDS procedure has been found to demonstrate a good degree of pitch-matching reliability as have the likeness tests; however, these methods are time-consuming; as such, they are rarely implemented in the clinical setting. Those conducting psychoacoustic evaluations should be mindful of the high level of pitch-matching variability across patients – whether it be due to daily fluctuations in tinnitus pitch, changes in pitch over time, or even difficulties in discerning an accurate and reflective pitch match – and consider the importance of obtaining several pitch-match replications (Tyler 2000).

3 Psychoacoustic Measures: Loudness

The two most common methods of defining the loudness of tinnitus are loudness matching and loudness rating. Although it has been suggested that loudness ratings are the more useful measure, loudness matching appears to be the more widely used technique, with loudness being determined by having the patient adjust an external pure tone stimuli so that it is equal in loudness to their tinnitus (Henry 2016; Moore 2014; Tyler 2000). Loudness rating is based on a more holistic approach, with ratings reflecting the impact of tinnitus rather than the perception itself (Henry 2016). Most of the measures available for the assessment of tinnitus pitch have been developed in such a way that they can be implemented for the determination of tinnitus loudness, including the adaptive, limits, and adjustment methods (Tyler and

Conrad-Armes 1983a), 2AFC method (Vernon and Fenwick 1984), and the FCDS protocol (Penner and Bilger 1992; Penner and Klafter 1992).

There appears to be a lack of correlation between loudness matching and loudness rating (Henry et al. 1999; Henry 2016). This is likely due to the two methods assessing slightly different aspects of loudness perception. Psychoacoustic matching is primarily based on the sensory judgment of tinnitus loudness, while loudness rating also depends on emotional and cognitive factors, perhaps being more reflective of what the subject is experiencing and feeling (Adamchic et al. 2012). Incorporation of cognitive and behavioral aspects by both the psychoacoustic match and ratings may be required to achieve effective evaluation of tinnitus loudness.

It is essential to validate a loudness measurement method that is accurate, reliable, and capable of detecting changes in tinnitus loudness. While tinnitus loudness matching typically demonstrates good reliability, with the great majority of loudness matches achieved between 0 and 20 dB SL (Goodwin and Johnson 1980; Graham and Newby 1962; Meikle and Taylor-Walsh 1984; Reed 1960; Ristovska et al. 2019; Roeser and Price 1980; Sandlin and Olsson 1999; Tyler and Conrad-Armes 1983b; Vernon 2000), it is not immune to the effects of interindividual variability (Burns 1984; Schechter and Henry 2002). Still, it must be noted that the reliability of loudness matching procedures is far superior to that observed in pitch matching (Penner 1983; Henry et al. 1999). Loudness ratings may be more easily influenced by factors such as annoyance or impact on quality of life than loudness matches (Henry 2016; Manning et al. 2019) but loudness matches are also not immune from psychological modifiers (Searchfield et al. 2012).

3.1 Tinnitus Loudness Matching and Choice of Units

There are a number of ways in which the magnitude of the matching tone can be specified, but there are questions as to their test–retest reliability (Hall et al. 2017). The simplest way of expressing tinnitus loudness is using intensity matches in either dB hearing level (HL) or dB sound pressure level (SPL); these reflect the dial values of equipment. The most common method is to use dB sensation level (SL), the difference between the loudness match and threshold to the same sound. The dB HL or dB SPL method is not independent of the listener's hearing threshold, so a person with hearing loss will have a higher dB HL match than a person with normal hearing, even if the perceived loudness were the same. The test–retest reliability of the SL measure may be less than SPL and HL methods, as it is the difference between the loudness match (dB HL on an audiometer) and auditory threshold (dB HL) meaning two measurements are required, increasing the chance of error. None of the intensity matches, whether it be dB HL, dB SPL, or dB SL, may truly reflect tinnitus loudness (Stevens and Davis 1938). Individuals can report different loudness to the same physical intensity of sounds, and loudness perception is influenced by context, memory, and personality (Searchfield et al. 2012).

In addition to the measure not representing loudness, Tyler and Conrad-Armes (1983b) found that the dB SL of a tinnitus loudness-matched tone depended on the pure-tone frequency used during matching. More specifically, tinnitus loudness matches are generally greater in frequency regions of normal hearing sensitivity, whereas loudness matches in frequency regions where hearing thresholds are elevated are often matched to only a few decibels above threshold (generally between 0 and 20 dB SL) (Goodwin and Johnson 1980; Graham and Newby 1962; Meikle and Taylor-Walsh 1984; Penner 1986; Reed 1960; Ristovska et al. 2019; Roeser and Price 1980; Sandlin and Olsson 1999; Tyler and Conrad-Armes 1983b; Vernon 2000). This may be explained by loudness recruitment (Tyler and Conrad-Armes 1983b). In an attempt to resolve this issue and offer a means of better understanding the loudness of tinnitus, Tyler and Conrad-Armes (1983b) converted dB SL measures into tinnitus loudness in sones, a conventional psychoacoustic unit of loudness. One sone is the loudness of a 1,000 Hz tone with a level of 40 dB SPL (Moore 2014). Tyler and Conrad-Armes (1983b) believed that there were several advantages to using the sone as a measure of tinnitus loudness: (1) it has diagnostic significance and can help identify those who complain of a very loud tinnitus but match their loudness to a soft tone, (2) presentation at the same sone for another listener (e.g. in demonstrating the percept to family and supplement counseling) should be the equivalent loudness, (3) it allows for comparisons to be made across patients, (4) it can be used as a quantitative measure to monitor changes in tinnitus with treatment, and (5) it offers a more meaningful psychoacoustic measure of the discomfort and annoyance that results from tinnitus (Tyler and Conrad-Armes 1983b).

Although Tyler and Conrad-Armes (1983b) suggested that using the sone was a more appropriate way of measuring tinnitus loudness, using the measure in the clinical setting presents two challenges. First, the sone scale is unfamiliar to many clinicians and as such makes it difficult for them to conceptualize the result (Matsuhira et al. 1992). This in turn presents a challenge with respect to not only validating and making sense of a patient's tinnitus percept, but providing the patient (and their family) with effective counseling. Secondly, the sone is based on loudness functions that represent complete loudness recruitment (see *Tinnitus loudness, recruitment, and hyperacusis* section); as such, the measure tends to result in the overestimation of tinnitus loudness as it assumes complete recruitment, which is not always the case in tinnitus patients (Matsuhira et al. 1992). In addition, loudness growth formulas are likely too general for application to specific individuals, in turn suggesting individualized functions need to be established at each loudness matching frequency (Henry and Meikle 2000).

Highlighting a number of these limitations, Matsuhira et al. (1992) proposed a method to account and correct for the effect of recruitment in tinnitus loudness matches using information obtained from standard clinical evaluation. The investigators devised an "average loudness function" which converted measures in dB SL into an estimate of the effective loudness level and corrected for abnormally rapid loudness growth by adjusting the mean loudness function for each participant using data generated by the individual (Henry and Meikle 2000; Matsuhira et al. 1992).

The measure was essentially the same as the phon scale, an alternative measure of loudness level, except for the difference in reference level (Matsuhira et al. 1992). However, the results of the study (Matsuhira et al. 1992) were highly variable, probably due to the large inter-subject variability in loudness recruitment between participants, even with the same level of hearing loss. The method has not been widely adopted.

Another method for quantifying tinnitus loudness is to use the personal loudness unit (PLU) developed by Hinchcliffe and Chambers (1983). They proposed calculating individualized loudness functions for each tinnitus patient and specifying the loudness match in terms of this loudness function. Much like the sone and phon measures, the PLU uses a loudness function at 1,000 Hz as the reference level; however, instead of using dB SL as unity, it employs the "most comfortable loudness level." Though the authors promoted the use of this method in the clinic, it has not been widely used likely due to clinical time constraints and difficulty in comprehension (Henry and Meikle 2000).

In summary, the measurement of dB SL has become the de facto unit for tinnitus loudness matching. It does have flaws, but its limitations should be considered in light of tinnitus being a complex concept. The need for precision in the measurement of loudness should also be balanced against time and benefits. At present: a precise loudness match is unnecessary for demonstrating the experience of tinnitus to a third party (i.e. family member/partner); loudness is not diagnostic or prognostic, and no treatment currently requires the measurement to be effective. As new treatments are developed precision of loudness match may become more important.

3.2 *Tinnitus Loudness, Recruitment, and Hyperacusis*

Tinnitus loudness tends to be matched to a relatively low intensity tone, often only a few decibels above threshold. The majority of matches have been reported to be within the 0–20 dB SL range (Goodwin and Johnson 1980; Graham and Newby 1962; Meikle and Taylor-Walsh 1984; Reed 1960; Ristovska et al. 2019; Roeser and Price 1980; Sandlin and Olsson 1999; Tyler and Conrad-Armes 1983a, b; Vernon 2000). Meikle and Taylor-Walsh (1984) evaluated the tinnitus percept of over 1,800 patients, reporting extensive fluctuations in loudness in 24% of cases (17% reported fluctuations from time to time, 55% reported a constant tinnitus, and 4% were not able to answer the question). Similar analyses were conducted by Stouffer and Tyler (1990) in 528 participants; loudness fluctuated in 56% of the tinnitus patients, changing either suddenly (25%) or gradually (31%). Further, half of the patients reported that the loudness of their tinnitus varied daily, while the remaining half did not notice daily fluctuation. Changes in tinnitus loudness since onset were also observed, with 33% of patients noticing an increase in loudness and 7% experiencing a decrease (there was no change for 60%).

Fowler, a pioneer in loudness measurement, was the first to note a paradoxical relationship between loudness matches as measured using psychoacoustic methods and subjective patient report (Fowler 1944). He found that although tinnitus loudness was matched only a few dB above the threshold of hearing, the subjectively perceived loudness of tinnitus was often been described by patients as being intolerable (Fowler 1944). Fowler (1944) suggested clinicians use the loudness matched tinnitus as a "factual foundation" for counseling patients, demonstrating that their perceived tinnitus is in fact a very soft sound. Fowler did not consider loudness recruitment, which has since been identified as a significant contributor to low-level loudness matches (Goodwin and Johnson 1980; Henry et al. 1999; Tyler and Conrad-Armes 1983b). Loudness recruitment is a phenomenon associated with hearing loss, in which there is disproportionately rapid loudness growth following increases in sound intensity (Goodwin and Johnson 1980; Penner 1986; Raj-Koziak et al. 2018; Tyler and Baker 1983). In turn what is found is that the growth of loudness for an external tone is more rapid when matches are made at the tinnitus frequency versus at frequencies outside of the tinnitus region (Eggermont and Roberts 2004; Mitchell et al. 1993; Penner 1986). As such, it comes as no surprise that large differences have been reported between loudness matches obtained at the tinnitus frequency, and at frequencies very different from the tinnitus pitch (Goodwin and Johnson 1980; Mitchell et al. 1993; Tyler and Conrad-Armes 1983b). Specifically, recruitment results in tinnitus being matched to a tone at a lower sensation level in regions affected by hearing loss (often the tinnitus frequency) than in regions of normal hearing (Penner 1986; Vernon and Fenwick 1984). In an attempt to avoid recruitment and better represent true tinnitus loudness, Vernon and Fenwick (1984) suggested routinely conducting loudness matches both at the tinnitus frequency and at a second frequency distinct from tinnitus in the normal hearing portion of the patient's audiogram. However, a number of researchers have noted that even when this method is applied, mean dB SL values are still too low to correspond to patient complaints (Henry and Meikle 1996; Jakes et al. 1986). A study by Hulshof (1986) considered the effects of recruitment by evaluating and comparing tinnitus loudness in those who are affected by the phenomenon versus those who are not (at least in one ear). The results demonstrated that although loudness recruitment has an effect on the measurement of tinnitus loudness, its effects are very small. Similarly, Henry and Meikle (1996) conducted measures of loudness growth at both reference and tinnitus frequencies, finding that the recruitment phenomenon is only responsible for 25% of the variability in loudness matching. The source of the remaining 75% of that variability remains unsolved.

Tinnitus is also associated with hyperacusis, a reduced tolerance to everyday sounds that cause significant discomfort, distress, and even pain (Baguley 2003; Moore 2014). It has been reported that approximately 85% of those with hyperacusis also suffer from tinnitus (Anari et al. 1999; Sheldrake et al. 2015). Often in cases where tinnitus is accompanied by hyperacusis, loudness discomfort measures are performed as part of the test battery. As in the case of psychoacoustic tinnitus measures, these testing protocols for hyperacusis evaluation vary and are not standardized (Goldstein and Shulman 1996). Loudness discomfort levels (LDL) are most frequently used to assess hyperacusis, defining the intensity level at which a sound is reported as being uncomfortable. Patients with hyperacusis will have lower LDLs than normal due to increased sensitivity to sound (Pienkowski et al. 2014). Generally, LDLs of 95 dB or greater are considered normal, whereas LDLs between 80 and 90 dB reflect mild hyperacusis, those between 65 and 75 dB imply moderate hyperacusis, and below 60 dB signify severe hyperacusis (Goldstein and Shulman 1996). More work is required to establish universally agreed upon frequencies for assessment and number of repetitions per judgment, as well as to determine norms for the range of LDLs for those with specific degrees and types of hearing loss (Goldstein and Shulman 1996; Pienkowski et al. 2014).

3.3 Tinnitus Loudness: Annoyance and Severity

A frequent complaint by tinnitus patients is the annoyance and distress experienced as a result of the perceived loudness of the tinnitus (Hallam et al. 1988). Several researchers have failed to identify a significant relationship between tinnitus loudness and annoyance, proposing that loudness is not a significant contributor to the perceived distress caused by tinnitus (Rosito et al. 2013; Sandlin and Olsson 1999). Andersson (2003) noted the lack of a relationship between loudness in dB SL and tinnitus annoyance, but reported a correlation when the loudness was expressed in dB HL, proposing that the degree of hearing loss was an important factor to consider when evaluating the impact of loudness on the perceived tinnitus distress. Meikle and Taylor-Walsh (1984) also reported that tinnitus severity was not correlated with loudness in dB SL, instead finding that loudness judgments are influenced by emotional factors. Loudness ratings, which reflect tinnitus impact (reactions) rather than actual loudness (percept), are found to significantly correlate with the severity of annoyance and distress reported by the patient (Henry 2016; Hiller and Goebel 2006; Schechter and Henry 2002; Stouffer and Tyler 1990; Tyler 2000; Ward and Baumann 2009). Factors other than loudness determine the perceived annoyance and in turn severity of tinnitus, including the duration since tinnitus onset (habituation factors) and the psychological state of the patient (Tyler 2000). Although tinnitus perceived at a greater loudness is generally more likely to be annoying, it does not necessarily mean that a softer tinnitus is any less severe of an issue for certain patients; in other words, it is often the case that the perceived intensity of tinnitus does not dictate how a patient reacts to their tinnitus and in turn how distressing they find it (Folmer et al. 1999; Tyler 2000). Ward and Baumann (2009) aimed to clarify the relationship between loudness and perceived distress using annoyance caused by aircraft noise near airports as an example. What is apparent is that although the relationship between aggregated loudness of flyovers and community annoyance is generally stable, there still remains a large amount of variability after the day-night noise level is accounted for (Ward and Baumann 2009). A subset of this variability is the result of differences in annoyance thresholds, which are also influenced by a number of variables such as fear of crashes and political interactions with airports; however, there are other unknown factors that are likely community-specific, as well as hypersensitive people who get annoyed by noises that are negligible to most people (Ward and Baumann 2009). As such, it appears that though loudness might influence tinnitus annoyance and distress, it is likely to be one of many factors defining overall tinnitus severity. For example, a low-level tinnitus might not be bothersome for one individual, but for another presenting with hyperacusis it might be highly disruptive (Hiller and Goebel 2006). The incorporation of cognitive and behavioral aspects (such as memory, attention, context, and personality) may provide a more meaningful understanding of a patient's tinnitus (Hiller and Goebel 2007; Searchfield et al. 2012; Welch and Dawes 2008).

3.4 Tinnitus Loudness, Magnitude, and the Adaptation Level Theory

Some of the variability in loudness matching could potentially be ascribed to auditory context, attention, and individual psychology (e.g. personality, memories, emotional state) (Searchfield et al. 2012). The adaptation level theory (ALT) of tinnitus (Searchfield et al. 2012) is founded in a psychoacoustical model proposing that stimuli do not act as singular entities, but instead interact with and influence each other (Helson 1964). In the context of tinnitus, it is proposed that the perceived tinnitus intensity is governed by several factors, not the least of which is the personality of the patient (Searchfield et al. 2012). The attitudes, ideals, experiences, learning, interpersonal relations, and intellectual and emotional behavior of an individual shape an individual's "frame of reference," which in turn dictates their response to stimuli. This is supported by findings suggesting a patient's response on a visual analogue scale (VAS) is correlated with the extent to which the individual is impacted by their tinnitus (Zenner et al. 2005).

Helson (1964) defined the adaptation level as the weighted product of focal, background, and residual stimuli, using a simple mathematical equation to demonstrate the ALT:

$$A = \overline{X}^p B^q R^r$$

where tinnitus audibility in the environment is the combined result of tinnitus magnitude ($\overline{X}$), background sound (e.g. sound therapy, B), and residual factors (R) such as personality, as influenced by weighting factors related to attention and auditory scene analysis (ASA). By adopting this theory, Searchfield et al. (2012) hypothesize that the variability in loudness matching and individual patient's overall response to tinnitus may be attributable to their internal reference for tinnitus loudness and be determined by interactions among many affecting factors.

A straightforward example of the ALT is its use in the context of chronic pain. Patients who suffer from chronic pain (Boureau et al. 1991; Rollman 1979) or have been severely injured in the past (Dar et al. 1995) have higher unpleasantness thresholds and find experimental pain less intense and more bearable than pain-free individuals. According to the ALT, chronic pain (as well as severe acute pain) can alter the internal anchor points for the subjective evaluation of pain; this in turn results in patients having a different adaptation level than a normal subject, which is demonstrated by the observed increase in pain threshold. Searchfield et al. (2012) considered the plausibility of this scenario in the context of tinnitus, using it to potentially explain the loudness match discrepancy. According to Searchfield et al. (2012), the experimental condition itself, as well as the introduction of a comparison sound (such as that used in loudness matching), can easily bias the adaptation level and in turn give rise to variability. Specifically, when a patient is asked to rate or describe the loudness of their tinnitus, they are often comparing their tinnitus to the quiet environment of the consultation room or research facility. However, when that same patient is instructed to perform tinnitus loudness matches using an external matching stimulus, they are no longer comparing loudness to the absence of sound, but rather to a new adaptation level which includes the test sound and the existing perceived tinnitus magnitude of tinnitus (Searchfield et al. 2012). In turn, it is the interaction between the stimulus (loudness matching sound) and the tinnitus itself which governs the overall magnitude of the tinnitus percept. Hence, adding the matching sound changes the internal anchor point for the subjective evaluation of tinnitus magnitude and results in tinnitus being matched to a tone at a level that is lower than expected (Searchfield et al. 2012). In addition to this interaction with the external matching stimulus, subjective loudness estimates vary from patient to patient as they are likely to have different concepts of tinnitus loudness relative to their own adaptation level (Mitchell et al. 1984).

It is possible that ALT could offer a more holistic approach to understanding tinnitus. For example, it could:

- Help determine the relative contributions of psychoacoustic, emotional, and cognitive aspects to tinnitus.
- Clarify why patients perceive their tinnitus in a certain way – what factors actively contribute to the intolerability of their tinnitus, and how these factors might be influenced in an attempt to "shift" the response to tinnitus.
- Identify attributes or elements that are potential risk factors for perceiving tinnitus in a negative and distressing manner, which could aid in grouping of tinnitus patients.
- Help form predictions for the success of tinnitus treatment for a given patient.
- Help identify the optimal route of treatment for a given patient, in turn not only forming a more rounded treatment approach, but also one tailored at the individual level.

4 Confusion of Pitch and Loudness: The Circular Problem

One of the most common issues that surfaces when performing psychoacoustic tinnitus matching is the confusion between pitch and loudness. Patients often find it difficult to conduct pitch matches if the matching sound differs in loudness from the tinnitus sound to which it is being matched; the same applies in the case of loudness matching to a sound distinct in pitch from the tinnitus (Mitchell et al. 1984; Moore 2014; Vernon and Fenwick 1984). Consequentially, this leads to patients deciding against a tentative tinnitus frequency match on the basis of loudness differences, or confirming an inappropriate pitch match simply because the loudness was comparable to the tinnitus (Vernon and Meikle 1981). This leads to the circular problem; in order to accurately measure pitch, the stimulus tone should be presented at the loudness of tinnitus, but in order to obtain an accurate measure of loudness, the tone should be presented at tinnitus pitch (Fowler 1940; Vernon and Meikle 1981). Further research disentangling pitch from loudness is needed.

5 The Minimum Masking Level

Tinnitus loudness and changes to it are often determined by the presence or absence of external auditory stimuli; for instance, tinnitus can be rather audible in quiet conditions, but less obvious in a noisy environment (Fowler 1944). Masking involves using an external sound to reduce or even fully conceal the tinnitus and is performed with the aim of determining how the addition of external stimuli might affect tinnitus and its perception. Often considered as the most critical aspect of psychoacoustic testing, masking can aid clinicians in deciding whether or not a patient will be a good candidate for sound therapy (Switalski and Sanchez 2019). It has been suggested that the lower the minimum level of broadband noise required to completely conceal a patient's tinnitus (known as the minimum masking level, MML), the more likely they are to benefit from masking therapy (Henry 2016).

Tinnitus is maskable in the large majority of cases (Roeser and Price 1980; Sandlin and Olsson 1999; Vernon and Meikle 2003), but tinnitus masking does not behave the same way as sound-on-sound masking does (Mitchell 1983; Mitchell et al. 1993; Searchfield et al. 2016; Tyler and Conrad-Armes 1984). There is a great degree of individual variability with respect to the frequency and sensation level (SL) required to mask tinnitus (Feldmann 1971; Tyler and Conrad-Armes 1984). A broadband sound cannot be masked by a pure tone, but a broadband tinnitus can be masked by a pure tone. Neither the pitch nor loudness of the tinnitus percept correlates with its maskability (Mitchell 1983). These findings have led to the realization that the neural processes underpinning tinnitus masking differ greatly from those responsible for the masking of external auditory stimuli (Vernon 2000) (see Table 3).

Table 3 Differences between tinnitus and sound-on-sound conventional masking (Mitchell 1983; Switalski and Sanchez 2019; Tyler and Conrad-Armes 1984; Vernon 2000)

	Sound-on-sound conventional masking	Tinnitus masking
Frequency	• It is difficult for a pure tone to mask a band of noise • There is an orderly frequency relationship: sounds that are higher in frequency than the masking sound are easier to mask than those below it	• Noise-like tinnitus can be masked by pure tones • Individual variability is high in terms of the frequency and intensity of sound required for effective masking of tinnitus
Critical band	• There is a "critical band" of frequencies surrounding the sound that is to be masked; frequencies within this band are effective maskers, while those outside are not	• There is no "critical band" for tinnitus masking – any frequency could be effective for a given patient
Beats	• In monoaural conventional masking, it is easy to generate the sensation of "beats" when two sounds of identical loudness are similar, but not exact, in regard to frequency and phase	• In the case of pure tone tinnitus it is very rare to generate "beats"
Sound presentation (ear)	• Contralateral masking has limited effect	• Strong contralateral masking is possible
Level of sound being masked	• Upon masking cessation, the sound being masked is perceived as being at its original (pre-masking) level unless auditory fatigue is experienced	• Upon masking cessation, the patient will often experience residual inhibition (a temporary reduction or even absence of their tinnitus sensation)

Tuning curves are an additional source of evidence that tinnitus differs from an external sound. While psychophysical tuning curves (PTC) depict the level of narrowband masker required to mask an acoustic narrowband signal of a fixed level (e.g. sinewave or narrow band of noise), tinnitus tuning curves (TTC) reveal the masker level that is needed to mask tinnitus (Moore 2014). For normal-hearing individuals, the PTC typically exhibits a V-shaped pattern, revealing a sharp tip at the signal frequency in normally hearing individuals; specifically, the closer the frequency of the masker is to the target, the lower the level required to mask the target (Fournier et al. 2019). This pattern is thought to arise from cochlear mechanisms; if tinnitus is processed in the same way as an external sound, a similar pattern would be observed. However, little similarity has been found between PTCs and TTCs; often the distinctive V-shape can only be observed for PTC, indicating tinnitus is unlikely to have a cochlear origin (Burns 1984; Penner 1987; Tyler and Conrad-Armes 1984). However, a recent study by Fournier et al. (2019) examining the shapes of PTCs and TTCs in 32 tinnitus patients found that 30% of cases demonstrated a V-shaped pattern for both PTC and TTC, suggesting that there is perhaps a subset of patients for whom tinnitus-related activity may share similar processing pathways with external sounds. The authors proposed that this might have implications for tinnitus research and treatment in terms of subtyping tinnitus and acoustic therapies (in particular those based on tinnitus frequency) (Fournier et al. 2019). Furthermore, the discrepancy between the conventional V-shape and the patterns observed in some TTC could simply reflect the fact that tinnitus, for some, is not narrowband in nature, but represents a broadband complex sound (Norena et al. 2002).

Fowler (1940) noted subgroups with respect to TTC profiles, claiming that his participants fell into one of the three groups: (1) tinnitus could be masked by tones at low SLs irrespective of its frequency, (2) high masker SPLs were required for all frequencies, and (3) tinnitus could not be masked at all. Feldmann (1971) replicated and extended these findings, exploring the effects that tones of selected frequencies had on the perception of tinnitus. Pure tones, narrow-band noise, and white noise were presented to 200 patients both ipsilaterally and contralaterally, plotting the intensities of the tones and noises just sufficient to mask the tinnitus so as to generate masking curves (Feldmann 1971). Feldmann identified five different tinnitus profiles based on masking pattern: convergent, divergent, congruent, distant, and persistent (Table 4). Results similar to that of Feldmann (1971) have been reported by others (Mitchell 1983; Penner 1987; Tyler and Conrad-Armes 1984). The heterogeneous nature of tinnitus, in particular with respect to its response to masking, suggests that tinnitus does not arise in the cochlea; it requires central involvement and higher-order processing (Tyler 2000).

6 Auditory Residual Inhibition

Auditory Residual Inhibition (ARI) reflects the temporary suppression or complete elimination of the tinnitus sensation that takes place following auditory stimulation (Henry 2016). Josephson (1931) was one of the first to describe what we now know to be ARI. The effect was initially named residual inhibition (RI) by Vernon and Schleuning (1978) in recognition of Feldmann's reporting that tinnitus "remains silent for a certain period of time after cessation of the inhibitory stimulus" (Feldmann 1971). To avoid confusion between acoustical RI and that observed in non-acoustical contexts, such as in neural transcranial magnetic stimulation (Vanneste and De Ridder 2012; van Zwieten et al. 2016), "auditory" is recommended to be routinely added to residual inhibition (auditory residual inhibition, ARI).

The clinical test for ARI is usually the presentation of broadband noise (2–12 kHz) binaurally to patients at 10 dB above their minimum masking level (MML). Exposure to the noise lasts 60 s before abrupt termination, at which time the participant is asked to report any perceived changes to their tinnitus percept (Henry 2016). In cases where tinnitus is suppressed, the patient is asked to describe any changes to the tinnitus sensation as it recovers and resumes its initial level. Changes to the tinnitus percept, as well as duration of the ARI, are noted and classed according to four categories (Henry 2016; Switalski and Sanchez 2019):

Table 4 Feldmann's tinnitus masking profiles and their prevalence among tinnitus patients as reported in several studies (Feldmann 1971 (n = 200), Mitchell 1983 (n = 32), and Tyler and Conrad-Armes 1984 (n = 10))

Masking pattern		Prevalence
Type I: Convergence	Common. Found in patients with high-pitch tinnitus and high-frequency hearing loss. Threshold and masking curves converge from low to high frequencies, meeting at the frequency corresponding to the tinnitus pitch and coinciding for higher frequencies. Occurs mostly in industrial deafness and sensorineural hearing loss associated with high-pitched tinnitus. Most like the masking of true sounds	• Feldmann (1971): 34% of patients • Mitchell (1983): 53% of patients • Tyler and Conrad-Armes (1984): 80% of patients
Type II: Divergence	Very rare. Defined by threshold and masking curve diverging from low to high frequencies. No clear pathologies associated	• Feldmann (1971): 3% of patients
Type III: Congruence	Common. Threshold and masking curve coincide within an intensity range of maximally 10 dB; any tone or narrowband noise raised at a level just above threshold will mask the tinnitus. Found in cases with a flat threshold curve, particularly in Meniere's disease, sudden deafness, and otosclerosis. Tinnitus may be tonal or noise-like	• Feldmann (1971): 32% of patients • Mitchell (1983): 19% of patients • Tyler and Conrad-Armes (1984): 10% of patients
Type IV: Distance	Relatively common. Masking sound has to be considerably louder than threshold in order to mask the tinnitus. Threshold curve and masking curve are therefore distant from each other. Present in cases of various pathologies of the middle and inner ear	• Feldmann (1971): 20% of patients • Mitchell (1983): 22% of patients • Tyler and Conrad-Armes (1984): 10% of patients
Type IVa: Dispersion	In type I–III the intensities required for masking with pure tones and narrowband noises are generally equal. In type IV however there can be differences in that higher intensities of pure tones are needed than for narrowband noises. This gives rise to type IVa	–
Type V: Persistence	Tinnitus cannot be masked irrespective of the stimulus. Often occurs in patients with severe sensorineural hearing loss or complete deafness	• Feldmann (1971): 11% of patients • Mitchell (1983): 6% of patients

1. Positive-complete: Tinnitus is entirely absent, ARI may vary from 1 s to several hours; ARI has been reported to last <2 min in 60% of patients, and <4 min in 80% of patients (Meikle et al. 2004).
2. Positive-partial: Tinnitus is still present, but less audible than prior to the testing procedure. Changes in the quality of the tinnitus might be reported.
3. Negative: No change in tinnitus loudness.

4. Rebound or Exacerbation: Increase in tinnitus loudness level in response to masker presentation (in these cases, the time taken for the tinnitus to return to its initial "original" level is recorded).

ARI has been reported in around 70% of tinnitus patients (Ristovska et al. 2019; Roberts et al. 2008), with some claiming an even higher prevalence of nearly 90% (Vernon 2000). However, results are variable. A study by Roeser and Price (1980) evaluated the efficacy of tinnitus masking, reporting either partial or complete ARI in around 64% of their sample, no effect in 23%, and exacerbation in around 5% (masking was ineffective in 8% of the sample). Mitchell et al. (1984) also examined changes in tinnitus following masking, observing ARI in only 42% of participants. Of those who did not experience ARI, 26% noted that the masking sound alleviated their tinnitus (while it was being presented), while the rest felt the sound was simply "one more noise on top of the tinnitus," in some cases even aggravating the tinnitus. Though the prevalence of ARI is somewhat variable, it appears that those who experience ARI tend to do so consistently (Henry et al. 2013).

The duration of the effect can last anywhere between 1 s and several hours, in some cases even days (Sandlin and Olsson 1999; Switalski and Sanchez 2019). The magnitude and duration of ARI are dictated by several factors, including the intensity, duration, and frequency of the masker (Terry et al. 1983; Vernon and Meikle 1981). High-intensity and long duration maskers have been found to produce longer relief from tinnitus through extending post-masking effects (ARI) (Tyler 2000). Terry et al. (1983) maintained that the greater the masker intensity, the greater the period of ARI. Evaluating the relationship between masker composition (frequency, bandwidth, intensity, and duration) and the magnitude and duration of ARI, the investigators found that ARI was proportional to the masker intensity given the tinnitus was fully masked (partial masking will result in little or no ARI). In addition, the time course of ARI demonstrates a linear increase as a function of the logarithm of masker duration for durations between 10 s and 10 min (Terry et al. 1983). The duration of the effect, measured as time taken to achieve complete recovery of tinnitus, increased to around 100 s for maskers presented for a period of 100 s, but only increased to 200 s for maskers presented at tenfold greater durations (Terry et al. 1983). Tyler and Conrad-Armes (1984) evaluated the perception of tinnitus following termination of a masker in 10 participants with sensorineural tinnitus, noting several different responses: low-level and short-duration maskers generally resulted in the tinnitus being heard immediately following masker termination, while higher-level and higher-duration masker presentation (and subsequent termination) led to (1) a silent period followed by an abrupt return to the pre-masking level, (2) a silent period followed by a more gradual return, (3) an increase in tinnitus loudness, (4) a reduction in tinnitus loudness, (5) a "wobbling" of the tinnitus, and (6) no ARI effect.

In addition to the well-known influence of masker intensity and duration on ARI, a number of studies have proposed that the post-masking effect is also, to some degree, frequency-dependent (Fournier et al. 2018; Sockalingam et al. 2007). Terry et al. (1983) found that ARI is in general maximal when the frequency of the masker is lower than the tinnitus frequency. Sockalingam et al. (2007) and Fournier et al. (2018) proposed that the closer the masker frequency is to the patient's tinnitus, the greater the ARI. Sockalingam et al. (2007) also noted that while the duration of ARI increased with increasing duration of frequency-matched stimuli, no such correlation was identified for non-frequency-matched stimuli.

6.1 *Cautions and Application*

There is limited evidence to suggest that temporary threshold shift (TTS) is produced during ARI (Terry et al. 1983; Vernon and Fenwick 1984); however, use of the test with persons who experience reduced sound tolerance is not recommended. ARI has some, limited, potential prognostic value in terms of masker-based therapies, determining the likelihood that a patient will benefit from sound therapy (Vernon 1977). In addition, it gives many individuals a feeling of renewed hope that their tinnitus can be managed and is not intractable. The demonstration that tinnitus can be altered – even if only for a brief moment – can be particularly rewarding if the patient is someone who has suffered from constant, unremitting tinnitus for a long period of time (Vernon 2000). Further, it shows patients that tinnitus relief might be possible by using external sounds to modify the percept.

7 The Future of Psychoacoustic Measures: Methods and Application to Therapy

7.1 *Methods*

Tinnitus evaluation is currently most commonly conducted using a standard audiometer primarily due to the availability, simplicity, and familiarity of the equipment. However, there are well-known limitations associated with assessing tinnitus using a conventional pure-tone audiometer (McFadden 1982; Vernon and Fenwick 1984). Audiometers are restricted in the range of frequencies they present, allowing only gross estimates of tinnitus to be obtained (Kostek and Poremski 2013). Frequencies above 8,000 Hz are often not available, resulting in those with tinnitus frequencies above 8,000 Hz not being appropriately and accurately pitch matched (McFadden 1982; Schechter and Henry 2002). Additionally, the typical 5 dB intensity level increments of audiometers may be too large to enable precise loudness matching of tinnitus (Kostek and Poremski 2013). While some audiometers allow for smaller increments (1 to 2 dB) to be used, this extends test duration (Kostek and Poremski 2013). The flexibility of digital platforms (software, apps) should free clinicians from the limitations of pure-tone audiometers, but with new innovative methods comes the potential for even less standardization. Kostek and Poremski (2013) evaluated the use of a multimedia-based synthesizer for measuring the psychoacoustical properties of tinnitus, noting its superiority over conventional audiometer use. The tinnitus synthesizer has many benefits above the use of an audiometer: (1) it obtains results more quickly, (2) it has a greater capacity for allocating the acoustic parameters of sound (in turn representing it more accurately), and (3) the participant does not have to be in close cooperation with the examiner or verbally describe the perceived listening experience, which is often challenging for a number of people.

The current pitch and loudness match of tinnitus offer a limited "cartoon-like" representation of the sensation. The use of pure tones as comparison stimuli has limitations, as the majority of individuals describe a broader more complex tinnitus spectra (Table 1). Patients and research participants are instructed to match to the prominent pitch of their tinnitus; however, this is often difficult to do (Henry et al. 2013; Moore 2014). Though the use of pure tones as comparison stimuli is not ideal, there are claims that it is still a reasonable methodological choice, having a number of advantages over the use of complex tones or noise bands (Norena et al. 2002). The pitch of pure tones is defined almost exclusively by their frequency; conversely, the perceptual attributes of complex tones and noise bands are defined by various physical parameters, including their fundamental frequency, center frequency, bandwidth, and spectral shape (which is in turn governed by the amplitude of their frequency constituents) (Norena et al. 2002).

Tinnitus likeness measures appear a compromise as they use tonal stimuli, but across a wide spectrum (Norena et al. 2002; Roberts et al. 2006). TLR have demonstrated a greater degree of reliability relative to several alternative methods (Kay and Searchfield 2008; Hébert 2018) and have tentatively been shown to improve the validity of pitch matching (Roberts et al. 2006). However, these methods still only offer a simplified representation of the global tinnitus experience, focusing solely on frequency and intensity components; two aspects that contribute to, but do not wholly define the sensation. They are also time-consuming. However, the use of complex and noise stimuli for tinnitus matching need not be a difficult and lengthy procedure. Kostek and Poremski (2013) recommended that for a patient reporting noise-like tinnitus a narrowband noise be presented with a center frequency equal to tonal pitch match. If the patient feels that the pure tone more closely resembles their tinnitus than the noise, this tone is the final match and no further testing needs to be conducted. If the patient feels the noise was a closer match to their tinnitus, the most appropriate form of noise should be determined by comparing broadband noise (speech noise or white noise) with narrowband noise.

More complex approaches such as tinnitus likeness ratings (TLR) are a step towards a more "complete" representation of the tinnitus sensation. Future methods of assessment may require more accurate and/or "realistic" replicas of tinnitus (Searchfield 2014). There appear to be two approaches to creating realistic copies of tinnitus. One is to start and build on basic building blocks of sound (frequency and intensity), the other is to start with complex sounds (real world) and modify them to match tinnitus (Kay and Searchfield 2008). In order to achieve appropriate tinnitus avatars (complex replicas of the tinnitus experience), future matches may need to incorporate several different sounds (Drexler et al. 2016), using real-world or complex sounds (Kay and Searchfield 2008), and defining tinnitus in 3-dimensional space (Searchfield et al. 2015).

7.2 Application to Treatment

Tinnitus heterogeneity may be responsible for the variable treatment responses seen among tinnitus patients (Cederroth et al. 2019; Simoes et al. 2019). Until recently, treatments have been largely independent of psychoacoustic measures. However, the increase in management strategies requiring accurate pitch and loudness matches including desynchronization with patterned tones (Reavis et al. 2010), tonotopic reorganization using sound and vagus nerve stimulation (De Ridder et al. 2015), active discrimination (Roberts and Bosnyak 2011; Wise et al. 2016), and categorization training tasks (Jepsen et al. 2010) has been a driving force for the need for accurate measurements. A treatment based on sound presentation in and around tinnitus pitch is likely to be compromised if the treatment sound is inaccurately prescribed.

Therapies currently attempting to personalize tinnitus therapy often only focus on one aspect of the percept such as pitch, loudness, sound preference, or maskability rather than considering the percept as the complex combination of these factors (Searchfield et al. 2017).

8 Summary

Psychoacoustic measures are crucial for characterizing and evaluating the perceptual properties of tinnitus. However, common approaches to the psychoacoustic matching of tinnitus are faced with a number of limitations, including:

- Lacking standardized protocols for psychoacoustic measures.
- Pitch-matching issues: equipment-based limitations, octave confusion, effects of cochlear dead regions, and tinnitus complexity.
- Loudness-matching issues: ample choice of loudness units of questionable relevance and appropriateness, effects of loudness recruitment.
- Masking based issues: variability and unpredictability of tinnitus behavior in response to an external sound.
- Residual inhibition-based issues: unpredictability in terms of whether or not a patient will demonstrate residual inhibition.

Despite these limitations, the current measurements are likely to be sufficiently accurate for counseling and as adjunct measures to questionnaires in research.

Throughout this chapter we have discussed and critiqued current psychoacoustic methods, offering suggestions for their improvement and a view of what successful tinnitus evaluation might look like and encompass. It is clear that a broader comprehension of tinnitus is required, taking into consideration not only its acoustic parameters and underlying pathophysiology, but also factors such as patient personality and activity within the neural networks (auditory, attention, memory, and emotion centers) affected by tinnitus. The heterogeneous nature of tinnitus should be taken into consideration at every stage of tinnitus assessment.

References

Adamchic I, Langguth B, Hauptmann C, Tass PA (2012) Psychometric evaluation of visual analog scale for the assessment of chronic tinnitus. Am J Audiol 21:215–225

Anari M, Axelsson A, Eliasson A, Magnusson L (1999) Hypersensitivity to sound: questionnaire data, audiometry and classification. Scand Audiol 28(4):219–230

Andersson G (2003) Tinnitus loudness matchings in relation to annoyance and grading of severity. Auris Nasus Larynx 30(2):129–133

Axelsson A, Ringdahl A (1989) Tinnitus – a study of its prevalence and characteristics. Br J Audiol 23(1):53–62

Baguley DM (2003) Hyperacusis. J R Soc Med 96(12):582–585

Baguley DM, Andersson G, McFerran D, McKenna L (2013) Tinnitus: a multidisciplinary approach. Wiley, Hoboken

Boureau F, Luu M, Doubrere JF (1991) Study of experimental pain measures and nociceptive reflex in chronic pain patients and normal subjects. Pain 44(2):131–138

Burns EM (1984) A comparison of variability among measurements of subjective tinnitus and objective stimuli. Audiology 23(4):426–440

Cederroth CR, Gallus S, Hall DA, Kleinjung T, Langguth B, Maruotti A et al (2019) Towards an understanding of tinnitus heterogeneity. Front Aging Neurosci 11:53

Dar R, Ariely D, Frenk H (1995) The effect of past-injury on pain threshold and tolerance. Pain 60 (2):189–193

De Ridder D, Kilgard M, Engineer N, Vanneste S (2015) Placebo-controlled vagus nerve stimulation paired with tones in a patient with refractory tinnitus: a case report. Otol Neurotol 36 (4):575–580

Douek E (1981) Classification of tinnitus. In: Tinnitus ciba foundation symposium, vol 85. pp. 4-15

Drexler D, López-Paullier M, Rodio S, González M, Geisinger D, Pedemonte M (2016) Impact of reduction of tinnitus intensity on patients' quality of life. Int J Audiol 55(1):11–19

Durai M, Sanders M, Kobayashi K, Searchfield GD (2019) Auditory streaming and prediction in tinnitus sufferers. Ear Hear 40(2):345–357

Eggermont JJ, Roberts LE (2004) The neuroscience of tinnitus. Trends Neurosci 27(11):676–682

Evered D, Lawrenson G (1981) Appendix II. Guidelines for recommended procedures in tinnitus testing. In: CIBA foundation symposium, vol 85. pp 303-306

Feldmann H (1971) Homolateral and contralateral masking of tinnitus by noise-bands and by pure tones. Audiology 10(3):138–144

Folmer RL, Griest SE, Meikle MB, Martin WH (1999) Tinnitus severity, loudness, and depression. Otolaryngol Head Neck Surg 121(1):48–51

Fournier P, Cuvillier AF, Gallego S, Paolino F, Paolino M, Quemar A et al (2018) A new method for assessing masking and residual inhibition of tinnitus. Trends Hear 22:1–19

Fournier P, Wrzosek M, Paolino M, Paolino F, Quemar A, Norena AJ (2019) Comparing tinnitus tuning curves and psychoacoustic tuning curves. Trends Hear 23:1–16

Fowler EP (1936) A method for the early detection of otosclerosis: a study of sounds well above threshold. Arch Otolaryngol 24(6):731–741

Fowler EP (1937) The diagnosis of diseases of the neural mechanism of hearing by the aid of sounds well above threshold (presidential address). Laryngoscope 47(5):289–300
Fowler EP (1940) Head noises: significance, measurement and importance in diagnosis and treatment. Arch Otolaryngol 32(5):903–914
Fowler EP (1944) Head noises in normal and in disordered ears: significance, measurement, differentiation and treatment. Arch Otolaryngol 39(6):498–503
Goldstein B, Shulman A (1996) Tinnitus – hyperacusis and the loudness discomfort level test – a preliminary report. Int Tinnitus J 2(1):83–89
Goodhill V (1952) A tinnitus identification test. Ann Otol Rhinol Laryngol 61(3):778–788
Goodwin PE, Johnson RM (1980) The loudness of tinnitus. Acta Otolaryngol 90(1–6):353–359
Graham JT, Newby HA (1962) Acoustical characteristics of tinnitus: an analysis. Arch Otolaryngol 75(2):162–167
Hall DA, Mehta RL, Fackrell K (2017) How to choose between measures of tinnitus loudness for clinical research? A report on the reliability and validity of an investigator-administered test and a patient-reported measure using baseline data collected in a phase IIa drug trial. Am J Audiol 26 (3):338–346
Hallam RS, Jakes SC, Hinchcliffe R (1988) Cognitive variables in tinnitus annoyance. Br J Clin Psychol 27(3):213–222
Hazell JWP, Wood SM (1981) Tinnitus masking – a significant contribution to tinnitus management. Br J Audiol 15(4):223–230
Hébert S (2018) Individual reliability of the standard clinical method vs patient-centered tinnitus likeness rating for assessment of tinnitus pitch and loudness matching. JAMA Otolaryngol Head Neck Surg 144(12):1136–1144
Helson H (1964) Adaptation-level theory: an experimental and systematic approach to behavior
Henry JA (2016) "Measurement" of tinnitus. Otol Neurotol 37(8):e276–e285
Henry JA, Meikle M (1996) Loudness recruitment only partially explains the small size of tinnitus loudness-matches. In: Reich GE, Vernon JA (eds.) Proceedings of the fifth international tinnitus seminar. American Tinnitus Association, Portland. pp 148-157
Henry JA, Meikle MB (2000) Psychoacoustic measures of tinnitus. J Am Acad Audiol 11 (3):138–155
Henry JA, Flick CL, Gilbert A, Ellingson RM, Fausti SA (1999) Reliability of tinnitus loudness matches under procedural variation. J Am Acad Audiol 10:502–520
Henry JA, Flick CL, Gilbert A, Ellingson RM, Fausti SA (2001) Comparison of two computer-automated procedures for tinnitus pitch matching. J Rehabil Res Dev 38(5):557–566
Henry JA, Flick CL, Gilbert A, Ellingson RM, Fausti SA (2004) Comparison of manual and computer-automated procedures for tinnitus pitch-matching. J Rehabil Res Dev 41(2)
Henry JA, Roberts LE, Ellingson RM, Thielman EJ (2013) Computer-automated tinnitus assessment: noise-band matching, maskability, and residual inhibition. J Am Acad Audiol 24 (6):486–504
Hiller W, Goebel G (2006) Factors influencing tinnitus loudness and annoyance. Arch Otolaryngol Head Neck Surg 132(12):1323–1330
Hiller W, Goebel G (2007) When tinnitus loudness and annoyance are discrepant: audiological characteristics and psychological profile. Audiol Neurotol 12(6):391–400
Hinchcliffe R, Chambers C (1983) Loudness of tinnitus: an approach to measurement. In: Aspects of modern otolaryngological practice, vol 29. Karger Publishers, pp 163–173
Hoare DJ, Edmondson-Jones M, Gander PE, Hall DA (2014) Agreement and reliability of tinnitus loudness matching and pitch likeness rating. PLoS One 9(12):e114553
Hulshof JH (1986) The loudness of tinnitus. Acta Otolaryngol 102(1–2):40–43
Jakes SC, Hallam RS, Chambers CC, Hinchcliffe R (1986) Matched and self-reported loudness of tinnitus: methods and sources of Error1. Audiology 25(2):92–100
Jepsen K, Sanders M, Searchfield G, Kobayashi K (2010) Perceptual training of tinnitus. In: Tinnitus discovery: Asia Pacific tinnitus symposium. NZMA

Jesteadt W (1980) An adaptive procedure for subjective judgments. Percept Psychophys 28 (1):85–88

Josephson EM (1931) A method of measurement of tinnitus aurium. Arch Otolaryngol 14 (3):282–283

Kay F, Searchfield GD (2008) Towards improving the assessment of tinnitus pitch. NZ Audiol Soc Bull 18(2):27–41

Kim TS, Yakunina N, Ryu YJ, Chung IJ, Nam EC (2017) Self-administered tinnitus pitch matching versus a conventional audiometric procedure. Audiol Neurotol 22(1):1–8

Korth D, Wollbrink A, Wunderlich R, Ivansic D, Guntinas-Lichius O, Salvari V et al (2020) One step closer towards a reliable tinnitus pitch-match frequency determination using repetitive recursive matching. Audiol Neurotol:1–10

Kostek B, Poremski T (2013) A new method for measuring the psychoacoustical properties of tinnitus. Diagn Pathol 8(1):209

Mahboubi H, Ziai K, Brunworth J, Djalilian HR (2012) Accuracy of tinnitus pitch matching using a web-based protocol. Ann Otol Rhinol Laryngol 121(10):671–674

Manning C, Grush L, Thielman E, Roberts L, Henry JA (2019) Comparison of tinnitus loudness measures: matching, rating, and scaling. Am J Audiol 28(1):137–143

Matsuhira T, Yamashita K, Yasuda M (1992) Estimation of the loudness of tinnitus from matching tests. Br J Audiol 26(6):387–395

McFadden D (1982) Tinnitus: facts, theories, and treatments. National Academies Press, Washington

Meikle MB, Taylor-Walsh E (1984) Characteristics of tinnitus and related observations in over 1800 tinnitus clinic patients. J Laryngol Otol 98(S9):17–21

Meikle MB, Creedon TA, Griest SE (2004) Tinnitus Archive, second edition. http://www.tinnitusarchive.org

Mitchell C (1983) The masking of tinnitus with pure tones. Audiology 22(1):73–87

Mitchell PL, Moffat DA, Fallside F (1984) Computer-aided tinnitus characterization. Clin Otolaryngol Allied Sci 9(1):35–42

Mitchell CR, Vernon JA, Creedon TA (1993) Measuring tinnitus parameters: loudness, pitch, and maskability. J Am Acad Audiol 4(3):139–151

Moore BCJ (2012) The psychophysics of tinnitus. In: Eggermont JJ, Zeng F, Popper AN, Fay RR (eds) Tinnitus. Springer, New York, pp 187–216

Moore BCJ (2014) Measuring the pitch and loudness of tinnitus. ENT Audiol News 22(6)

Moore BCJ, Vinay SN (2010) The relationship between tinnitus pitch and the edge frequency of the audiogram in individuals with hearing impairment and tonal tinnitus. Hear Res 261(1–2):51–56

Nageris BI, Attias J, Raveh E (2010) Test-retest tinnitus characteristics in patients with noise-induced hearing loss. Am J Otolaryngol 31(3):181–184

Neff P, Langguth B, Schecklmann M, Hannemann R, Schlee W (2019) Comparing three established methods for tinnitus pitch matching with respect to reliability, matching duration, and subjective satisfaction. Trends Hear 23:1–9

Norena AJ (2011) An integrative model of tinnitus based on a central gain controlling neural sensitivity. Neurosci Biobehav Rev 35(5):1089–1109

Norena AJ, Micheyl C, Chéry-Croze S, Collet L (2002) Psychoacoustic characterization of the tinnitus spectrum: implications for the underlying mechanisms of tinnitus. Audiol Neurotol 7 (6):358–369

Ohsaki K, Fujimura T, Sugiura T, Tamura K, Nakagiri S, Zheng HX et al (1990) Reproducibility of pitch-matching test for tinnitus, using a heptatonic scale. Scand Audiol 19(2):123–126

Pan T, Tyler RS, Ji H, Coelho C, Gehringer AK, Gogel SA (2009) The relationship between tinnitus pitch and the audiogram. Int J Audiol 48(5):277–294

Penner MJ (1983) Variability in matches to subjective tinnitus. J Speech Lang Hear Res 26 (2):263–267

Penner MJ (1986) Magnitude estimation and the" paradoxical" loudness of tinnitus. J Speech Lang Hear Res 29(3):407–412

Penner MJ (1987) Masking of tinnitus and central masking. J Speech Lang Hear Res 30(2):147–152
Penner MJ, Bilger RC (1992) Consistent within-session measures of tinnitus. J Speech Lang Hear Res 35(3):694–700
Penner MJ, Klafter EJ (1992) Measures of tinnitus: step size, matches to imagined tones, and masking patterns. Ear Hear 13(6):410–416
Pienkowski M, Tyler RS, Roncancio ER, Jun HJ, Brozoski T, Dauman N et al (2014) A review of hyperacusis and future directions: part II. Measurement, mechanisms, and treatment. Am J Audiol 23(4):420–436
Plack CJ, Barker D, Prendergast G (2014) Perceptual consequences of "hidden" hearing loss. Trends Hear 18:2331216514550621
Raj-Koziak D, Gos E, Swierniak W, Karpiesz L, Niedzialek I, Wlodarczyk E et al (2018) Relationship between tinnitus loudness measure by visual analogue scale and psychoacoustic matching of tinnitus loudness. Otol Neurotol 40(1):16–21
Ratnayake SAB, Jayarajan V, Bartlett J (2009) Could an underlying hearing loss be a significant factor in the handicap caused by tinnitus? Noise Health 11(44):156
Reavis KM, Chang JE, Zeng FG (2010) Patterned sound therapy for the treatment of tinnitus. Hear J 63(11):21–22
Reed GF (1960) An audiometric study of two hundred cases of subjective tinnitus. AMA Arch Otolaryngol 71(1):84–94
Ristovska LE, Jachova Z, Stojcheska V (2019) Psychoacoustic characteristics of tinnitus in relation to audiometric profile. Arch Acoust 44
Roberts LE, Bosnyak DJ (2011) Auditory training in tinnitus. In: Møller AR, Langguth B, DeRidder D, Kleinjung T (eds) Textbook of tinnitus. Springer-Verlag, New York, pp 563–573
Roberts LE, Moffat G, Bosnyak DJ (2006) Residual inhibition functions in relation to tinnitus spectra and auditory threshold shift. Acta Oto-Laryngol 126(sup556):27–33
Roberts LE, Moffat G, Baumann M, Ward LM, Bosnyak DJ (2008) Residual inhibition functions overlap tinnitus spectra and the region of auditory threshold shift. J Assoc Res Otolaryngol 9 (4):417–435
Roeser RJ, Price DR (1980) Clinical experience with tinnitus maskers. Ear Hear 1(2):63–68
Rollman GB (1979) Signal detection theory pain measures: empirical validation studies and adaptation-level effects. Pain 6(1):9–21
Rosito LS, Bonfitto M, Deutsch KM, Bombardell K, Butzke B, Silva AL, Dalligna C (2013) Is pitch and loudness associated with tinnitus annoyance? Otolaryngol Head Neck Surg 149(2):229–229
Sandlin RE, Olsson RJ (1999) Evaluation and selection of maskers and other devices used in the treatment of tinnitus and hyperacusis. Trends Amplif 4(1):6–26
Schaette R, Kempter R (2009) Predicting tinnitus pitch from patients' audiograms with a computational model for the development of neuronal hyperactivity. J Neurophysiol 101 (6):3042–3052
Schechter MA, Henry JA (2002) Assessment and treatment of tinnitus patients using a. J Am Acad Audiol 13(10):545–558
Schecklmann M, Vielsmeier V, Steffens T, Landgrebe M, Langguth B, Kleinjung T (2012) Relationship between audiometric slope and tinnitus pitch in tinnitus patients: insights into the mechanisms of tinnitus generation. PLoS One 7(4)
Searchfield GD (2014) Tinnitus what and where: an ecological framework. Front Neurol 5:271
Searchfield GD, Kobayashi K, Sanders M (2012) An adaptation level theory of tinnitus audibility. Front Syst Neurosci 6(46):1–9
Searchfield GD, Kobayashi K, Proudfoot K, Tevoitdale H, Irving S (2015) The development and test–retest reliability of a method for matching perceived location of tinnitus. J Neurosci Methods 256:1–8
Searchfield GD, Kobayashi K, Hodgson SA, Hodgson C, Tevoitdale H, Irving S (2016) Spatial masking: development and testing of a new tinnitus assistive technology. Assist Technol 28 (2):115–125

Searchfield GD, Durai M, Linford T (2017) A state-of-the-art review: personalization of tinnitus sound therapy. Front Psychol 8:1599

Sereda M, Hall DA, Bosnyak DJ, Edmondson-Jones M, Roberts LE, Adjamian P, Palmer AR (2011) Re-examining the relationship between audiometric profile and tinnitus pitch. Int J Audiol 50(5):303–312

Sheldrake J, Diehl PU, Schaette R (2015) Audiometric characteristics of hyperacusis patients. Front Neurol 6(105):1–7

Simoes JP, Neff P, Schoisswohl S, Bulla J, Schecklmann M, Harrison SS et al (2019) Towards personalized tinnitus treatment: an exploratory study based on internet crowdsensing. Front Public Health 7:157

Sockalingam R, Dunphy L, Nam KE, Gulliver M (2007) Effectiveness of frequency-matched masking and residual inhibition in tinnitus therapy: a preliminary study. Audiol Med 5 (2):92–102

Spaulding AJ (1903) Tinnitus with a plea for its more accurate musical notation. Arch Otolaryngol 32:263–272

Stephens SDG (1984) The treatment of tinnitus—a historical perspective. J Laryngol Otol 98 (10):963–972

Stevens SS, Davis H (1938) Hearing: its psychology and physiology. Wiley, New York

Stouffer JL, Tyler RS (1990) Characterization of tinnitus by tinnitus patients. J Speech Hear Disord 55(3):439–453

Switalski W, Sanchez C (2019) Tinnitus assessment: the key to successful tinnitus patient management [white paper]. https://madsen.hu/pdf/tajekoztato/Astera_Tinnitus_Whitepaper_7-26-9058-EN_00_STD.pdf

Terry AMP, Jones DM, Davis BR, Slater R (1983) Parametric studies of tinnitus masking and residual inhibition. Br J Audiol 17(4):245–256

Tyler RS (2000) The psychoacoustical measurement of tinnitus. In: Tyler RS (ed) Tinnitus handbook. Singular Publishing Group, California, pp 149–180

Tyler RS, Baker LJ (1983) Difficulties experienced by tinnitus sufferers. J Speech Hear Disord 48 (2):150–154

Tyler RS, Conrad-Armes D (1983a) Tinnitus pitch: a comparison of three measurement methods. Br J Audiol 17(2):101–107

Tyler RS, Conrad-Armes D (1983b) The determination of tinnitus loudness considering the effects of recruitment. J Speech Lang Hear Res 26(1):59–72

Tyler RS, Conrad-Armes D (1984) Masking of tinnitus compared to masking of pure tones. J Speech Lang Hear Res 27(1):106–111

van Zwieten G, Smit JV, Jahanshahi A, Temel Y, Stokroos RJ (2016) Tinnitus: is there a place for brain stimulation? Surg Neurol Int 7(S4):125–129

Vanneste S, De Ridder D (2012) Noninvasive and invasive neuromodulation for the treatment of tinnitus: an overview. Neuromodul Technol Neural Interf 15(4):350–360

Vanneste S, Song JJ, De Ridder D (2013) Tinnitus and musical hallucinosis: the same but more. NeuroImage 82:373–383

Vernon JA (1977) Attempts to relieve tinnitus. J Am Audiol Soc 2(4):124–131

Vernon JA, Meikle MB (2000) Tinnitus masking. In: Tyler RS (ed) Tinnitus handbook. Singular Publishing Group, San Diego, pp 313–355

Vernon JA, Fenwick J (1984) Identification of tinnitus: a plea for standardization. J Laryngol Otol 98(S9):45–53

Vernon JA, Meikle MB (1981) Tinnitus masking: unresolved problems. In: CIBA foundation symposium, vol 85. pp 239–256

Vernon JA, Meikle MB (2003) Masking devices and alprazolam treatment for tinnitus. Otolaryngol Clin N Am 36(2):307–320

Vernon J, Schleuning A (1978) Tinnitus: a new management. Laryngoscope 88(3):413–419

Ward LM, Baumann M (2009) Measuring tinnitus loudness using constrained psychophysical scaling. Am J Audiol 18:119–128

Wegel RL (1931) A study of tinnitus. Arch Otolaryngol 14(2):158–165
Welch D, Dawes PJ (2008) Personality and perception of tinnitus. Ear Hear 29(5):684–692
Wise K, Kobayashi K, Magnusson J, Welch D, Searchfield GD (2016) Randomized controlled trial of a perceptual training game for tinnitus therapy. Games Health J 5(2):141–149
Wunderlich R, Stein A, Engell A, Lau P, Waasem L, Shaykevich A et al (2015) Evaluation of iPod-based automated tinnitus pitch matching. J Am Acad Audiol 26(2):205–212
Yost WA (2015) Psychoacoustics: a brief historical overview. Acoust Today 11(3):46–53
Zenner PH, De Maddalena H, Zalaman IM (2005) Validity and reliability study of three tinnitus self-assessment scales: loudness, annoyance and change. Acta Otolaryngol 125(11):1184–1188

Emerging Topics in the Behavioral Neuroscience of Tinnitus

Grant D. Searchfield, Jinsheng Zhang, Roshni Biswas, Dirk De Ridder, Brian Deutsch, Deborah A. Hall, Sylvie Hébert, Tobias Kleinjung, Maria Kleinstäuber, Berthold Langguth, Jose Antonio Lopez-Escamez, Michael R. D. Maslin, Birgit Mazurek, Jay F. Piccirillo, Richard Salvi, Winfried Schlee, Abraham Shulman, Susan Shore, Agnieszka J. Szczepek, Paul F. Smith, Sarah M. Theodoroff, Dunja Vajsakovic, Cornelia Weise, and Yiwen Zheng

Contents

The original version of this chapter was revised. A correction to this chapter can be found at https://doi.org/10.1007/7854_2021_230

G. D. Searchfield (✉)
Audiology, School of Population Health, Faculty of Medical and Health Sciences, The University of Auckland, Auckland, New Zealand

Eisdell Moore Centre, The University of Auckland, Auckland, New Zealand

Centre for Brain Research, The University of Auckland, Auckland, New Zealand
e-mail: g.searchfield@auckland.ac.nz

J. Zhang
Department of Otolaryngology-Head and Neck Surgery, Wayne State University, Detroit, MI, USA

Department of Communication Sciences and Disorders, Wayne State University, Detroit, MI, USA

R. Biswas
School of Medicine, The University of Nottingham, Nottingham, UK

D. De Ridder
Neurosurgery, Department of Surgical Sciences, University of Otago, Dunedin, New Zealand

B. Deutsch
Icahn School of Medicine at Mount Sinai, New York, NY, USA

Curr Topics Behav Neurosci (2021) 51: 461–484
https://doi.org/10.1007/7854_2020_217
Published Online: 5 March 2021

Abstract This volume has highlighted the many recent advances in tinnitus theory, models, diagnostics, therapies, and therapeutics. But tinnitus knowledge is far from complete. In this chapter, contributors to the Behavioral Neuroscience of Tinnitus consider emerging topics and areas of research needed in light of recent findings. New research avenues and methods to explore are discussed. Issues pertaining to current assessment, treatment, and research methods are outlined, along with recommendations on new avenues to explore with research.

D. A. Hall
Hearing Sciences, Division of Clinical Neuroscience, School of Medicine, University of Nottingham, Nottingham, UK

National Institute for Health Research (NIHR) Nottingham Biomedical Research Centre, Ropewalk House, Nottingham, UK

University of Nottingham Malaysia, Semenyih, Selangor Darul Ehsan, Malaysia

S. Hébert
École d'orthophonie et d'audiologie, Université de Montréal, Montréal, QC, Canada

T. Kleinjung
Department of Otorhinolaryngology, Head and Neck Surgery, University Hospital of Zurich, University of Zurich, Zurich, Switzerland

M. Kleinstäuber
Department of Psychological Medicine, Otago Medical School, University of Otago, Dunedin, New Zealand

B. Langguth and W. Schlee
Department of Psychiatry and Psychotherapy, University of Regensburg, Regensburg, Germany

J. A. Lopez-Escamez
Otology and Neurotology Group CTS495, Department of Otolaryngology, Instituto de Investigación Biosanitaria, ibs.Granada, Hospital Universitario Virgen de las Nieves, Universidad de Granada, Granada, Spain

M. R. D. Maslin
Eisdell Moore Centre, The University of Auckland, Auckland, New Zealand

School of Psychology, Speech and Hearing, University of Canterbury, Christchurch, New Zealand

B. Mazurek
Charité Universitätsmedizin Berlin, Tinnitus Center, Berlin, Germany

J. F. Piccirillo
Department of Otolaryngology-Head and Neck Surgery, Washington University School of Medicine, St. Louis, MO, USA

Keywords Diagnostic · Discovery · Future · Mechanism · Methods · Rehabilitation · Tinnitus · Treatment

1 Introduction

In this chapter, we consider the current status of tinnitus research with a focus on new and emerging topics. Advances in research hypothesized to be essential for future research are discussed. The literature reviewed is primarily from the last decade.

2 Tinnitus Is, What Tinnitus Is

Tinnitus is a complex experience and its management is made challenging by individual responses to treatments. Simply stated, tinnitus is composed of its perception (hearing a sound) and its reaction (the degree to which hearing the sound is a

R. Salvi
Center for Hearing and Deafness, University at Buffalo, Buffalo, NY, USA

A. Shulman
Otolaryngology, SUNY Downstate, Brooklyn, NY, USA

S. Shore
Kresge Hearing Research Institute, Ann Arbor, MI, USA

A. J. Szczepek
Department of Otorhinolaryngology, Head and Neck Surgery, Charité – Universitätsmedizin Berlin, Corporate Member of Freie Universität Berlin, Humboldt- Universität zu Berlin, and Berlin Institute of Health, Berlin, Germany

Faculty of Medicine and Health Sciences, University of Zielona Góra, Zielona Góra, Poland

P. F. Smith and Y. Zheng
Eisdell Moore Centre, The University of Auckland, Auckland, New Zealand

Department of Pharmacology and Toxicology, School of Biomedical Sciences and the Brain Health Research Centre, University of Otago, Dunedin, New Zealand

S. M. Theodoroff
Veterans Affairs (VA) Rehabilitation Research and Development, National Center for Rehabilitative Auditory Research, VA Portland Health Care System, Portland, OR, USA

Department of Otolaryngology/Head and Neck Surgery, Oregon Health and Science University, Portland, OR, USA

D. Vajsakovic
Audiology, School of Population Health, Faculty of Medical and Health Sciences, The University of Auckland, Auckland, New Zealand

Centre for Brain Research, The University of Auckland, Auckland, New Zealand

Esidell Moore Centre, The University of Auckland, Auckland, New Zealand

C. Weise
Department of Psychology, Philipps-University of Marburg, Marburg, Germany

problem for the person). Advanced neuroimaging technology reveals tinnitus-associated changes in neuronal activity and connectivity involving multiple neural networks, both in human patients (Boyen et al. 2014; Husain and Schmidt 2014; Kraus and Canlon 2012; Leaver et al. 2011; Maudoux et al. 2012; Song et al. 2012; Vanneste et al. 2011; Vanneste and De Ridder 2012) and animal models (Chen et al. 2014). It has been increasingly recognized by the international tinnitus research community that disabling tinnitus is unlikely to originate from a single pathological source, but rather complex network changes involving not only the auditory system but also other sensory systems and systems related to memory, emotion, attention, and stress (Henry et al. 2014; Knobel and Sanchez 2008; Leaver et al. 2016; Roberts et al. 2010; Simonetti and Oiticica 2015).

A more complete understanding of tinnitus will emerge when we can tie tinnitus-related behaviors with neuroscience and this must consider both intrinsic and extrinsic factors that shape an individual, their perception and reaction (Searchfield 2014). Tinnitus is difficult to define, but it can be viewed as a false perception of sound, which can cause emotional, cognitive, behavioral, and autonomic reactions, leading to functional and behavioral changes, potentially disabling the person experiencing it. Tinnitus is perceived as a sound, but it is the creation of central processing, not a sensation of the external world. Our response to real-world sounds is based on prior knowledge and experience of the sound; the same cannot be said for tinnitus. Normally, our perception and reaction to sound are determined by its source, which tinnitus lacks. The absence of an external source introduces a sense of discomfort, for the individual cannot locate where the sound is coming from and what is causing it, challenging the listener's reality of sound perception (Feldmann 1992). Brefczynski-Lewis and Lewis (2017) proposed a neurobiological model of auditory perception describing the interaction between bottom-up and top-down influences. This model classifies sounds according to three basic categories of sound-source: (1) action sounds produced by living things; (2) action sounds produced by non-living things; and (3) vocalizations. The effect and attention capturing characteristics of tinnitus could be examined by extending the model from real-world natural sounds (Brefczynski-Lewis and Lewis 2017) to include false sounds such as tinnitus and hallucinations.

Calls to recognize the heterogeneity of tinnitus have become more frequent, and over the last decade, this heterogeneity has led to an increased drive for the subtyping of tinnitus and the development of personalized therapeutic approaches (Cederroth et al. 2019a; McFerran et al. 2019; Simoes et al. 2019; Van de Heyning et al. 2015). Therefore, in the next decade, we will most likely see advancement in at least two broad areas, individualized identification of tinnitus-related network activity and individualized treatment targeting those networks (e.g., emotion, attention, hearing).

3 The Pathway to Precision in Assessment

No two patients are likely to experience tinnitus in precisely the same way; their percept will differ relative to its psychoacoustic characteristics, related comorbidities, their personality, and psychological reaction to the tinnitus

(Searchfield 2014), individual genetics (Maas et al. 2017; Veile et al. 2018), and even their lifestyle (Veile et al. 2018). It is clear that in order to make progress in tinnitus assessment, we must adopt a holistic stance. A broader comprehension of tinnitus is required. Accumulating evidence suggests complex interactions between genetic, demographic, lifestyle, and other environmental factors influence tinnitus. It can be assumed that epigenetics may also become involved in the genetic and environmental interaction that leads to tinnitus (Lopez-Escamez et al. 2016). Request has been made for reinforced efforts to create large datasets that incorporate a broader spectrum of information from each participant to understand further tinnitus heterogeneity (Schlee et al. 2018). As part of these efforts, a multi-omics approach to bridge tinnitus-related metabolomic data with proteomic, transcriptomic, and genomic data combined with data on a wide array of individual, lifestyle and environmental factors, is needed. Furthermore, the question needs to be addressed whether environmental influences translate epigenomics to different tinnitus phenotypes. These types of information are often missing from current datasets, but are nevertheless important for unravelling the biological basis of tinnitus.

Another challenge for precise assessment is the temporal dynamic of the subjective tinnitus perception. Patients often report that their perception of tinnitus fluctuates between days, and also within days (Probst et al. 2017). These patients report that there are moments with loud and pronounced tinnitus perception, but also moments with reduced tinnitus perception. Early work in this line of research demonstrated substantial variability in the tinnitus loudness as well as the tinnitus distress measures (Henry et al. 2012; Schlee et al. 2016; Wilson et al. 2015). The underlying neurobiological mechanisms of this moment-to-moment fluctuation are largely unknown. Yet, a better understanding of these mechanisms could reveal innovative ways for clinical interventions.

3.1 Biomarkers of Tinnitus

The search for physiological and behavioral markers of tinnitus has a long history (Ciba Foundation Symposium 85 1981), but methods are now emerging that suggest the distinction between tinnitus and non-tinnitus is becoming possible by examining cortical/subcortical morphology (Liu et al. 2019) and patterns of EEG activity (Vanneste et al. 2018). Imaging and EEG methods, with the aid of machine learning algorithms, offer new possibilities to accurately distinguish tinnitus from non-tinnitus neural activity at an individual level (Durai et al. 2020; Han et al. 2019).

The investigation of genetic and blood-based biomarkers for tinnitus is at an exploratory phase (Haider et al. 2020; Szczepek et al. 2014). While platelet volume and distribution (Ulusoy et al. 2018) and activity of circulating proteasomes (Yun et al. 2020) may be indicative of tinnitus in humans, they must be differentiated from other concomitant pathologies such as mild cognitive impairment (Yun et al. 2020). Acoustic trauma in rats changes several metabolic pathways (He et al. 2017), some of which may also be associated with tinnitus, and highlights the importance of networks of biochemical variables in determining changes in auditory function (He et al. 2017). Transient tinnitus induced in rats using salicylate induces dysregulation of cytokines and N-methyl D-aspartate receptor subunit 2A genes (Chen and Zheng 2017). The genotype A/T at glutamate metabotropic receptor 7 (GRM7) (Haider et al. 2017) may be a marker for tinnitus severity in humans. Evaluation of tinnitus heritably in twins and careful recording and control of phenotypes in studies will aid the development of genetic markers of tinnitus, which may in turn guide drug development (Lopez-Escamez et al. 2016). In a similar approach, large-scale metabolomics analysis in noise trauma has been postulated to be able to diagnose tinnitus in rats (He et al. 2017).

Molecular genetics studies are needed in humans to define the genes and biological processes involved in tinnitus severity. The mechanisms leading to severe tinnitus seem to be independent of hearing loss (Lopez-Escamez and Amanat 2020). Evidence is beginning to emerge to support a genetic predisposition to develop severe tinnitus. Concordance studies in twins and adoptees support heritability in bilateral tinnitus (Cederroth et al. 2019b; Maas et al. 2017) and familial aggregation studies reveal that severe tinnitus clusters in families and this effect is stronger in women (Trpchevska et al. 2020).

3.2 Acute, Chronic, Bothersome, and Disabling Tinnitus

A potentially fruitful avenue for increasing our understanding of tinnitus is the differentiation of tinnitus based on its etiology (e.g., noise, ototoxicity, or blast trauma), stage of pathogenesis (acute, chronic), and severity of symptoms (non-bothersome, bothersome, disabling) which may aid in the discovery of therapies best attuned to different mechanisms. For example, in the management of pain, there has been considerable progress in efforts to predict the risk and mitigate the transition from acute to chronic pain (George et al. 2020; Traeger et al. 2016). The development of similar frameworks may aid tinnitus management across its progression from acute to chronic, perhaps informing practices that could prevent chronification.

Animal models are important in research due to the ability to control the mode of tinnitus induction, the duration of pathology, and the use of more invasive measurement techniques to assess structural and functional changes linked to structure and function. Animal models have helped progress knowledge from a gross understanding of pathophysiology (e.g., from peripheral to central mechanisms) to single-cell populations (e.g., fusiform cells in the DCN (Shore and Wu 2019) and the role of receptors (e.g., $GABA_A$ receptor function in thalamic circuits (Caspary and Llano 2017) and targets for therapy (e.g., attention (Brozoski and Bauer 2016)). Basic research in animals is also contributing to understanding the roles of different pathologies in explaining some of the heterogeneity in tinnitus, such as distinguishing neural changes due to hearing loss from tinnitus (Shore and Wu

2019), chronic noise-induced tinnitus from that resulting from salicylate ototoxicity (Eggermont 2016) and blast trauma (Zhang 2019). The effects of stress are also now being modelled in animals (Jiang et al. 2017) in an attempt to mimic the stress-tinnitus axis seen in humans. Clues as to the role of stress and tinnitus in humans are being explored through indicators of stress, such as salivary markers, blood pressure, and heart rate (Alsalman et al. 2016; Aydin and Searchfield 2019; Betz et al. 2017). Stress markers may clarify how the acute onset of the false perception of sound can evolve into disabling chronic tinnitus. However, the auditory system of different species (Szczepek et al. 2018) and of different sexes (Keesom et al. 2018; Willott 2009) may respond differently to stress. Thus, fundamental work on tinnitus-related stress responses in animals as well as in humans must be pursued.

Acute and chronic, mild and disabling tinnitus may have distinct pathophysiological markers in humans. Resting-state fMRI from persons with mild acute tinnitus appears similar to controls (Wineland et al. 2012). Resting-state fMRI to distinguish tinnitus chronicity from tinnitus severity has revealed disrupted connectivity between the precuneus and other default mode regions, which could be an indicator of long-term tinnitus, with the strength of the disruption correlated with tinnitus severity (i.e., with more bothersome tinnitus demonstrating more substantial decreases) (Schmidt et al. 2017). Individual differences in auditory and non-auditory systems may impact how tinnitus is perceived and how "bothersome" it is.

EEG studies also demonstrate that, in some patients with tinnitus, cognitive functions may worsen with activity changes in the hippocampus, the pregenual and subgenual anterior cingulate cortex extending into the right insula (Vanneste et al. 2016). A careful review found mixed support for the claim that tinnitus impairs working memory, executive attention, and selective attention (Mohamad et al. 2016). However, as they are all complex concepts in themselves, the relationship of cognition and executive function to tinnitus requires further evaluation. The continued development of biomarkers indicative of sequelae and different dimensions of tinnitus may lead to treatments that may be more effective at one given phase of tinnitus development than another.

3.3 *Psychoacoustic Measures*

Tinnitus is a more complicated perception than the relatively simple psychoacoustic matches used today imply. Psychoacoustic assessments can be improved both in their reliability and efficiency (Hebert and Fournier 2017). Tinnitus patients describe their perception in a number of ways, most commonly as a ringing or hissing sensation, but in some cases as more complex sounds such as crickets and even music (Meikle et al. 2004). Most methods of tinnitus synthesis are based on the adjustment of a pure tone to match the dominant pitch of tinnitus (Bertet et al. 2013; Burns 1984; Henry and Meikle 2000; Mitchell et al. 1993; Penner and Klafter 1992). More complex approaches such as Tinnitus Likeness Ratings are a step towards a broader and more "complete" representation of the tinnitus sensation through the demonstration of its spectral constituents (Basile et al. 2013). But even these advanced matching methods do not capture all of the acoustic parameters that likely define tinnitus, for example timing and location (Searchfield et al. 2015). There is still a need to develop accurate and realistic tinnitus avatars (complex replicas of the tinnitus experience). There appear to be two approaches to creating realistic copies of tinnitus, one is to build on basic building blocks of sound (frequency and intensity), the other is to start with complex sounds (real world) and modify them to match tinnitus. The resources and time required for more involved measures (cost-benefit) need to be ascertained. The need for greater accuracy in pitch-matching is becoming urgent, as new therapies based around sounds presented in or around tinnitus pitch become available (Korth et al. 2020).

3.4 Standardized Outcome Measures

One of the major methodological limitations for cross-study comparisons is the lack of agreed standards for how some of the most fundamental variables related to tinnitus are defined. This hampers cross-study comparisons and data pooling. In order to pool data, it is essential to have standardized methods (Langguth et al. 2007). The call for standardization is not new (Ciba Foundation Symposium 85 1981) but recently increased efforts have been made to develop and adopt a core set of measures for human participant research (Hall et al. 2018). A key component to this is settling on standard questionnaires with high-quality translations/adaptations into different languages so that data can be reliably pooled across countries. Much has already been debated on the utility of questionnaires to assess tinnitus symptom severity and to formulate treatment (Langguth et al. 2007). There may be diminished value in developing new composite measures of tinnitus symptom severity as existing questionnaires (e.g., Tinnitus Handicap Inventory (Newman et al. 1996), Tinnitus Functional Index (Meikle et al. 2012), Tinnitus Primary Function Index (Richard Tyler et al. 2014) have become accepted norms. However, the diagnosis of comorbidities (e.g., depression, anxiety) using questionnaires is becoming more and more important. Moreover, sex-specific aspects will increasingly play a role (Lugo et al. 2019; Niemann et al. 2020a; Van der Wal et al. 2020; Vanneste et al. 2012). It may be useful to continue the development of measures that are more specifically tuned to ascertain distinct health concepts (e.g., illness beliefs, impact on concentration, or tinnitus intrusiveness) or mechanisms of effect, expectations, and benefit (Probst et al. 2019), for example: information from a client-oriented perspective may enable individuals to prioritize their treatment needs (e.g., COSIT (Searchfield 2019)). There are also new opportunities to develop and test methods that can capture tinnitus in greater temporal resolution, possibly in real-time, through momentary ecological evaluation (Goldberg et al. 2017; Schlee et al. 2016). Momentary evaluation samples subjective states regularly, up to several times daily with recording facilitated through the use of the omnipresent smartphone

(Goldberg et al. 2017; Schlee et al. 2016). This methodology allows for the incorporation of patient- and environment-specific factors shown to affect the severity of tinnitus. The most sensitive measures, duration, and regularity of assessment need to be determined, and considered alongside any unintended side effects that may occur from regularly attending to tinnitus. The increasing depth of understanding of tinnitus opens up prognostic opportunities that have not previously existed; these in turn increase the potential of applying precision medicine to tinnitus.

3.5 Predicting Success

There is a new wave of tinnitus research focused on predicting treatment outcomes. Tinnitus case history information (Simoes et al. 2019) and assessment of personality (Durai et al. 2017; Kleinstäuber et al. 2018), tinnitus severity (Mazurek et al. 2006), hearing level and localization of tinnitus (Theodoroff et al. 2014), and gender aspects (Lugo et al. 2019; Niemann et al. 2020a; Van der Wal et al. 2020; Sven Vanneste et al. 2012), may all be markers that can be used to predict therapy outcomes. Physiological measures such as MRI and EEG, in highly controlled circumstances, also appear to accurately predict therapy outcomes (Durai et al. 2020; Han et al. 2019). Artificial intelligence and machine learning algorithms may mean that successful treatments will be selected a priori for patients (James et al. 2017; Niemann et al. 2020b). Drawing inspiration from other domains that have developed novel approaches to predict the successful therapeutic outcomes would also be worthy. For instance, some studies have reported that initial symptom profiles of patients (Uckelstam et al. 2019) and patient language (Goodwin et al. 2019) can predict the rate of change in psychotherapy, allowing professionals to anticipate prognosis and length of therapy, and to adapt their intervention. Other studies have used machine learning approaches to examine behavioral indices not under the control of patients, such as facial movements and expressions (Anis et al. 2018), to better measure psychopathology and predict a successful outcome.

4 The Research Pathway to Precision Therapy

Many health disciplines are moving towards personalized medicine that is focused on tailored diagnosis and treatment for an individual (Schleidgen et al. 2013; Tutton 2012), including tinnitus (Tzounopoulos et al. 2019). After the first sequencing of the human genome in 2000, the idea that the individual's genome could influence therapy led to the concept of personalized genomic medicine (Ginsburg and Willard 2009). Yet genomic data can only be applied at a group level, and thus the term precision medicine was coined (Juengst et al. 2016; Roden and Tyndale 2013). The fundamental difference between personalized medicine and precision medicine is that personalized medicine treats an individual patient, whereas precision medicine targets a disease specified by genomic data (Juengst et al. 2016; Roden and Tyndale 2013). The genomic information need to be integrated with proteomes, creating interactomes, if one wants to move again from precision medicine to individualized medicine (Zhang et al. 2015).

The development of new assessment methods and machine learning application may identify targets optimized for more individualized therapy. These new methods may lead to a re-evaluation of treatments that have fallen out of favor due to large variability in benefit between individuals, such as neurofeedback (Guntensperger et al. 2017; Jensen et al. 2020) or re-invigorate therapies that are widely used but criticized for lack of demonstrated benefit in comparison with control groups such as sound therapy (Brennan-Jones et al. 2020). Knowing who may benefit most from treatment should lead to greater efficiency and a reduction in the overall cost of tinnitus healthcare.

4.1 Network-Based Models

In many individuals, complete elimination of tinnitus perception may be analogous to the Greek myth of Hercules defeating the multiheaded Hydra serpent (Ogden 2013). In the legend, the Hydra would die only if all its heads were removed, but two heads grew in place of each one removed. Within a tinnitus neural network, the suppression or elimination of tinnitus-related activity may lead to aberrant activity emerging elsewhere within the network and so tinnitus persistence. With this in mind, we may need to prescribe multiple therapeutics, each with an individual target within the network, develop multicomponent therapeutics, or design a single therapeutic with multiple targets (Hopkins 2008). Developers of future tinnitus treatments should consider where in the sequelae of tinnitus development individual's sit and consider the tinnitus network biology and poly-therapeutics that aim to treat the tinnitus-causing networks rather than a single target within the network (e.g., (Jun He et al. 2017)). Treatments that combine several therapeutic actions, multimodal therapies, may become more common. The combination of auditory stimulation with counselling can be viewed as acting on bottom-up processing and top-down control of tinnitus networks (Searchfield this volume). In a different vein, bimodal auditory and somatosensory (dorsal column and trigeminal) activation with specific bimodal intervals based on underlying brain stem circuitry has shown effectiveness in reducing tinnitus loudness and distress (Marks et al. 2018). In different approaches, auditory combined with tongue stimulation has shown effectiveness on tinnitus distress (Conlon et al. 2020) sound combined with vagal nerve stimulation (Tyler et al. 2017) has shown some promising results. The ultimate therapeutic effectiveness of these new approaches remains to be determined. The combination of stimulation parameters derived from precise study of underlying circuitry would appear a key to success using multimodal therapies (Marks et al. 2018). However, there is some ambiguity as to what the optimal parameters for success (e.g. Conlon et al. 2020), indicating future research needs.

4.1.1 Polytherapeutics

For the treatment of HIV-positive patients, there are six different classes of drugs (Pau and George 2014). It was shown that triple therapy was more effective than dual therapy, and that dual therapy was more effective than monotherapy in first line treatment, dramatically increasing the success of AIDS treatment (Jordan et al. 2002). Yet combining four drugs does not seem to add any more benefit (Feng et al. 2019). A question is whether and how combination therapy may also benefit tinnitus?

Emerging drugs for tinnitus have mostly targeted the cochlear genesis of tinnitus, intending to block aberrant afferent activity. Altering cochlear-neural afferent drive with a pharmaceutical is expected to be most effective in an acute phase, prior to reactive plasticity creating tinnitus in-related central networks (Langguth et al. 2019, Kleinjung and Langguth this volume). Other drugs aim to affect central processes, for example oxytocin increases the salience of sound through inhibition (Marlin et al. 2015) with preliminary benefits for tinnitus (Azevedo et al. 2017). Similarly methylenedioxymethamphetamine (MDMA) has been shown to create an indifference to negative sounds, by mediating serotonergic signaling (Kuypers et al. 2018) and preliminary investigations of MDMA as a potential tinnitus therapy have begun (Searchfield et al. 2020a). There is no conclusive evidence yet that these drugs will be effective clinically, but the mechanisms putatively modified by these drugs are worth exploring further.

Several potential therapeutics have actions throughout putative tinnitus networks, and as such, have some of the characteristics potentially needed in a network therapeutic, examples are nitrous oxide (an N-Methyl-D-aspartate (NMDA) antagonist) and BGG492 (an α-amino-3-hydroxy-5-methyl-4-isoxazolepropionic acid (AMPA) antagonist) (Langguth et al. 2019). However, nitrous oxide was found no more effective than a placebo (Hong et al. 2018), and BGG492 development was cancelled for tinnitus, possibly due to unwanted side effects (Cederroth et al. 2018). One class of medicines aiming at achieving network modulation that has already been used in clinical practice for hundreds of years is the Traditional Chinese Medicines (TCMs). TCM uses a combination of several herbs acting on different organs/systems, especially for complex chronic diseases (Li and Zhang 2013). Traditionally, TCM takes a network approach in treating hearing loss and tinnitus by restoring kidney function (Dong and Shi 2012). A study reviewed 150 TCM prescriptions used to treat deafness over the past dynasties (Zhu et al. 1996). It was found that a total of 162 individual Chinese medicines were used, and on average, each prescription consisted of 16.3 components (Zhu et al. 1996). Although the scientific evidence for TCM is limited, a multi-site modulation approach combined with the increasingly popular network pharmacology approach for drug discovery might identify potential tinnitus treatments (Elgoyhen et al. 2014; Hopkins 2008).

An effective tinnitus medication may need to block or activate homeostatic processes to maintain the temporary shifts in excitation and inhibition seen with current therapies (Tetteh et al. 2018). This may require medication with multiple actions over time – an adaptive therapeutic – or the prescription of primary and then secondary pharmaceutical products based on careful monitoring of intervention effect.

4.1.2 From Polypharmacology to Multimodal Treatment

If a poly-therapeutic drug is not possible or insufficiently successful in alleviating the tinnitus, medications may also need to be paired with therapies that may potentiate each therapy's action (such as attempts to enhance auditory training through the prescription of fluoxetine (Searchfield et al. 2020b) or that target residual tinnitus mechanisms that may remain following the therapeutic's effect, for example cognitive-behavioral therapy (CBT) to manage remaining negative reactions to tinnitus not completely eliminated, or sound therapy modifies detection while the drug modifies emotion. Real-time monitoring of tinnitus symptoms and potentially physiology, through momentary evaluation online or through apps, may improve the therapy process by informing the medication dosage that is needed or to change the tinnitus drug therapy. An extension of the poly-therapeutic concept might also include interventions aligned with the ecology, context, of the individual's tinnitus experience, perhaps with a focus on attention (Searchfield 2014). This ecological approach would require rapid, easy access to therapy. Tools that might suit this might include internet-based CBT (Beukes et al. 2019), counselling, mindfulness, stress reduction, and sound-based apps (Mehdi et al. 2020; Sereda et al. 2019) as well as serious-games (Wise et al. 2016). These may be part of a self-directed therapeutic approach to tinnitus, not necessarily independent of other therapies (e.g., a tinnitus drug may be the primary therapy), but as at-home adjuncts to clinic-based approaches.

5 Improving Research to Improve Outcomes

There is limited funding available for medical research, as such, we should make the most of the resources available (Kleinjung and Langguth 2020). To do this, we need to optimize research methods and the reporting of results (Kleinjung and Langguth 2020). Efficacious research requires the selection and application of the best methods, with the least waste of effort. For all the advancements described in this volume, there remains a critical need to improve our understanding of what tinnitus is and to accurately evaluate proposed treatments.

5.1 Diversity

One of the great strengths of the tinnitus research community is its multidisciplinary membership. The International Tinnitus Seminars (since 1979) and Tinnitus Research Initiative conferences (since 2006) have showcased the diverse nature of tinnitus research and clinical practice. The movement towards a greater multi- and Inter-disciplinary approach has been a great achievement of the last decade. However, while diversity in knowledge and technology for discovery has increased, it would be fair to say that the quantitative tradition of research has dominated tinnitus research. More extensive use of qualitative and mixed methods designs may address gaps in understanding that cannot be filled by quantitative research (Durai and Searchfield 2017; Heinrich et al. 2016; Pryce et al. 2018; Taylor et al. 2020). This requires an understanding of researchers' worldviews and encouraging greater diversity in approaches. The inclusion of researchers trained in qualitative techniques to the current cadre of quantitative researchers from multidisciplinary backgrounds will help expand our understanding of tinnitus.

Worldviews, or positioning, are philosophies of research that can be categorized as post-positivist, constructivist, transformative, and pragmatic (Creswell 2014). As summarized by Creswell: Post-positivism is the conventional "scientific method" of reductionism, theories are reduced to testable ideas based on hypothesis testing. Constructionist views are often associated with qualitative research, where interest is on the meaning of the subject being studied, questions are open-ended. The transformative view extends Constructionism to include political change. The pragmatic worldview uses multiple methods, the "mixed methods" approach, for knowledge (Creswell 2014). In addition, understanding of the world (and reaction to therapies within it) may differ from culture to culture; indigenous worldviews are currently absent in our field (Curtis 2016).

Most tinnitus methods literature, to this point, has focused on the need for standardization in measures and rigor in research trials, that is, the post-positivist worldview. In addition to the need to improve the quality of trial designs, the field should consider alternative methods that may be better suited to answering questions about tinnitus.

5.2 Clinical Research Design and Methods

Most researchers are familiar with the hierarchy of research evidence that ranges from low level (expert opinion) to moderate (randomized controlled trials) to high level "filtered" information in the form of systematic reviews. A considerable body of tinnitus research has generated informative insights and raised some intriguing suggestions for interventions, but too often it has been unable to generate definitive knowledge. "Definitive" in this context means a finding obtained from an appropriately powered study focusing on the effectiveness or efficacy of intervention of interest and replicability. Many of the Cochrane systematic reviews assessing the evidence for various tinnitus interventions conclude that the evidence has a "moderate to high risk of bias" and/or has only "low to moderate certainty" (e.g. (Brennan-Jones et al. 2020)). Prospective large-scale studies recruiting many hundreds of participants and retrospective data mining of pooled datasets are both important directions for identifying patterns in the variability in tinnitus. Randomization of participants, with blinding to intervention, and where possible blinding of researchers, minimizes the risk of bias and increases the certainty in findings. RCTs do have shortcomings: they can be impractical due to cost, time from recruitment to publication, and require rigid protocols (Pham et al. 2016). Cross-over designs, in which participants are their own controls, can use fewer participants while maintaining statistical power. But such designs require a greater commitment from participants (essentially twice the time and testing) than equivalent parallel-arm studies. Higher quality evidence is more likely to be associated with strong recommendations, but, sometimes, even low or very low-quality evidence can lead to strong recommendation (Balshem et al. 2011). Small proof of concept and feasibility trials can be valuable to explore ideas. If the results of such pilot studies are promising, high quality RCTs can follow as a second step.

Case-reports can have strong internal validity for assessing causal relationships between interventions and outcomes (Lobo et al. 2017). In the so-called single case designs, the dependent variable is measured repeatedly across time with varying interventions or levels of intervention, termed phases, allowing for fine-grained time-series analysis (Lobo et al. 2017). Close examination of individual differences may identify factors that contribute to tinnitus therapy success in some patients and not others. Findings from these studies are informative to develop a feasibility trial to help us to: (1) determine if it is feasible to conduct a multicenter study; (2) optimize the design of a future definitive RCT; and (3) inform which outcome(s) is/are relevant for patients. A feasibility trial itself requires substantial effort and external funding.

The reductionist perspective inevitably means that the scope and questions asked in RCTs are narrow. This is fine if the understanding of parameters is well developed, but if not, a mixed methods design that includes broad questions to capture unanticipated responses should be considered (Christ 2014). Qualitative designs allow for a broad capture of experiences (Christ 2014). RCTs in areas of rapid technology development would have greater value and currency if they focused on evaluating intervention principles, the goals of the technology rather than the technology version (Mohr et al. 2015). For example, adaptive quantitative designs such as "Continuous Evaluation of Evolving Behavioral Intervention Technologies (CEEBIT)" gather information about an original app alongside version iterations (Mohr et al. 2013). Such a design accounts for continuous improvement in the technology within the trial, that is to avoid the situation in which an mHealth app version chosen for a trial is no longer current on trial completion. Once RCTs have been undertaken, validation in the real-world application is valuable, as participant behavior in the less constrained clinical practice environment may change. Despite the critique offered here, RCTs are the preferred research design for determining relationships between interventions and tinnitus measures. A critical aspect of the scientific approach is replication. The replication of findings, repeating research to determine the extent to which findings generalize across time and situations, is one of the defining hallmarks of quantitative research. Once an RCT has been undertaken, the findings need to be replicated; for research integrity and transparency, this would ideally be undertaken by researchers independent of the original RCT and free of conflicts of interest.

Researchers planning clinical trials should consider the various recommendations for trials that have been published. In addition, the FDA (USA) gives explicit guidance on what it means to run an adequate and well-controlled clinical trial, whether it is a drug trial or a medical device trial. These provide a clear framework for trial development. Trials should have:

- A clear statement of the objectives of the investigation and a summary of the proposed methods of analysis in the protocol.
- A design that permits a valid comparison with control to provide a quantitative assessment of the effect.
- Participant selection that provides adequate assurance that the participant has the disease or condition that the treatment is directed at.
- A method of assigning patients to treatment and control groups that minimizes bias and assures comparability of the groups.
- Adequate measures to minimize bias, by the participants, observers, and the data analysts.
- Well-defined and reliable assessment of participants' response.
- Analysis of the results that is adequate to assess the effect of the drug or device.

5.3 *Tinnitus Research Community and Cooperation*

Tinnitus research is becoming more interdisciplinary than ever. Collaboration maximizes the resources of geographically dispersed and discipline-specific expertise. There have been attempts to support such endeavors in sensory aging research, tinnitus and psychology. For example, in sensory aging research, the Sense Network (https://www.sensenetwork.org/) aims to create a diverse and inclusive network and encourages researchers from all backgrounds and under-represented groups. Likewise, the Tinnitus Research Initiative (https://www.tinnitusresearch.net) aims to support scientific and clinical research on collaborative, multidisciplinary, and international projects which are planned to lead to novel, effective therapies for the treatment of tinnitus. TRI's database project (https://tinnitusresearch.net/index.php/for-clinicians/database) is open to everyone willing to collect data according to a predefined protocol. The sharing of data between different research groups can only accelerate tinnitus research progress (Landgrebe et al. 2010).

A fairly unique model is the Psychological Science Accelerator (PSA, https://psysciacc.org/). This global network works to accelerate the accumulation of reliable and generalizable evidence in psychological science through a distributed laboratory network in which research is ongoing (as opposed to time or task limited), diverse (both in terms of human participants and participating researchers), and inclusive (welcoming ideas, contributions, study proposals, or other input from anyone in the field of psychology). The PSA embraces principles of diversity and inclusion (endeavoring to achieve cultural and geographic diversity among participants and researchers, as well as a diversity of research topics), decentralized authority (policies and procedures are set by committees with international representations, in conjunction with the membership at large), and openness to criticism (integrating critical assessment of its policies and research products). Consistent with these principles, the PSA has a globally distributed and democratically appointed leadership. There may be potential for such an umbrella group in the tinnitus field. Collaboration, especially with trial methodologists, and specialist clinical trial units, may jointly raise the quality and impact of tinnitus research. An inclusive research community could help foster and grow quality work, enable global research, and translate good ideas into research programs.

Research communities can support quality research at all levels of evidence. Although the qualitative research evidence hierarchy places systematic reviews and RCT as the highest forms of evidence, not all researchers can (through limited resources and/or funding), or are motivated to undertake systematic reviews and large RCTs. Journals and institutions often view systematic reviews and RCTs as more impactful than other research. But, it is incredibly important for the tinnitus field (and others) that basic knowledge and early-discovery research is published and publicized. Such research forms the foundations that other research builds upon, and without innovation, the field will not move forward.

6 Summary

This chapter has brought together the views of contributing authors on topics that they see as emerging and important for tinnitus research. The ideas and developments discussed should offer hope to tinnitus sufferers who report frustration at the speed at which advancements in tinnitus are made. Collectively we are optimistic about developing a range of therapies, some of which may require personalization that will be more effective than those currently at hand.

References

Alsalman OA, Tucker D, Vanneste S (2016) Salivary stress-related responses in tinnitus: a preliminary study in young male subjects with tinnitus. Front Neurosci 10:338

Anis K, Zakia H, Mohamed D, Jeffrey C (2018) Detecting depression severity by interpretable representations of motion dynamics. Paper presented at the 2018 13th IEEE international conference on automatic face & gesture recognition (FG 2018)

Aydin N, Searchfield GD (2019) Changes in tinnitus and physiological biomarkers of stress in response to short-term broadband noise and sounds of nature. Complement Ther Med 46:62–68

Azevedo AA, Figueiredo RR, Elgoyhen AB, Langguth B, Penido NO, Schlee W (2017) Tinnitus treatment with oxytocin: a pilot study. Front Neurol 8:494. https://doi.org/10.3389/fneur.2017.00494

Balshem H, Helfand M, Schünemann HJ, Oxman AD, Kunz R, Brozek J et al (2011) GRADE guidelines: 3. Rating the quality of evidence. J Clin Epidemiol 64(4):401–406

Basile C-É, Fournier P, Hutchins S, Hébert S (2013) Psychoacoustic assessment to improve tinnitus diagnosis. PLoS One 8(12):e82995

Bertet S, Baskind A, Londero A, Bonfils L, Viaud-Delmon I, Warusfel O (2013) Design and evaluation of tinnitus synthesis methods: from spectral to spatial matching. Am J Otolaryngol 34 (2):121–132. https://doi.org/10.1016/j.amjoto.2012.09.014

Betz LT, Mühlberger A, Langguth B, Schecklmann M (2017) Stress reactivity in chronic tinnitus. Sci Rep 7(1):1–9

Beukes EW, Manchaiah V, Allen PM, Baguley DM, Andersson G (2019) Internet-based interventions for adults with hearing loss, tinnitus, and vestibular disorders: a systematic review and meta-analysis. Trends Hear 23:2331216519851749

Boyen K, de Kleine E, van Dijk P, Langers DRM (2014) Tinnitus-related dissociation between cortical and subcortical neural activity in humans with mild to moderate sensorineural hearing loss. Hear Res 312:48–59. https://doi.org/10.1016/j.heares.2014.03.001

Brefczynski-Lewis JA, Lewis JW (2017) Auditory object perception: a neurobiological model and prospective review. Neuropsychologia 105:223–242

Brennan-Jones CG, Thomas A, Hoare DJ, Sereda M (2020) Cochrane corner: sound therapy (using amplification devices and/or sound generators) for tinnitus. Int J Audiol 59(3):161–165

Brozoski TJ, Bauer CA (2016) Animal models of tinnitus. Hear Res 338:88–97

Burns EM (1984) A comparison of variability among measurements of subjective tinnitus and objective stimuli. Audiology 23(4):426–440. https://doi.org/10.3109/00206098409081535

Caspary DM, Llano DA (2017) Auditory thalamic circuits and GABAA receptor function: putative mechanisms in tinnitus pathology. Hear Res 349:197–207

Cederroth CR, Dyhrfjeld-Johnsen J, Langguth B (2018) An update: emerging drugs for tinnitus. Expert Opin Emerg Drugs 23(4):251–260

Cederroth CR, Gallus S, Hall DA, Kleinjung T, Langguth B, Maruotti A et al (2019a) Towards an understanding of tinnitus heterogeneity. Front Aging Neurosci 11:53

Cederroth CR, PirouziFard M, Trpchevska N, Idrizbegovic E, Canlon B, Sundquist J et al (2019b) Association of genetic vs environmental factors in Swedish adoptees with clinically significant tinnitus. JAMA Otolaryngol Head Neck Surg 145(3):222–229

Chen XH, Zheng LL (2017) Expression of pro-inflammatory cytokines in the auditory cortex of rats with salicylate-induced tinnitus. Mol Med Rep 16(4):5643–5648. https://doi.org/10.3892/mmr.2017.7235

Chen Y-C, Wang J, Jiao Y, Zang F-C, Yang M, Tong J-X et al (2014) Changes in resting-state fMRI activity during salicylate-induced tinnitus and sound stimulation. 8th international TRI tinnitus conference, Abstract, p 106

Christ TW (2014) Scientific-based research and randomized controlled trials, the "gold" standard? Alternative paradigms and mixed methodologies. Qual Inq 20(1):72–80

Ciba Foundation Symposium 85 (1981) Appendix II guidelines for recommended procedures in tinnitus testing. In: Tinnitus. Pitman Books, Ltd., London, pp 303–306

Conlon B, Langguth B, Hamilton C, Hughes S, Meade E, Connor CO et al (2020) Bimodal neuromodulation combining sound and tongue stimulation reduces tinnitus symptoms in a large randomized clinical study. Sci Transl Med 12(564). https://doi.org/10.1126/scitranslmed.abb2830

Creswell JW (2014) Research design: qualitative, quantitative and mixed methods approaches, 4th edn. Sage, Thousand Oaks

Curtis E (2016) Indigenous positioning in health research: the importance of Kaupapa Māori theory-informed practice. AlterNative Int J Indigenous Peoples 12(4):396–410

Dong Y, Shi JR (2012) Biological research evaluating the Chinese medical theory of the association of the kidney with the ears. Zhong Xi Yi Jie He Xue Bao 10(2):128–134. https://doi.org/10.3736/jcim20120202

Durai M, Searchfield GD (2017) A mixed-methods trial of broad band noise and nature sounds for tinnitus therapy: group and individual responses modeled under the adaptation level theory of tinnitus. Front Aging Neurosci 9:44

Durai M, O'Keeffe MG, Searchfield GD (2017) The personality profile of tinnitus sufferers and a nontinnitus control group. J Am Acad Audiol 28(4):271–282. https://doi.org/10.3766/jaaa.15103

Durai M, Sanders P, Doborjeh Z, Doborjeh M, Kasabov N, Searchfield G (2020) Prediction of tinnitus masking benefit within a case series using a spiking neural network model. In: Progress in brain research, Elsevier, pp 1–17

Eggermont JJ (2016) Can animal models contribute to understanding tinnitus heterogeneity in humans? Front Aging Neurosci 8:265

Elgoyhen AB, Langguth B, Nowak W, Schecklmann M, De Ridder D, Vanneste S (2014) Identifying tinnitus-related genes based on a side-effect network analysis. CPT Pharmacometrics Syst Pharmacol 3(1):1–10

Feldmann H (1992) Tinnitus – reality or phantom. In: Aran JM, Dauman R (eds) Tinnitus 91: proceedings of the fourth international Tinnitus seminar. Kugler Publications, Amsterdam

Feng Q, Zhou A, Zou H, Ingle S, May MT, Cai W et al (2019) Quadruple versus triple combination antiretroviral therapies for treatment naive people with HIV: systematic review and meta-analysis of randomised controlled trials. BMJ 366:l4179. https://doi.org/10.1136/bmj.l4179

George SZ, Lentz TA, Beneciuk JM, Bhavsar NA, Mundt JM, Boissoneault J (2020) Framework for improving outcome prediction for acute to chronic low back pain transitions. Pain Rep 5(2): e809. https://doi.org/10.1097/PR9.0000000000000809

Ginsburg GS, Willard HF (2009) Genomic and personalized medicine: foundations and applications. Transl Res 154(6):277–287. https://doi.org/10.1016/j.trsl.2009.09.005

Goldberg RL, Piccirillo ML, Nicklaus J, Skillington A, Lenze E, Rodebaugh TL et al (2017) Evaluation of ecological momentary assessment for tinnitus severity. JAMA Otolaryngol Head Neck Surg 143(7):700–706

Goodwin BJ, Constantino MJ, Westra HA, Button ML, Antony MM (2019) Patient motivational language in the prediction of symptom change, clinically significant response, and time to response in psychotherapy for generalized anxiety disorder. Psychotherapy 56(4):537

Guntensperger D, Thuring C, Meyer M, Neff P, Kleinjung T (2017) Neurofeedback for tinnitus treatment – review and current concepts. Front Aging Neurosci 9:386. https://doi.org/10.3389/fnagi.2017.00386

Haider HF, Flook M, Aparicio M, Ribeiro D, Antunes M, Szczepek AJ et al (2017) Biomarkers of Presbycusis and tinnitus in a Portuguese older population. Front Aging Neurosci 9:346. https://doi.org/10.3389/fnagi.2017.00346

Haider HF, Ribeiro SF, Martins C, Ribeiro D, Trigueiros N, Szczepek AJ et al (2020) Tinnitus, hearing loss and inflammatory processes in an older Portuguese population. Int J Audiol 59 (5):323–332

Hall DA, Smith H, Hibbert A, Colley V, Haider HF, Horobin A et al (2018) The COMiT'ID study: developing Core Outcome domains sets for clinical trials of sound-, psychology-, and pharmacology-based interventions for chronic subjective Tinnitus in adults. Trends Hear 22:2331216518814384. https://doi.org/10.1177/2331216518814384

Han L, Na Z, Chunli L, Yuchen C, Pengfei Z, Hao W et al (2019) Baseline functional connectivity features of neural network nodes can predict improvement after sound therapy through adjusted narrow band noise in tinnitus patients. Front Neurosci:13

He J, Zhu Y, Aa J, Smith PF, De Ridder D, Wang G, Zheng Y (2017) Brain metabolic changes in rats following acoustic trauma. Front Neurosci 11:148

Hebert S, Fournier P (2017) Clinical validation of a new tinnitus assessment technology. Front Neurol 8:38. https://doi.org/10.3389/fneur.2017.00038

Heinrich S, Rozental A, Carlbring P, Andersson G, Cotter K, Weise C (2016) Treating tinnitus distress via the internet: a mixed methods approach of what makes patients seek help and stay motivated during internet-based cognitive behavior therapy. Internet Interv 4:120–130

Henry JA, Meikle MB (2000) Psychoacoustic measures of tinnitus. J Am Acad Audiol 11(3):138–155

Henry JA, Galvez G, Turbin MB, Thielman EJ, McMillan GP, Istvan JA (2012) Pilot study to evaluate ecological momentary assessment of tinnitus. Ear Hear 32(2):179

Henry JA, Roberts LE, Caspary DM, Theodoroff SM, Salvi RJ (2014) Underlying mechanisms of tinnitus: review and clinical implications. J Am Acad Audiol 25(1):5–22.; quiz 126. https://doi.org/10.3766/jaaa.25.1.2

Hong HY, Karadaghy O, Kallogjeri D, Brown FT, Yee B, Piccirillo JF, Nagele P (2018) Effect of nitrous oxide as a treatment for subjective, idiopathic, nonpulsatile bothersome tinnitus: a randomized clinical trial. JAMA Otolaryngol Head Neck Surg 144(9):781–787

Hopkins AL (2008) Network pharmacology: the next paradigm in drug discovery. Nat Chem Biol 4 (11):682–690. https://doi.org/10.1038/nchembio.118

Husain FT, Schmidt SA (2014) Using resting state functional connectivity to unravel networks of tinnitus. Hear Res 307:153–162. https://doi.org/10.1016/J.Heares.2013.07.010

James GA, Thostenson JD, Brown G, Carter G, Hayes H, Tripathi SP et al (2017) Neural activity during attentional conflict predicts reduction in tinnitus perception following rTMS. Brain Stimul 10(5):934–943. https://doi.org/10.1016/j.brs.2017.05.009

Jensen M, Hüttenrauch E, Schmidt J, Andersson G, Chavanon M-L, Weise C (2020) Neurofeedback for tinnitus: study protocol for a randomised controlled trial assessing the specificity of an alpha/delta neurofeedback training protocol in alleviating both sound perception and psychological distress in a cohort of chronic tinnitus sufferers. Trials 21:1–13

Jiang C, Luo B, Manohar S, Chen GD, Salvi R (2017) Plastic changes along auditory pathway during salicylate-induced ototoxicity: hyperactivity and CF shifts. Hear Res 347:28–40. https://doi.org/10.1016/j.heares.2016.10.021

Jordan R, Gold L, Cummins C, Hyde C (2002) Systematic review and meta-analysis of evidence for increasing numbers of drugs in antiretroviral combination therapy. BMJ 324(7340):757. https://doi.org/10.1136/bmj.324.7340.757

Juengst E, McGowan ML, Fishman JR, Settersten RA Jr (2016) From "Personalized" to "Precision" medicine: the ethical and social implications of rhetorical reform in genomic medicine. Hastings Cent Rep 46(5):21–33. https://doi.org/10.1002/hast.614

Keesom SM, Morningstar MD, Sandlain R, Wise BM, Hurley LM (2018) Social isolation reduces serotonergic fiber density in the inferior colliculus of female, but not male, mice. Brain Res 1694:94–103. https://doi.org/10.1016/j.brainres.2018.05.010

Kleinjung T, Langguth B (2020) Avenue for future tinnitus treatments. Otolaryngol Clin North Am 53(4):667–683

Kleinstäuber M, Weise C, Andersson G, Probst T (2018) Personality traits predict and moderate the outcome of internet-based cognitive behavioural therapy for chronic tinnitus. Int J Audiol 57 (7):538–544

Knobel KA, Sanchez TG (2008) Influence of silence and attention on tinnitus perception. Otolaryngol Head Neck Surg 138(1):18–22. https://doi.org/10.1016/j.otohns.2007.09.023

Korth D, Wollbrink A, Wunderlich R, Ivansic D, Guntinas-Lichius O, Salvari V et al (2020) One step closer towards a reliable tinnitus pitch-match frequency determination using repetitive recursive matching. Audiol Neurotol:1–10

Kraus KS, Canlon B (2012) Neuronal connectivity and interactions between the auditory and limbic systems. Effects of noise and tinnitus. Hear Res 288(1–2):34–46. https://doi.org/10.1016/J.Heares.2012.02.009

Kuypers KP, De La Torre R, Farre M, Pizarro N, Xicota L, Ramaekers JG (2018) MDMA-induced indifference to negative sounds is mediated by the 5-HT 2A receptor. Psychopharmacology (Berl) 235(2):481–490

Landgrebe M, Zeman F, Koller M, Eberl Y, Mohr M, Reiter J et al (2010) The tinnitus research initiative (TRI) database: a new approach for delineation of tinnitus subtypes and generation of predictors for treatment outcome. BMC Med Inform Decis Mak 10(1):42

Langguth B, Goodey R, Azevedo A, Bjorne A, Cacace A, Crocetti A et al (2007) Consensus for tinnitus patient assessment and treatment outcome measurement: tinnitus research initiative meeting, Regensburg, July 2006. Prog Brain Res 166:525–536

Langguth B, Elgoyhen AB, Cederroth CR (2019) Therapeutic approaches to the treatment of tinnitus. Annu Rev Pharmacol Toxicol 59:291–313

Leaver AM, Renier L, Chevillet MA, Morgan S, Kim HJ, Rauschecker JP (2011) Dysregulation of limbic and auditory networks in tinnitus. Neuron 69(1):33–43

Leaver AM, Seydell-Greenwald A, Rauschecker JP (2016) Auditory-limbic interactions in chronic tinnitus: challenges for neuroimaging research. Hear Res 334:49–57. https://doi.org/10.1016/j.heares.2015.08.005

Li S, Zhang B (2013) Traditional Chinese medicine network pharmacology: theory, methodology and application. Chin J Nat Med 11(2):110–120. https://doi.org/10.1016/S1875-5364(13)60037-0

Liu Y, Niu H, Zhu J, Zhao P, Yin H, Ding H et al (2019) Morphological neuroimaging biomarkers for tinnitus: evidence obtained by applying machine learning. Neural Plast 2019

Lobo MA, Moeyaert M, Cunha AB, Babik I (2017) Single-case design, analysis, and quality assessment for intervention research. J Neurol Phys Ther 41(3):187

Lopez-Escamez JA, Amanat S (2020) Heritability and genetics contribution to tinnitus. Otolaryngol Clin North Am 53(4):501–513

Lopez-Escamez JA, Bibas T, Cima RF, Van de Heyning P, Knipper M, Mazurek B et al (2016) Genetics of tinnitus: an emerging area for molecular diagnosis and drug development. Front Neurosci 10:377. https://doi.org/10.3389/fnins.2016.00377

Lugo A, Trpchevska N, Liu X, Biswas R, Magnusson C, Gallus S, Cederroth CR (2019) Sex-specific association of tinnitus with suicide attempts. JAMA Otolaryngol Head Neck Surg 145(7):685–687

Maas IL, Brüggemann P, Requena T, Bulla J, Edvall NK, vB Hjelmborg J et al (2017) Genetic susceptibility to bilateral tinnitus in a Swedish twin cohort. Genet Med 19(9):1007–1012

Marks KL, Martel DT, Wu C, Basura GJ, Roberts LE, Schvartz-Leyzac KC, Shore SE (2018) Auditory-somatosensory bimodal stimulation desynchronizes brain circuitry to reduce tinnitus in guinea pigs and humans. Sci Transl Med 10(422)

Marlin BJ, Mitre M, D'Amour JA, Chao MV, Froemke RC (2015) Oxytocin enables maternal behaviour by balancing cortical inhibition. Nature 520(7548):499–504. https://doi.org/10.1038/nature14402

Maudoux A, Lefebvre P, Cabay JE, Demertzi A, Vanhaudenhuyse A, Laureys S, Soddu A (2012) Connectivity graph analysis of the auditory resting state network in tinnitus. Brain Res 1485:10–21. https://doi.org/10.1016/j.brainres.2012.05.006

Mazurek B, Fischer F, Haupt H, Georgiewa P, Reisshauer A, Klapp BF (2006) A modified version of tinnitus retraining therapy: observing long-term outcome and predictors. Audiology and Neurotology 11(5):276–286

McFerran DJ, Stockdale D, Holme R, Large CH, Baguley DM (2019) Why is there no cure for tinnitus? Front Neurosci 13:802

Mehdi M, Riha C, Neff P, Dode A, Pryss R, Schlee W et al (2020) Smartphone apps in the context of tinnitus: systematic review. Sensors 20(6):1725

Meikle MB, Henry JA, Griest SE, Stewart BJ, Abrams HB, McArdle R et al (2012) The tinnitus functional index: development of a new clinical measure for chronic, intrusive tinnitus. Ear Hear 33(2):153–176

Meikle MB, Creedon TA, Griest SE. (2004) Tinnitus Archive, second edition. http://www.tinnitusArchive.org/

Mitchell CR, Vernon JA, Creedon TA (1993) Measuring tinnitus parameters: loudness, pitch, and maskability. J Am Acad Audiol 4(3):139–151

Mohamad N, Hoare DJ, Hall DA (2016) The consequences of tinnitus and tinnitus severity on cognition: a review of the behavioural evidence. Hear Res 332:199–209

Mohr DC, Cheung K, Schueller SM, Hendricks Brown C, Duan N (2013) Continuous evaluation of evolving behavioral intervention technologies. Am J Prev Med 45(4):517–523. https://doi.org/10.1016/j.amepre.2013.06.006

Mohr DC, Schueller SM, Riley WT, Brown CH, Cuijpers P, Duan N et al (2015) Trials of intervention principles: evaluation methods for evolving behavioral intervention technologies. J Med Internet Res 17(7):e166. https://doi.org/10.2196/jmir.4391

Newman CW, Jacobson GP, Spitzer JB (1996) Development of the tinnitus handicap inventory. Arch Otolaryngol Head Neck Surg 122(2):143–148

Niemann U, Boecking B, Brueggemann P, Mazurek B, Spiliopoulou M (2020a) Gender-specific differences in patients with chronic tinnitus – baseline characteristics and treatment effects. Front Neurosci 14:487

Niemann U, Boecking B, Brueggemann P, Mebus W, Mazurek B, Spiliopoulou M (2020b) Tinnitus-related distress after multimodal treatment can be characterized using a key subset of baseline variables. PLoS One 15(1):e0228037

Ogden D (2013) Drakon: dragon myth and serpent cult in the Greek and Roman worlds. Oxford University Press, Oxford

Pau AK, George JM (2014) Antiretroviral therapy: current drugs. Infect Dis Clin North Am 28 (3):371–402. https://doi.org/10.1016/j.idc.2014.06.001

Penner MJ, Klafter EJ (1992) Measures of tinnitus: step size, matches to imagined tones, and masking patterns. Ear Hear 13(6):410–416

Pham Q, Wiljer D, Cafazzo JA (2016) Beyond the randomized controlled trial: a review of alternatives in mHealth clinical trial methods. JMIR Mhealth Uhealth 4(3):e107. https://doi.org/10.2196/mhealth.5720

Probst T, Pryss RC, Langguth B, Rauschecker JP, Schobel J, Reichert M et al (2017) Does tinnitus depend on time-of-day? An ecological momentary assessment study with the "TrackYourTinnitus" application. Front Aging Neurosci 9:253

Probst T, Weise C, Andersson G, Kleinstäuber M (2019) Differences in baseline and process variables between non-responders and responders in internet-based cognitive behavior therapy for chronic tinnitus. Cogn Behav Ther 48(1):52–64

Pryce H, Durand M-A, Hall A, Shaw R, Culhane B-A, Swift S et al (2018) The development of a decision aid for tinnitus. Int J Audiol 57(9):714–719

Roberts LE, Eggermont JJ, Caspary DM, Shore SE, Melcher JR, Kaltenbach JA (2010) Ringing ears: the neuroscience of tinnitus. J Neurosci 30(45):14972–14979. https://doi.org/10.1523/JNEUROSCI.4028-10.2010

Roden DM, Tyndale RF (2013) Genomic medicine, precision medicine, personalized medicine: what's in a name? Clin Pharmacol Ther 94(2):169–172. https://doi.org/10.1038/clpt.2013.101

Schlee W, Pryss RC, Probst T, Schobel J, Bachmeier A, Reichert M, Langguth B (2016) Measuring the moment-to-moment variability of tinnitus: the TrackYourTinnitus smart phone app. Front Aging Neurosci 8:294

Schlee W, Hall DA, Canlon B, Cima RF, de Kleine E, Hauck F et al (2018) Innovations in doctoral training and research on tinnitus: the European school on interdisciplinary tinnitus research (ESIT) perspective. Front Aging Neurosci 9:447

Schleidgen S, Klingler C, Bertram T, Rogowski WH, Marckmann G (2013) What is personalized medicine: sharpening a vague term based on a systematic literature review. BMC Med Ethics 14 (1):55

Schmidt SA, Carpenter-Thompson J, Husain FT (2017) Connectivity of precuneus to the default mode and dorsal attention networks: a possible invariant marker of long-term tinnitus. Neuroimage Clin 16:196–204

Searchfield GD (2014) Tinnitus what and where: an ecological framework. Front Neurol 5:271. https://doi.org/10.3389/fneur.2014.00271

Searchfield GD (2019) A client oriented scale of improvement in tinnitus for therapy goal planning and assessing outcomes. J Am Acad Audiol 30(4):327–337

Searchfield G, Kobayashi K, Proudfoot K, Tevoitdale H, Irving S (2015) The development and test–retest reliability of a method for matching perceived location of tinnitus. J Neurosci Methods 256:1–8

Searchfield GD, Poppe T, Durai M, Jensen M, Kennedy MA, Maggo S et al (2020a) A proof-of-principle study of the short-term effects of 3,4-methylenedioxymethamphetamine (MDMA) on tinnitus and neural connectivity. Int J Neurosci 130(7):671–682. https://doi.org/10.1080/00207454.2019.1702544

Searchfield G, Spiegel D, Poppe T, Durai M, Jensen M, Kobayashi K et al (2020b) A proof-of-concept study comparing tinnitus and neural connectivity changes following multisensory perceptual training with and without a low-dose of fluoxetine. Int J Neurosci:1–12

Sereda M, Smith S, Newton K, Stockdale D (2019) Mobile apps for management of tinnitus: users' survey, quality assessment, and content analysis. JMIR Mhealth Uhealth 7(1):e10353

Shore SE, Wu C (2019) Mechanisms of noise-induced tinnitus: insights from cellular studies. Neuron 103(1):8–20

Simoes JP, Neff P, Schoisswohl S, Bulla J, Schecklmann M, Harrison SS et al (2019) Towards personalized tinnitus treatment: an exploratory study based on internet crowdsensing. Front Public Health 7:157

Simonetti P, Oiticica J (2015) Tinnitus neural mechanisms and structural changes in the brain: the contribution of neuroimaging research. Int Arch Otorhinolaryngol 19(3):259–265. https://doi.org/10.1055/s-0035-1548671

Song JJ, De Ridder D, Van de Heyning P, Vanneste S (2012) Mapping tinnitus-related brain activation: an activation-likelihood estimation metaanalysis of PET studies. J Nucl Med 53 (10):1550–1557. https://doi.org/10.2967/jnumed.112.102939

Szczepek AJ, Haupt H, Klapp BF, Olze H, Mazurek B (2014) Biological correlates of tinnitus-related distress: an exploratory study. Hear Res 318:23–30

Szczepek AJ, Dietz GPH, Reich U, Hegend O, Olze H, Mazurek B (2018) Differences in stress-induced modulation of the auditory system between Wistar and Lewis rats. Front Neurosci 12:828. https://doi.org/10.3389/fnins.2018.00828

Taylor JA, Thompson DM, Hall DA, Walker D-M, McMurran M, Casey A et al (2020) The TinMan study: feasibility trial of a psychologically informed, audiologist-delivered, manualised intervention for tinnitus. Int J Audiol:1–10

Tetteh H, Lee M, Lau CG, Yang S, Yang S (2018) Tinnitus: prospects for pharmacological interventions with a seesaw model. Neuroscientist 24(4):353–367. https://doi.org/10.1177/1073858417733415

Theodoroff SM, Schuette A, Griest S, Henry JA (2014) Individual patient factors associated with effective tinnitus treatment. J Am Acad Audiol 25(7):631–643

Traeger AC, Henschke N, Hubscher M, Williams CM, Kamper SJ, Maher CG et al (2016) Estimating the risk of chronic pain: development and validation of a prognostic model (PICKUP) for patients with acute low Back pain. PLoS Med 13(5):e1002019. https://doi.org/10.1371/journal.pmed.1002019

Trpchevska N, Bulla J, Prada Hellberg M, Edvall NK, Lazar A, Mehraei G et al (2020) Sex-dependent aggregation of tinnitus in Swedish families. J Clin Med 9(12):3812

Tutton R (2012) Personalizing medicine: futures present and past. Soc Sci Med 75(10):1721–1728

Tyler R, Ji H, Perreau A, Witt S, Noble W, Coelho C (2014) Development and validation of the tinnitus primary function questionnaire. Am J Audiol 23(3):260–272

Tyler R, Cacace A, Stocking C, Tarver B, Engineer N, Martin J et al (2017) Vagus nerve stimulation paired with tones for the treatment of tinnitus: a prospective randomized double-blind controlled pilot study in humans. Sci Rep 7(1):11960. https://doi.org/10.1038/s41598-017-12178-w

Tzounopoulos T, Balaban C, Zitelli L, Palmer C (2019) Towards a mechanistic-driven precision medicine approach for Tinnitus. J Assoc Res Otolaryngol 20(2):115–131. https://doi.org/10.1007/s10162-018-00709-9

Uckelstam C-J, Philips B, Holmqvist R, Falkenström F (2019) Prediction of treatment outcome in psychotherapy by patient initial symptom distress profiles. J Couns Psychol

Ulusoy B, Bozdemir K, Akyol M, Mise HI, Kutluhan A, Korkmaz MH (2018) Investigation of neutrophil-to-lymphocyte ratio, platelet-to-lymphocyte ratio and mean platelet volume in patients with tinnitus. J Laryngol Otol 132(2):129–132. https://doi.org/10.1017/S0022215117002481

Van de Heyning P, Gilles A, Rabau S, Van Rompaey V (2015) Subjective tinnitus assessment and treatment in clinical practice: the necessity of personalized medicine. Curr Opin Otolaryngol Head Neck Surg 23(5):369–375. https://doi.org/10.1097/MOO.0000000000000183

Van der Wal A, Luyten T, Cardon E, Jacquemin L, Vanderveken OM, Topsakal V et al (2020) Sex differences in the response to different tinnitus treatment. Front Neurosci 14:422

Vanneste S, De Ridder D (2012) The auditory and non-auditory brain areas involved in tinnitus. An emergent property of multiple parallel overlapping subnetworks. Front Syst Neurosci 6:31. https://doi.org/10.3389/fnsys.2012.00031

Vanneste S, de Heyning PV, De Ridder D (2011) The neural network of phantom sound changes over time: a comparison between recent-onset and chronic tinnitus patients. Eur J Neurosci 34 (5):718–731. https://doi.org/10.1111/J.1460-9568.2011.07793.X

Vanneste S, Joos K, De Ridder D (2012) Prefrontal cortex based sex differences in tinnitus perception: same tinnitus intensity, same tinnitus distress, different mood. PLoS One 7(2): e31182

Vanneste S, Faber M, Langguth B, De Ridder D (2016) The neural correlates of cognitive dysfunction in phantom sounds. Brain Res. https://doi.org/10.1016/j.brainres.2016.03.016

Vanneste S, Song JJ, De Ridder D (2018) Thalamocortical dysrhythmia detected by machine learning. Nat Commun 9(1):1103. https://doi.org/10.1038/s41467-018-02820-0

Veile A, Zimmermann H, Lorenz E, Becher H (2018) Is smoking a risk factor for tinnitus? A systematic review, meta-analysis and estimation of the population attributable risk in Germany. BMJ Open 8(2)

Willott JF (2009) Effects of sex, gonadal hormones, and augmented acoustic environments on sensorineural hearing loss and the central auditory system: insights from research on C57BL/6J mice. Hear Res 252(1–2):89–99. https://doi.org/10.1016/j.heares.2008.12.002

Wilson MB, Kallogjeri D, Joplin CN, Gorman MD, Krings JG, Lenze EJ et al (2015) Ecological momentary assessment of tinnitus using smartphone technology: a pilot study. Otolaryngol Head Neck Surg 152(5):897–903

Wineland AM, Burton H, Piccirillo J (2012) Functional connectivity networks in nonbothersome tinnitus. Otolaryngol Head Neck Surg 147(5):900–906

Wise K, Kobayashi K, Magnusson J, Welch D, Searchfield GD (2016) Randomized controlled trial of a perceptual training game for tinnitus therapy. Games Health J 5(2):141–149

Yun Y, Lee S-Y, Hoon Choi W, Park J-C, Han Lee D, Kim YK et al (2020) Proteasome activity in the plasma as a novel biomarker in mild cognitive impairment with chronic tinnitus. J Alzheimer's Dis:1–11

Zhang J (2019) Blast-induced tinnitus: animal models. J Acoust Soc Am 146(5):3811. https://doi.org/10.1121/1.5132551

Zhang X, Kuivenhoven JA, Groen AK (2015) Forward individualized medicine from personal genomes to interactomes. Front Physiol 6:364. https://doi.org/10.3389/fphys.2015.00364

Zhu M-J, Tan J-S, Li F-C, Xu S-Q (1996) [Analysis of the frequency and pattern of the individual Chinese medicine usage in 150 traditional Chinese medicine prescriptions for treating deafness in the past dynasties] (in Chinese). Hunan Zhong Yi Xue Yuan Xue Bao 16(2):16–17

Correction to: Prevalence, Incidence, and Risk Factors for Tinnitus

Roshni Biswas and Deborah A. Hall

Correction to:
Chapter "Epidemiology of Tinnitus" in: Roshni Biswas and Deborah A. Hall, Curr Topics Behav Neurosci, https://doi.org/10.1007/7854_2020_154

The original version of this chapter unfortunately contained errors in chapter title and keywords.

The incorrect chapter title "Epidemiology of Tinnitus" is now corrected and updated as "Prevalence, Incidence, and Risk Factors for Tinnitus".

The incorrect keywords "Incidence · Measures of effects · Population · Prevalence · Risk factors" are corrected and updated as "Adults · Epidemiology · Measures of effects · Population · Tinnitus".

The updated online version of this chapter can be found at
https://doi.org/10.1007/7854_2020_154

Curr Topics Behav Neurosci (2021) 51: 485–486
https://doi.org/10.1007/7854_2020_188
Published Online: 3 April 2021

Correction to: Emerging Topics in the Behavioral Neuroscience of Tinnitus

Grant D. Searchfield, Jinsheng Zhang, Roshni Biswas, Dirk De Ridder, Brian Deutsch, Deborah A. Hall, Sylvie Hébert, Tobias Kleinjung, Maria Kleinstäuber, Berthold Langguth, Jose Antonio Lopez-Escamez, Michael R. D. Maslin, Birgit Mazurek, Jay F. Piccirillo, Richard Salvi, Winfried Schlee, Abraham Shulman, Susan Shore, Agnieszka J. Szczepek, Paul F. Smith, Sarah M. Theodoroff, Dunja Vajsakovic, Cornelia Weise, and Yiwen Zheng

Correction to:
Chapter "Emerging Topics in the Behavioral Neuroscience of Tinnitus" in:
Grant D. Searchfield et al., Curr Topics Behav Neurosci, https://doi.org/10.1007/7854_2020_217

The original version of this chapter unfortunately contained two errors: in author name and order of author. These two errors has been corrected and the below are the updated correction:

1. The author name "Sylvie Hall Hébert" is changed to "Sylvie Hébert".
2. The order of author name Deborah A. Hall is listed before Sylvie Hébert.

The updated online version of this chapter can be found at https://doi.org/10.1007/7854_2020_217

Curr Topics Behav Neurosci (2021) 51: 487–488
https://doi.org/10.1007/7854_2021_230
Published Online: 11 June 2021

The manufacturer's authorised representative in the EU is Springer Nature Customer Service Centre GmbH, Europaplatz 3, 69115 Heidelberg, Germany. If you have any concerns regarding our products, please contact ProductSafety@springernature.com

Printed and bound by CPI Group (UK) Ltd, Croydon, CR0 4YY
30/09/2026
02225687-0013